NUMERICAL METHODS
FOR ENGINEERING AND SCIENCE

SAUMYEN GUHA

Professor
Department of Civil Engineering
IIT Kanpur

RAJESH SRIVASTAVA

Professor
Department of Civil Engineering
IIT Kanpur

OXFORD
UNIVERSITY PRESS

OXFORD
UNIVERSITY PRESS

Oxford University Press is a department of the University of Oxford.
It furthers the University's objective of excellence in research, scholarship,
and education by publishing worldwide. Oxford is a registered trade mark of
Oxford University Press in the UK and in certain other countries.

Published in India by
Oxford University Press
22 Workspace, 2nd Floor, 1/22 Asaf Ali Road, New Delhi 110002, India

First Edition published XXXX
Second Edition published XXXX

Digitally Printed in 2023

ISBN-13: 978-0-19-569348-5
ISBN-10: X-XX-XXXXXX-X

eISBN-13 (eBook): XXX-X-XX-XXXXXX-X
eISBN-10 (eBook): X-XX-XXXXXX-X

Typeset in XXXXX xXXXXX
by XXXXXx XXXXXXXXX
Printed at Manipal Technologies Limited, Manipal
Cover image: © XXXXXXXX/Shutterstock

For product information and current price, please visit www.india.oup.com

Preface

Numerical methods attempt to approximate mathematical problems using simple arithmetic operations. Even before the calculator was invented, many problems such as measuring the circumference of a circle involving the approximation of π, linear interpolation, solution of non-linear algebraic equations, and linear simultaneous equations were solved using large computation tables. Later, mechanical calculators (e.g., addition, multiplication and division machines, slide rule, etc.) followed by electronic calculators made these solutions much easier. The biggest leap in this field came about with the advent of computers in the late 1940s. This led to attempts for the solution of mathematical problems from diverse fields using approximate methods, since it was felt that a *good* approximation is good enough for most fields. This also called for more complicated and sophisticated approximations. Since the true solutions to the problems were mostly unknown, the question that cropped up was: How do we know that an approximation is *better* or how do we choose the *best* approximation from a set of *good* approximations for a given problem? This triggered large leaps in the area of *mathematical analyses* of *numerical methods*, often called *numerical analyses*.

In today's world, numerical methods and analyses are the backbone of computer simulation, which is fast replacing all forms of destructive laboratory and pilot scale testing (nuclear bomb testing, space shuttle launch and landing, aircraft design, car design, large buildings and bridges, etc., to name a few). Therefore, it becomes necessary for the scientists and engineers of this generation to become familiar with the fundamentals of numerical methods and analyses. Beyond the boundaries of science, numerical methods are steadily increasing their application base in the spheres of arts and commerce as well. With this background, we describe the intent and outline of this book.

ABOUT THE BOOK

This book is the outcome of our experience of nearly a decade of teaching the course Computational Methods in Engineering at IIT Kanpur. There are plenty of very good textbooks on the subject but when we looked for a book to recommend, no single book satisfied what we were teaching in the class. They either fell short on derivation and analysis of the methods, or contained too much for an undergraduate course, or did not cover all the topics that we were teaching in the course.

While writing this book, we have assumed that the students have the background of at least three mathematics courses covering calculus, complex variable, linear algebra, ordinary differential equations, and partial differential equations.

We have followed the principle of science-based engineering in this book. Students are expected to not only know a method but also understand its derivation, assumptions, and limitations. This book is an attempt to strike a balance between numerical methods and their analyses so that it remains legible to undergraduate students as well as serve as a basic textbook for both undergraduate and graduate courses in engineering. Keeping in mind the diverse background of students, we have kept the course content general and explained the principles in such a manner that the methods can be applied to any discipline.

Most of the methods presented in this book have been derived and analysed for convergence. A close link has been maintained throughout between a method and its mathematical background. To clarify the concepts better, relevant mathematical theorems along with their proofs have been sprinkled throughout the text. This not only gives a complete picture but also enables the students to visualize the assumptions and approximations that went into the method. Solved examples have been included in all the chapters to reinforce the working of the methods discussed.

The use of computer programs and pseudocodes has been deliberately avoided, since the emphasis is on the concepts and not on providing recipes.

An attempt has been made to integrate the concepts presented in various chapters by cross-referencing between them. However, it is easy to go overboard with this by playing squash with the reader by tossing them everywhere in the book while they are trying to concentrate on understanding a concept in one chapter. In order to reduce this, the chapter numbers are often given in order to give the idea that the concepts in that chapter are useful. It is not to say that the reader should stop reading at that point and go to the referred chapter. If the concept in a different place is very crucial for understanding the material, it has been rephrased and presented briefly so that the reader can keep reading. They can go back to the referred section of the book at a later time and revise. However, in order to keep the book at a reasonable length and avoid repetition, the proofs of theorems and examples are presented at only one place and are referred to whenever needed.

We have used the fictitious character Batman in all the chapters in order to integrate the applicability aspect of different numerical methods presented in various chapters. The concept at a glance may appear to be inane but we wanted to show that, (i) a practical problem often requires meaningful approximations and robust mathematical formulation, (ii) different aspects of the same problem may require more than one numerical method to obtain a meaningful picture, (iii) there are many ways to model the same problem, and (iv) it's ok to be silly if it helps to clarify a concept.

CONTENT AND STRUCTURE

The book starts by introducing numerical method as an approximate method for solving a mathematical problem by using basic arithmetic operations. Thereafter a direct link is established between a *mathematical problem* and its equivalent *numerical algorithm* separated by an *error*. This is followed by the introduction of various kinds of errors as well as concepts of *condition* of a numerical problem and algorithm. Then the idea of convergent and divergent algorithms is introduced. At this point, we tell the students how easy it is to produce completely worthless results using numerical methods and caution them with a few examples to be skeptical of their results when using numerical methods and impress upon them the importance of analysis. This constitutes *Chapter 1* of the book. Concepts presented here have been referred to in all the chapters of the book.

Chapter 2 begins with a brief recapitulation and visualization of some basic concepts of linear algebra. For the solution of linear system of equations, one direct method namely Gauss elimination is derived and extended to derive other direct methods suitable for different types of applications. Similarly, two iterative methods are derived using a common framework and extended to derive more efficient method using their analyses. For eigenvalue estimation, methods have been presented for single as well as multiple eigenvalues.

Chapter 3 presents the solution of non-linear equations. The chapter starts with presentation of bracketing and open methods for a single equation. It goes on to generalize the methods in order to be able to derive higher order methods. Special methods are presented for finding real and complex roots of polynomials. Finally, the methods for single equation have been generalized to transform a non-linear system of equations to a linear system and thereby establishing a link with the previous chapter.

Chapter 4 deals with the approximation theory for functions in a generalized framework. Weirstrass approximation theorem is presented for generalized applicability. This establishes least square problems and interpolation problems in the same theoretical framework thereby enabling us to deal with continuous functions and discrete data vectors in a similar manner by simply altering the appropriate definitions of inner products and norms. Various types of basis functions and their specialized applications in different least square and interpolation problems are then tackled. Finally, Fourier approximation for periodic functions and data vectors is taken up. The error bounds of approximations have been presented throughout the chapters. Methods of Chapter 2 are used to solve the least square problems.

In *Chapter 5*, the approximations presented in Chapter 4 are used to derive the approximations for derivatives and integrals. These are applicable equally well to discrete data vectors as well as continuous functions. The formulae for derivatives using the Taylor's series approximations are also derived. Quadrature formulae for integration are derived for integration using the concepts of polynomial approximation. The concept of truncation error introduced in Chapter 1 is used to define order of approximation. Moreover, the concept of Richardson's extrapolation is introduced to obtain successively higher order approximations using a lower order method. Concept of phase error is also briefly introduced.

Chapter 6 starts with the solution of initial-value problems (IVPs). Solution is first motivated by graphical means and single step implicit and explicit methods are introduced graphically and also derived from Taylor series. A common framework using Taylor series is established for derivation of multi-step implicit and explicit methods of arbitrary order as well as backward difference formulae (BDFs). Similarly, another common framework is established for derivation of arbitrary order explicit Runge-Kutta methods. Concepts of consistency, stability, and convergence are introduced. Formal analyses methods are introduced for truncation error, phase error, and stability of a method. Combination methods such as predictor-corrector formulae are described. A generalization leads to solution and analysis of systems of initial-value problems using methods of Chapter 2. Stiff systems are introduced and Gear's method is described briefly. Nonlinear IVPs are addressed and linked to the concepts presented in Chapters 2 *and* 3. Higher order boundary-value problems (BVPs) are first solved by decomposing into a system of IVPs and solving by using secant method of Chapter 3. Direct methods are then discussed for boundary-value problems.

Solution of partial differential equations (PDEs) in *Chapter 7* relies heavily on the previous chapter. In effect, parabolic PDEs are decomposed into a system of IVPs and elliptic methods are decomposed into a system of linear algebraic equations using the concepts of direct methods for BVPs. Some discretisation schemes are also derived for hyperbolic PDEs. Von Neumann linear stability analysis is introduced.

Some special methods are taken up and described in *Chapter 8*. These methods include rational interpolation, fast Fourier transforms, finite element method, and the multigrid method. These are not covered for the regular courses but we often find some advanced students asking questions which can be answered better by presenting these methods. The chapter could very well constitute material for a graduate course that may use this book as text.

Online Resources

The Online Resource Centre of the book contains several solved examples of varying degrees of difficulty and using several means of solution (Microsoft Excel, MATLAB, and C). We hope that it would be useful for those who learn best by doing rather than (or, in addition to) thinking!

ACKNOWLEDGEMENTS

Lastly, this preface cannot be completed without paying due regards to numerous unnamed students who have been asking intelligent questions in the class, sitting and listening to our lectures, willingly and unknowingly subjecting themselves to our numerous little experiments to make the understandings of various concepts easier. The comments received on the draft manuscript by Dr S.K. Gupta and Dr S.N. Tripathi were very helpful in improving the presentation of the matter and are gratefully acknowledged. Last but not least, thanks to our beloved student Shivam Gupta for writing and testing the programs in the online resources.

Saumyen Guha
Rajesh Srivastava

Brief Contents

Detailed Contents

8. ADVANCED TOPICS

Introduction

1.1 INTRODUCTION

Many real world problems, from computing the income tax to the design of the nuclear bomb, depend on simulation and prediction through complex mathematical models and a large amount of data. Mathematical modelling utilizes physical laws to develop equations representing system behaviour. In some cases, a phenomenon may be very difficult or hazardous or expensive to model experimentally but may yield to reasonably accurate mathematical modelling.

The mathematical model representing a physical system may be solved using experimental, analytical, and numerical methods or a combination of them. However, rarely does one obtain an *exact* solution because of the approximations introduced in the process at various stages.

The mathematical model may, knowingly or inadvertently, involve some simplifying assumptions; the experimental procedure may have limitations in the accuracy of measurement; the analytical solution may involve irrational numbers or infinite series; and the numerical solution invariably contains errors due to limited precision of computers.

The objective, therefore, is to obtain a *reasonably accurate* solution with *optimum use of resources*. The desired accuracy and efficiency would, of course, depend on the physical problem and must take into account both the nature of the problem and the intended use of the answer.

In experimental methods, we reproduce the system using full-size or scaled models and obtain the solution by measuring and (if needed) scaling the relevant variables or parameters. There is generally an excellent representation of the physical system but the process is typically slow and often expensive. Scaling effects are sometimes non-linear and complicated that need to be properly accounted for in the interpretation of results. Measurement accuracy, cost, and scalability often restrict the application of such models.

Analytical solutions, if possible, are very useful for understanding the response of the system. However, they are often limited to very simple geometry and finding a solution can often be extremely difficult. Sometimes, the analytical solution may involve an infinite series, which may require summation of a large number of terms for desired accuracy and may be time consuming to evaluate.

More often than not, it is impossible to obtain a closed form solution of the complex mathematical models. Instead, one relies on a large number of numerical calculations to obtain an approximation to the solution of the model. The branch of mathematics that deals with transformation of the mathematical problems to numerical calculations is often termed as *Numerical Methods*. Essentially, a numerical method transforms a mathematical problem into a set of computations involving addition, subtraction, multiplication, and division. Mathematical analysis that helps determine the proximity of the approximate solution, obtained using a numerical method, to the exact solution is termed as *Numerical Analysis*.

Numerical methods use numbers to simulate mathematical processes and hence, may be used to solve extremely complex systems. The solution is often quick and is easily adapted for parametric sensitivity studies. With the rapid advancement of computers, numerical methods have become an invaluable engineering tool. Some situations in which numerical methods would be the method of choice are: integration of a function for which either the integral cannot be expressed as analytical expressions or it is too cumbersome and time consuming to evaluate, solution of differential equations for complicated geometry, and/or boundary conditions, large systems of equations, repeated solution of the same system under changing conditions, etc. Moreover, as an engineer, one is likely to come across many software packages to perform numerical simulations, the use of which would probably be the most efficient means for solving engineering problems. One may be tempted to use these packages as *black-box*es but a meaningful use of these would require thorough understanding of the underlying numerical methods.

Sometimes, an engineer may have to develop his or her own software for numerical simulations due to the prohibitive cost, lack of flexibility, and poor computational efficiency of available packages. Even when the available packages are suitable for a particular problem, a working knowledge of numerical methods is valuable in case any difficulties or unreasonable results are encountered. We may also come across new mathematical models, which require a new software for solutions. The use of a numerical method is equally an art and a science (or shall we say, mathematics!). It is desirable to have the ability to select (and possibly modify) a numerical method for a specific problem.

We will try to illustrate the basic concepts with an example. Let us assume *Joker* drops the wheel of *Bat-mobile* from the top of the *Vampire State Building* and *Batman* is looking out of the window at the 13th floor, which is 666 m below the top. As the wheel passes by the window he reports to the police. How long after *Joker* drops the wheel, the police are going to receive the call?

Transforming any practical problem into a mathematical model requires a series of assumptions, such as:

1. Assume the initial downward velocity to be zero (*reasonable assumption*).
2. Ignore all resistances due to air, wind, boundary layer around the building, etc. (*questionable*).
3. Assume that the acceleration due to gravity remains constant from the top of the building to the ground level (*reasonable*).
4. Assume the time taken between viewing the object and making the call to be zero (*reasonable,* don't forget we are talking about *Batman!*).
5. We do not worry about what the *Joker* was doing at the top of the *Vampire State Building* with the wheel.

Using assumptions 2 and 3, the model for the problem can be formulated as a differential equation as follows:

$$\frac{d^2h}{dt^2} = g \tag{1.1}$$

where h is the distance travelled in time t and g is the gravitational acceleration.

Integrating this equation and using assumptions 1 and 4, we obtain an algebraic equation as

$$t = \sqrt{\frac{2h}{g}} \tag{1.2}$$

where t now represents the time when the police receives the call. If g is taken as 9.81 m/sec^2, one can compute t (in sec) for $h = 666$ m. Now, let us try to calculate this time, which is $\sqrt{2 \times 666/9.81}$. After performing one multiplication and one division, one arrives at approximately $\sqrt{135.77982}$. Since one cannot write or store (in a computer) infinitely many digits after the decimal, we need to *chop* or *round-off* this number.

We intend to apply a *numerical method* and use a computer to compute the square root of the rounded-off number. Recall, a numerical method as well as a computer only does simple algebraic operations. How does one compute square root of a number (rational or irrational) by addition, subtraction, multiplication, and division? We invoke our mathematical knowledge to transform the square root problem into a language that computer can understand, namely, $+$, $-$, $\times$, and $\div$.

It is possible to generate a sequence that converges to the square root of a number. One can then get arbitrarily close to the required value by progressing further and further in the sequence. For example, to compute the square root of a number a $(a > 0)$, one can write two different iterative sequences as follows:

Sequence 1:
$$x_{n+1} = \frac{a}{x_n}$$
(1.3)

Sequence 2:
$$x_{n+1} = \frac{1}{2}\left(x_n + \frac{a}{x_n}\right)$$
(1.4)

It is easy to see that *if the sequences converge*, their limits (l) will be equal to $\sqrt{a}$.[1] However, the primary concern is whether they converge? One is often interested to know that before one does a few iterations and discovers that the sequence is not converging. For example, for any initial guess $x_0 > 0$, *Sequence 1* oscillates between x_0 and a/x_0 and does not converge to $\sqrt{a}$.

On the other hand, for *Sequence 2*:
For any initial guess $x_0 > 0$,

$$2x_{n+1}x_n = x_n^2 + a \geq 2x_n\sqrt{a}$$

since, $(x_n - \sqrt{a})^2 \geq 0$. This means $x_{n+1} \geq \sqrt{a}$. So, the sequence is bounded.
Also, for $n \geq 2$,

$$x_n - x_{n+1} = \frac{1}{2}\left(\frac{x_n^2 - a}{x_n}\right) \geq 0$$

since, $x_{n+1} \geq \sqrt{a}$, $\forall n$ and $x_0 > 0$. So, the sequence is decreasing.

This analysis shows that *Sequence 2* is a monotonically decreasing sequence that is bounded below (*Cauchy* sequence) and thus convergent. For example, to find the square root of 135.77982, we start with an initial guess of the square root as $x_0 = 12$ (we know that the square of 12 is 144, therefore the desired root should be close to 12). Subsequent iterations (with an eight digit accuracy) using Eq. (1.4) result in the values $x_1 = 11.657493$, $x_2 = 11.652461$, $x_3 = 11.652460$, $x_4 = 11.652460$. Thus, the square root of 135.77982 is obtained in only 4 iterations

---

[1] Put $x_{n+1} = x_n = l$.

as 11.652460. Assuming that we have no idea about the approximate answer, we may start from a very different value, say, $x_0 = 1$. It would take us 10 iterations to arrive at the root. Similarly, starting from an initial guess of 100, we would again converge to the correct answer in 10 iterations.

We have now seen two computational schemes, Eqs (1.3) and (1.4), for finding the square root of a number. Subsequent mathematical analysis showed that one scheme, Eq. (1.4), converges to the true solution, whereas the other one, Eq. (1.3), does not. The computational scheme of Eq. (1.4) that converges to the true solution can be termed as a *numerical method* for computing the square root of a number. The mathematical analysis conducted to determine whether the computational scheme converges or not is called *numerical analysis*.

Using the numerical method described above, one can solve the algebraic form of the model for the problem of *Joker* dropping the wheel [Eq. (1.2)]. However, *Batman* being a very precise person wants to know the precision of your calculation. So, we need to know where the errors are and how to estimate them.

1.2 ERRORS

In any problem-solving exercise, one encounters errors in many forms and shapes. Some errors may add-up and some may cancel each other to give a net error in the final result. Different forms of errors can be classified as follows:
1. Error in the model or model error.
2. Error in the data or data error.
3. Truncation error.
4. Round-off error.

Let us discuss these errors with the help of the example cited in the previous section.

Mathematical model of a physical process is more often than not a *spherical cow*. One can minimize assumptions and make it a *cylindrical horse* at best but rarely, if ever, they represent the true physical process. This is because, the physical processes are often too complex or some of the processes cannot be characterized. For example, the assumptions 1 and 2 may not be valid for the problem of *Joker* dropping the wheel. Errors imparted in the final result due to these assumptions or as a result of approximations in the model formulation are termed as *model errors*.

The value of gravitational acceleration (g) was taken as 9.81 m/sec^2. The value of g at the site may be 9.80897653879 m/sec^2, which was approximated as 9.81 m/sec^2. Similarly, the distance between the window and the rooftop was taken as 666 m. If one measures more accurately, it may be 665.99 m or 666.04 m. These errors in the data also lead to some error in the final result, which is known as *data error*.

Many of the processes of mathematics, such as differentiation, integration, and the evaluation of series, imply the use of a limit which is an infinite process. The machine has finite speed and can only do a finite number of operations in a finite length of time. This leads to a *truncation error* in the process. To illustrate truncation error, let us consider the Taylor's series expansion of a function $h(t)$ at $(t + \Delta t)$,

$$h(t + \Delta t) = h(t) + \Delta t \frac{dh}{dt} + \frac{\Delta t^2}{2!} \frac{d^2 h}{dt^2} + \frac{\Delta t^3}{3!} \frac{d^3 h}{dt^3} + \frac{\Delta t^4}{4!} \frac{d^4 h}{dt^4} + \cdots \qquad (1.5)$$

Using this, one can express $h(t)$ at $(t + 2\Delta t)$ and at $(t - \Delta t)$ as follows:

$$h(t + 2\Delta t) = h(t) + (2\Delta t) \frac{dh}{dt} + \frac{(2\Delta t)^2}{2!} \frac{d^2 h}{dt^2} + \frac{(2\Delta t)^3}{3!} \frac{d^3 h}{dt^3} + \frac{(2\Delta t)^4}{4!} \frac{d^4 h}{dt^4} + \cdots \qquad (1.6)$$

$$h(t - \Delta t) = h(t) - \Delta t \frac{dh}{dt} + \frac{\Delta t^2}{2!} \frac{d^2 h}{dt^2} - \frac{\Delta t^3}{3!} \frac{d^3 h}{dt^3} + \frac{\Delta t^4}{4!} \frac{d^4 h}{dt^4} - \cdots \qquad (1.7)$$

Using various combinations of Eqs (1.5)–(1.7), one can approximate the second derivative in the differential form of the model problem Eq. (1.1) in different ways:

$$\frac{d^2 h}{dt^2} = \frac{h(t + \Delta t) - 2h(t) + h(t - \Delta t)}{\Delta t^2} - \frac{\Delta t^2}{12} \frac{d^4 h}{dt^4} - \cdots \qquad (1.8)$$

$$\frac{d^2 h}{dt^2} = \frac{h(t + 2\Delta t) - 2h(t + \Delta t) + h(t)}{\Delta t^2} - \Delta t \frac{d^3 h}{dt^3} - \frac{7\Delta t^2}{12} \frac{d^4 h}{dt^4} \cdots \qquad (1.9)$$

If we consider the first term on the right-hand side as the approximation of the second derivative, we make some error. This error is due to approximation of an infinite series by a finite number of terms or in other words, due to truncation of the series. Error encountered in this way is termed as *truncation error*. As seen from these equations, the truncation error is proportional to $(\Delta t)^2$ in the first form [Eq. (1.8)] and to Δt in the second [Eq. (1.9)]. For small Δt, therefore, the first form will have a much smaller error. Further discussion of the truncation error and how to solve the differential form of the model problem using these approximations of the second derivative will be discussed in a later chapter. Here we illustrate the truncation error by using the example of the well-known infinite series for the sine function.

Example 1.1   The sine of an angle x is evaluated using an infinite series given by

$$\sin x = \sum_{n=1}^{\infty} \frac{(-1)^{n-1} x^{2n-1}}{(2n - 1)!}$$

Show a plot of the truncation error over the interval $(0, \pi)$ with the series truncated after 1, 2, 3, 5, 10, and 100 terms.

Solution

The following plot (Fig. 1.1) shows the computed values obtained using 1, 2, 3, and 5 terms of the series (beyond 5 terms the curves are not shown since they plot on the same line):

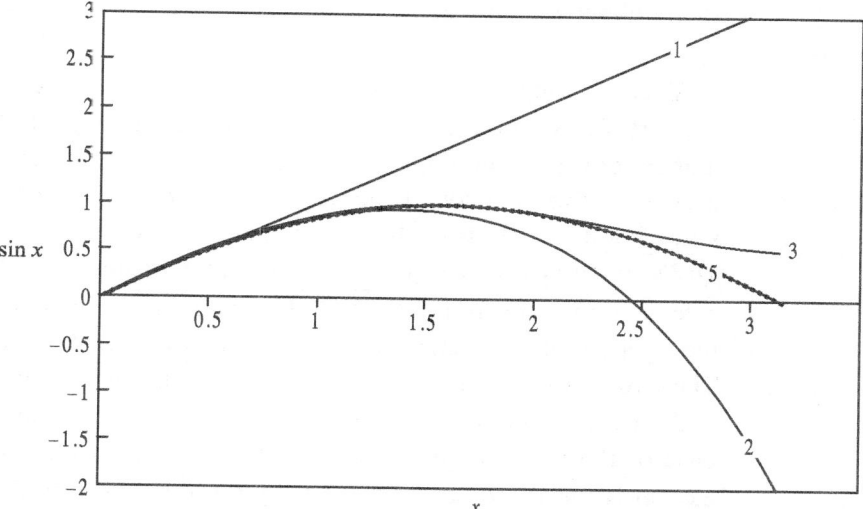

Fig. 1.1 The sine series truncated after a few terms. Symbols show the exact values and the lines show the computed values using the number of terms mentioned on the curves.

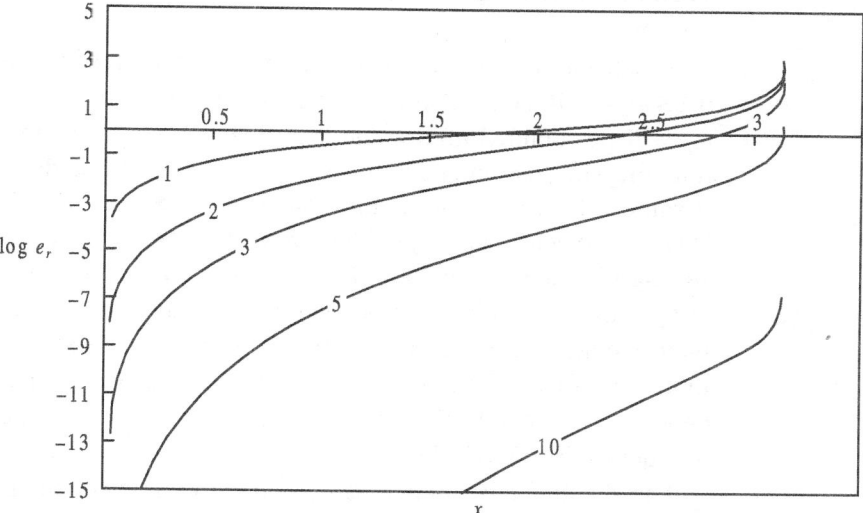

Fig. 1.2 The truncation error in the sine series computed with a few terms. The lines show the log of the relative error.

One may compute the errors in each of the truncated series. Figure 1.2 shows the errors in the computed values obtained using 1, 2, 3, 5, and 10 terms of the series (for 100 terms the error was negligible). The error, e_r, is the error relative to the true value of $\sin x$ (naturally, the points $x = 0$ and π have been excluded).

It is apparent that the error is very small for small x but increases significantly as x approaches π. However, even with only 10 terms, the error near $x = \pi$ is quite small (about 10^{-7}).

Since all computing machines have finite maximum length for storing a number, they are always rounded-off after a few significant digits following decimal. The number of significant digits is often dependent on the machine and precision one uses to perform the computations. For example, if one uses a precision of four significant digits, time for the phone call in the model problem will be $\sqrt{(2\times 666)/(9.81)} = \sqrt{135.8}$. But the same problem with double precision calculation will result in $\sqrt{135.77982}$. Now, if one uses *Sequence* 2 to compute the square root, the number will be rounded-off in each iteration of the sequence. The error in the final result due to these rounding-off is termed as *round-off error*.

A floating point representation of a number in a computer is mb^p where b is the base of the number system, e.g., 2 (binary), 10 (decimal), 16 (hexadecimal), etc., m is the mantissa given by $(1/b) \le m < 1$, and p is the power which is an integer. Let us illustrate the round-off error by using decimal system. Computers use the binary system but humans understand the decimal system much better. Therefore, the examples used here assume the numbers to be stored in the decimal system.

A real number (a) in a decimal system is represented as $a = m_a \times 10^p$ where $0.1 \le m_a < 1$. If the number is irrational and we try to store this number in a computer, m_a will be rounded after some significant digits following decimal. Instead of rounding (i.e., incrementing the last digit by 1 if the next digit is 5 or larger, keeping it same otherwise), one could also use chopping (not changing the last digit irrespective of the next digit). However, rounding is more accurate, though it requires storing a few additional digits. Let us assume that our machine can store a maximum of t places after decimal, i.e., the number is rounded-off after t decimal places. The intermediate computations are generally carried out using a *few* (typically 2 or 3) more significant digits than t and the final result is rounded-off to t digits. We denote the rounded-off mantissa as $\tilde{m}_a$ and the corresponding approximate real number as $\tilde{a} = \tilde{m}_a \times 10^p$.

The absolute error in the mantissa due to round-off can be written as,

$$\left| m_a - \tilde{m}_a \right| \le 0.5 \times 10^{-t}$$

The relative error in real number a due to round-off is given by (note that $0.1 \le m_a < 1$)

$$\left| \frac{a - \tilde{a}}{a} \right| \le \frac{0.5 \times 10^{-t} \times 10^p}{m_a \times 10^p} \le 0.5 \times 10^{1-t} = u \text{ (say)} \quad (1.10)$$

Since t will vary depending on the machine one uses and the precision of calculation (single vs. double), u is called the *machine precision unit* or simply *round-off unit*. For a few isolated computations, round-off error would be very small. However, most practical computations involve a series of steps with each step using the computed values from the previous one. This may cause an unacceptable accumulation of errors in the final result. To analyze this behaviour, we look at the propagation of error and the related concepts of condition of problems and stability of algorithms.

1.3 ERROR PROPAGATION AND CONDITION NUMBER OF A PROBLEM

In the previous section, we saw that the errors in variables are introduced by several means, e.g., data error, round-off error, etc. One would then be interested in estimating the error introduced in the final computation of the functional value as a result of these errors in the variables. In other words, how the errors in variables propagate into the function, whether they grow or decay as the computations proceed. For example, in surveying, if one has measured two sides of a triangle and the included angle, and computes the area of the triangle, what is the likely error in the area due to measurement errors in the sides and the angle? In this section, we will derive an approximate expression that will enable us to calculate the error in the functional value if one has the knowledge of the function and the error in the independent variables.

Let y be a function of n independent variables, $x_1, x_2, \ldots, x_n$, which are written compactly as a vector X. To find the error in y as a result of errors in the data, ΔX, we write $y = f(X)$ and the approximate vector, $\tilde{X} = X - \Delta X$, in which $\Delta x_i = x_i - \tilde{x}_i$.

Due to this error in the variables, approximate functional value is given by $\tilde{y} = f(\tilde{X})$ while the true functional value is $y = f(X)$. So, the error in the functional value is expressed as

$$\Delta y = y - \tilde{y} = f(X) - f(\tilde{X}) \quad (1.11)$$

Since $\Delta X = X - \tilde{X}$, i.e., $X = \tilde{X} + \Delta X$, we use the multivariate Taylor's series and write

$$f(X) = f(\tilde{X} + \Delta X) = f(\tilde{X}) + \sum_{i=1}^{n} \Delta x_i \left. \frac{\partial f}{\partial x_i} \right|_{\tilde{x}_i} + \cdots \text{(higher-order terms)} \quad (1.12)$$

The errors are likely to be small and the *higher-order terms* consisting of errors raised to the power 2 or more can, therefore, be neglected. Then, from Eqs (1.11) and (1.12), we have

$$\Delta y = \sum_{i=1}^{n} \Delta x_i \left.\frac{\partial f}{\partial x_i}\right|_{\bar{x}_i} \tag{1.13}$$

Following example illustrates the use of Eq. (1.13).

Example 1.2   The gravitational acceleration is computed by observing the time, t, it takes for an object to fall a certain height, h. The height measurement is accurate to a cm and the time measurement to a 100th of a second. What error is expected in the computed value of g, if the measured height is 666.00 m and the time 11.65 sec?

Solution   Accounting for the error, the actual values of h and t may be written as 666.00 ± 0.01 m and 11.65 ± 0.01 sec, respectively. Writing $g = f(h,t) = \dfrac{2h}{t^2}$, we have $\dfrac{\partial f}{\partial h} = \dfrac{2}{t^2}$ and $\dfrac{\partial f}{\partial t} = -\dfrac{4h}{t^3}$. From Eq. (1.13), therefore,

$$\Delta g = \frac{2}{\tilde{t}^2}\Delta h - \frac{4\tilde{h}}{\tilde{t}^3}\Delta t = 0.014736\Delta h - 1.6848\Delta t$$

Clearly, the errors due to error in h and t may offset each other if the errors in both h and t are of the same sign (e.g., if we have made positive errors and underestimated both the distance and time). However, we would consider the worst case scenario of making the maximum error (0.01 m and 0.01 s) in both h and t and with opposite signs and obtain an upper bound on the magnitude of the error in g as $|\Delta g| = 0.016995$ m/sec^2. In other words, we will consider the absolute values of all the terms in Eq. (1.13).

The error is generally written as

True error (e) = True value − Approximate value

Sometimes the error is written as (approximate value − true value). We will, however, follow the more common representation.

Note that this definition of error is contingent on the *true value* being known. In most cases where we apply the numerical method, the true value is not known and we seek an approximate value which is *close* to the true value. Typically it is done by generating a sequence [see Eq. (1.4)] which approaches the true value. Therefore, it is customary to define an approximate error as

Approximate error (ε) = Current approximation − Previous approximation

In Example 1.2, both h and t had a maximum measurement error of 0.01 (m and sec, respectively). However, since the magnitude of h was very large compared to t, the error in t is much more detrimental to the accuracy of the computed g. The error in g was obtained as about 0.017 m/sec^2. Whether this is acceptable or not will depend on the expected value of g. For example, if g is close to 10 m/sec^2, the error may be acceptable. But if g is, say, about 0.1 m/sec^2, the error is not acceptable. Therefore, instead of the *absolute error* in the functional value and the variables, it may be better to define a *relative error* as

$$\text{True relative error } (e_r) = \frac{\text{True value} - \text{Approximate value}}{\text{True value}}$$

and

$$\text{Approximate relative error } (\varepsilon_r) = \frac{\begin{array}{c}\text{Current approximation}\\ -\text{ Previous approximation}\end{array}}{\text{Current approximation}}$$

One could then ask whether the *relative* error in the variables will be magnified or reduced in the functional value following computations. In other words, if we make a small relative error in the variable, how much relative error in the function would it cause? This sensitivity of the function to small errors in the variables is defined through a *condition number* (C_P) of a problem.

To describe the relative error in $f(x)$ for a small relative error in x, we define the condition number as the ratio of the relative change in the function $f(x)$, to the relative change in the variable x for a small perturbation Δx,

$$C_P = \frac{\left| \dfrac{f(x + \Delta x) - f(x)}{f(x)} \right|}{\left| \dfrac{\Delta x}{x} \right|} \qquad (1.14)$$

For arbitrarily small Δx or as $\Delta x \to 0$, the above expression reduces to,

$$C_P = \left| \frac{xf'(x)}{f(x)} \right| \qquad (1.15)$$

It is easy to see from the definition that, if $C_P = 1$, the relative error in the functional value remains same as that in the variable. When $C_P > 1$, the error is magnified in the function and the problem is called *ill-conditioned*. When $C_P < 1$, the error is attenuated and the problem is *well-conditioned* (depending on the type of problem, sometimes ill-conditioning is assumed to occur when C_P is larger than a specified number greater than 1. For example, we may accept relative errors of, say, 5%, in the function for a 1% error in the variable). The condition number of a problem is illustrated in Example 1.3.

Example 1.3   The equation for a hyperbola is given as $xy - y + 1 = 0$. The abscissa is measured at some locations and the ordinate is computed as $y = 1/(1 - x)$. It is known that the measurement of x involves some error. Find the condition number of the problem (computing y for a measured x). For what values of x would it be well-conditioned?

Solution   Using Eq. (1.15),

$$C_P = \left| \frac{xf'(x)}{f(x)} \right| = \left| \frac{x/(1-x)^2}{1/(1-x)} \right| = \left| \frac{x}{1-x} \right|$$

For the problem to be well-conditioned, therefore, x should be less than 0.5. At $x = 0.5$, $y = 2$, and if we change x by 1% to 0.505, we get $y = 2.0202$, a relative change of about 1%. This indicates a condition number of 1 (which is also obtained from the expression above). Similarly, for $x = 0.98$, $y = 50$, and if we change x by 1% to 0.9898, we get $y = 98.04$, a relative change of about 96% (the condition number from the equation comes out to be 49 at $x = 0.98$ and 97 at $x = 0.9898$).

If the problem is ill-conditioned, there is not much we can do about it except to use higher precision in the computations. On the other hand, a well-conditioned problem can often be solved using more than one *numerical method* or, in any *numerical method*, by more than one algorithm. Each algorithm will have a specific set of floating point operations that will lead to round-off errors. For the same problem, one can have different round-off errors by using different algorithms. So, even for a well-conditioned problem, the errors may be very different for different algorithms, as shown in the next example.

Example 1.4   A well-conditioned problem is to find the positive root of the equation $y^2 + 2y - x = 0$ for x close to zero. The evaluation of $y = \sqrt{1 + x} - 1$ has a condition number equal to 1 for x approaching zero. Let the computation be done by an algorithm which uses the following three steps:
(a) compute $1 + x$,
(b) perform the square root operation on the result of (a), and
(c) subtract 1 from the result of (b).
Find the condition number of each step.

Solution   The condition number of step (a) is $\left| \dfrac{x \times 1}{1 + x} \right|_{x \to 0} = 0$ and the result (x_a) is close to 1 as x approaches zero.

The condition number of step (b) is

$$\left| \frac{x_a \times \dfrac{1}{2\sqrt{x_a}}}{\sqrt{x_a}} \right|_{x_a \to 1} = \frac{1}{2}$$

and the result (x_b) is close to 1 as x_a approaches 1.
The condition number of step (c) is

$$\left| \frac{x_b \times 1}{x_b - 1} \right|_{x_b \to 1} = \infty$$

and its result (y) is close to zero as x_b approaches 1.

Due to this ill-conditioned step, the algorithm will result in large relative error in y computed with small errors in x, and the above algorithm is *unstable*. Since the problem is well-conditioned, we should be able to devise an algorithm which would be stable, i.e., all steps would be well-conditioned. The reader should verify that the equivalent algorithm given by $y = \dfrac{x}{\sqrt{1 + x} + 1}$ is stable.

1.4 CONDITION NUMBER OF AN ALGORITHM

One can also define a condition number for an algorithm, which takes into account propagation of errors through a set of numerical computation in an algorithm. This is done through a *backward error analysis*, which determines the change necessary in the input variable (or data) to explain the final error in the output (or result) obtained through errors introduced at various stages of calculation. A measure of this change required in the input data to explain the errors in the output result is termed the condition number of the algorithm (C_A). Denoting the *exact* input data by x and the corresponding function value by y, let us assume that an algorithm A operating on x produces the result y_A on a machine with precision unit u [see Eq. (1.10)]. We look for the input data, x_A, which would result in the exact function value, y, if computations are done with infinite precision. The condition number of the algorithm is then defined through

$$\frac{|x - x_A|}{|x|} \leq C_A u \tag{1.16}$$

For example, a multiplication between two real numbers $x = 0.12$ and $y = 0.1554$ would yield 0.018648. We assume that x is not subjected to round-off errors and is exactly known. If we want to carry out a floating point operation ($x \times y$) with precision of four decimal places ($u = 0.5 \times 10^{-3}$), the result, generally written as, $fl(x \times y)$, is 0.1865×10^{-1}. However, we know that the true result $(x \times y)_T$ is 0.18648×10^{-1}. We can obtain the relative error in this computation as follows:

$$\frac{\left| fl(x \times y) - (x \times y)_T \right|}{\left| (x \times y)_T \right|} = 0.10725 \times 10^{-3} \leq u \qquad (1.17)$$

As pointed out in the previous equation, u is an upper bound for the relative error. If we did a multiplication between $x = 0.12$ and $y = 0.1554167$ with infinite precision, we would obtain the same result (0.01865) as we did using floating point operation. If we denote the new value of y required to explain the final result as $y\,(1 + \delta)$, then one can write,

$$fl(x \times y) = x \times y(1 + \delta) \qquad (1.18)$$

So, the final error in the result is equivalent to introducing an error in y. In this case, $\delta = 0.10725 \times 10^{-3}$. By combining Eqs (1.17) and (1.18), it is easily seen that u is an upper bound for δ as well, i.e., $\left| \delta \right| \leq u$.

This process of estimating perturbation (δ) required in the input variable (y) to explain the final result is called *backward error analysis*. An obvious advantage of using the backward error analysis is that we do not need to know the true value of the result, which was needed in a forward error analysis and which is generally not known. The illustration above was for one operation only. It is easy to imagine that, in an algorithm, a series of floating point operation will be required to obtain the final result. In the process, the round-off error will accumulate or may sometimes get cancelled. In the worst case of accumulating round-off error, a larger perturbation will be required in the input variable to explain the final error. This required perturbation in the input data is expressed as $C_A u$, where C_A is the condition number of the algorithm. For the multiplication problem, therefore, the condition number of the algorithm is equal to 1, as u is the upper bound for the relative change in the input variable. On the other hand, if we consider the algorithm

$$f(x) = \sqrt{1 + x^2} - 1$$

and evaluate the function using four significant digits (without using additional buffer digits) for $x = 0.4980 \times 10^{-1}$, we get a value of 0.1000×10^{-2}. This function value corresponds to an exact computation with $x_A = 0.447325 \times 10^{-1}$, a relative error of 0.10176. With a u value of 0.5×10^{-3}, the condition number C_A is obtained as 204, indicating the instability of the algorithm.

One should note that we have used the absolute values in Eqs (1.14)–(1.16) indicating that there is a single input variable, x. If there are several input variables subjected to round-off errors, we would replace the modulus with some indicator of the size (norm) of the vector, X. Although we assume reader's familiarity with vector algebra, we briefly mention here various vector norms, which could be used to represent the *magnitude* of a vector. These would also be useful in Chapter 2 when we discuss the matrix norms.

The L_p norm of an n-dimensional vector, X, i.e. $(x_1, x_2, \ldots, x_n)$, is given by

$$\|X\|_p = \left(|x_1|^p + |x_2|^p + |x_3|^p + \ldots + |x_n|^p\right)^{1/p}, \qquad p \geq 1 \qquad (1.19)$$

The *Euclidean norm*, representing the *length* of the vector in the n-dimensional space is obtained with $p = 2$ and is, therefore, called the L_2 norm. The L_∞ norm or the maximum norm, representing the maximum distance travelled along any direction (i.e., $\max_i |x_i|$) is obtained as $p \to \infty$. And the L_1 norm represents the total distance travelled *along the individual directions*, i.e., $\sum_{i=1}^{n} |x_i|$. Some properties of the vector norms[2] are (α is a scalar):

$$\|X\| = 0 \text{ only if } X \text{ is a null vector; otherwise } \|X\| > 0$$
$$\|\alpha X\| = |\alpha| \|X\| \qquad (1.20)$$
$$\|X_1 + X_2\| \leq \|X_1\| + \|X_2\|$$

The discussion of the condition number of a problem and the condition number of an algorithm, combined with the norm of a vector, allows us to write the error in the *actual computation* of a function as follows:

Let X and Y represent the exact input and output vectors, respectively, of the function $f(X)$. Let $\tilde{X}$ represent the actual input and let Y_A be the result of an algorithm A to evaluate the function, $f(X)$, applied to this input on a machine with precision unit u. Also, let X_A be an input which would result in the same output, Y_A, if computations are performed with infinite precision. The error in the output is then $Y - Y_A$, and the relative error could be written as

$$\begin{aligned}
\frac{\|Y - Y_A\|}{\|Y\|} = \frac{\|f(X) - f_A(\tilde{X})\|}{\|f(X)\|} &\leq \frac{\|f(X) - f(\tilde{X})\|}{\|f(X)\|} + \frac{\|f(\tilde{X}) - f_A(\tilde{X})\|}{\|f(X)\|} \\
&\simeq \frac{\|f(X) - f(\tilde{X})\|}{\|f(X)\|} + \frac{\|f(\tilde{X}) - f_A(\tilde{X})\|}{\|f(\tilde{X})\|} \\
&= \frac{\|f(X) - f(\tilde{X})\|}{\|f(X)\|} + \frac{\|f(\tilde{X}) - f(X_A)\|}{\|f(\tilde{X})\|} \qquad (1.21) \\
&\leq C_P \frac{\|X - \tilde{X}\|}{\|X\|} + \tilde{C}_P \frac{\|\tilde{X} - X_A\|}{\|\tilde{X}\|} \\
&\leq C_P(r + C_A u)
\end{aligned}$$

---

[2] True for all L_p norms.

where r is the relative error in the input and $\tilde{C}_P$ represents the condition number of the problem of evaluating $f(X)$ at $\tilde{X}$, which is approximately equal to C_P at X. Note that we have used the properties of the norm, the definition of the condition number of a problem, and the definition of the condition number of an algorithm to arrive at the final result for the upper bound of the error in an actual computation.

Example 1.5   The computation of $y = f(x) = 2x$ is to be performed for $x = 0.045676$ on a computer using 4 significant digits. What would be an upper bound on the relative error?

Solution   Since we use only 4 significant digits, the exact value of x would be rounded-off to $\tilde{x} = 0.4568 \times 10^{-1}$, a relative error in the input, r, of 8.7573×10^{-5}. The condition number of the Problem is 1, using Eq. (1.15). Condition number of the algorithm for floating point multiplication was obtained earlier as 1. Hence, the upper bound on relative error would be $(r + u) = 5.8757 \times 10^{-4}$.

While the description of different errors and their analyses is interesting and provides us an overview of possible errors in computations and their propagation, from a practical viewpoint, we are more interested in the errors which are in our control while developing or applying a numerical method. The round-off errors could often be minimized by writing the algorithm in a slightly different form and the truncation errors could be reduced by using more steps in the computation. However, generally the reduction of error comes at the cost of more computation time and proper application of a numerical method boils down to an optimum balance between the error and effort. The discussion of various topics in the subsequent chapters of this book, therefore, stresses these two aspects. For example, when acceptable errors in the result are specified, we would look to minimize the computational effort to achieve the desired accuracy. And when different methods with similar computational efforts are compared, we would look to minimize the errors.

EXERCISE 1.1

1. An instrument gives result y in terms of the independent variable x according to
   $y(x) = x + \dfrac{a^3}{x^2}$, where $a = 6.8704294974$ is an instrument constant. We need to compute the difference $\Delta y = y(x_1) - y(x_2)$ for two subsequent measurements with $x_1 = 8.5834541395$ and $x_2 = 8.7282534483$.
   Compute Δy from the definition of y using floating point operations rounding

mantissa to 5 decimals (5-digit precision). Use double precision calculation to obtain a more accurate estimate and treat this as true solution. Estimate the relative error.

2. In Problem 1, express Δy in terms of products and ratios involving a, x_1, and x_2. Recompute using the same 5-digit precision and estimate the relative error. Comment on the differences observed.

3. It is required to compute the roots of the quadratic equation $x^2 - x - 2\varepsilon^2$ for $\varepsilon = 0.001$. Perform the computation with 5-digit precision and estimate the corresponding relative error, performing operations by rounding all mantissas to six decimals. Use double precision calculation to obtain the true solution that is needed to calculate the relative errors.

4. Express the roots by employing a Taylor series expansion. Now compute taking only 2 terms of the Taylor series and estimate the corresponding relative error.

5. On a triangular plot of land, two sides were measured as $a = 100.0 \pm 0.1$ m and $b = 101.0 \pm 0.1$ m. The included angle between the two measured sides was estimated using a theodolite as $C = 58.00° \pm 0.01°$. How accurately is it possible to estimate the third side c? What is the range of error in the estimation of the area of the plot?

6. What is meant by the condition number of a problem? The following set of equations is solved to get the value of x for a given ε. For what values of ε will this problem be well-conditioned?

$$x + y = 2$$
$$x + (1 - \varepsilon)y = 1$$

OUTLINE OF THE BOOK

In the next chapter, we look at the methods of solving a set of linear equations because these are quite frequently encountered during application of several numerical methods. We look at the conditions under which a solution exists, and discuss various techniques of solution, broadly classified as direct or iterative methods, depending on whether we get the solution in a definite number of operations or whether we keep on going till the desired accuracy is achieved. Some aspects of the ill-conditioned system are discussed. Finally, the methods for computation of the eigenvalues are presented.

Chapter 3 starts with the solution of non-linear equations in a single variable. These are also called root-finding problems since the given non-linear equation could be written as $f(x) = 0$ and the solution of the equation amounts to finding the zeros of $f(x)$. Various methods, their errors, and the rate of convergence are described. The methods have been broadly classified in bracketing and open methods depending on whether, to start the process, we need to obtain a closed interval containing the root or not. The extension of these methods to non-linear equations involving multiple variables is also discussed.

The next two chapters (4 and 5) describe the approximations of functions (given either as a continuous function or in discrete form as a set of values), their derivatives, and integrals. Chapter 4 describes various techniques of interpolation (when the approximate function passes through all the given data points) and regression (where the approximation captures the general trend of the function or data), and the errors associated with these. Numerical differentiation and integration is discussed in Chapter 5 for both continuous and discrete cases.

Chapters 6 and 7 discuss various techniques of numerically solving differential equations, starting with the easier problem of ordinary differential equations (ODE) in Chapter 6, and moving on to partial differential equations (PDE) in Chapter 7. Both explicit and implicit methods have been described for solution of ODEs. The classification of PDEs into elliptic, parabolic, and hyperbolic, and the solution techniques for each have been discussed. Error and stability analysis for the differential equations is emphasized.

Finally, in Chapter 8, some advanced topics are briefly covered to introduce the readers to techniques which are generally not mentioned in an introductory course on numerical analysis. The treatment of these topics is, by choice, rather superficial and the interested reader is directed to appropriate resources for further details.

2

System of Linear Equations

2.1 INTRODUCTION

The *Joker* escaped because the *Batmobile* lost the wheel while chasing. To avoid the adverse publicity, *Barrel Motors* decided to provide *Batman* with their new and exclusive model *FruitBat NSX 2009*. Before setting out to chase the *Joker* around, *Batman* decided to explore his new car. To his surprise, he soon found out that the engine does not provide a steady acceleration in the initial period. From zero to a certain unknown speed u, the engine delivers an uneven acceleration. However, that point onwards, the engine provides a steady thrust and a constant acceleration of f. Let us denote the distance covered by the *Batmobile* to reach a speed of u from rest as S_0. Batman knew enough, physics to write his distance of chase S in terms of f, u, S_0, and time t as

$$S = S_0 + ut + \frac{1}{2}ft^2 \qquad (2.1)$$

Since, each batmobile engine is custom manufactured, the f, u, and S_0 are engine specific constants. These constants need to be calibrated after every 1000 km. It is easy to see why it is so important for the Batman to calibrate these parameters precisely. Without these parameters, he will not be able to estimate the precise time his chase will take once his global positioning system (GPS) provides the distance S of the target.

For the first calibration run, he located three cathedrals at distances of 439, 640, and 875 m from his starting position. He timed the chases with a precision of 0.01 sec as 6.20, 14.20, and 22.40 sec, respectively. So, the calibration equations become

$$S_0 + 6.2u + 19.22f = 439$$
$$S_0 + 14.2u + 100.82f = 640 \qquad (2.2)$$
$$S_0 + 22.4u + 250.88f = 875$$

In the matrix form, the set of equations can be written as

$$
\begin{bmatrix} 1 & 6.2 & 19.22 \\ 1 & 14.2 & 100.82 \\ 1 & 22.4 & 250.88 \end{bmatrix}
\begin{bmatrix} S_0 \\ u \\ f \end{bmatrix} =
\begin{bmatrix} 439 \\ 640 \\ 875 \end{bmatrix} \qquad (2.3)
$$

Batman needs to program the solution of this system of equations into his *Batmobile* computer. For recalibration of the constants, he will then need to make three trial runs to preselected targets and measure the travel times. The computer can then take the distance and time data, re-evaluate the constants and store them for future use.

For an actual chase, the GPS picks up the actual distance of the target object and using the constants evaluated above, the *Batmobile* computer displays the precise time of chase. However, Batman did not know how to program the computer to solve the set of equations (2.3). He enquired around and was directed to the *experts* and we gave him this chapter! It starts with some mathematical background of linear algebra which is useful in visualization of a matrix equation and determining whether a solution to any given system of equations exists or not, and, if it does, is it a unique solution or would a different set of values of S_0, u, and f, also satisfy the equations? This is followed by the two main sections of numerical linear algebra, namely, solution of systems of linear equation and estimation of eigenvalues.

2.1.1 Properties of Matrices

A large number of engineering problems involve solution of a system of linear algebraic equations with several independent variables. These equations can be written in the following form:

$$a_{11}x_1 + a_{12}x_2 + a_{13}x_3 + \ldots + a_{1n}x_n = b_1$$
$$a_{21}x_1 + a_{22}x_2 + a_{23}x_3 + \ldots + a_{2n}x_n = b_2$$
$$a_{31}x_1 + a_{32}x_2 + a_{33}x_3 + \ldots + a_{3n}x_n = b_3$$
$$\ldots \qquad \ldots$$
$$a_{m1}x_1 + a_{m2}x_2 + a_{m3}x_3 + \ldots + a_{mn}x_n = b_m$$

In short, the above system of equations can be written as:

$$\sum_{j=1}^{n} a_{ij}x_j = b_i, \quad \text{where } i = 1, 2, 3, \cdots, m \qquad (2.4)$$

Alternatively, the system can be expressed in terms of two vectors (b and x) and a matrix (A), as follows:

$$Ax = b \tag{2.5}$$

where b is a vector of dimension m, x is a vector of dimension n, and A is a matrix of dimension ($m \times n$). The set of equations is as follows:

$$
\begin{bmatrix}
a_{11} & a_{12} & \cdots & a_{1n} \\
a_{21} & a_{22} & \cdots & a_{2n} \\
a_{31} & a_{32} & \cdots & a_{3n} \\
\cdots & \cdots & \cdots & \cdots \\
a_{m1} & a_{m2} & \cdots & a_{mn}
\end{bmatrix}
\begin{bmatrix}
x_1 \\ x_2 \\ x_3 \\ \cdots \\ x_n
\end{bmatrix}
=
\begin{bmatrix}
b_1 \\ b_2 \\ b_3 \\ \cdots \\ b_m
\end{bmatrix}
\tag{2.6}
$$

Let us now try to visualize what this equation means. The matrix A transforms an n-dimensional vector x to an m-dimensional vector b. If x_i's and b_i's are real numbers indicating the position vectors, the matrix A can be seen as a linear transformation $\Re^n \mapsto \Re^m$.

A *linear transformation* (T) is a mapping between two vector spaces X and Y, $T : X \mapsto Y$, if the following holds:

$$T(\alpha x + \beta y) = \alpha T x + \beta T y, \quad \forall x, y \in X \text{ and } \forall \alpha, \beta \in \Re \tag{2.7}$$

In the case of linear transformation $T : \Re^n \mapsto \Re^m$, let $\{e_1, e_2, \cdots, e_n\}$ be the standard basis for $\Re^n$, every vector x in $\Re^n$ can be expressed as a linear combination of the basis vectors (if you are not sure why, please see Box 2.1 for a brief discussion on vector space or consult other books on linear algebra) as

$$x = x_1 e_1 + x_2 e_2 + x_3 e_3 + \cdots + x_n e_n \tag{2.8}$$

Since T is linear, $\quad Tx = x_1 T e_1 + x_2 T e_2 + x_3 T e_3 + \cdots + x_n T e_n \tag{2.9}$

Notice that by definition of the linear transformation T, each Te_1 to Te_n are vectors of dimension m in $\Re^m$. Let us define,

$$u_1 = Te_1, u_2 = Te_2, u_3 = Te_3, \cdots, u_n = Te_n \tag{2.10}$$

Now, construct the $m \times n$ matrix A with u_i's as its columns:

$$A = [u_1 \ u_2 \ u_3 \ \cdots \ u_n] \tag{2.11}$$

Using the relations of Eqs (2.10) and (2.11) in Eq. (2.9), we can say

$$Tx = Ax \tag{2.12}$$

In summary, one can say that a matrix is a representation of the linear transform with a particular set of bases. It is important to note that the composition of A is not arbitrary but depends on the basis $\{e_1, e_2, ..., e_n\}$. For each $x \in X$, the product Ax is a vector in Y. So, it follows that $A(\alpha x + \beta y) = \alpha Ax + \beta Ay$. If the bases are

BOX 2.1 Vector Space, Linear Independence, and Basis

1. Let X be a collection of vectors. If for every x and y in X, a and b in $\Re$ (set of real numbers), there is a unique vector $ax + by$ that also belongs to X (it is closed under addition and scalar multiplication), then X is called a *real vector space*. If a and b are complex numbers, it is called a *complex vector space*. Examples of real vector space include $\Re^2, \Re^3, \Re^4, \ldots, \Re^n$. Any non-empty subset of a vector space that is closed under addition and scalar multiplication is a *subspace* of the vector space. Example: $\Re^2$ is a subspace of $\Re^3$.

2. A vector x in X is a *linear combination* of a set of vectors $\{u_1, u_2, \ldots, u_n\}$ in X if it can be expressed as $x = a_1u_1 + a_2u_2 + \cdots + a_nu_n$ for some scalars $a_1, a_2, \ldots, a_n$.

3. A set of vectors $\{u_1, u_2, \ldots, u_n\}$ in X are *linearly independent* if $a_1u_1 + a_2u_2 + \ldots + a_nu_n = 0$ only when all a_i's are zero. If the summation is zero with some non-zero a_i's, the vectors are linearly dependent.

4. A set of vectors $\{u_1, u_2, \ldots, u_n\}$ is said to *span* a vector space X if X contains all linear combinations of the vectors. Furthermore, if the set of vectors are linearly independent and span the vector space, they form a *basis* for the vector space. For example, [1, 0, 0], [0, 1, 0], and [0, 0, 1] form a basis for $\Re^3$. So, is [1, 0, 0], [1, 1, 0], and [1, 1, 1]. But, [1, 1, 0], [0, 1, 1], and [1, 2, 1] do not.

5. Number of vectors in a basis for a given vector space is constant. It is the minimum number of vectors required to span the space and at the same time, maximum number of independent vectors possible. Number of vectors in the basis is termed as the *dimension* of the vector space. We will denote the vectors $e_1 = [1, 0, \ldots, 0]$, $e_2 = [0, 1, \ldots, 0], \ldots, e_n = [0, 0, \ldots, 1]$, as the *standard basis* for $\Re^n$.

changed for either of the vector spaces, the matrix A is changed as well and a new matrix represents the same linear transformation. Uniqueness of matrix A for a given set of bases is shown in Theorem 2.1. Example 2.1 shows how a change in basis affects the matrix A.

Theorem 2.1

Let $\{u_1, u_2, \cdots, u_n\}$ be a basis for vector space X and $\{v_1, v_2, \cdots, v_m\}$ be a basis for vector space Y. For a linear transformation $T: X \mapsto Y$ with these bases, there exists a unique $m \times n$ matrix A, such that,

$$x = \sum_{i=1}^{n} \xi_i u_i \Rightarrow Tx = \sum_{j=1}^{m} \zeta_j v_j \text{ with } \zeta = A\xi$$

Sketch of Proof: If the j column of matrix A is the coordinate vector of Tu_j with respect to the basis $\{v_1, v_2, \cdots, v_m\}$, then it follows:

$$Tu_j = \sum_{i=1}^{m} a_{ij} v_i, \text{ for } j = 1, 2, \cdots, n \tag{2.13}$$

where a_{ij} are the elements of matrix A.

Since, $\{v_1, v_2, \cdots, v_m\}$ form a basis for the subspace Y and Tu_j is an arbitrary vector in Y, the coefficients a_{ij} are unique.

For an arbitrary vector x in X, $x = \sum_{j=1}^{n} \xi_j u_j$, we can write,

$$Tx = \sum_{j=1}^{n} \xi_j Tu_j = \sum_{j=1}^{n} \xi_j \sum_{i=1}^{m} a_{ij} v_i = \sum_{i=1}^{m} \zeta_i v_i \tag{2.14}$$

Thus, $\zeta_i = \sum_{j=1}^{n} a_{ij} \xi_j$ for $i = 1, 2, \cdots, m$. This essentially means, $\zeta = A\xi$.

Example 2.1 Let us consider a linear transformation $T_1 : \Re^2 \mapsto \Re^3$ with standard bases for $\Re^2$ and $\Re^3$. If the transformation maps the vector $[1, 0]$ to $[1, 2, 3]$ and $[0, 1]$ to $[3, 2, 1]$, the matrix A_1 representing the transformation T_1 can be written as:

$$A_1 = \begin{bmatrix} 1 & 3 \\ 2 & 2 \\ 3 & 1 \end{bmatrix}$$

Let us change the basis for $\Re^2$ to $[1, 0]$ and $[1, 1]$ but keep the standard basis for $\Re^3$. The same transformation T_1 maps $[1, 0]$ to $[1, 2, 3]$ and $[1, 1]$ to $[4, 4, 4]$. So, the new matrix B_1 representing the transformation is

$$B_1 = \begin{bmatrix} 1 & 4 \\ 2 & 4 \\ 3 & 4 \end{bmatrix}$$

If we formulate a new matrix C_1 with the new set of basis for $\Re^2$ as $C_1 = \begin{bmatrix} 1 & 1 \\ 0 & 1 \end{bmatrix}$, note that $B_1 = A_1 C_1$.

Instead, if we changed the basis for $\Re^3$ and kept the standard basis for $\Re^2$, the new matrix representing the same linear transformation could be obtained as $B_1 = D_1^{-1}A_1$ where D_1 is a square matrix of dimension 3 consisting of the new basis vectors as columns. Inverse will always exist since basis vectors by definition are linearly independent. So, in summary, if both bases are changed, the new matrix is given by $B_1 = D_1^{-1}A_1C_1$.

At this point, one can visualize the unknown solution x as the vector that undergoes a linear transformation represented by the matrix A. The image after transformation is the vector b. Before we embark on solving the system of equation, we will try to understand the circumstances under which a solution will exist. We can re-write the system of Eq. (2.6) as a linear combination of column vectors:

$$x_1 \begin{bmatrix} a_{11} \\ a_{21} \\ a_{31} \\ \vdots \\ a_{m1} \end{bmatrix} + x_2 \begin{bmatrix} a_{12} \\ a_{22} \\ a_{32} \\ \vdots \\ a_{m2} \end{bmatrix} + x_3 \begin{bmatrix} a_{13} \\ a_{23} \\ a_{33} \\ \vdots \\ a_{m3} \end{bmatrix} + \cdots + x_n \begin{bmatrix} a_{1n} \\ a_{2n} \\ a_{3n} \\ \vdots \\ a_{mn} \end{bmatrix} = \begin{bmatrix} b_1 \\ b_2 \\ b_3 \\ \vdots \\ b_m \end{bmatrix} \tag{2.15}$$

The above form is similar to linear transformation described by Eq. (2.9). In this form, one can easily see that the solution of the system is finding suitable coefficients (x_i's) corresponding to the column vectors that result in the right-hand side vector b. Each column of the matrix A is a vector of length m. The column vectors would span an n-dimensional vector space if they are linearly independent. The vector space spanned by the columns is often referred to as the *column* space of A. So, a solution of the system will only exist if the right-hand side vector b belongs to this vector space spanned by the columns of A. If some of the vectors in the columns of A are linear combination of other columns such that there are only r linearly independent columns, the *column space* is of dimension r. This r is known as the *rank* of matrix A. When, $r = n$, the columns span the complete n-dimensional vector space. In this case, each vector b can be represented as a linear combination of the columns and there exists at least one solution.

The other $(n - r)$ columns can be expressed as a linear combination of the r linearly independent columns and therefore, add nothing new to the *column space*. Certain set of x_i values may lead to the right-hand vector to be zero. In other words, they provide the solution to equation $Ax = 0$. Collection of all solutions to this equation forms the *null space* of A. Notice that each vector in the *null space* has n elements and lies in n-dimensional vector space. However, since there are not sufficiently many vectors to span the full n-dimensional vector space, the *null space* has a dimension less than n. Actually, it will have a dimension of $n - r$.

Notice that the null space always contains the zero vector $x = 0$. If it also contains some non-zero vectors, from the definition of linear independence of column vectors in Eq. (2.15), we can conclude that the columns of A are not independent.

For a square matrix A ($m = n$), if the rank of the matrix $r = n$ then the columns of the matrix are linearly independent. This set then forms a basis for an n-dimensional vector space. Since, any vector in that space can be represented as a *unique* linear combination of the bases, the system of Eq. (2.15) will have a unique solution for any arbitrary right-hand side vector b. Uniqueness of the solution is shown in Theorem 2.2. In this case, null space contains only $x = 0$. There would also exist another $n \times n$ matrix B such that $BA = I$. That is to say that $B = A^{-1}$ or the matrix A is *invertible*. If the columns are independent, the rows will also be independent in a square matrix. We leave it to the reader to explore why! Concepts of column space and the null space are depicted in Example 2.2.

Theorem 2.2

If $\{u_1, u_2, \cdots, u_n\}$ form a basis for vector space X, every vector in X has a unique representation as a linear combination of the basis vectors.

Proof: Let us choose an arbitrary vector x in X and assume that it has two representations as follows:

$$x = a_1 u_1 + a_2 u_2 + \cdots + a_n u_n \text{ and } x = b_1 u_1 + b_2 u_2 + \cdots + b_n u_n$$

Since, $x - x = 0$, we can write:

$$x - x = 0 = (a_1 - b_1) u_1 + (a_2 - b_2) u_2 + \cdots + (a_n - b_n) u_n$$

By definition of the basis of a vector space $\{u_1, u_2, \cdots, u_n\}$ are a linearly independent set of vectors. Therefore, the above is only possible if all the coefficients are zero, i.e., $a_1 = b_1, a_2 = b_2, \cdots, a_n = b_n$.

Example 2.2   Let us consider the matrix $A = \begin{bmatrix} 1 & -1 & 0 \\ 0 & -2 & 1 \\ -1 & -1 & 1 \end{bmatrix}$.

It is easy to see that the columns are not independent. If we represent the column vectors as u_1, u_2, and u_3, we observe that $u_1 + u_2 + 2u_3 = 0$. Thus, the columns do not span the complete three-dimensional vector space or in other words columns do not form the basis for a three-dimensional vector space. Therefore, solution to equation of the form $Ax = b$ involving this matrix will not exist for any arbitrary right-hand side vector b since it will not be possible to express it as a linear combination of the column vectors. For example, the readers can try to solve the system for $b = [1, 1, 1]^T$ and see for themselves that a solution does not exist. Now, question is, does it have solution for some right-hand side vectors b? If yes, then which ones?

Note that, any two columns of the matrix are linearly independent. Thus, the rank of the matrix is two, dimension of the null space is 1. Therefore, columns of matrix A span a two-dimensional vector space (a plane), which is a subspace of the larger three-dimensional vector space. If we can get some insight into the plane they span, then the solution will exist for any b lying on this plane. In order to construct a vector b, we notice that the column vectors of matrix A can be represented as $[c, c + d, d]^T$, i.e., second element is summation of the first and the third elements. Therefore, any linear combination of them will also inherit this property and the solution will exist for any b that has this property. The plane is the one perpendicular to the vector $[1, -1, 1]^T$ and passes through the origin. The system has a solution for $b = [1, 2, 1]^T$ and the solution is $[0, -1, 0]^T$. It is also easy to check that $Ax = 0$ for the vector $[1, 1, 2]^T$. So, this vector lies in the null space of the matrix A. Let us consider two solution vectors x and y for the same right-hand side vector b. Since, $Ax = b$ and $Ay = b$, $A(x - y) = 0$. This means that the difference between two solutions lies in the null space, i.e., it is a vector in the null space. For this example, one solution is $[0, -1, 0]^T$ and $[1, 1, 2]^T$ is a vector in the null space. Thus, more solutions can be generated by adding the vector in the null space to one of the solution. This is to say, $[1, 0, 2]^T$, $[2, 1, 4]^T$, $[3, 2, 6]^T$, etc. are also solution to the system of equation with the same right-hand side vector.

The solution of the set of Eq. (2.5) is of course $x = A^{-1}b$ provided the A^{-1} exists. If it exists, one can compute the inverse and multiply with b vector to obtain the solution vector x. Although, the solution of the system of Eq. (2.5) can be obtained using A^{-1}, it is rarely, if ever, computed for the engineering problems because more efficient methods of solution are available. Instead, one relies on transforming the set of Eq. (2.6) into a form that can be solved easily. Various *numerical methods* exist for such solutions and form the focus of this chapter.

At this point, we will assume that the reader is familiar with basic matrix algebra such as addition, multiplication, and transpose of a matrix. We will also assume familiarity with computation of the trace and determinant of a matrix. These will not be discussed in this book. Unless otherwise mentioned, we will assume real matrices in $\Re^{m \times n}$ and real vectors in $\Re^n$. Many theorems discussed here are equally applicable for the complex matrices in $C^{m \times n}$ and complex vectors in C^n. We will mention them at appropriate places. To refresh some concepts, here we mention some terms related to matrices.

A real square matrix is *symmetric* if $A = A^T$ and *skew symmetric* if $A^T = -A$. For a complex square matrix, it is *hermitian* if $A = A^H$ and *skew symmetric* if $A^H = -A$. A^T and A^H are the *transpose* and *conjugate transpose* of matrix A. A square matrix is *normal* if $AA^T = A^TA$ for real A and $AA^H = A^HA$ for complex A. A matrix is called *orthogonal* if $AA^T = I$ and *unitary* if $AA^H = I$. A real matrix A is *positive definite* if $x^TAx > 0 \ \forall x \neq 0$. They are also characterized by positive *eigenvalues* and positive *pivots*. In terms of application, we will use these properties for derivation of some of the numerical methods. However, the readers have to wait a few more sections for clear implication of pivots and eigenvalues. The *pivots* will first appear when we perform Gauss elimination and we discuss the eigenvalues in the next section.

2.1.2 Eigenvalues and Eigenvectors

For a matrix A, if there exist a vector x and a scalar λ, such that $Ax = \lambda x$, x is called an eigenvector and λ is the corresponding eigenvalue. Thus, it follows from the definition:

$$Ax = \lambda x \quad \text{and} \quad A^kx = \lambda^kx \tag{2.16}$$

Therefore, eigenvalues are the roots of the polynomial given by $\det(A - \lambda I) = 0$. The resulting polynomial is known as the *characteristic polynomial*. Then it automatically follows that

$$\det(A) = \prod_{i=1}^{n} \lambda_i$$

From properties of the polynomials and coefficient of the term λ^{n-1} in characteristic polynomial, one can also write,

$$\text{trace}(A) = \sum_{i=1}^{n} \lambda_i \tag{2.17}$$

Once eigenvalues are determined, the eigenvectors can be obtained by solving the set of equations $(A - \lambda I) x = 0$. Note that $(A - \lambda I)$ is a singular matrix since its determinant is zero and there would not be a unique solution for x (we will get a direction of the eigenvector, its magnitude can be arbitrary).

Before we proceed to further mathematical interpretation, let us spend some time on visualization of the concept of eigenvalue and eigenvector. For this purpose, let us focus on a square matrix A of size 2 and a vector of size 2 so that the results can be plotted within the realm of the pages of this book.

A vector, when multiplied by a matrix experiences rotation and/or stretching or contraction. However, the equation $Ax = \lambda x$ implies that the eigenvectors do not experience rotation. They simply stretch or contract with a ratio of λ. In order to visualize this, let us consider a unit vector x,

$$x = \begin{bmatrix} \cos\theta \\ \sin\theta \end{bmatrix}, \quad \theta \in (0, 2\pi) \tag{2.18}$$

Next consider a matrix

$$A = \begin{bmatrix} 3 & 1 \\ 2 & 2 \end{bmatrix} \tag{2.19}$$

If we plot the position vector x for all values of θ, it traces a circle of unit radius (Fig. 2.1). In the same plot, the position vector of Ax is shown as the ellipse. The circle is transformed to an ellipse due to rotation and elongation of vector x as a result of multiplication by the matrix A.

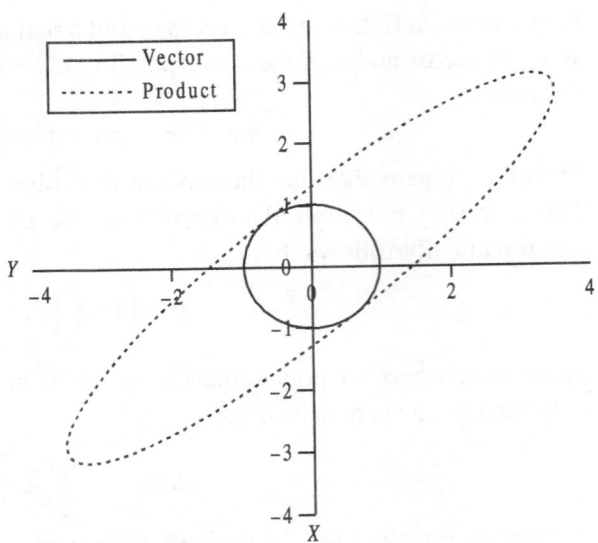

Fig. 2.1 Traces of vectors x and Ax of Eqs (2.18) and (2.19).

By equating the determinant of the matrix $(A - \lambda I)$ to zero, one obtains the eigenvalues as 4 and 1. The corresponding eigenvectors are given by (1,1) and (−0.5, 1), respectively. By normalizing the eigenvectors in order to correspond to the circle of x one obtains, $(1/\sqrt{2}, 1/\sqrt{2})$ and $(-1/\sqrt{5}, 2/\sqrt{5})$. The reader can easily check that these two vector directions do not change as a result of multiplication by matrix A. The eigenvector $(1/\sqrt{2}, 1/\sqrt{2})$ corresponding to eigenvalue $\lambda = 4$ elongates by a factor of 4. The eigenvector $(-1/\sqrt{5}, 2/\sqrt{5})$ corresponding to eigenvalue $\lambda = 1$ remains unaltered. At this point, the readers may also note that if x is an eigenvector, −x will also be one (in fact any scalar multiple of x would be an eigenvector).

Let us look at an enlarged portion of the second quadrant (Fig. 2.2). Vector marked as 1 is the eigenvector corresponding to $\lambda = 1$. Vector 2 becomes Vector 3 as a result of multiplication by matrix A although no elongation or contraction takes place (both lie on the circle). All vectors other than the eigenvectors will rotate as a result of the multiplication by a matrix. They may or may not experience elongation or contraction. The eigenvectors on the other hand will not rotate by definition. They will experience elongation or contraction in the same proportion as the corresponding eigenvalue. Mathematically speaking, the vectors x and Ax will belong to the same subspace which is *invariant* for matrix A. A subspace $X \subseteq C^n$ is *invariant* for matrix A iff $x \in X$ implies $Ax \in X$.

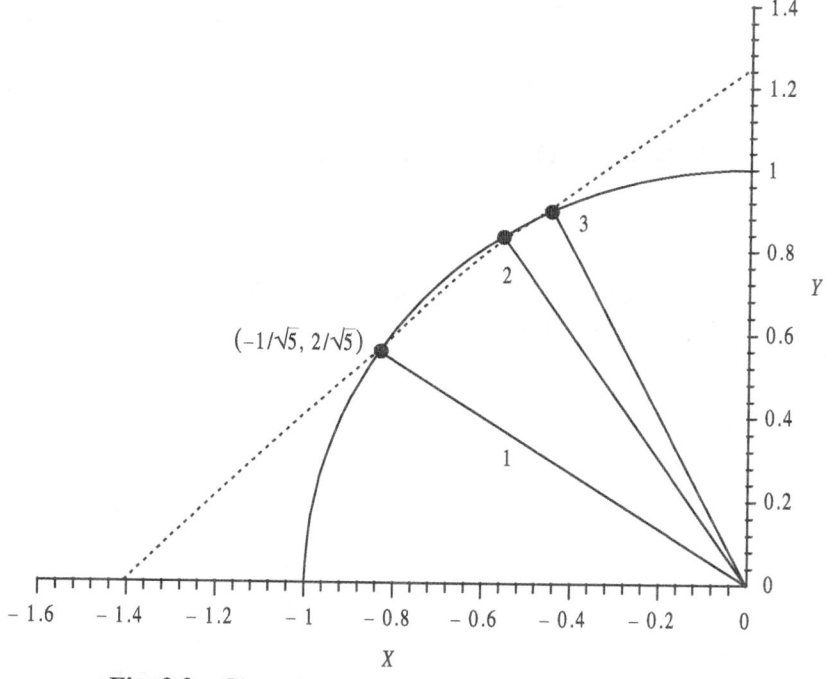

Fig. 2.2   Closer look at the second quadrant of Fig. 2.1.

There may be more than one linearly independent eigenvectors corresponding to an eigenvalue. For a simple example, let us consider a diagonal matrix,

$$A = \begin{bmatrix} 2 & 0 \\ 0 & 2 \end{bmatrix} \tag{2.20}$$

The loci of x and Ax are now concentric circles (Fig. 2.3). It is easy to see that both the eigenvalues are equal to 2. Every vector x is an eigenvector.

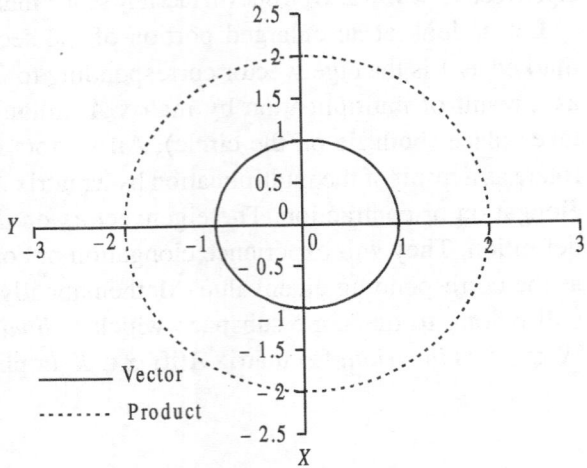

Fig. 2.3 Traces of vectors *x* and *Ax* of Eqs (2.18) and (2.20).

If the diagonal matrix is changed to

$$A = \begin{bmatrix} 2 & 0 \\ 0 & 1 \end{bmatrix} \tag{2.21}$$

once again we get back two unique eigenvectors corresponding to eigenvalues 2 and 1 lying along the *x* and *y* axes, respectively (Fig. 2.4).

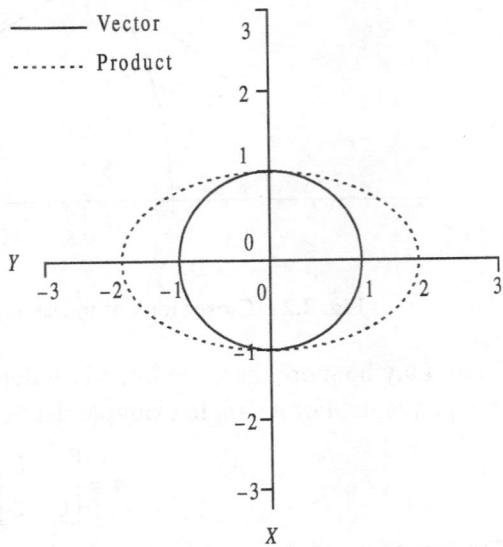

Fig. 2.4 Traces of vectors *x* and *Ax* of Eqs (2.18) and (2.21).

We have seen in the previous examples that an eigenvalue can occur more than once in a matrix. If an eigenvalue occurs only once in a matrix, it is known as a *simple eigenvalue*. If an eigenvalue occurs n times in a matrix, it is said to have an *algebraic multiplicity* of n. For example, the eigenvalue $\lambda = 2$ in the matrix in Eq. (2.20), has an algebraic multiplicity of 2. An eigenvalue with algebraic multiplicity >1 may or may not have multiple linearly independent eigenvectors associated with them. The matrix in Eq. (2.20) has two linearly independent eigenvectors for the eigenvalue with algebraic multiplicity of 2. Now consider the following matrix:

$$A = \begin{bmatrix} 3 & 1 \\ -1 & 1 \end{bmatrix} \tag{2.22}$$

It is easy to verify that for this matrix, the eigenvalue $\lambda = 2$ has an algebraic multiplicity of 2. However, there exists only one linearly independent eigenvector associated with it which is $(1, -1)$. The number of linearly independent eigenvectors associated with an eigenvalue is known as its *geometric multiplicity*. If the *algebraic multiplicity* of an eigenvalue is greater than its *geometric multiplicity*, it is known as a *defective* eigenvalue. The matrix with a defective eigenvalue is a *defective* matrix. Otherwise, the matrix is *non-defective*. This classification has important implications when the eigenvalues are estimated using diagonalization (Section 2.3).

From a computational point of view, we would be interested in the condition of a matrix (see discussion of condition number in Chapter 1) in terms of whether it will significantly amplify small round-off errors or not. For this purpose, we should define some measure of the size of a matrix. Similar to vector norms, described in Chapter 1, we now define matrix norms.

2.1.3 Matrix Norms

A matrix norm is useful to have a sense of the *size* of a given matrix. This *size* is not related to the number of rows or columns or elements in a matrix but represents the magnitude of the effect it creates when operated on a vector. There are many ways to define such a norm for a matrix. Before we get on with the definitions of various matrix norms, let us first list a set of properties that any defined norm must satisfy and have a visual sense of what matrix norm might mean. Following on the discussion of vector norm (in Chapter 1, Section 1.4), one may list the following properties a matrix norm must satisfy:

- $\|A\| > 0 \quad \forall A \neq 0$ and $\|A\| = 0$ iff $A = 0$ (2.23)

- $\|\alpha A\| = |\alpha| \|A\|$ for all scalar α (2.24)

- $\|A + B\| \leq \|A\| + \|B\|$ (2.25)

- $\|AB\| \leq \|A\|\|B\|$ (2.26)

- $\|Ax\| \leq \|A\|\|x\|$ for the matrix norm to be consistent with the norm
  of vector x (2.27)

Let us take a re-look at Fig. 2.1 for a sense of this *size*. All the unit vectors contained in the circle remain within the closed curve traced by Ax. It may be an ellipse (Fig. 2.1) or a circle (Fig. 2.3) depending on the eigenvalues. Thus, if we can find the maximum magnitude of extension a unit vector experiences as a result of the multiplication, expansion of all other vectors would be less than or equal to that. If this magnitude of maximum expansion is defined as the *size* or norm of the matrix, its effect is to stretch any vector by the same factor or less. The only job then is to find this maximum extension. Intuitively, one may think that it is in the direction of the major axis of ellipse but it is not always so. For example, in Fig. 2.1, the vector along the major axis of the ellipse, which is also an eigenvector, stretches by a factor of 4, same as the corresponding eigenvalue. It will be easy for the reader to check that a larger extension occurs for the vector (0.86491, 0.50193), which is transformed to (3.0967, 2.7337) and experiences an elongation to the extent of 4.131. This vector not being an eigenvector also experiences a rotation. The task is then to find the magnitude of this maximum extension in a logical manner. For an arbitrary $m \times n$ matrix A, it is easy to verify that $A^T A$ is a square and symmetric matrix. The square root of the maximum eigenvalue of $A^T A$ provides this magnitude of maximum extension for A. A brief outline of the derivation is shown below.

Let us consider a matrix A that transforms the unit vector x ($x^T x = 1$) to a vector b. We are interested in finding the direction x for which norm of vector b is maximum. Amount of maximum extension using Euclidean norm for vectors can then be calculated as $\|b\| = \sqrt{b^T b}$. This is a maximization problem which can be stated as, maximize $b^T b$ subject to $x^T x = 1$. Using Lagrange multiplier (λ), one may write the following:

$$\nabla\left[b^T b - \lambda\left(x^T x - 1\right)\right] = 0 \qquad (2.28)$$

Since $Ax = b$,

$$\nabla\left[x^T A^T A x - \lambda\left(x^T x - 1\right)\right] = 0 \text{ or } A^T A x = \lambda x \qquad (2.29)$$

Therefore, the maximum extension is obtained when the multiplier λ (of maximum absolute magnitude) is an eigenvalue of the matrix $A^T A$ and the magnitude of this extension is $\|b\| = \sqrt{b^T b} = \sqrt{x^T A^T A x} = \sqrt{x^T \lambda x} = \sqrt{\lambda}$. This quantity is known as the *spectral norm* of matrix A. Spectral norm is typically denoted as $\|A\|_2$ and formally defined as

$$\|A\|_2 = \left[\max_{i=1,\,n} |\lambda_i| \right]^{1/2} \tag{2.30}$$

where λ_i's are eigenvalues of the square symmetric matrix $A^T A$.

Example 2.3   Find the spectral norm of the matrix in Eq. (2.19).

Solution

$$A = \begin{bmatrix} 3 & 1 \\ 2 & 2 \end{bmatrix}, \quad A^T = \begin{bmatrix} 3 & 2 \\ 1 & 2 \end{bmatrix}, \quad A^T A = \begin{bmatrix} 13 & 7 \\ 7 & 5 \end{bmatrix} \tag{2.31}$$

The largest eigenvalue of $A^T A$ is 17.06226 and its square root is 4.13065, which is the maximum extension for any vector due to multiplication by A.

Similar to *p-norm* for a vector (Section 1.4), a matrix norm can also be defined as follows:

$$\|A\|_p = \max_{x \neq 0} \frac{\|Ax\|_p}{\|x\|_p} \tag{2.32}$$

Numerator is the norm of the transformed vector and the denominator is the original vector. The above norm gives a measure of the maximum extension ratio attainable in the transformation. Using this definition, the following two norms can be easily computed:

In Eq. (2.32), if one is interested to see how much a unit vector would expand due to linear transformation given by matrix A, one can use the standard basis (Box 2.1) for a vector space as unit vectors. In this case, if one uses L_1 norm (Chapter 1) for computing the norm for the vectors, one obtains the *column-sum norm* as follows:

$$\|A\|_1 = \max_{1 \le j \le n} \sum_{i=1}^{m} |a_{ij}| \tag{2.33}$$

Alternatively, one can consider extension for a vector in which all the components are unity. For this vector, the L_∞ norm (Chapter 1) is unity. In this case, the L_∞ norm of the resulting vector gives the *row-sum norm* as follows:

$$\|A\|_\infty = \max_{1 \le i \le m} \sum_{j=1}^{n} |a_{ij}| \tag{2.34}$$

Frobenius norm for a matrix is defined as follows:

$$\|A\|_F = \sqrt{\sum_{i=1}^{m}\sum_{j=1}^{n} a_{ij}^2} = \sqrt{\text{trace}(A^T A)} \tag{2.35}$$

For a matrix A with p distinct eigenvalues, the largest absolute eigenvalue is known as the *spectral radius* and is typically denoted as $\rho(A)$. That is,

$$\rho(A) = \max_{1 \le i \le p} |\lambda_i| \tag{2.36}$$

The *spectral radius* provides a lower bound for the matrix norms. In order to see this, let us consider a eigenvalue–eigenvector pair,

$$Ax_j = \lambda_j x_j \text{ so } \|Ax_j\| = |\lambda_j|\|x_j\|; \text{ but } \|Ax_j\| \le \|A\|\|x_j\|$$

This essentially means

$$|\lambda_j| \le \|A\| \; \forall j \quad \text{and therefore} \quad \rho(A) = \max_{1 \le j \le n} |\lambda_j| \le \|A\| \tag{2.37}$$

We will end this section by stating a relationship between the matrix norms and the spectral radius. For any norm of matrix A,

$$\rho(A) = \lim_{p \to \infty} \|A^p\|^{1/p} \tag{2.38}$$

Thus, spectral radius can be interpreted as the asymptotic growth rate of the norm of the power sequence of matrix A. The above relation is known as the *Gelfand's formula* [Gelfand (1941)].

With the background described in this section, we are now ready to proceed to numerical linear algebra. In the following sections, we will cover two topics, methods for solution of systems of linear equations and estimation of eigenvalues of a matrix.

EXERCISE 2.1

1. (a) Show that all the eigenvalues of a positive-definite matrix are positive.
    (b) Show that the spectral radius of a square matrix cannot exceed the largest sum of the absolute values of the elements of the matrix along any row.
2. (a) For $x, a \in \Re$ but $\ne 0$, show that the sequence $x_{k+1} = x_k + x_k(1 - ax_k)$ converges to $1/a$ if and only if $|1 - ax_0| < 1$.
    (b) Let us consider the matrix analogy of this sequence for a nonsingular $n \times n$ matrix A, and an arbitrary initial $n \times n$ matrix X_0. We define a sequence of matrices as
    $$X_{k+1} = X_k + X_k(I - AX_k), \quad k = 0, 1, 2, \ldots$$

Show that $\lim_{k \to \infty} X_k = A^{-1}$ if and only if the special radius $\rho(I - AX_0) < 1$.

(c) Compute the inverse of matrix $\begin{bmatrix} 4 & 1 \\ 1 & 4 \end{bmatrix}$ using this iterative scheme.

3. Derive Eqs (2.33) and (2.34) from Eq. (2.32).
4. (a) We have seen (in Problem 3) that the 1-norm is the *column sum norm* and the ∞-norm is the *row sum norm*. Show from Definition (2.32) for the matrix norm that the 2-norm is same as the *spectral norm.*
   (b) For a symmetric matrix, prove that the *spectral norm* and the *spectral radius* are equal.

2.2 SOLUTION OF LINEAR SYSTEMS

When the solution of a system exists, one is often interested in computing it in the most economical fashion. Economy in this context refers to minimum number of floating point computation. Although, solution of the system is $x = A^{-1}b$, computation of inverse of the coefficient matrix A requires large number of computations and is best avoided for solving a system of equation. We will see the computational requirements later in the chapter. So, goal of all the *numerical methods* in this direction is to minimize the number of computations to solve a system. The existing methods can broadly be classified into two distinct categories. These are as follows:

Direct Methods This group of methods obtains the solution of a system in a finite number of computations or steps. Final errors in the solution are mainly due to round-off error. In this group, we will outline the *Gauss elimination, Gauss Jordon,* and *LU decomposition* for general coefficient matrices and *Thomas algorithm* for a special group of matrix.

Indirect Methods In these methods, an approximate solution is assumed and iteratively improved to obtain desired accuracy. Since, the solution vector in this method is approached in a sequential manner, the final error can be arbitrary. However, the minimum error or the lower bound of the error is the accumulation through round-off error, which in turn, is a function of the number of iterations. We will outline three methods in this group: *Jacobi iteration, Gauss Seidel iteration,* and a modification of Gauss-Seidel in the form of *successive over-relaxation (SOR).*

2.2.1 Direct Methods

We will consider equations of the form (2.5) with square coefficient matrix A of dimension $(n \times n)$ with rank n, such that a unique solution exists. If the coefficient matrix has only non-zero elements along the major diagonal (A) and all other off-diagonal elements are zero, i.e., A is a *diagonal matrix*, the solution vector x can be obtained directly by dividing the elements of b vector by the diagonal element in the same row.

If we have a coefficient matrix which is an upper triangular matrix, i.e., all the entries below the main diagonal are zero, the set of equation can be represented as follows:

$$
\begin{bmatrix}
a_{11} & a_{12} & \cdots & \cdots & a_{1n} \\
0 & a_{22} & \cdots & \cdots & a_{2n} \\
0 & 0 & a_{33} & \cdots & a_{3n} \\
\cdots & \cdots & \cdots & \cdots & \cdots \\
0 & 0 & 0 & \cdots & a_{nn}
\end{bmatrix}
\begin{bmatrix}
x_1 \\ x_2 \\ x_3 \\ \cdots \\ x_n
\end{bmatrix}
=
\begin{bmatrix}
b_1 \\ b_2 \\ b_3 \\ \cdots \\ b_n
\end{bmatrix}
\tag{2.39}
$$

The above set of equations can be solved directly starting from the last equation. Solution of the last equation or nth equation is

$$
x_n = \frac{b_n}{a_{nn}}
\tag{2.40}
$$

Using the value of x_n, one can now obtain the value of x_{n-1} from the $(n-1)$th equation, as

$$
x_{n-1} = \frac{b_{n-1} - a_{n-1n} x_n}{a_{n-1n-1}}
\tag{2.41}
$$

Knowing the values of x_n and x_{n-1}, one can now obtain x_{n-2} from the $(n-2)$th equation. This way one can continue up to the first equation and obtain all the x_i's provided the diagonal elements or a_{ii}'s are non-zero for all values of i. This method of solving an upper triangular system is known as *back substitution*. The algorithm for the *back substitution* can be written as follows:

$$
x_n = \frac{b_n}{a_{nn}} \text{ and } x_i = \frac{\left(b_i - \displaystyle\sum_{j=i+1}^{n} a_{ij} x_j \right)}{a_{ii}}; \quad i = (n-1), \ldots, 3, 2, 1
\tag{2.42}
$$

If the coefficient matrix is a lower triangular matrix, i.e., the entries above the main diagonal are zero, the system of equation can be written as follows:

$$
\begin{bmatrix}
a_{11} & 0 & \cdots & \cdots & 0 \\
a_{21} & a_{22} & 0 & \cdots & 0 \\
a_{31} & a_{32} & a_{33} & \cdots & 0 \\
\cdots & \cdots & \cdots & \cdots & \cdots \\
a_{n1} & a_{n2} & a_{n3} & \cdots & a_{nn}
\end{bmatrix}
\begin{bmatrix}
x_1 \\ x_2 \\ x_3 \\ \vdots \\ x_n
\end{bmatrix}
=
\begin{bmatrix}
b_1 \\ b_2 \\ b_3 \\ \vdots \\ b_n
\end{bmatrix}
\tag{2.43}
$$

This lower triangular system can be solved similarly starting from the first equation and sequentially substituting the known values in the subsequent equations. In this way, one can obtain all the x_i's progressing from the first equation down to the last equation provided all the diagonal elements are non-zero. This method of solving a lower triangular system is called the *forward substitution*. The algorithm can be written as follows:

$$x_1 = \frac{b_1}{a_{11}} \text{ and } x_i = \frac{\left(b_i - \displaystyle\sum_{j=1}^{i-1} a_{ij} x_j \right)}{a_{ii}}, \ i = 2, 3, 4, \ldots, n \qquad (2.44)$$

All the direct methods attempt to transform the coefficient matrix A into an upper triangular or a lower triangular or a diagonal matrix or a combination of these with non-zero diagonal elements through a finite number of operations.

Number of floating point operations required to solve a triangular system can be calculated using Eq. (2.42) or Eq. (2.44). In Eq. (2.44), evaluation of x_1 requires only one division. Evaluation of x_2 requires one subtraction, one multiplication, and one division. Proceeding similarly, one can easily see that to obtain x_i one requires to perform $(i-1)$ addition, $(i-1)$ multiplication, one subtraction, and one division, i.e., total $2i$ operations. Thus, in order to solve the complete system one needs to perform $n(n+1)$ operations. For large n, one can say that the number of operations required is approximately of the order n^2. If one assumes that addition and subtraction take less CPU time compared to multiplication and division, the number of effective computation becomes half. We first describe the most common method for transforming the coefficient matrix to an upper triangular matrix.

2.2.1.1 *Gauss Elimination*

This is one of the most popular methods for solving a linear system of equation of the form of Eq. (2.6). In this section, we will assume that the coefficient matrix has all sorts of nice properties, such as, square $(n \times n)$ with rank n such that we do not have to worry about the existence of the solution. In this method, the coefficient matrix will be reduced to an upper triangular matrix through a series of row operations. Our starting matrix A is the one shown in Eq. (2.6) with $m = n$. Although, we will mainly operate on the rows, the elements below the leading diagonal will be reduced to zero column-wise. As we proceed to make sub-diagonal elements of each column to zero, we will count these as *steps*.

Step 1: The goal is to reduce the elements of 1st column to zero except the element a_{11}. In order to do that, we will keep the 1st row unaltered and define a multiplying factor for each row. For example, to make a_{21} zero, we can multiply the 1st row by (a_{21}/a_{11}) and subtract from the 2nd row. So, (a_{21}/a_{11}) is the multiplying factor for the 2nd row. Similarly, multiplying 1st row by the multiplying factor for the 3rd row (a_{31}/a_{11}) and deducting it from 3rd row will make a_{31} zero. Proceeding this way, we can reduce a_{i1} to zero by multiplying the 1st row by the multiplying factor for the ith row (a_{i1}/a_{11}) and deducting it from the ith row. This sequence of operation is equivalent to multiplying the first equation by a constant and deducting the resulting equation from all subsequent equations in order to make the coefficient of the variable x_1 to be zero, i.e., eliminate the variable. So, similar operations also need to be conducted for the right-hand vector b. These operations on b can be carried out independently or alternatively, the vector can be added as the $(n + 1)$th column of the matrix A.

We will denote the multiplying factor of ith row as $l_{i1} = (a_{i1}/a_{11})$. Note that the second index on the multiplying factor as well as 1 in the indices on a correspond to the step number. The algorithm for *Step 1* can be written as follows:

$$l_{i1} = \frac{a_{i1}}{a_{11}}, \quad a_{ij} = a_{ij} - l_{i1}a_{1j}, \text{ and } b_i = b_i - l_{i1}b_1 \text{ for } i = 2, \ldots, n \text{ and } j = 1, \ldots, n \ (2.45)$$

If the vector b is augmented as the $(n + 1)$th column of the matrix A

$$l_{i1} = \frac{a_{i1}}{a_{11}}, \quad a_{ij} = a_{ij} - l_{i1}a_{1j} \text{ for } i = 2 \text{ to } n \text{ and } j = 1 \text{ to } (n + 1) \tag{2.46}$$

We notice that augmenting the vector b essentially results in the same operation to be computed with range of j extended to $(n + 1)$ in place of n. This point onwards, the algorithm for the augmented matrix will not be shown explicitly.

It is easy to see that the entire operation will fail if the element a_{11} is zero or very close to zero, since it appears in the denominator of the multiplying factor. It is known as the *pivot* of *Step 1*. Also note that while a multiplying factor is defined for each row, it is always multiplied to the first row. At the end of *Step 1*, only first row remained unchanged. All other elements in the matrix in rows 2 to n are the modified values.

After the first step, the modified matrix A is of the following form:

$$\begin{bmatrix} a_{11} & a_{12} & \cdots & \cdots & a_{1n} \\ 0 & a_{22} & \cdots & \cdots & a_{2n} \\ 0 & a_{32} & \cdots & \cdots & a_{3n} \\ \cdots & \cdots & \cdots & \cdots & \cdots \\ 0 & a_{n2} & \cdots & \cdots & a_{nn} \end{bmatrix} \tag{2.47}$$

Notice that all the elements except for the first row are modified.

Step 2: In this step, we will reduce sub-diagonal elements of column 2 $(a_{32}, a_{42}, ..., a_{n2})^T$ to zero. For the third row, we define a multiplying factor $l_{32} = (a_{32} / a_{22})$, multiply the second row with it and deduct from the third row. Similarly, for the *i*th row, we define the multiplying factor as $l_{i2} = (a_{i2} / a_{22})$, multiply the second row with it and deduct from the *i*th row. Once again note that the second index on the multiplying factor as well as 2 in the indices of *a* correspond to the step number. The computations are done on rows 3 to *n* and in each row, on columns 2 to *n*. So, the algorithm for *Step 2* can be written as,

$$l_{i2} = \frac{a_{i2}}{a_{22}}, \quad a_{ij} = a_{ij} - l_{i2}a_{2j} \quad \text{and} \quad b_i = b_i - l_{i2}b_2 \quad \text{for } i = 3 \text{ to } n, \; j = 2 \text{ to } n \quad (2.48)$$

At the end of this step, the modified matrix is as follows:

$$\begin{bmatrix} a_{11} & a_{12} & \cdots & \cdots & a_{1n} \\ 0 & a_{22} & \cdots & \cdots & a_{2n} \\ 0 & 0 & \cdots & \cdots & a_{3n} \\ \cdots & \cdots & \cdots & \cdots & \cdots \\ 0 & 0 & \cdots & \cdots & a_{nn} \end{bmatrix} \quad (2.49)$$

Continuing in this fashion, one can write the computations of the *k*th step as follows:

$$l_{ik} = \frac{a_{ik}}{a_{kk}}, \quad a_{ij} = a_{ij} - l_{ik}a_{kj} \quad \text{and} \quad b_i = b_i - l_{ik}b_k$$

$$\text{for } i = (k + 1) \text{ to } n, \, j = k \text{ to } n \quad (2.50)$$

After completion of *Step k*, the coefficient matrix is of the following form:

$$\begin{bmatrix} a_{11} & a_{12} & a_{13} & a_{14} & \cdots & a_{1k} & a_{1k+1} & \cdots & a_{1n} \\ 0 & a_{22} & a_{23} & a_{24} & \cdots & a_{2k} & a_{2k+1} & \cdots & a_{2n} \\ 0 & 0 & a_{33} & a_{34} & \cdots & a_{3k} & a_{3k+1} & \cdots & a_{3n} \\ 0 & 0 & 0 & a_{44} & \cdots & a_{4k} & a_{4k+1} & \cdots & a_{4n} \\ \cdots & \cdots & \cdots & \cdots & \cdots & \cdots & \cdots & \cdots & \cdots \\ 0 & 0 & 0 & 0 & \cdots & a_{kk} & a_{kk+1} & \cdots & a_{kn} \\ 0 & 0 & 0 & 0 & \cdots & 0 & a_{k+1k+1} & \cdots & a_{k+1n} \\ \cdots & \cdots & \cdots & \cdots & \cdots & \cdots & \cdots & \cdots & \cdots \\ 0 & 0 & 0 & 0 & \cdots & 0 & a_{nk+1} & \cdots & a_{nn} \end{bmatrix} \quad (2.51)$$

To obtain an upper triangular matrix, we have to reduce the sub-diagonal elements of first $(n-1)$ columns to zero. Since, in the k step, the sub-diagonal elements of column k are reduced to zero, we only need to proceed for $(n-1)$ steps. So, Eq. (2.50) is the algorithm for Gauss elimination with $k = 1$ to $(n-1)$. The full algorithm becomes,

$$l_{ik} = \frac{a_{ik}}{a_{kk}}, \; a_{ij} = a_{ij} - l_{ik}a_{kj}, \; \text{and} \; b_i = b_i - l_{ik}b_k \qquad (2.52)$$

for $k = 1$ to $(n-1)$, $i = (k+1)$ to n, and $j = k$ to n.

If the vector b is added as the $(n+1)$ column of the matrix A, the operations could be extended as in Eq. (2.46). The computations can be done on j starting from $(k+1)$ to avoid computing the zeros on the subdiagonal elements on kth column. However, in that case, subdiagonal elements on kth column have to be set to zero to avoid retention of older values.

The above is known as *forward elimination* of the Gauss elimination procedure. At this point, one can easily obtain the solution of the system of equation using *back substitution*. So, the combination of *forward elimination* and *back substitution* provides the required solution of the set of equations and is known as *Gauss elimination* procedure.

As mentioned earlier, the process will fail if any of the elements a_{kk} for $k = 1$ to $(n-1)$ become zero. These elements are called *pivots* and at each step k, the kth row and column are known as *pivotal row* and *pivotal column*, respectively. We will discuss the situations when pivots are very small or zero, in a latter section.

Using Eq. (2.52), one can calculate the number of computations required to reduce a full system of equation to a triangular system using Gauss elimination. At each k, we require $(n-k)$ computations for l_{ik}'s and $2(n-k)(n-k+1)$ for a_{ij}'s. Thus, total number of computation is

$$\sum_{k=1}^{n-1}\left[(n-k)+2(n-k)(n-k+1)\right] \text{ or } \left(\frac{2n^3}{3}+\frac{n^2}{2}-\frac{7n}{6}\right)$$

In order to solve the system completely, one needs to add the number of computations required to solve a triangular system which was calculated as $n(n+1)$. So, overall computation requirement is $(2n^3/3 + 3n^2/2 - n/6)$. For large n, it is safe to say that Gauss elimination require approximately $2n^3/3$ operations. If we ignore the additions and subtractions, the approximate number of computations are $n^3/3$.

Example 2.4   Solve Batman's problem given by Eq. (2.3) using Gauss elimination.

Solution   The augmented matrix is as follows:

$$\begin{bmatrix} 1 & 6.2 & 19.22 & 439 \\ 1 & 14.2 & 100.82 & 640 \\ 1 & 22.4 & 250.88 & 875 \end{bmatrix}$$

$k = 1$, pivot $a_{11} = 1$, define the multiplying factors as $l_{21} = 1/1 = 1$ and $l_{31} = 1/1 = 1$. Using these multiplying factors and performing the row operations by Eq. (2.52), we obtain

$$\begin{bmatrix} 1 & 6.2 & 19.22 & 439 \\ 0 & 8 & 81.6 & 201 \\ 0 & 16.2 & 231.66 & 436 \end{bmatrix}$$

$k = 2$, pivot $a_{22} = 8$, define the multiplying factor $l_{32} = 16.2 / 8 = 2.025$. Once again performing the row operations:

$$\begin{bmatrix} 1 & 6.2 & 19.22 & 439 \\ 0 & 8 & 81.6 & 201 \\ 0 & 0 & 66.42 & 28.975 \end{bmatrix}$$

Now, applying the backward elimination algorithm of Eq. (2.42), we obtain the solutions as $f = 28.975/66.42 = 0.436239$, $u = (201 - 81.6 \times 0.436239) / 8 = 20.67536$, and $S_0 = (439 - 19.22 \times 0.436239 - 6.2 \times 20.67536) / 1 = 302.42825$.

2.2.1.2 *Gauss Jordon Elimination*

In this method, the matrix A is reduced to an identity matrix through a series of operations. This is easily accomplished by minor modification of the Gauss elimination algorithm described in the previous section. The modifications are as follows:

- At each step, first the pivotal element is made unity by dividing the pivotal row with the pivotal element. So, at *Step k*, the *k*th row is divided by a_{kk}. As a result, the multiplication factor for *i*th row at *Step k* becomes $l_{ik} = a_{ik}$.
- In addition to the sub-diagonal elements, the elements above the diagonal are also made zero. So, at *Step k*, the row operations are conducted for all rows except the pivotal row. At this step, the columns 1 to $k - 1$ have already been operated to resemble part of the identity matrix. So, the operations are done on columns k to n.

As a result, the algorithm becomes

For $k = 1, \ldots, n$

$$a_{kj} = \frac{a_{kj}}{a_{kk}}, \ b_k = \frac{b_k}{a_{kk}}, \ a_{ij} = a_{ij} - a_{ik}a_{kj}, \text{ and } b_i = b_i - a_{ik}b_k \qquad (2.53)$$

for $i = 1$ to n but $\neq k$, and $j = k$ to n.

If the vector b is augmented as the $(n + 1)$ column of the matrix A

$$a_{kj} = \frac{a_{kj}}{a_{kk}} \text{ and } a_{ij} = a_{ij} - a_{ik}a_{kj}$$

for $k = 1$ to n, $i = 1$ to n but $\neq k$, $j = k$ to $(n + 1)$. $\qquad (2.54)$

It is easy to see that the Gauss Jordon elimination process can be used to obtain the inverse of a matrix. An identity matrix of size $(n \times n)$ can be augmented to the existing $(n \times n)$ matrix A such that the resultant matrix becomes $n \times 2n$, as follows:

$$\begin{bmatrix} a_{11} & a_{12} & \cdots & \cdots & a_{1n} & 1 & 0 & 0 & \cdots & 0 \\ a_{21} & a_{22} & \cdots & \cdots & a_{2n} & 0 & 1 & 0 & \cdots & 0 \\ a_{31} & a_{32} & \cdots & \cdots & a_{3n} & 0 & 0 & 1 & \cdots & 0 \\ \cdots & \cdots & \cdots & \cdots & \cdots & \cdots & \cdots & \cdots & \cdots & \cdots \\ a_{n1} & a_{n2} & \cdots & \cdots & a_{nn} & 0 & 0 & 0 & \cdots & 1 \end{bmatrix} \qquad (2.55)$$

Now one can perform the Gauss Jordon elimination on the augmented matrix such that the original A matrix becomes an identity matrix. The original identity matrix will transform in the process to give A^{-1}. The algorithm can be written as follows:

$$a_{kj} = \frac{a_{kj}}{a_{kk}} \text{ and } a_{ij} = a_{ij} - a_{ik}a_{kj}, \text{ for } k = 1 \text{ to } n, i = 1 \text{ to } n \text{ but } \neq k, \text{ and } j = k \text{ to } 2n$$

$$(2.56)$$

We leave it to the readers to calculate the number of computations required to invert a matrix using *Gauss Jordon* method.

Example 2.5   Solve Batman's problem given by Eq. (2.3) using Gauss Jordon elimination.

Solution   We will use the algorithm of augmented matrix given by Eq. (2.54). The augmented matrix is given by

$$\begin{bmatrix} 1 & 6.2 & 19.22 & 439 \\ 1 & 14.2 & 100.82 & 640 \\ 1 & 22.4 & 250.88 & 875 \end{bmatrix}$$

Steps of the Gauss Jordon iterations are shown in the matrix form below:

$$\text{Step 1:} \begin{bmatrix} 1 & 6.2 & 19.22 & 439 \\ 0 & 8 & 81.6 & 201 \\ 0 & 16.2 & 231.66 & 436 \end{bmatrix}$$

$$\text{Step 2:} \begin{bmatrix} 1 & 0 & -44.02 & 283.225 \\ 0 & 1 & 10.2 & 25.125 \\ 0 & 0 & 66.42 & 28.975 \end{bmatrix}$$

$$\text{Step 3:} \begin{bmatrix} 1 & 0 & 0 & 302.4282 \\ 0 & 1 & 0 & 20.6754 \\ 0 & 0 & 1 & 0.4362 \end{bmatrix}$$

Note that the augmented right-hand side column gives the solution vector.

2.2.1.3 *LU Decomposition*

In many engineering problems, one requires to solve a system repeatedly where coefficient matrix (A) remains the same but the right-hand side vector (b) changes. Let us say the solutions are required for right-hand side vectors $b_1, b_2, b_3, \ldots, b_m$. If these vectors are mutually independent, one can augment all the columns with the coefficient matrix and obtain the solutions together. However, most often, vector b_2 will depend upon the solution vector obtained from b_1 and so on. In this case, b_2 is not known until $Ax = b_1$ is solved. So, one needs to run the Gauss elimination repeatedly even though the coefficient matrix remained same. We will illustrate it through an example.

Example 2.6   Let us consider a series of adjacent reactors. Each reactor is completely mixed such that the concentration within each reactor is uniform. However, there can be exchange of flow in both directions between two adjacent reactors. Such a system is shown in Fig. 2.5.

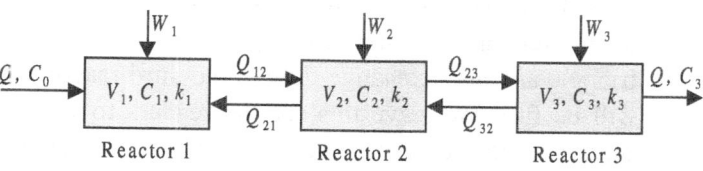

Fig. 2.5

In the figure, Q_i's are the flow rates, C_i's are concentrations, V_i's are reactor volumes and k_i's are the first order reaction rates and W_i's are the external loads in the reactors. All the Q_i's, V_i's, W_i's, k_i's, and C_0 are known. Formulate the set of equations to calculate C_1, C_2, and C_3 at steady state.

Solution

The mass balance equations for the above reactors can be written as:

Reactor 1: $\quad V_1 \dfrac{dC_1}{dt} = QC_0 + W_1 + Q_{21}C_2 - Q_{12}C_1 - k_1V_1C_1$

Reactor 2: $\quad V_2 \dfrac{dC_2}{dt} = Q_{12}C_1 + W_2 + Q_{32}C_3 - Q_{21}C_2 - Q_{23}C_2 - k_2V_2C_2$

Reactor 3: $\quad V_3 \dfrac{dC_3}{dt} = Q_{23}C_2 + W_3 - Q_{32}C_3 - k_3V_3C_3 - QC_3$

At steady state, the left-hand side of all the equations is zero. Thus, the set of equations to solve is

$$(-Q_{12} - k_1V_1)C_1 + Q_{21}C_2 = (-QC_0 - W_1)$$
$$Q_{12}C_1 + (-Q_{21} - Q_{23} - k_2V_2)C_2 + Q_{32}C_3 = -W_2$$
$$Q_{32}C_2 + (-Q_{32} - k_2V - Q)C_3 = -W_3$$

This set of equations can be solved to obtain the C_i's. However, the concentrations in each reactor as well as the final concentration C_3 are often governed by guideline values. The coefficient matrix will remain unaltered as long as the reaction rates, volumes, and the flow rates remain constant. The W_i's are often adjusted depending on the values of C_i's so that the required concentrations are obtained. In addition, the initial concentration C_0 may also have diurnal and seasonal variations. This changes the right-hand side vector, one as a function of C_i's and the other as a function of time. Under these circumstances, it is very handy to have a *LU* decomposition of the coefficient matrix so that the solution can be obtained quickly for any right-hand side vector. So, now we introduce *LU* decomposition.

In the above example, the coefficient matrix is tri-diagonal. Notice that the reactor 3 has exchange flow only with reactor 2. If we place three reactors as vertices of a triangle and add exchange flow to the third reactor from both 1 and 2, the matrix will be full. We leave this for the readers to work out. This example has high practical significance since, lakes, rivers and estuaries are often simulated by dividing into small reactors with exchange flows. Division is made into horizontal segments as well as into vertical segments.

In summary, the method of *LU* decomposition provides an alternative by which one can avoid repeated operation on the same coefficient matrix A. For large matrices, this saves significant amount of computation time. Any square matrix, on which Gauss elimination will work, can be expressed as a product of a lower triangular matrix L and an upper triangular matrix U. More specifically, the condition for such a decomposition is given by Theorem 2.3 also known as the *LU theorem*.

Once such a decomposition is obtained, one can express the equation of the form $Ax = b$ as $LUx = b$. The latter is equivalent to two triangular systems of equations: $Ux = y$ and $Ly = b$. One can easily obtain y by solving the triangular system $Ly = b$ by *forward substitution* and subsequently obtain the solution vector x by solving the other triangular system $Ux = y$ by *back substitution*.

In the enlarged form, *LU* decomposition looks as follows:

$$
\begin{bmatrix}
a_{11} & a_{12} & \cdots & a_{1n} \\
a_{21} & a_{22} & \cdots & a_{2n} \\
a_{31} & a_{32} & \cdots & a_{3n} \\
\cdots & \cdots & \cdots & \cdots \\
a_{11} & a_{12} & \cdots & a_{nn}
\end{bmatrix}
=
\begin{bmatrix}
l_{11} & 0 & \cdots & \cdots & 0 \\
l_{21} & l_{22} & 0 & \cdots & 0 \\
l_{31} & l_{32} & l_{33} & \cdots & 0 \\
\cdots & \cdots & \cdots & \cdots & \cdots \\
l_{n1} & l_{n2} & l_{n3} & \cdots & l_{nn}
\end{bmatrix}
\begin{bmatrix}
u_{11} & u_{12} & \cdots & \cdots & u_{1n} \\
0 & u_{22} & u_{23} & \cdots & u_{2n} \\
0 & 0 & u_{33} & \cdots & u_{3n} \\
\cdots & \cdots & \cdots & \cdots & \cdots \\
0 & 0 & 0 & \cdots & u_{nn}
\end{bmatrix}
$$

$$(2.57)$$

Reader should note at this point that for a given matrix A, all a_{ij}'s are known. On the right-hand side of the equations all l_{ij}'s and u_{ij}'s are unknown. Since there are $n(n + 1)/2$ non-zero elements in a triangular matrix, overall there are $(n^2 + n)$ unknowns. By simple matrix multiplication, one can obtain n^2 equation since there are n^2 numbers of known a_{ij} values. As a result, values of n unknowns have to be set arbitrarily. This is accomplished by setting either l_{ii} or $u_{ii} = 1$ for $i = 1, 2, \ldots, n$. This allows unique determination of the remaining n^2 unknowns.

Theoretically, one can obtain these values by writing these equations explicitly through matrix multiplication and solving them. This is often simpler for a small system, for example a 2×2 or 3×3 matrix. However, for a large system, this becomes quite cumbersome and a more systematic method is called for. In order to do that, we will use Gauss elimination steps and algorithm to show that it is equivalent to *LU* decomposition [Dahlquist and Bjorck (1969)]. In the process we will derive the general algorithm for the *LU* decomposition.

Theorem 2.3

Let A_k be a sequence of matrices formed by first k rows and k columns of a square matrix A. If $\det(A_k) \neq 0$ for $k = 1, 2, \ldots, (n-1)$, then there exist an upper triangular matrix U and a lower triangular matrix L such that, $A = LU$. Furthermore, if the diagonal elements of either L or U are unity, i.e., l_{ii} or $u_{ii} = 1$ for $i = 1, 2, \ldots, n$, both L and U are unique.

Proof: For $k = 1$, there is nothing to prove. Let us assume that the theorem is valid up to k, i.e., $A_k = L_k U_k$. We will check for the validity for A_{k+1}:

$$A_{k+1} = \begin{bmatrix} A_k & \bar{a} \\ \bar{b}^T & a_{k+1k+1} \end{bmatrix} \tag{2.58}$$

where $\bar{a}$ and $\bar{b}$ are column vectors.

If this has an LU decomposition, they will be of the form:

$$L_{k+1} = \begin{bmatrix} L_k & 0 \\ \bar{\beta}^T & 1 \end{bmatrix} \text{ and } U_{k+1} = \begin{bmatrix} U_k & \bar{\alpha} \\ 0 & u_{kk} \end{bmatrix} \tag{2.59}$$

where $\bar{\alpha}$ and $\bar{\beta}$ are column vectors.

By multiplying and equating the elements, we obtain

$$A_k = L_k U_k \tag{2.60}$$

$$L_k \bar{\alpha} = \bar{a} \tag{2.61}$$

$$\bar{\beta}^T U_k = \bar{b}^T \text{ or } U_k \bar{\beta} = \bar{b} \tag{2.62}$$

$$\bar{\beta}^T \bar{\alpha} + u_{kk} = a_{kk} \tag{2.63}$$

Since the theorem is valid up to A_k, decomposition in Eq. (2.59) is valid and $\det(A_k) \neq 0$. Moreover, determinant of a triangular matrix (upper or lower) is the product of the diagonal entries. Thus, $\det(A_k) = \det(L_k)\det(U_k) \neq 0$. So, Eqs (2.61) and (2.62) have unique solution vectors, $\bar{\beta}$ and $\bar{\alpha}$. Then, by Eq. (2.63), u_{kk} is also uniquely determined.

Since it is difficult to visualize a $n \times n$ matrix, let us first try to visualize the steps of Gauss elimination with a 5×5 matrix:

$$
\begin{bmatrix}
a_{11}^0 & a_{12}^0 & a_{13}^0 & a_{14}^0 & a_{15}^0 \\
a_{21}^0 & a_{22}^0 & a_{23}^0 & a_{24}^0 & a_{25}^0 \\
a_{31}^0 & a_{32}^0 & a_{33}^0 & a_{34}^0 & a_{35}^0 \\
a_{41}^0 & a_{42}^0 & a_{43}^0 & a_{44}^0 & a_{45}^0 \\
a_{51}^0 & a_{52}^0 & a_{53}^0 & a_{54}^0 & a_{55}^0
\end{bmatrix}
\xrightarrow{\text{Step 1}}
\begin{bmatrix}
a_{11}^0 & a_{12}^0 & a_{13}^0 & a_{14}^0 & a_{15}^0 \\
0 & a_{22}^1 & a_{23}^1 & a_{24}^1 & a_{25}^1 \\
0 & a_{32}^1 & a_{33}^1 & a_{34}^1 & a_{35}^1 \\
0 & a_{42}^1 & a_{43}^1 & a_{44}^1 & a_{45}^1 \\
0 & a_{52}^1 & a_{53}^1 & a_{54}^1 & a_{55}^1
\end{bmatrix}
\xrightarrow{\text{Step 2}}
$$

$$
\xrightarrow{\text{Step 2}}
\begin{bmatrix}
a_{11}^0 & a_{12}^0 & a_{13}^0 & a_{14}^0 & a_{15}^0 \\
0 & a_{22}^1 & a_{23}^1 & a_{24}^1 & a_{25}^1 \\
0 & 0 & a_{33}^2 & a_{34}^2 & a_{35}^2 \\
0 & 0 & a_{43}^2 & a_{44}^2 & a_{45}^2 \\
0 & 0 & a_{53}^2 & a_{54}^2 & a_{55}^2
\end{bmatrix}
\xrightarrow{\text{Step 3}}
\begin{bmatrix}
a_{11}^0 & a_{12}^0 & a_{13}^0 & a_{14}^0 & a_{15}^0 \\
0 & a_{22}^1 & a_{23}^1 & a_{24}^1 & a_{25}^1 \\
0 & 0 & a_{33}^2 & a_{34}^2 & a_{35}^2 \\
0 & 0 & 0 & a_{44}^3 & a_{45}^3 \\
0 & 0 & 0 & a_{54}^3 & a_{55}^3
\end{bmatrix}
\xrightarrow{\text{Step 4}}
$$

$$
\xrightarrow{\text{Step 4}}
\begin{bmatrix}
a_{11}^0 & a_{12}^0 & a_{13}^0 & a_{14}^0 & a_{15}^0 \\
0 & a_{22}^1 & a_{23}^1 & a_{24}^1 & a_{25}^1 \\
0 & 0 & a_{33}^2 & a_{34}^2 & a_{35}^2 \\
0 & 0 & 0 & a_{44}^3 & a_{45}^3 \\
0 & 0 & 0 & 0 & a_{55}^4
\end{bmatrix}
\tag{2.64}
$$

Note that the superscripts represent number of times the value has been actively modified during the step operations according to Eq. (2.52) in Gauss elimination. This process can be thought of as a sequence of matrices generated from the original matrix by the steps of Gauss elimination and can be written as:

$$
A^{(1)} \xrightarrow{\text{Step 1}} A^{(2)} \xrightarrow{\text{Step 2}} A^{(3)} \xrightarrow{\text{Step 3}} A^{(4)} \xrightarrow{\text{Step 4}} A^{(5)}
\tag{2.65}
$$

Furthermore, let us define the following notation at the kth step for the elements in the matrix,

$$
A^{(k)} = \left[a_{ij}^{(k)} \right]
\tag{2.66}
$$

The superscripts in the parentheses in Eqs (2.65) and (2.66) are different from those in Eq. (2.64). In Eq. (2.64), superscripts represent the number of times a non-zero entry has been actively modified during the Gauss elimination steps. While the superscript in the parenthesis (k) in Eqs (2.65) and (2.66) correspond to the elements at the beginning of *Step k* of Gauss elimination [Eq. (2.52)]. Let us focus on the entries at or above the main diagonal $(i \leq j)$ and below the main diagonal $(i > j)$ separately, for the number of times they have been modified (look at Eq. (2.64) for example):

- For $i \leq j$, the entries (a_{ij}'s) are actively modified for $(i - 1)$ steps. For rest of the steps, they remain unaltered, i.e., $a_{ij}^{(i)} = a_{ij}^{(i+1)} = \cdots = a_{ij}^{(n)}$.
- For $i > j$, the entries (a_{ij}'s) are actively modified for j steps. For rest of the steps, they are zero, i.e., $a_{ij}^{(j+1)} = a_{ij}^{(j+2)} = \cdots = a_{ij}^{(n)} = 0$.

The above two statements are mathematically equivalent to saying that any entry a_{ij} is altered for m steps, where $m = \min(i - 1, j)$. At each step of the Gauss elimination, the elements are modified according to Eq. (2.52), which can now be written following the additional superscript convention of Eq. (2.66), as follows:

$$a_{ij}^{(k+1)} = a_{ij}^{(k)} - l_{ik} a_{kj}^{(k)}, \text{ where } l_{ik} = \frac{a_{ik}^{(k)}}{a_{kk}^{(k)}} \qquad (2.67)$$

Notice that the step number corresponds to the pivotal index as in the Gauss elimination. Since, any element is modified for m steps, we can sum Eq. (2.67) over m steps to get the original elements back,

$$\sum_{k=1}^{m} \left(a_{ij}^{(k+1)} - a_{ij}^{(k)} \right) = -\sum_{k=1}^{m} l_{ik} a_{kj}^{(k)} \text{ or } a_{ij}^{(m+1)} - a_{ij}^{(1)} = -\sum_{k=1}^{m} l_{ik} a_{kj}^{(k)} \qquad (2.68)$$

Note that $a_{ij}^{(1)}$'s are the original elements of matrix A. Expanding Eq. (2.68) for the original elements (a_{ij}'s), one can write more specifically for values of $m = \min(i - 1, j)$ for entries above and below the diagonal:

$$a_{ij} = \begin{cases} a_{ij}^{(i)} + \displaystyle\sum_{k=1}^{i-1} l_{ik} a_{kj}^{(k)}, & i \leq j \\[4mm] 0 + \displaystyle\sum_{k=1}^{j} l_{ik} a_{kj}^{(k)}, & i > j \end{cases} \qquad (2.69)$$

If we define, $l_{ii} = 1$, we can write Eq. (2.69) simply as

$$a_{ij} = \sum_{k=1}^{t} l_{ik} a_{kj}^{(k)}, \text{ where } t = \min(i, j) \qquad (2.70)$$

This is nothing but product of two matrices L and U whose elements are given by,

$$l_{ij} = \begin{cases} l_{ij}, & \text{for } i > j \\ 1, & \text{for } i = j \\ 0, & \text{for } i < j \end{cases} \quad \text{and} \quad u_{ij} = \begin{cases} a_{ij}^{(i)}, & \text{for } i \leq j \\ 0, & \text{for } i > j \end{cases} \tag{2.71}$$

Thus, Gauss elimination is equivalent to LU decomposition where the final upper triangular matrix is the U matrix and one can construct the L matrix simply by saving the multiplying factor l_{ik} at each step of Eq. (2.52).

Equations (2.69) and (2.70) are two alternative forms of the general formulation for LU decomposition. They provide n^2 independent equations for the determination of the elements of L and U. It is a small step to derive different forms of LU decomposition algorithms using these general equations, which allows systematic calculation of the elements of L and U.

Doolittle's Algorithm We can rewrite Eq. (2.69) using the definition of Eq. (2.71) as:

$$k = 1, 2..., n$$
$$l_{kk} = 1$$

$$u_{kj} = a_{kj} - \sum_{m=1}^{k-1} l_{km} u_{mj}, \quad j = k, k+1, \cdots, n \tag{2.72}$$

$$l_{ik} = \frac{a_{ik} - \sum_{m=1}^{k-1} l_{im} u_{mk}}{u_{kk}}, \quad i = k+1, \cdots, n$$

Crout's Algorithm Define $u_{kj} = a_{kj}^{(k)}$ and $u_{kk} = 1$ in Eq. (2.69) to obtain

$$k = 1, 2..., n$$
$$u_{kk} = 1$$

$$l_{ik} = a_{ik} - \sum_{m=1}^{k-1} l_{im} u_{mk}, \quad i = k, k+1, \cdots, n \tag{2.73}$$

$$u_{kj} = \frac{a_{kj} - \sum_{m=1}^{k-1} l_{km} u_{mj}}{l_{kk}}, \quad j = k+1, \cdots, n$$

Cholesky Algorithm The LU decomposition for a symmetric and positive definite matrix A can be computed much more efficiently if we change the composition of matrices L and U somewhat. The decomposition in this case can be expressed as $A = LL^T$. Then one needs to compute the elements of L and the U will be L^T. This substantially reduces the computational effort. Why we specifically require a symmetric and positive definite matrix for this will be clear once we go through the details of the derivation. At this point, let us take a closer look at the U matrix. The upper triangular matrix from LU decomposition can be further decomposed into a diagonal matrix and an upper triangular matrix with diagonal elements as 1. For example:

$$\begin{bmatrix} d_1 & u_{12} & \cdots & \cdots & u_{1n} \\ 0 & d_2 & \cdots & \cdots & u_{2n} \\ 0 & 0 & d_3 & \cdots & u_{3n} \\ \cdots & \cdots & \cdots & \cdots & \cdots \\ 0 & 0 & 0 & \cdots & d_n \end{bmatrix} = \begin{bmatrix} d_1 & 0 & \cdots & \cdots & 0 \\ 0 & d_2 & \cdots & \cdots & 0 \\ 0 & 0 & d_3 & \cdots & 0 \\ \cdots & \cdots & \cdots & \cdots & \cdots \\ 0 & 0 & 0 & \cdots & d_n \end{bmatrix} \times \begin{bmatrix} 1 & u_{12}/d_1 & \cdots & \cdots & u_{1n}/d_1 \\ 0 & 1 & \cdots & \cdots & u_{2n}/d_2 \\ 0 & 0 & 1 & \cdots & u_{3n}/d_3 \\ \cdots & \cdots & \cdots & \cdots & \cdots \\ 0 & 0 & 0 & \cdots & 1 \end{bmatrix}$$

$$(2.74)$$

Theorem 2.4

Let A be a $n \times n$ invertible matrix then there exists a decomposition of the form $A = LDU$ where L is a $n \times n$ lower triangular matrix with diagonal elements as 1, U is a $n \times n$ upper triangular matrix with diagonal elements as 1, and D is a diagonal matrix.

We leave this proof to the readers.

Thus, instead of LU, one can talk about a more general LDU decomposition of a matrix. The statement for existence of such a decompostion is given by Theorem 2.4.

Let us highlight the derivation of Cholesky method with an example. Notice that the diagonal elements of the U matrix, which are originally the pivots of the Gauss elimination are retained in the diagonal matrix [Eq. (2.74)]. So, for a positive definite matrix, all $d_i > 0$. Now let us look at the LDU decomposition of a symmetric matrix. We use a 3×3 matrix for illustration:

$$\begin{bmatrix} d_1 & a & b \\ a & d_2 & c \\ b & c & d_3 \end{bmatrix} = \begin{bmatrix} 1 & 0 & 0 \\ l_{21} & 1 & 0 \\ l_{31} & l_{32} & 1 \end{bmatrix} \begin{bmatrix} p_1 & 0 & 0 \\ 0 & p_2 & 0 \\ 0 & 0 & p_3 \end{bmatrix} \begin{bmatrix} 1 & u_{12} & u_{13} \\ 0 & 1 & u_{23} \\ 0 & 0 & 1 \end{bmatrix} \qquad (2.75)$$

We can write the l_{ij}'s and u_{ij}'s either using the Gauss elimination or Doolittle's algorithm and modify the u_{ij}'s according to Eq. (2.74). These are as follows:

$$l_{21} = u_{12} = \frac{a}{d_1} = \alpha \text{ (say)}$$

$$l_{31} = u_{13} = \frac{b}{d_1} = \beta \text{ (say)}$$

$$l_{32} = u_{23} = \frac{cd_1 - ab}{d_1 d_2 - a^2} = \gamma \text{ (say)}$$

$$p_1 = d_1$$

$$p_2 = \frac{d_1 d_2 - a^2}{d_1} \tag{2.76}$$

$$p_3 = \frac{(d_1 d_2 - a^2)(d_1 d_3 - b^2) - (d_1 c - ab)^2}{(d_1 d_2 - a^2)d_1}$$

So, the above equation can be written as

$$\begin{bmatrix} d_1 & a & b \\ a & d_2 & c \\ b & c & d_3 \end{bmatrix} = \begin{bmatrix} 1 & 0 & 0 \\ \alpha & 1 & 0 \\ \beta & \gamma & 1 \end{bmatrix}\begin{bmatrix} p_1 & 0 & 0 \\ 0 & p_2 & 0 \\ 0 & 0 & p_3 \end{bmatrix}\begin{bmatrix} 1 & \alpha & \beta \\ 0 & 1 & \gamma \\ 0 & 0 & 1 \end{bmatrix} \tag{2.77}$$

Notice that in this case $U = L^T$. In fact, if you pay close attention to the process of Gauss elimination, you will notice, this will always be the case for a symmetric matrix. Therefore, for a symmetric matrix, the decomposition can be written as $A = LDL^T$.

The pivots (p_i's) of Gauss elimination are in the diagonal matrix D. Since, all pivots are positive for a positive definite matrix, their square roots are real. Square root of a diagonal matrix is a diagonal matrix with the square root of the diagonal elements. Thus, the above decomposition for a positive definite matrix can be written as

$$A = LDL^T = (LD^{1/2})(D^{1/2}L^T) = L_C L_C^T \tag{2.78}$$

where
$$L_C = LD^{1/2} \tag{2.79}$$

This is known as *Cholesky* decomposition. In the above example of 3 × 3 matrix, the decomposition is as follows:

$$\begin{bmatrix} d_1 & a & b \\ a & d_2 & c \\ b & c & d_3 \end{bmatrix} = \begin{bmatrix} \sqrt{p_1} & 0 & 0 \\ \alpha\sqrt{p_1} & \sqrt{p_2} & 0 \\ \beta\sqrt{p_1} & \gamma\sqrt{p_2} & \sqrt{p_3} \end{bmatrix}\begin{bmatrix} \sqrt{p_1} & \alpha\sqrt{p_1} & \beta\sqrt{p_1} \\ 0 & \sqrt{p_2} & \gamma\sqrt{p_2} \\ 0 & 0 & \sqrt{p_3} \end{bmatrix} \tag{2.80}$$

Once, you follow the logic of the Cholesky decomposition described above, the elementwise algorithm can be written easily using Eq. (2.70). We present the algorithm below and leave the reader for checking how it can be derived similar to the LU algorithm,

$$k = 1, 2, ..., n$$

$$l_{kk} = \left(a_{kk} - \sum_{p=1}^{k-1} l_{kp}^2 \right)^{\frac{1}{2}} \tag{2.81}$$

$$l_{ik} = \frac{a_{ik} - \sum_{p=1}^{k-1} l_{ip} l_{kp}}{l_{kk}}, \quad i = k+1, ..., n$$

Example 2.7 Solve the Batman problem of Eq. (2.2) by performing an LU decomposition of the coefficient matrix using, (a) Gaussian elimination, (b) Doolittle algorithm, (c) Crout's algorithm.

Solution (a) *Gaussian elimination*
The problem was solved by Gauss elimination in Example 2.4. We use those results to assemble the L and U matrices. Recall that the U matrix is the same as the final upper triangular matrix obtained in Gauss elimination:

$$U = \begin{bmatrix} 1 & 6.2 & 19.22 \\ 0 & 8 & 81.6 \\ 0 & 0 & 66.42 \end{bmatrix}$$

Using the values of $l_{21} = 1$, $l_{31} = 1$, and $l_{32} = 2.025$, we can assemble the L matrix as:

$$L = \begin{bmatrix} 1 & 0 & 0 \\ 1 & 1 & 0 \\ 1 & 2.025 & 1 \end{bmatrix}$$

Now, we can first solve the equation $Ly = b$:

$$\begin{bmatrix} 1 & 0 & 0 \\ 1 & 1 & 0 \\ 1 & 2.025 & 1 \end{bmatrix} \begin{bmatrix} y_1 \\ y_2 \\ y_3 \end{bmatrix} = \begin{bmatrix} 439 \\ 640 \\ 875 \end{bmatrix}$$

Using forward substitution of Eq. (2.44), we obtain the solution vector y as,

$$\begin{bmatrix} y_1 \\ y_2 \\ y_3 \end{bmatrix} = \begin{bmatrix} 439 \\ 201 \\ 28.975 \end{bmatrix}$$

Notice that the above is same as the augmented right-hand side vector in Example 2.4 after the Gauss elimination. In order to obtain the solution vector x we now have to solve the following:

$$\begin{bmatrix} 1 & 6.2 & 19.22 \\ 0 & 8 & 81.6 \\ 0 & 0 & 66.42 \end{bmatrix} \begin{bmatrix} x_1 \\ x_2 \\ x_3 \end{bmatrix} = \begin{bmatrix} 439 \\ 201 \\ 28.975 \end{bmatrix}$$

The above can be solved by the back substitution of Eq. (2.42). Since, the equations are same as in Example 2.5, we obtain the same solutions as, $x_1 = S_0 = 302.42824$, $x_2 = u = 20.67536$ and $x_3 = f = 0.436$.

(b) *Doolittle's method*

The coefficient matrix is $A = \begin{bmatrix} 1 & 6.2 & 19.22 \\ 1 & 14.2 & 100.82 \\ 1 & 22.4 & 250.88 \end{bmatrix}$

Now, we evaluate the elements of matrices L and U using the set of Eq. (2.72). The sequence in which the elements can be evaluated are

$$i = 1, j = 1, u_{11} = a_{11} = 1$$
$$i = 1, j = 2, u_{12} = a_{12} = 6.2$$
$$i = 1, j = 3, u_{13} = a_{13} = 19.22$$
$$j = 1, i = 2, l_{21} = 1/1 = 1$$
$$j = 1, i = 3, l_{31} = 1/1 = 1$$
$$i = 2, j = 2, u_{22} = 14.2 \, (- (1 \times 6.2) = 8$$
$$i = 2, j = 3, u_{23} = 100.82 \, (- (1 \times 19.22) = 81.6$$
$$j = 2, i = 3, l_{32} = (22.4 - (1 \times 6.2))/8 = 2.025$$
$$i = 3, j = 3, u_{33} = 250.88 - (1 \times 19.22) - (2.025 \times 81.6) = 66.42$$

Although it may appear from Eq. (2.72) that the elements of matrix U and L can be calculated independently, it is not so. Notice in the above example, the first row of matrix U is first calculated followed by the first column of L. This is important because in calculation of second row of the matrix U, the elements of the first column of L were required.

So, the sequence of computation in Doolittle's method can be presented as

1st Row of U $\Rightarrow$ 1st Column of L $\Rightarrow$ 2nd Row of U $\Rightarrow$ 2nd Column of L $\Rightarrow$ so on...

Following the computation of the elements, the L and U matrices of the Doolittle method can be synthesized as follows:

$$U = \begin{bmatrix} 1 & 6.2 & 19.22 \\ 0 & 8 & 81.6 \\ 0 & 0 & 66.42 \end{bmatrix} \text{ and } L = \begin{bmatrix} 1 & 0 & 0 \\ 1 & 1 & 0 \\ 1 & 2.025 & 1 \end{bmatrix}$$

(c) *Crout's method*

The elements of the Crout's decomposition can be calculated using Eq. (2.73) in the same way as was done in Doolittle's method. However, recall that the difference is that the diagonal elements of U matrix is now unity. The computation sequence is also changed accordingly. The computation sequence of Crout's is

1st Column of L $\Rightarrow$ 1st Row of U $\Rightarrow$ 2nd Column of L $\Rightarrow$ 2nd Row of U $\Rightarrow$ so on...

Using the above computation sequence in combination with Eq. (2.73), the L and U matrices can be computed as

$$L = \begin{bmatrix} 1 & 0 & 0 \\ 1 & 8 & 0 \\ 1 & 16.2 & 66.42 \end{bmatrix} \text{ and } U = \begin{bmatrix} 1 & 6.2 & 19.22 \\ 0 & 1 & 10.2 \\ 0 & 0 & 1 \end{bmatrix}$$

Using the LU decompositions in (b) and (c), the solutions can be computed similarly as was done in the case of (a).

Example 2.8 Perform a Cholesky decomposition of the following positive definite matrix:

$$\begin{bmatrix} 4 & 2 & 1 \\ 2 & 5 & 1 \\ 1 & 1 & 4.3125 \end{bmatrix}$$

Solution Using Eq. (2.81), the elements of matrix L can be computed as follows:

$$k = 1, \, l_{11} = \sqrt{4} = 2$$
$$k = 1, \, i = 2, \, l_{21} = 2/2 = 1$$
$$k = 1, \, i = 3, \, l_{31} = 1/2 = 0.5$$
$$k = 2, \, l_{22} = \sqrt{4} = 2$$
$$k = 2, \, i = 3, \, l_{32} = (1 - (1 \times 0.5))/2 = 0.25$$
$$k = 3, \, l_{33} = \sqrt{4.3125 - 0.5^2 - 0.25^2} = 2$$

So, the matrix L can be synthesized as

$$L = \begin{bmatrix} 2 & 0 & 0 \\ 1 & 2 & 0 \\ 0.5 & 0.25 & 2 \end{bmatrix}$$

The Cholesky decomposition is then given by $A = LL^T$ as follows:

$$\begin{bmatrix} 4 & 2 & 1 \\ 2 & 5 & 1 \\ 1 & 1 & 4.3125 \end{bmatrix} = \begin{bmatrix} 2 & 0 & 0 \\ 1 & 2 & 0 \\ 0.5 & 0.25 & 2 \end{bmatrix} \begin{bmatrix} 2 & 1 & 0.5 \\ 0 & 2 & 0.25 \\ 0 & 0 & 2 \end{bmatrix}$$

2.2.1.4 *Banded Matrices and Thomas Algorithm*

Advantage of the direct methods described so far is that the solution can be obtained in finite number of steps and as a result, the quantum of round-off errors can be estimated. However, if the co-efficient matrix contains a lot of zeros as elements, it is easy to see that many computations will involve operation on zero. It is too cumbersome to eliminate these computations on zeros within the framework of the methods described above. At the same time, one wishes to eliminate the computations involving zeros thereby making the solution procedure more efficient. Before proceeding with the methods, let us categorize a few types of matrices based on the location of zeros in the matrix.

A matrix will be called *dense* if most of the elements are non-zero. Alternatively, if the matrix contains a large proportion of the elements as zero, it will be called a *sparse* matrix. A special kind of *sparse* matrix contains non-zero elements only along the main diagonal and lines parallel to the diagonal elements, both above and below the main diagonal of a matrix. These are known as *banded* matrix and one such is shown in Fig. 2.5. *Bandwidth*(s) of a banded matrix shown in Fig. 2.5 is $s = (a + b - 1)$. So, for a banded matrix $a_{ij} = 0$ for $j \geq (i + a)$ or $i \geq (j + b)$. A special case is when $a = b = 2$ and the bandwidth $s = 3$. The matrix is called *tri-diagonal*.

Thomas Algorithm for Tri-diagonal Matrix This special kind of sparse matrix often appears in the computation as a result of simulation of data using splines (Chapter 4) or finite difference solution of differential equations (Chapters 6 and 7) in one-dimensional problems (and sometimes, in 2-dimensional problems with special algorithms). Solution of linear systems involving these matrices can be accomplished using the direct methods described before. However, most of the computations would involve zero. Instead of matrix operation, a tri-diagonal system can be solved by vector operations. Thomas algorithm is one such procedure.

$$
\begin{array}{c}
\leftarrow \; a \; \rightarrow \\
\begin{array}{c} \uparrow \\ \\ b \\ \downarrow \\ \\ \\ \\ \end{array}
\begin{bmatrix}
x & x & x & 0 & 0 & 0 & \cdots & 0 \\
x & x & x & x & 0 & 0 & \cdots & 0 \\
x & x & x & x & x & 0 & \cdots & 0 \\
x & x & x & x & x & x & \cdots & 0 \\
0 & x & x & x & x & x & \cdots & 0 \\
0 & 0 & x & x & x & x & \cdots & 0 \\
\cdots & \cdots & \cdots & \cdots & \cdots & \cdots & \cdots & \cdots
\end{bmatrix}
\end{array}
$$

Fig. 2.5 A banded matrix of bandwidth $(a+b-1)$. 'x' denotes positions that may contain non-zero entries.

Let us consider the following system of linear equations with tri-diagonal matrix:

$$
\begin{bmatrix}
d_1 & u_1 & 0 & \vdots & 0 & 0 \\
l_2 & d_2 & u_2 & \vdots & 0 & 0 \\
0 & l_3 & d_3 & \vdots & 0 & 0 \\
\cdots & \cdots & \cdots & \cdots & \cdots & \cdots \\
0 & 0 & 0 & l_{n-1} & d_{n-1} & u_{n-1} \\
0 & 0 & 0 & 0 & l_n & d_n
\end{bmatrix}
\begin{bmatrix} x_1 \\ x_2 \\ x_3 \\ \cdots \\ x_{n-1} \\ x_n \end{bmatrix}
=
\begin{bmatrix} b_1 \\ b_2 \\ b_3 \\ \cdots \\ b_{n-1} \\ b_n \end{bmatrix}
\tag{2.82}
$$

The above system of equations can be stored in four vectors of length n. These are vector $\boldsymbol{d}$ consisting of the diagonal elements, vector $\boldsymbol{u}$ consisting of the upper diagonal elements, vector $\boldsymbol{l}$ consisting of the lower diagonal elements and the right-hand side vector $\boldsymbol{b}$. The ith equation can be written as follows:

$$
l_i x_{i-1} + d_i x_i + u_i x_{i+1} = b_i \tag{2.83}
$$

Note that l_1 and u_n are zero. Let us define two new vectors, $\boldsymbol{\alpha}$ and $\boldsymbol{\beta}$ of length n and initialize them as $\alpha_1 = d_1$ and $\beta_1 = b_1$. Using these, the first two equations can be written as follows:

$$
\begin{aligned}
\alpha_1 x_1 + u_1 x_2 &= \beta_1 \\
l_2 x_1 + d_2 x_2 + u_2 x_3 &= b_2
\end{aligned}
\tag{2.84}
$$

Multiplying the first equation by l_2/α_1 and deducting from the second equation, we obtain

$$\alpha_2 x_2 + u_2 x_3 = \beta_2 \text{ where } \alpha_2 = d_2 - \left(\frac{l_2}{\alpha_1}\right) u_1 \text{ and } \beta_2 = b_2 - \left(\frac{l_2}{\alpha_1}\right) \beta_1 \quad (2.85)$$

Similarly, we can subsequently remove x_2, x_3 and so on. Note that the vectors α and β generated during the process of elimination can be calculated by the iterative formulae:

$$\alpha_i = d_i - \left(\frac{l_i}{\alpha_{i-1}}\right) u_{i-1} \text{ and } \beta_i = b_i - \left(\frac{l_i}{\alpha_{i-1}}\right) \beta_{i-1}, \quad i = 2, \ldots, n \quad (2.86)$$

At the $(i-1)$ stage, the resulting equation can be written as

$$\alpha_i x_i + u_i x_{i+1} = \beta_i \quad (2.87)$$

After the completion of $(n-1)$ elimination steps, the last equation becomes $\alpha_n x_n = \beta_n$ which leads to the solution $x_n = \beta_n / \alpha_n$. All other x_i's can easily be calculated using Eq. (2.87) as follows:

$$x_i = \frac{\beta_i - u_i x_{i+1}}{\alpha_i}, \quad i = n-1, \ldots, 1 \quad (2.88)$$

Example 2.9   Solve the following tri-diagonal system of equation using Thomas algorithm:

$$\begin{bmatrix} -2 & 1 & 0 & 0 \\ 1 & -4 & 1 & 0 \\ 0 & 1 & -4 & 1 \\ 0 & 0 & 1 & -2 \end{bmatrix} \begin{bmatrix} x_1 \\ x_2 \\ x_3 \\ x_4 \end{bmatrix} = \begin{bmatrix} 3 \\ 1 \\ 2 \\ -2 \end{bmatrix}$$

Solution   Using the same notation for the lower and upper diagonal and the diagonal elements as in Eq. (2.82), we can compute α_i's and β_i's using Eq. (2.86). Finally the solution vector was computed using Eq. (2.88). Table 2.1 shows the computations.

Table 2.1

Index	l	d	u	b	α	β	x
1		-2	1	3	-2	3	-1.93333
2	1	-4	1	1	-3.5	2.5	-0.86667
3	1	-4	1	2	-3.71429	2.714286	-0.53333
4	1	-2		-2	-1.73077	-1.26923	0.733333

2.2.1.5 *Stability of the Direct Methods and Pivoting*

In the context of Gauss elimination, what we mean by stability is that the round-off errors do not grow. Round-off error incurred in any element will grow only if it is multiplied by a number of magnitude larger than unity at any time during the process of elimination. Looking at the process of elimination, it is easy to see that at each step, elements are multiplied by the factor l_{ik}. So, the condition of stability is

$$|l_{ik}| \leq 1 \tag{2.89}$$

Since, $l_{ik} = a_{ik}/a_{kk}$ for $i = (k+1)$ to n, it follows that $|a_{kk}| \geq |a_{ik}| \ \forall i = (k+1) \cdots n$. In other words, at any step k, the absolute value of the pivot should be larger than the absolute values of the elements in the pivotal column of the subsequent rows.

If at any step, the above condition is not satisfied, it can be easily achieved by interchanging of rows. In order to illustrate this, let us have a re-look at the Gauss elimination matrix in the beginning of the kth step [Eq. (2.52)]

$$\begin{bmatrix}
a_{11} & a_{12} & a_{13} & a_{14} & \cdots & a_{1k} & a_{1k+1} & \cdots & a_{1n} \\
0 & a_{22} & a_{23} & a_{24} & \cdots & a_{2k} & a_{2k+1} & \cdots & a_{2n} \\
0 & 0 & a_{33} & a_{34} & \cdots & a_{3k} & a_{3k+1} & \cdots & a_{3n} \\
0 & 0 & 0 & a_{44} & \cdots & a_{4k} & a_{4k+1} & \cdots & a_{4n} \\
\cdots & \cdots & \cdots & \cdots & \cdots & \cdots & \cdots & \cdots & \cdots \\
0 & 0 & 0 & 0 & \cdots & a_{kk} & a_{kk+1} & \cdots & a_{kn} \\
0 & 0 & 0 & 0 & \cdots & a_{k+1k} & a_{k+1k+1} & \cdots & a_{k+1n} \\
\cdots & \cdots & \cdots & \cdots & \cdots & \cdots & \cdots & \cdots & \cdots \\
0 & 0 & 0 & 0 & \cdots & a_{nk} & a_{nk+1} & \cdots & a_{nn}
\end{bmatrix} \tag{2.90}$$

Before we perform the elimination operations of the kth step, we search for $\max\limits_{k \leq i \leq n} |a_{ik}|$. If this occurs at $i = p$, we interchange kth row with the pth row (of course, accompanied by the interchange of entries in the right-hand side vector as well, i.e., interchange b_k with b_p. This is done automatically if one works with the augmented matrix.) It is equivalent to changing the position or sequence of the equations in the set and thus has no bearing on the final solution. The operation is known as *partial pivoting*.

You may have already guessed that there is also something called the *full pivoting*. In this, the search is made for the largest absolute valued element in rows as well as columns between k and n, i.e, search for $\max\limits_{\substack{k \leq i \leq n \\ k \leq j \leq n}} |a_{ij}|$. If this occurs in the pth row and *s*th column, interchange *k*th row with the *p*th row and *k*th column with *s*th column such that a_{ps} becomes the pivot. While partial pivoting had no bearing on the final solution, exchange of columns in full pivoting amounts to renaming of the variable. So, this should be properly accounted for while interpreting the final solution vector.

We have already shown that the *LU* decomposition is nothing but Gauss elimination. So, the same stability criteria and pivoting scheme are applicable there as well. The same stability criteria and partial pivoting scheme would also work for Gauss Jordon and Thomas algorithm.

Example 2.10   Perform Gauss elimination on the matrix $\begin{bmatrix} 2 & 1 & -1 \\ 1 & -0.5 & -1 \\ 2 & 4 & 1 \end{bmatrix}$

Solution   After the first step of Gauss elimination, the matrix is

$$\begin{bmatrix} 2 & 1 & -1 \\ 0 & 0 & -0.5 \\ 0 & 3 & 2 \end{bmatrix}$$

We run into a problem to proceed further with elimination as the pivot $a_{22} = 0$. However, if we switch the third row with the second, we obtain

$$\begin{bmatrix} 2 & 1 & -1 \\ 0 & 3 & 2 \\ 0 & 0 & -0.5 \end{bmatrix}$$

This is already an upper triangular matrix.

2.2.1.6 *Perturbation Analysis*

In many engineering applications, one is interested in finding the effect of small changes in some elements of the coefficient matrix and/or in the elements of the right-hand side vector on the solution vector. These small changes in the coefficient matrix or in the right-hand side vector often represent small perturbations in the physical process. In this section, we will try to establish a relation between the perturbation quantities and the resulting changes in the solution vector. More precisely, we would like to obtain upper bounds for the changes in the solution vector as a function of the perturbation quantities. This analysis is immensely important from the application point of view where one is often interested in the possibility of failure as a result of small perturbation. An analysis of upper bounds of the effect will enable an engineer to restrict the perturbations within certain limits to avoid failure.

We first explore the effect of perturbation in the co-efficient matrix A keeping b unaltered. Let us assume that the matrix A has been given a perturbation of δA. As a result the solution vector changes by an amount δx. Notice that if the matrix A is a $(n \times n)$ matrix, δA is also one with at least one non-zero element which is small compared to the corresponding element in A. If more than one element in δA are non-zero, all such elements are small compared to the corresponding elements in A. So, the resulting equation to solve now is as follows:

$$(A + \delta A)(x + \delta x) = b \tag{2.91}$$

Since, $Ax = b$, the above equation can be simplified to yield the following:

$$A\delta x + \delta A(x + \delta x) = 0 \quad \text{or} \quad \delta x = -A^{-1}\delta A(x + \delta x) \tag{2.92}$$

Now, we take the norm in order to get the upper bound. Notice that the negative sign disappears due to this.

$$\|\delta x\| = \|A^{-1}\delta A(x + \delta x)\| \le \|A^{-1}\|\|\delta A\|\|x + \delta x\| \le \|A^{-1}\|\|\delta A\|\|x\| + \|A^{-1}\|\|\delta A\|\|\delta x\| \tag{2.93}$$

Second term on the right-hand side involve product of two perturbation quantities which is likely to be much smaller compared to the first term. So, ignoring this term and carrying out some algebraic manipulation, one obtains,

$$\frac{\|\delta x\|}{\|x\|} \le \|A\|\|A^{-1}\|\frac{\|\delta A\|}{\|A\|} \tag{2.94}$$

If we did not ignore the product term, the relationship would be,

$$\frac{\|\delta x\|}{\|x + \delta x\|} \le \|A\|\|A^{-1}\|\frac{\|\delta A\|}{\|A\|} \tag{2.95}$$

Notice that both the Eqs (2.94) and (2.95) are essentially saying the same thing, that is, the relative change in vector x is proportional to the relative change in the matrix A with the proportionality constant $\|A\|\|A^{-1}\|$. One expresses the relative change in x with respect to the old vector x and the other with respect to the new vector $(x + \delta x)$. Since, the perturbation quantity is likely to be smaller compared to the original vector, these ratios are unlikely to make much difference. In other words, ignoring the term involving product of two perturbation quantities in Eq. (2.93) does not have much effect on the final results. One may use either of the two equations.

The proportionality constant $\|A\|\|A^{-1}\|$ is known as the *condition number* of the matrix A. We will denote the condition number as $C(A)$. It is easy to see, that if $C(A)$ is very large, the error is amplified considerably and for small $C(A)$, the error stays within acceptable limits. Depending on this, the matrix can be categorized as *ill-conditioned* if $C(A)$ is large and *well-conditioned* otherwise. How large is large depends on the type of problem and the matrix norm chosen to evaluate the condition number.

Now, let us apply a small perturbation δb to the right-hand side vector b. Let us assume that the resulting change in the solution vector x is δx. The equation set can now be written as

$$A(x + \delta x) = (b + \delta b) \tag{2.96}$$

Once again using the equality $Ax = b$, one obtains,

$$\delta x = A^{-1}\delta b \tag{2.97}$$

Taking the norms, we get

$$\|\delta x\| \le \|A^{-1}\|\|\delta b\| \tag{2.98}$$

Multiplying the numerator and denominator of the right-hand side by the norm of vector b,

$$\|\delta x\| \le \|A^{-1}\|\frac{\|b\|}{\|b\|}\|\delta b\| \tag{2.99}$$

Using the equality $Ax = b$,

$$\|\delta x\| \le \|A^{-1}\|\frac{\|Ax\|}{\|b\|}\|\delta b\| \le \|A^{-1}\|\|A\|\|x\|\frac{\|\delta b\|}{\|b\|} \tag{2.100}$$

$$\frac{\|\delta x\|}{\|x\|} \le \|A^{-1}\|\|A\|\frac{\|\delta b\|}{\|b\|} \tag{2.101}$$

Once again, we observe that relative change in the norms of solution vector x is proportional to the relative perturbation in the norms of the vector b and the proportionality constant is the *condition number* of matrix A. Thus, the sensitivity of the solution to perturbation in either the coefficient matrix A or the right hand side vector b depend on the condition number of matrix A.

Example 2.11   In the Batman problem of (2.2), if the distances were measured with an accuracy of ±0.1 m, what is the maximum relative error in the estimation of parameters?

Solution   Using L_1 norm for the vectors and column sum norm $\left(\|A\|_1\right)$ for the matrices, we get

$$\|\delta b\| = 0.1, \|b\| = 875, \|A\|_1 = 370.92$$

Using Gauss Jordon elimination, we can compute the inverse as

$$A^{-1} = \begin{bmatrix} 2.454 & -2.117 & 0.663 \\ -0.282 & 0.436 & -0.154 \\ 0.015 & -0.03 & 0.015 \end{bmatrix}$$

The column sum norm of the inverse is given by the sum of the absolute values of the first column as $\|A^{-1}\|_1 = 2.751$. The condition number of the matrix A is $370.92 \times 2.751 = 1020.4$ and the upper bound of the relative error in estimation is $\dfrac{0.1 \times 1020.4}{875} = 0.1166$ or 11.66%.

2.2.1.7 Iterative Improvement of Solution by Direct Methods

In the cases of moderate *ill-conditioning*, an iteration scheme can be applied to improve the solution using the direct methods. In order to see this, let us assume that we have computed an approximate solution vector $\tilde{x}$ by a direct method. We can then compute a residual vector r,

$$r = b - A\tilde{x} \tag{2.102}$$

The residual vector gives the error vector with the approximate solution. If x is the true solution vector, $(Ax - b)$ is zero. Adding $(Ax - b)$ to the right-hand side of the above equation, we obtain,

$$A(x - \tilde{x}) = r \tag{2.103}$$

Define $(x - \tilde{x}) = \delta x$ as the improvement in solution. The vector δx can be calculated by solving the above equation. However, since this will also not be calculated exactly with infinite precision, an improvement of the existing solution can be calculated as $(\tilde{x} + \delta x)$. Using this improved approximate solution a new residual vector can be calculated. Thus, Eqs (2.102) and (2.103) in combination define an iterative process to improve an approximate solution. Since, the coefficient matrix A remains constant, application of LU decomposition makes the process more efficient. The only thing to keep in mind here is that, the residual vector is a difference of two almost equal quantities, and it should be computed with a higher precision. All other quantities, however, may be computed with normal precision. If r is not computed using higher precision, there is no gain. On the other hand, if the whole computation is done with higher precision, it significantly increases storage as well as CPU time. Iterative improvement is fast and less computation intensive once the *LU* decomposition of the matrix *A* is computed. In the next section, we outline the iterative methods for solution of a system of linear equations.

2.2.2 Iterative Methods

Any $n \times n$ matrix A can be written as a summation of three matrices as follows:

$$
\begin{bmatrix}
a_{11} & a_{12} & \cdots & \cdots & a_{1n} \\
a_{21} & a_{22} & \cdots & \cdots & a_{2n} \\
a_{31} & a_{32} & \cdots & \cdots & a_{3n} \\
\cdots & \cdots & \cdots & \cdots & \cdots \\
a_{n1} & a_{n2} & \cdots & \cdots & a_{nn}
\end{bmatrix}
=
\begin{bmatrix}
0 & 0 & \cdots & \cdots & 0 \\
a_{21} & 0 & 0 & \cdots & 0 \\
a_{31} & a_{32} & 0 & 0 & 0 \\
\cdots & \cdots & \cdots & \cdots & \cdots \\
a_{n1} & a_{n2} & a_{n3} & \cdots & 0
\end{bmatrix}
+
$$

$$
\begin{bmatrix}
a_{11} & 0 & 0 & \cdots & 0 \\
0 & a_{22} & 0 & 0 & 0 \\
0 & 0 & a_{33} & 0 & 0 \\
\cdots & \cdots & \cdots & \cdots & \cdots \\
0 & 0 & 0 & \cdots & a_{nn}
\end{bmatrix}
+
\begin{bmatrix}
0 & a_{12} & a_{13} & \cdots & a_{1n} \\
0 & 0 & a_{23} & \cdots & a_{2n} \\
0 & 0 & 0 & \cdots & a_{3n} \\
\cdots & \cdots & \cdots & \cdots & \cdots \\
0 & 0 & 0 & \cdots & 0
\end{bmatrix}
$$

In the matrix form, this can be written as

$$A = L + D + U \tag{2.104}$$

We define a *strictly lower triangular matrix* as the one that contains non-zero values below the leading diagonal. All the entries on and above the leading diagonal are zero. Notice that *L* is a *strictly lower triangular matrix*. A *diagonal matrix* is the one where all entries other than the leading diagonal are zero. The *D* is a diagonal matrix. Similarly, a *strictly upper triangular matrix* is the one that has non-zero values above the leading diagonal. All the entries on and below the leading diagonal are zero. The *U* is such a matrix.

Using Eq. (2.104), a set of linear simultaneous equation given by $Ax = b$ can be written as follows:

$$(L + D + U)x = b \tag{2.105}$$

Recall that if a_{ij}'s are the elements of an $n \times n$ matrix A, in the expanded form the set of equation is still represented as:

$$\sum_{j=1}^{n} a_{ij} x_j = b_i, \quad i = 1, \ 2, \dots, n \tag{2.106}$$

Therefore,

$$l_{ij} = \begin{cases} a_{ij}, & \text{for } i > j \\ 0, & \text{for } i \le j \end{cases}$$

$$d_{ij} = \begin{cases} a_{ij}, & \text{for } i = j \\ 0, & \text{for } i \ne j \end{cases}$$

$$u_{ij} = \begin{cases} a_{ij}, & \text{for } i < j \\ 0, & \text{for } i \ge j \end{cases}$$

From Eq. (2.105), we can derive two iterative methods by algebraic rearrangements. These methods are outlined below.

Jacobi Iteration For an iteration counter k,

$$Dx^{(k+1)} = -(U + L)x^{(k)} + b \quad \text{or} \quad x^{(k+1)} = -D^{-1}(U + L)x^{(k)} + D^{-1}b \qquad (2.107)$$

In the expanded form, it is equivalent to

$$x_i^{(k+1)} = \frac{b_i - \sum_{j=1, j \ne i}^{n} a_{ij} x_j^{(k)}}{a_{ii}}, \quad i = 1, 2, \dots, n \qquad (2.108)$$

Gauss Seidel Iteration During Jacobi iteration, some of the x_i values were known in the new iteration step but they were not used in calculation, e.g., $x_1^{(k+1)}$ was known while calculating $x_2^{(k+1)}$ but was not used in computation. To make convergence faster, Gauss Seidel iteration scheme uses the known values of x vector at new iterations. For an iteration counter k, the scheme is given as follows:

$$(L + D)x^{(k+1)} = -Ux^{(k)} + b \quad \text{or} \quad x^{(k+1)} = -(L + D)^{-1}Ux^{(k)} + (L + D)^{-1}b \qquad (2.109)$$

In the expanded form, it is equivalent to

$$x_i^{(k+1)} = \frac{b_i - \sum_{j=1}^{i-1} a_{ij} x_j^{(k+1)} - \sum_{j=i+1}^{n} a_{ij} x_j^{(k)}}{a_{ii}}, \quad i = 1, 2, \dots, n \qquad (2.110)$$

Both the schemes require an initial x vector to be given for the start-up. It is a common practice to give a zero vector for start-up. However, often in engineering problems, the variables have some physical significance and one knows the order of magnitude values of the variables. This information can be put in the start-up vector to make the convergence faster.

Both the schemes also require a termination criterion in order to be able to stop the iteration. If the true solution vector is x and the approximate solution vector at the kth iteration is $x^{(k)}$, the true error vector is given by $e^{(k)} = x - x^{(k)}$. However, since the true solution is not known, we will define the approximate error as the improvement between iterations, i.e., $\varepsilon^{(k)} = x^{(k+1)} - x^{(k)}$.

An upper bound or tolerance can now be defined on the norm of the error vector $\varepsilon^{(k)}$. Any vector norm defined in Section 1.4 can be used for this purpose. Similarly, one can also work with approximate relative error. The error expressions for absolute and relative errors are as follows:

$$\varepsilon^{(k)} = \left\| x^{(k+1)} - x^{(k)} \right\| \text{ and } \varepsilon_r^{(k)} = \frac{\left\| x^{(k+1)} - x^{(k)} \right\|}{\left\| x^{(k+1)} \right\|}$$

In the next section, we will try to understand under what condition the iterative methods do not converge.

Example 2.12 Solve the Batman Problem (2.2) using: (a) Gauss Seidel method and (b) Jacobi method with an absolute error of 0.001 or less.

Solution (a) *Gauss Seidel method*
We assume an initial solution vector as,

$$\begin{bmatrix} S_0 \\ u \\ f \end{bmatrix} = \begin{bmatrix} 0 \\ 0 \\ 0 \end{bmatrix}$$

Now, using Eq. (2.110), we can compute the new solutions. We tabulate the values in Table 2.2.

Table 2.2

Iteration no.	S_0	u	f	$\varepsilon^{(k)}$
	0	0	0	
1	439	14.15493	0.47405	439
2	342.1282	17.61113	0.551588	96.8718
3	319.2094	18.67461	0.547988	22.91876
4	312.6851	19.15963	0.530689	6.52436
5	310.0105	19.47081	0.513566	2.674627
6	308.4102	19.70507	0.499028	1.600213
7	307.2372	19.8909	0.487112	1.173016
8	306.3141	20.04051	0.477433	0.923099

(contd.)

(contd.)

9	305.5726	20.16145	0.469591	0.741564
10	304.9735	20.25932	0.46324	0.599104
11	304.4887	20.33855	0.458098	0.484749
12	304.0963	20.40269	0.453936	0.392382
13	303.7787	20.45461	0.450566	0.31765
14	303.5215	20.49665	0.447838	0.257158
15	303.3133	20.53068	0.445629	0.208188
16	303.1448	20.55823	0.443841	0.168544
17	303.0083	20.58054	0.442393	0.136449
18	302.8979	20.59859	0.441221	0.110465
19	302.8084	20.61321	0.440273	0.08943
20	302.736	20.62505	0.439505	0.0724
21	302.6774	20.63463	0.438883	0.058613
22	302.63	20.64238	0.438379	0.047452
23	302.5916	20.64866	0.437972	0.038416
24	302.5605	20.65375	0.437642	0.0311
25	302.5353	20.65786	0.437375	0.025178
26	302.5149	20.6612	0.437158	0.020384
27	302.4984	20.66389	0.436983	0.016502
28	302.485	20.66608	0.436842	0.01336
29	302.4742	20.66784	0.436727	0.010816
30	302.4655	20.66928	0.436634	0.008756
31	302.4584	20.67044	0.436559	0.007089
32	302.4526	20.67137	0.436498	0.005739
33	302.448	20.67213	0.436449	0.004646
34	302.4442	20.67275	0.436409	0.003761
35	302.4412	20.67325	0.436376	0.003045
36	302.4387	20.67365	0.43635	0.002465
37	302.4367	20.67397	0.436329	0.001996
38	302.4351	20.67424	0.436312	0.001616
39	302.4338	20.67445	0.436298	0.001308
40	302.4327	20.67463	0.436287	0.001059
41	302.4319	20.67477	0.436278	0.000857

So, the solution vector is

$$\begin{bmatrix} S_0 \\ u \\ f \end{bmatrix} = \begin{bmatrix} 302.4319 \\ 20.67477 \\ 0.436278 \end{bmatrix}$$

You may want to compare this solution with the one obtained using Gauss elimination in Example 2.4.

(b) *Jacobi iteration*

If we start with the same initial solution vector and perform the iterations using Eq. (2.108), we will see that the method does not converge for this problem. The solution oscillates and the oscillation increases with iterations. We tabulate the first 25 iterations in Table 2.3.

Table 2.3

Iteration no.	S_0	u	f	$\varepsilon^{(k)}$
	0	0	0	
1	439	45.07042	3.487723	439
2	92.52934	−10.6079	−2.28626	346.4707
3	548.71097	54.78673	4.066038	456.1816
4	21.172992	−22.4401	−3.59109	527.538
5	647.14926	69.07614	5.406906	625.9763
6	−93.19279	−38.8925	−5.25931	740.342
7	781.21737	88.97437	7.331732	874.4102
8	−253.557	−62.0002	−7.57033	1034.774
9	968.90277	116.6759	10.03412	1222.46
10	−477.2462	−94.4044	−10.7918	1446.149
11	1231.7256	155.301	13.81898	1708.972
12	−789.4667	−139.786	−15.2881	2021.192
13	1599.507	209.2119	19.11537	2388.974
14	−1225.511	−203.29	−21.5675	2825.018
15	2113.9254	284.5032	26.52347	3339.436
16	−1834.701	−292.114	−30.3404	3948.626
17	2833.2502	389.6915	36.88241	4667.951
18	−2685.967	−416.319	−42.5994	5519.217
19	3838.9405	536.6788	51.36527	6524.908
20	−3875.649	−589.971	−59.7319	7714.59
21	5244.8676	742.1002	71.61192	9120.517
22	−5538.402	−832.731	−83.6771	10783.27
23	7210.2066	1029.206	99.91461	12748.61
24	−7862.437	−1172.08	−117.145	15072.64
25	9957.4566	1430.495	139.4775	17819.89

2.2.2.1 *Convergence of the Iterative Methods*

From Eqs (2.107) and (2.109), it is easy to see that both the methods can be expressed as

$$x^{(k+1)} = Sx^{(k)} + c \tag{2.111}$$

where the matrix S and the vector c are given as follows:

For *Jacobi* iteration:

$$S = -D^{-1}(L + U) \quad \text{and} \quad c = D^{-1}b \tag{2.112}$$

For *Gauss Seidel* iteration:

$$S = -(L + D)^{-1}U \quad \text{and} \quad c = (L + D)^{-1}b \tag{2.113}$$

In both the methods, S and c, as well as form of the iteration Eq. (2.111) remain unchanged for all the iterations during the entire solution process. These are known as *stationary iteration methods* and the matrix S is called the *iteration matrix*. Note that for a coefficient matrix A of size $n \times n$, the iteration matrix S is also $n \times n$ and c is a vector of size n.

If we denote the true solution by x, error vector $e^{(k)}$ at kth iteration is given by

$$e^{(k)} = x - x^{(k)} \qquad (2.114)$$

Needless to say that the true solution vector satisfies the iteration Eq. (2.111), i.e.,

$$x = Sx + c \qquad (2.115)$$

Deducting Eq. (2.111) from Eq. (2.115) and using the definition of Eq. (2.114), one obtains

$$e^{(k+1)} = Se^{(k)} \qquad (2.116)$$

The above equation means that the error vector satisfies the homogenous form of the iteration Eq. (2.111). If at the start of the iteration process, the error vector is given by $e^{(0)}$, error at kth iteration will be given by,

$$e^{(k)} = S^k e^{(0)} \qquad (2.117)$$

Thus, the methods will converge if the error vector approaches zero with the progression of the iteration, i.e.,

$$\lim_{k \to \infty} e^{(k)} = 0 \qquad (2.118)$$

Since $e^{(0)}$ is a constant vector, it is easy to see from Eq. (2.117), that Eq. (2.118) is possible only if,

$$\lim_{k \to \infty} S^k = 0 \qquad (2.119)$$

Let us assume that the original linear system of equations and the corresponding coefficient matrix A is such that the iteration matrix S has a set of linearly independent eigenvectors, which form a basis for an n-dimensional vector space. If we denote the eigenvectors of S as $\{v_j\}_{j=1}^n$ and the eigenvalues as $\{\lambda_j\}_{j=1}^n$, any n-dimensional vector can be expressed as a linear combination of these vectors.

$$e^{(0)} = \sum_{j=1}^n C_j v_j \qquad (2.120)$$

Using Eq. (2.117), we can write, $e^{(k)} = \sum_{j=1}^n C_j \lambda_j^k v_j$.

From the above equation one can easily see that in order to satisfy the convergence criterion of Eq. (2.118), the largest eigenvalue or the *spectral radius* (Section 2.1.3) of the iteration matrix S must be strictly less than unity, i.e.,

$$\rho(S) < 1 \tag{2.121}$$

This is the necessary condition for the methods to converge. However, this condition is very hard to use in application. It will be easier, if we can derive a condition based on the original coefficient matrix A. For this purpose, we recall the relation derived in Eq. (2.37), $\rho(A) \le \|A\|$ for arbitrary matrix A. Using this relation in conjunction with the necessary condition of Eq. (2.121), one can get a sufficient condition as,

$$\|S\| < 1 \tag{2.122}$$

Now, for *Jacobi* iteration: $S = -D^{-1}(L + U)$. The elements (s_{ij}) of the iteration matrix are related to a_{ij}'s and can be written as follows:

$$s_{ij} = \begin{cases} -\dfrac{a_{ij}}{a_{ii}}, & \text{for } i \ne j \\[2ex] 0, & \text{for } i = j \end{cases} \tag{2.123}$$

If we use, *row-sum norm* [Eq. (2.34)] for the iteration matrix,

$$\|S\| = \max_{1 \le i \le n} \sum_{j=1}^{n} |s_{ij}| = \max_{1 \le i \le n} \sum_{j=1}^{n} \left| \frac{a_{ij}}{a_{ii}} \right| \tag{2.124}$$

Application of the sufficient condition [Eq. (2.122)] translates to,

$$|a_{ii}| > \sum_{j=1, j \ne i}^{n} |a_{ij}|, \quad i = 1, 2, \ldots, n \tag{2.125}$$

This means that the *Jacobi* method surely converges if A is strictly diagonally dominant. Using the definition of the iteration matrix in Eq. (2.113) and expanded form of Eq. (2.110), one can derive a similar condition for the *Gauss Seidel* method. This is left for the reader as an exercise.

2.2.2.2 *Rate of Convergence of the Iterative Methods*

From Eqs (2.116) and (2.120), for large k, one can write:

$$\frac{\left\| e^{(k+1)} \right\|}{\left\| e^{(k)} \right\|} \cong \rho(S) \tag{2.126}$$

If the error is reduced by a factor of 10^{-m} in k iterations,

$$\rho(S)^k \le 10^{-m} \quad \text{or} \quad k \ge -\frac{m}{\log_{10} \rho(S)} \tag{2.127}$$

The quantity $R = \log_{10} \rho\,(S)$ is called the asymptotic rate of convergence. It is also evident that $(1/R)$ represents the minimum number of iteration required to reduce the error by one order of magnitude.

If the eigenvalue corresponding to the spectral radius is denoted as λ_{max} such that, $\rho(S) = |\lambda_{max}|$, one may write Eq. (2.126) as,

$$e^{(k+1)} \cong \lambda_{max} e^{(k)} \quad \text{or} \quad e^{(k+1)} - e^{(k)} \cong \lambda_{max}(e^{(k)} - e^{(k-1)}) \tag{2.128}$$

In an iterative methods, the solution vector $x^{(k)}$ is modified in each iteration. Let us denote the modification vector as $d^{(k)}$ and define as follows:

$$x^{(k+1)} = x^{(k)} + d^{(k)} \tag{2.129}$$

Using the definition of error vectors from Eq. (2.114) into Eq. (2.128) and combining with the relation defined by Eq. (2.129), we obtain

$$d^{(k)} \cong \lambda_{max} d^{(k-1)} \tag{2.130}$$

Now, recall Eq. (2.110) of Gauss Seidel iteration and rewrite as,

$$x_i^{(k+1)} = x_i^{(k)} + \frac{b_i - \sum_{j=1}^{i-1} a_{ij} x_j^{(k+1)} - \sum_{j=i}^{n} a_{ij} x_j^{(k)}}{a_{ii}}, \quad i = 1, 2, \ldots, n \tag{2.131}$$

or
$$x_i^{(k+1)} = x_i^{(k)} + d_i^{(k)}, \quad i = 1, 2, \ldots, n \tag{2.132}$$

This is same as Eq. (2.129). This means that, at every iteration, the solution vector is modified by a displacement vector d. Furthermore, implication of Eq. (2.130) is that, for a $\lambda_{max} > 0$, the direction of the displacement vector does not change from iteration $(k-1)$ to iteration (k) for large enough value of k. It is only the magnitude of displacement that changes and the magnitude is approximately given by the largest eigenvalue of the iteration matrix. This observation suggests relaxation procedure as follows:

$$x_i^{(k+1)} = x_i^{(k)} + \omega d_i^{(k)}, \quad i = 1, 2, \ldots, n, \ \omega > 0 \tag{2.133}$$

where ω is a relaxation parameter. Depending on the value of ω, the following relaxation procedures can be defined:

$$0 < \omega < 1 : \text{Under relaxation}$$
$$\omega = 1 : \text{Gauss Seidel}$$
$$1 < \omega < 2 : \text{Over relaxation}$$

The values of $\omega > 2$ is not commonly used.

Using Eq. (2.133) in conjunction with Eq. (2.131), a modified Gauss Seidel iteration scheme can be written as follows:

$$x_i^{(k+1)} = (1-\omega)x_i^{(k)} + \omega \frac{b_i - \sum_{j=1}^{i-1} a_{ij} x_j^{(k+1)} - \sum_{j=i+1}^{n} a_{ij} x_j^{(k)}}{a_{ii}}, \quad i = 1, 2, \dots, n \quad (2.134)$$

When over relaxation is done in succession, the method is commonly known as method of *successive over relaxation (SOR)*.

Example 2.13   Solve the Batman's problem of (2.2) using successive over relaxation (SOR) with ω as 1.1, 1.4, and 1.6. Compare the number of iterations required for computing the solution with an absolute error of 0.001 or less.

Solution   We assume an initial solution vector as,

$$\begin{bmatrix} S_0 \\ u \\ f \end{bmatrix} = \begin{bmatrix} 0 \\ 0 \\ 0 \end{bmatrix}$$

and perform SOR iterations using Eq. (2.134). Tables 2.4 – 2.6 show the computations for three values of ω = 1.1, 1.4, and 1.6.

$\omega = 1.1$

Table 2.4

Iteration no.	S_0	u	f	$\varepsilon^{(k)}$
0	0	0		
1	482.9	12.16971831	0.523948265	482.9
2	340.5352069	17.88896915	0.534049181	142.3647931
3	315.5528419	19.17340953	0.51642553	24.98236495
4	309.6637942	19.63880005	0.498300819	5.889047728
5	307.4619283	19.90438206	0.483683569	2.201865869
6	306.1798835	20.09129792	0.472408698	1.282044872
7	305.2716952	20.2310157	0.463795915	0.908188321
8	304.5917302	20.33698309	0.457231032	0.679964961
9	304.0758238	20.41762267	0.452229589	0.515906405
10	303.6831931	20.47903504	0.448419676	0.392630719
11	303.3841729	20.52581276	0.445517501	0.299020154
12	303.1564087	20.56144468	0.443306802	0.227764216
13	302.982914	20.58858677	0.441622829	0.173494683

(contd.)

(contd.)

14	302.8507569	20.60926191	0.440340084	0.13215707
15	302.750088	20.62501093	0.439362969	0.100668927
16	302.6734047	20.63700755	0.438618665	0.076683277
17	302.6149922	20.64614582	0.4380517	0.058412518
18	302.5704972	20.65310679	0.437619821	0.044494999
19	302.5366037	20.65840921	0.437290843	0.033893504
20	302.5107858	20.66244827	0.437040249	0.02581795
21	302.4911193	20.66552497	0.436849361	0.019666498
22	302.4761386	20.66786861	0.436703955	0.014980707
23	302.4647272	20.66965385	0.436593194	0.011411364
24	302.4560347	20.67101373	0.436508823	0.008692463
25	302.4494134	20.6720496	0.436444554	0.006621374
26	302.4443696	20.67283866	0.436395599	0.005043748
27	302.4405276	20.67343972	0.436358307	0.003842012
28	302.437601	20.67389757	0.436329901	0.002926604
29	302.4353717	20.67424633	0.436308263	0.002229304
30	302.4336736	20.674512	0.43629178	0.001698144
31	302.43238	20.67471436	0.436279225	0.00129354
32	302.4313947	20.67486851	0.436269661	0.000985338

$\omega = 1.4$

Table 2.5

Iteration no.	S_0	u	f	$\varepsilon^{(k)}$
	0	0	0	
1	614.6	2.504225352	1.140096831	614.6
2	316.3455984	19.57533625	0.214535317	298.2544016
3	312.3751257	22.3384284	0.26153006	3.97047272
4	288.7151404	23.0987384	0.27972458	23.6599853
5	291.0900655	22.37961332	0.349084477	2.374925134
6	294.5157651	21.64008095	0.394665437	3.42569956
7	298.3381337	21.10596591	0.421867251	3.822368647
8	300.7133584	20.81504881	0.434096561	2.37522471
9	301.9593627	20.68701081	0.438256437	1.246004283
10	302.4603968	20.64747911	0.438738	0.5010341
11	302.5901604	20.64571146	0.438042204	0.129763622
12	302.5723207	20.65509359	0.437247309	0.017839755
13	302.5194088	20.66445866	0.436689901	0.052911896
14	302.4742835	20.67070224	0.436384231	0.045125299
15	302.4463642	20.67399577	0.436250608	0.027919254
16	302.4325397	20.67536956	0.436209479	0.013824551
17	302.4272517	20.67575021	0.436207858	0.005287986

(contd.)

(*contd.*)

18	302.4261065	20.67572697	0.436217802	0.001145238
19	302.4264987	20.67559876	0.436227662	0.000392209

$\omega = 1.6$

Table 2.6

Iteration no.	S_0	u	f	$\varepsilon^{(k)}$
	0	0	0	
1	702.4	−7.030985915	2.105191865	702.4
2	285.96852	20.19448294	−0.391462937	416.43148
3	342.5278855	25.84831394	−0.061870437	56.55936544
4	242.3706341	29.99716864	−0.213561485	100.1572514
5	265.9731494	26.57162777	0.216289879	23.60251532
6	272.5742165	23.0000586	0.426504586	6.601067058
7	297.5790198	19.9375184	0.578412096	25.00480328
8	308.2850768	18.84305692	0.575340804	10.70605707
9	312.8129488	19.02444092	0.522394845	4.527871952
10	309.9250905	19.84246911	0.455718715	2.8878583
11	305.5933902	20.59717194	0.415535342	4.33170027
12	301.9414774	21.01231649	0.403629262	3.651912832
13	300.3805269	21.07436457	0.411863939	1.560950444
14	300.4483474	20.93594804	0.426264394	0.13841653
15	301.3379043	20.75517703	0.437775357	0.889556896
16	302.2434335	20.63084364	0.442855634	0.905529178
17	302.7772746	20.58758063	0.442583299	0.533841049
18	302.8945138	20.6034221	0.439735934	0.117239254
19	302.7545851	20.6420299	0.436821355	0.139928723
20	302.545182	20.67556955	0.435114203	0.209403079
21	302.3906089	20.6922557	0.434740556	0.154573133
22	302.3293165	20.69339481	0.435192909	0.061292346
23	302.3408812	20.68626956	0.435865636	0.01156467
24	302.3839372	20.67805115	0.436361467	0.043055993
25	302.4243825	20.67279235	0.436557284	0.040445273
26	302.4462608	20.67125797	0.436519459	0.021878379
27	302.449518	20.67224129	0.436380909	0.003257136
28	302.44207	20.67406445	0.436251087	0.007448019
29	302.4324453	20.67552979	0.436181027	0.009624704
30	302.4258383	20.67619091	0.436170754	0.006606914
31	302.4235601	20.67616765	0.436194771	0.002278219
32	302.4244193	20.67581197	0.436225693	0.000859146
33	302.4264812	20.67544177	0.436246875	0.002061944
34	302.428265	20.67522227	0.436254146	0.001783799
35	302.4291486	20.67517181	0.436251357	0.000883544

From the three tables, it appears that there is an optimum value of ω for this problem. We plot the number of computations required to attain the same accuracy with different ω values (Fig. 2.6).

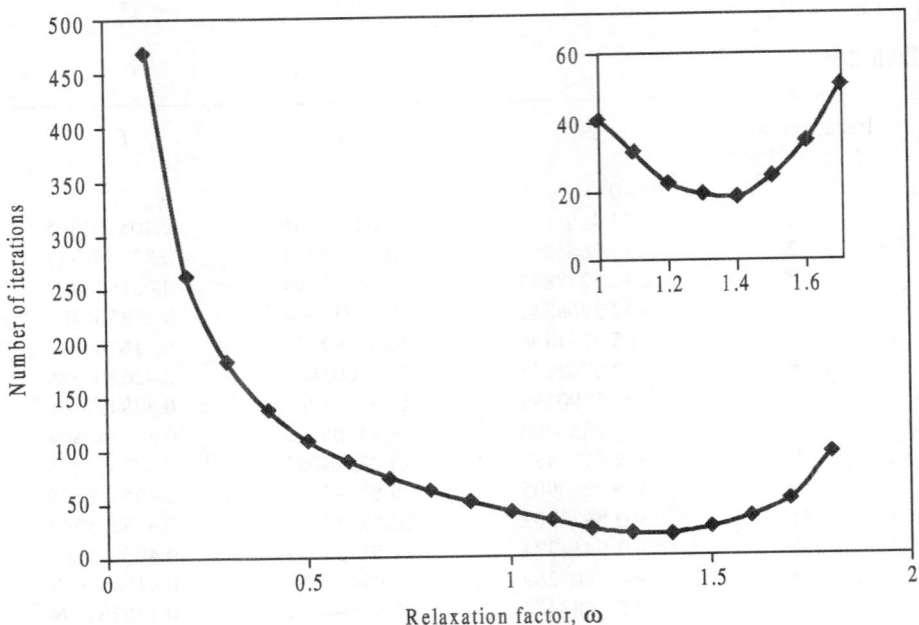

Fig. 2.6 Number of iterations required vs. relaxation factor. Inset shows the optimum near 1.4.

Beyond the value of 1.8 for ω, the solution starts to oscillate. In general, for a problem where Gauss Seidel oscillates (but converges), under relaxation will help to stabilize the oscillation. Over relaxation almost always speeds up the convergence for the problems where Gauss Seidel has a steady convergence. Number of iterations required was half with $\omega = 1.4$ compared to Gauss Seidel method for the same problem (Example 2.12). There is an optimum value of the relaxation factor beyond which the convergence slows down and eventually leads to oscillation. The optimum value depends on the specific matrix.

2.2.3 Scaling and Equilibration

In a system of linear equation of engineering interest, unknowns are physical quantities. These can be expressed in various units. If the units of the variables are unmatched, we may have a coefficient matrix with values differing in orders of magnitude. The same may happen if one has done measurements for parameters with values differing by orders of magnitudes.

For example, let us consider the following problem:

$$\begin{bmatrix} 0.003 & 1.45 & 0.3 \\ 0.00002 & 0.0096 & 0.0021 \\ 0.0015 & 0.966 & 0.201 \end{bmatrix} \begin{bmatrix} x_1 \\ x_2 \\ x_3 \end{bmatrix} = \begin{bmatrix} 11 \\ 0.12 \\ 19 \end{bmatrix}$$

It is evident by looking at the coefficient matrix that the elements in the first column are about 3 orders of magnitude smaller compared to the other elements. These elements correspond to the variable x_1. This may sometimes appear because of a different unit in x_1. For example, x_1 may have been entered in centimetre while the other variables were entered in metres. The second row elements are about 2 orders of magnitude smaller compared to the other elements. This may have happened due to a smaller range of measurement for the second equation compared to the other two equations. If one wants to solve this equation by the method of Gauss elimination, after 1st elimination step, we obtain the following coefficient matrix:

$$\begin{bmatrix} 0.003 & 1.45 & 0.3 \\ 0 & 0.0000667 & 0.0001 \\ 0 & 0.241 & 0.041 \end{bmatrix}$$

If one is doing the calculation by 4 digits, the second pivot will be approximated as 0.0001. Carrying on, one finds the solution vector as [−22328.9, −35.49, 431.5]. The true solution vector is [−22400, −37.454545, 441.69697]. However, if we *scale* the variable x_1 as $x_1 = 10^3 \times x_1'$ and replace x_1 by x_1', the set of equation becomes

$$\begin{bmatrix} 3 & 1.45 & 0.3 \\ 0.02 & 0.0096 & 0.0021 \\ 1.5 & 0.966 & 0.201 \end{bmatrix} \begin{bmatrix} x_1' \\ x_2 \\ x_3 \end{bmatrix} = \begin{bmatrix} 11 \\ 0.12 \\ 19 \end{bmatrix}$$

Now, we can multiply the second equation by 100 in order to bring the values in the coefficient matrix to the same order of magnitude as the rest of the values. This operation is called *equilibration*.

Following the equilibration step, the equation becomes

$$\begin{bmatrix} 3 & 1.45 & 0.3 \\ 2 & 0.96 & 0.21 \\ 1.5 & 0.966 & 0.201 \end{bmatrix} \begin{bmatrix} x_1' \\ x_2 \\ x_3 \end{bmatrix} = \begin{bmatrix} 11 \\ 12 \\ 19 \end{bmatrix}$$

Now, if one solves the set of equation with 4 digit arithmetic, one obtains the solution vector as [−22.456, −37.336, 441.685]. To get the original solution, we have to multiply the first element with the scaling factor. We observe that the solution improved. Essentially, we reduced round-off error by making the order of magnitudes of the elements much closer compared to the original form.

For incorporation of these operations in matrix form into an algorithm, let us make some observations. We will consider the original set of equations of the form $Ax = b$.

Scaling operation essentially involves multiplying each column by a constant factor. Some columns may have this factor as unity and some may have a scale. However, this essentially translates into multiplying the scaled variable with a diagonal scaling matrix to yield the original variable, i.e., $x = Sx'$. For the above example, it was,

$$\begin{bmatrix} x_1 \\ x_2 \\ x_3 \end{bmatrix} = \begin{bmatrix} 10^3 & 0 & 0 \\ 0 & 1 & 0 \\ 0 & 0 & 1 \end{bmatrix} \begin{bmatrix} x_1' \\ x_2 \\ x_3 \end{bmatrix}$$

Notice that scale factors of corresponding elements are along the diagonals of matrix S. Using this relation, the original set of equation becomes $Ax = ASx' = b$. The modified equation gives the solution vector for x' and the original vector can be obtained by using the relation $x = Sx'$. Notice that the new coefficient matrix is A post-multiplied by the diagonal matrix S. This yields the required effect of multiplying the columns with related scale factors. For our example:

$$\begin{bmatrix} 0.003 & 1.45 & 0.3 \\ 0.00002 & 0.0096 & 0.0021 \\ 0.0015 & 0.966 & 0.201 \end{bmatrix} \begin{bmatrix} 10^3 & 0 & 0 \\ 0 & 1 & 0 \\ 0 & 0 & 1 \end{bmatrix} = \begin{bmatrix} 3 & 1.45 & 0.3 \\ 0.02 & 0.0096 & 0.0021 \\ 1.5 & 0.966 & 0.201 \end{bmatrix}$$

Similarly, equilibration is equivalent to premultiplying the coefficient matrix with a diagonal matrix containing the equilibration factors along the diagonal. So, the equation after scaling and equilibration would become,

$$EASx' = Eb \tag{2.135}$$

where E and S are diagonal matrices containing the equilibration and scaling factors on the respective diagonals.

Notice that the right-hand side vector also changed due to equilibration. We have seen this for our example problem. So, for our example problem, final coefficient matrix is

$$\begin{bmatrix} 1 & 0 & 0 \\ 0 & 10^2 & 0 \\ 0 & 0 & 1 \end{bmatrix} \begin{bmatrix} 0.003 & 1.45 & 0.3 \\ 0.00002 & 0.0096 & 0.0021 \\ 0.0015 & 0.966 & 0.201 \end{bmatrix} \begin{bmatrix} 10^3 & 0 & 0 \\ 0 & 1 & 0 \\ 0 & 0 & 1 \end{bmatrix} = \begin{bmatrix} 3 & 1.45 & 0.3 \\ 2 & 0.96 & 0.21 \\ 1.5 & 0.966 & 0.201 \end{bmatrix}$$

Final right-hand side vector is

$$\begin{bmatrix} 1 & 0 & 0 \\ 0 & 10^2 & 0 \\ 0 & 0 & 1 \end{bmatrix} \begin{bmatrix} 11 \\ 0.12 \\ 19 \end{bmatrix} = \begin{bmatrix} 11 \\ 12 \\ 19 \end{bmatrix}$$

We see that we need to decide on scaling factors and equilibration factors before we start our solution procedure. This is true for both direct and iterative methods. Once decided, it can be easily implemented by incorporating a few matrix multiplication steps.

Scaling and equlibration do not eliminate the need for pivoting operation. You will notice in the Example that after Step 1, the pivot becomes small. Exchanging row 2 and 3 in a partial pivoting operation eliminates division by a small number. This is regardless of whether scaling and/or equlibration was done or not. Moreover, while scaling and equilibration helps in reducing round-off errors, it does not transform an *ill-conditioned* matrix into a *well-conditioned* one.

EXERCISE 2.2

1. Solve the following system of equations by Gauss elimination and Gauss Jordon method:

    (a) $x_1 + x_2 - x_3 = 0.5$
        $2x_1 - x_2 + 3x_3 = 5.5$
        $3x_1 + 2x_2 - 2x_3 = 2.0$

    (b) $x_1 + 2x_2 - 6x_3 = 9.0$
        $x_1 - 4x_2 + 4x_3 = -7.0$
        $3x_1 + 8x_2 - 2x_3 = 13.0$

    (c) $8x_1 + x_2 + 2x_3 - x_4 = 3$
        $x_1 + 2x_2 - 2x_3 + x_4 = 3$
        $-2x_1 - x_2 + 5x_3 + 3x_4 = 0$
        $2x_1 + 3x_2 - 2x_3 - 6x_4 = -11$

    (d) $x_1 + 2x_2 - x_4 = 2$
        $x_2 + 2x_3 = 3$
        $-x_3 + 2x_4 = 1$
        $3x_1 + 8x_2 - 2x_3 = 9$

2. Solve the following system of equations using Jacobi and Gauss Seidel methods and compare the number of iterations required for solution:

    (a) $-5x_1 + 2x_2 + x_3 = 2$
    $x_1 - 8x_2 + 3x_3 = -6$
    $3x_1 + x_2 - 7x_3 = -16$

    (b) $-x_1 + x_3 = 3$
    $x_1 + 8x_2 + 3x_3 = 14$
    $x_2 - 2x_3 = -5$

    (c) $x_1 + 2x_2 - x_4 = 1$
    $x_2 + 2x_3 = 1.5$
    $-x_3 - 2x_4 = 1.5$
    $x_1 + 2x_3 - x_4 = 2$

3. Solve the following system of equations by Gauss elimination, Doolittle's method, Crout's method, and Cholesky decomposition:

$$\begin{bmatrix} 9.3746 & 3.0416 & -2.4371 \\ 3.0416 & 6.1832 & 1.2163 \\ -2.4371 & 1.2163 & 8.4429 \end{bmatrix} \begin{bmatrix} x_1 \\ x_2 \\ x_3 \end{bmatrix} = \begin{bmatrix} 9.67685 \\ 6.74135 \\ 2.3925 \end{bmatrix}$$

Can you assemble the *LU* decomposition of Doolittle from the steps of Gauss elimination?

4. Solve the following sets of equations using (a) Doolittle and Crout decomposition, (b) Thomas algorithm, and (c) Cholesky decomposition.

    (i)
$$\begin{bmatrix} -2 & 1 & 0 & 0 \\ 1 & -4 & 1 & 0 \\ 0 & 1 & -4 & 1 \\ 0 & 0 & 1 & -2 \end{bmatrix} \begin{bmatrix} x_1 \\ x_2 \\ x_3 \\ x_4 \end{bmatrix} = \begin{bmatrix} 3 \\ 1 \\ 2 \\ -2 \end{bmatrix}$$

    (ii) $4x_1 + x_2 = 6$
    $x_1 + 4x_2 + x_3 = 12$
    $x_2 + 4x_3 = 14$

5. For the circuit shown in the figure, find the currents through the elements using (a) Gauss elimination, (b) Jacobi iteration, (c) Gauss Seidel method, (d) SOR method. Compare the number of iterations taken by methods (b), (c), and (d). (Use $R_1 = 10\ \Omega$; $R_2 = 20\ \Omega$; $R_3 = 40\ \Omega$; $V_A - V_C = 200$ V; $V_B - V_C = 100$ V)

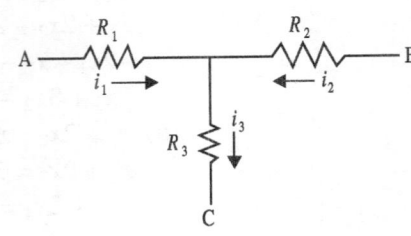

6. In the above problem, how much will be the change in current in R_3 due to unit change in voltage difference $(V_A - V_C)$? Which of the currents is the most sensitive to a change in voltage at B?

7. Compute the inverse of the following matrices using Gauss Jordan elimination:

   (a) $\begin{bmatrix} -3 & 1 \\ 1 & -2 \end{bmatrix}$

   (b) $\begin{bmatrix} -4 & 1 & -1 \\ 1 & -2 & 1 \\ -1 & 1 & -3 \end{bmatrix}$

   (c) $\begin{bmatrix} 4 & -3 & 2 & 0 \\ -3 & 8 & -2 & -1 \\ -1 & -2 & 8 & -3 \\ 0 & -3 & 2 & 4 \end{bmatrix}$

8. Consider the following matrix:

$$A = \begin{bmatrix} 9 & 3 & -2 \\ 3 & 6 & 1 \\ -2 & 1 & 9 \end{bmatrix}$$

   (a) Reduce it to an upper triangular matrix using Gauss elimination procedure.
   (b) Synthesize a lower triangular matrix L and an upper triangular matrix U from the steps of (a) above such that $A = LU$.
   (c) Compute A^{-1} using the LU decomposition obtained in (b).
   (d) Using the above results, compute the determinant and condition number of A. Use row sum norms for the matrices.

9. The following set of equations were originally constructed by T.S. Wilson [Hildebrand, 1974]:

$$\begin{bmatrix} 10 & 7 & 8 & 7 \\ 7 & 5 & 6 & 5 \\ 8 & 6 & 10 & 9 \\ 7 & 5 & 9 & 10 \end{bmatrix} \begin{bmatrix} x_1 \\ x_2 \\ x_3 \\ x_4 \end{bmatrix} = \begin{bmatrix} 32 \\ 23 \\ 33 \\ 31 \end{bmatrix}$$

An approximation to the x-values as [−7.2, 14.6, −2.5, 3.1] yields the right-hand side vector as [31.9, 23.1, 32.9, 31.1]. A very different set of x-values [0.18, 2.36, 0.65, 1.21] also yields a very close right-hand side vector as [31.99, 23.01, 32.99, 31.01]. It is not clear whether any of the x-values are close to the true solution. Use Crout's decomposition and improve the solution starting from each of the above approximations of x-values.

10. Consider the following set of equations:

$$\begin{bmatrix} 10^{-5} & 10^{-5} & 1 \\ 10^{-5} & -10^{-5} & 1 \\ 1 & 1 & 2 \end{bmatrix} \begin{bmatrix} x_1 \\ x_2 \\ x_3 \end{bmatrix} = \begin{bmatrix} 2 \times 10^{-5} \\ -2 \times 10^{-5} \\ 1 \end{bmatrix}$$

(a) Solve the system using Gaussian elimination, without pivoting, using 3-digit floating-point arithmetic with round-off.
(b) Perform complete pivoting and carry out Gaussian elimination steps once again using 3-digit floating-point arithmetic with round-off.
(c) Rewrite the set of equations after scaling and equilibration. Solve the system using Gaussian elimination without pivoting using 3-digit floating-point arithmetic with round-off.
(d) Explain your results with respect to effect of pivoting, scaling, and equilibration.

11. Consider the following matrix:

$$A = \begin{bmatrix} 10 & 7 & 8 & 7 \\ 7 & 5 & 6 & 5 \\ 8 & 6 & 10 & 9 \\ 7 & 5 & 9 & 10 \end{bmatrix}$$

(a) Compute A^{-1} using Cholesky decomposition.
(b) Find the solution of $Ax = b$ where $b = [4\ 3\ 3\ 1]^1$.
(c) If b is perturbed by δb such that $\|\delta b\|_\infty = 0.01$, find an upper bound for the corresponding perturbation vector $\|\delta b\|_\infty$.
(d) Compute the condition number $C(A)$ and check the inequality

$$\frac{\|\delta x\|_\infty}{\|x\|_\infty} \bigg/ \frac{\|\delta b\|_\infty}{\|b\|_\infty} \leq C(A).$$

12. Consider the family of so-called Hilbert matrices, given by ($n = 1, 2, \ldots$):

$$H_n = \begin{bmatrix} 1 & 1/2 & 1/3 & 1/4 & \cdots & 1/n \\ 1/2 & 1/3 & 1/4 & 1/5 & \cdots & 1/(n+1) \\ 1/3 & 1/4 & 1/5 & 1/6 & \cdots & 1/(n+2) \\ 1/4 & 1/5 & 1/6 & 1/7 & \cdots & 1/(n+3) \\ \cdots & \cdots & \cdots & \cdots & \cdots & \cdots \\ 1/n & 1/(n+1) & 1/(n+2) & 1/(n+3) & \cdots & 1/(2n-1) \end{bmatrix}$$

(a) Solve the system of equations $H_3 x = b$, where x is a column vector of 3 variables and $b^T = [11/6,\ 13/12,\ 47/60]$.

(b) Write the LU decomposition of H_3 from Gaussian elimination steps.

(c) What is the condition number of H_3?

13. The following system of five equations (not a tri-diagonal) can be solved using Thomas algorithm for tri-diagonal matrices:

$$\begin{array}{cc} A & \quad b \qquad x \qquad d \\[4pt] \left[\begin{array}{cccc|c} 4 & 1 & 0 & 0 & -2 \\ -2 & 10 & 1 & 0 & -1 \\ 0 & -3 & 8 & 1 & -2 \\ \hline 0 & 0 & 1 & 5 & 3 \\ -3 & 2 & 1 & -5 & 13 \end{array}\right] & \begin{bmatrix} x_1 \\ x_2 \\ x_3 \\ x_4 \\ x_5 \end{bmatrix} = \begin{bmatrix} 4.75 \\ -1.15 \\ 0.55 \\ 12.35 \\ -4.15 \end{bmatrix} \\[6pt] c^T & \quad a_{55} \qquad x_5 \qquad d_5 \end{array}$$

The system can be partitioned as shown above and expressed as

$$\begin{bmatrix} A & b \\ c^T & a_{55} \end{bmatrix} \begin{bmatrix} x \\ x_5 \end{bmatrix} = \begin{bmatrix} d \\ d_5 \end{bmatrix}$$

Answer the following:

(a) If vector x is decomposed into two parts as $x = x^H + x^I$ such that, vector x^H is the solution of tri-diagonal system $Ax^H = d$, prove the following:

$$x^I = -x_5 y$$

$$x_5 = \frac{d_5 - c^T x^H}{a_{55} - c^T y}$$

where the vector y is the solution of tri-diagonal system of equations $Ay = b$.

(b) Using the results of (a), compute the solution consisting of vector x and variable x_5.

14. Show that a square matrix of size n with n independent columns will have a decomposition of the form $A = LDU$.

15. Derive a sufficient condition for the convergence of Gauss Seidel iteration method similar to Eq. (2.125) for Jacobi iteration.

2.3 COMPUTATION OF EIGENVALUES

We may recall that the eigenvalues of a matrix are the roots of the polynomial $\det(A - \lambda I) = 0$. The eigenvalues are important in many engineering problems. The problem of estimation of eigenvalues also appears from a group of ordinary differential equations known as eigenvalue problems.

Sometimes, it is sufficient to get the range within which the eigenvalues lie. For example, to compute the condition number of a symmetric matrix, we only need the largest and the smallest eigenvalues. In some engineering analysis, one requires precise estimation of all the eigenvalues. In this section, we will first discuss a simple iterative process known as *power method* to estimate the largest eigenvalue. A variation of this method can also be used to determine the smallest eigenvalue. Then we will present a method to determine all the eigenvalues from the first principle by solving the characteristic polynomial. At the end we will discuss a method of triangulation/diagonalization by which all the eigenvalues can be estimated simultaneously.

2.3.1 Power Method

If an $n \times n$ matrix A has a unique eigenvalue of maximum absolute magnitude (λ_1) such that, $|\lambda_1| > |\lambda_2| \geq |\lambda_3| \geq ... \geq |\lambda_n|$ and n independent eigenvectors $\{x_1, x_2, ..., x_n\}$, this method provides an easy way to estimate the maximum eigenvalue. Starting from an arbitrary non-zero vector $(z^{(0)})$ of length n, the method generates a sequence of vectors $(z^{(k)})$ through an iterative process that converges to the eigenvector corresponding to the maximum eigenvalue. The iteration scheme is given by

$$z^{(k+1)} = Az^{(k)} = A^{k+1}z^{(0)} \tag{2.136}$$

The proof that the vector $z^{(k)}$ will converge to the eigenvector corresponding to the maximum eigenvalue involves expressing the arbitrary initial vector $z^{(0)}$ as a linear combination of n linearly independent eigenvectors as in Eq. (2.120) followed by application of the iterative scheme described above by Eq. (2.136).

$$z^{(0)} = \alpha_1 x_1 + \alpha_2 x_2 + \cdots + \alpha_n x_n \tag{2.137}$$

where x_i is the eigenvector corresponding to the eigenvalue λ_i and α_i's are the constants.

Since, by definition of eigenvalue $Ax_j = \lambda_j x_j$, therefore, it follows:

$$z^{(k)} = A^k z^{(0)} = A^k \sum_{i=1}^{n} \alpha_i x_i = \sum_{i=1}^{n} \lambda_i^k \alpha_i x_i = \lambda_1^k \left(\alpha_1 x_1 + \sum_{j=2}^{n} \left(\frac{\lambda_j}{\lambda_1} \right)^k \alpha_j x_j \right) \tag{2.138}$$

Since λ_1 is a unique maximum eigenvalue satisfying $|\lambda_1| > |\lambda_2| \geq |\lambda_3| \geq \ldots \geq |\lambda_n|$, the term $(\lambda_j / \lambda_1)^k$ approaches zero for large k. This essentially means that the vector $z^{(k)}$ approaches the same direction as x_1. Since, the magnitude of the maximum eigenvalue λ_1 are often different from unity, the elements of $z^{(k)}$ may approach zero (for $\lambda_1 < 1$) or become very large (for $\lambda_1 > 1$) with large k because of the factor λ_1^k. Therefore, it is a good practice to normalize the vector $z^{(k)}$ after every iteration with its norm. Any suitable vector norm can be used for this purpose. The L_2 norm is shown below for illustration of the final iteration scheme:

$$y^{(k)} = \frac{z^{(k)}}{\left\| z^{(k)} \right\|_2} \quad \text{and} \quad z^{(k+1)} = Ay^{(k)} \tag{2.139}$$

Termination criteria will be when $y^{(k)}$ becomes stationary or the variation is within a chosen tolerance. Once an eigenvector (x_i) is known, the corresponding eigenvalue (λ_i) can be estimated using the *Rayleigh quotient* given by

$$\lambda_i = \frac{x_i^T A x_i}{x_i^T x_i}$$

An approximation of λ_1 at kth iteration can therefore be obtained as follows:

$$\lambda_1^{(k)} = \frac{y^{(k)^T} Ay^{(k)}}{y^{(k)^T} y^{(k)}} = y^{(k)^T} z^{(k+1)} , \text{ since } y^{(k)^T} y^{(k)} = 1 \text{ and } z^{(k+1)} = Ay^{(k)} \tag{2.140}$$

Termination criterion can either be applied on the eigenvalue $\lambda_1^{(k)}$ or on the eigenvector $y^{(k)}$. We can set a tolerance for either absolute approximate error or relative approximate error. If applied on eigenvalue, one can work with absolute value of the difference between successive iterations. For the eigenvector, one will have to choose a vector norm. Any norm described in Chapter 1 can be used. However, Euclidean or the maximum norm are mostly used. The error expressions are shown below:

For eigenvalue: $\varepsilon^{(k)} = \left| \lambda_1^{(k+1)} - \lambda_1^{(k)} \right|$ or $\varepsilon_r^{(k)} = \left| \frac{\lambda_1^{(k+1)} - \lambda_1^{(k)}}{\lambda_1^{(k+1)}} \right|$

For eigenvectors: $\varepsilon^{(k)} = \left\| y^{(k+1)} - y^{(k)} \right\|$ or $\varepsilon_r^{(k)} = \dfrac{\left\| y^{(k+1)} - y^{(k)} \right\|}{\left\| y^{(k+1)} \right\|}$

Example 2.14 Compute the largest eigenvalue and the corresponding eigenvector of the following coefficient matrix of Example 2.9 using the power method with a relative error of 0.1% *or less* on the eigenvalue.

$$\begin{bmatrix} 2 & -1 & 0 & 0 \\ -1 & 4 & -1 & 0 \\ 0 & -1 & 4 & -1 \\ 0 & 0 & -1 & 2 \end{bmatrix}$$

Solution We assume an initial vector $z_0 = [1\ 0\ 0\ 0]^T$. We now apply the algorithm given by Eqs (2.139) and (2.140). The computations are tabulated in Table 2.7.

Table 2.7

Iteration no.	y	z	λ	ε (%)
1	1	2	2	
	0	−1		
	0	0		
	0	0		
2	0.894427	2.236068	3.2	60
	−0.44721	−2.68328		
	0	0.447214		
	0	0		
3	0.635001	2.032002	4.354839	36.08871
	−0.762	−3.81		
	0.127	1.270001		
	0	−0.127		
4	0.451287	1.748735	4.848051	11.32561
	−0.84616	−4.11799		
	0.282054	2.002584		
	−0.02821	−0.33846		
5	0.355916	1.549959	5.073126	4.642583
	−0.83813	−4.116		
	0.407582	2.53734		
	−0.06889	−0.54536		

(contd.)

(contd.)

6	0.303509 −0.80599 0.496856 −0.10679	1.413004 −4.02431 2.900198 −0.71044	5.18924	2.288802
7	0.271393 −0.77294 0.557036 −0.13645	1.315727 −3.92019 3.137536 −0.82994	5.248124	1.134739
8	0.250299 −0.74576 0.596874 −0.15788	1.246363 −3.83023 3.291145 −0.91264	5.276903	0.548367
9	0.23601 −0.72529 0.623209 −0.17282	1.19731 −3.76038 3.390941 −0.96884	5.290633	0.260201
10	0.226225 −0.7105 0.6407 −0.18306	1.162954 −3.70894 3.456363 −1.00682	5.297102	0.122255
11	0.219508 −0.70006 0.65239 −0.19004	1.13908 −3.67215 3.499661 −1.03246	5.300129	0.057163

The error has become less than the specified 0.1%. So, the iterations may be terminated. The maximum eigenvalue is therefore 5.300129 and the corresponding eigenvector is given by the final z or by normalizing, we obtain

$$[0.214898 \quad -0.69279 \quad -0.660245 \quad -0.19478]^T$$

2.3.2 Inverse Power Method

If a $(n \times n)$ matrix A has an unique eigenvalue of minimum absolute magnitude, it can also be obtained using the same principle as the power method described above. If λ_i's are the eigenvalues of matrix A, the inverse of the matrix have the eigenvalues as $1/\lambda_i$. Thus, the unique eigenvalue of minimum magnitude of the matrix A will translate into a unique eigenvalue of maximum magnitude for the matrix A^{-1}. Thus, application of the power method on A^{-1} will converge to the inverse of the minimum eigenvalue of matrix A. The algorithm can be written similar to Eq. (2.139) as follows:

$$y^{(k)} = \frac{z^{(k)}}{\left\| z^{(k)} \right\|_2}; \quad z^{(k+1)} = A^{-1} y^{(k)} \tag{2.141}$$

where $z^{(0)}$ is an arbitrary non-zero vector of length n. One can avoid inversion of matrix A. Instead, an LU decomposition can be carried out for A which can then be used repeatedly to solve for $z^{(k+1)}$ with known right-hand side vector $y^{(k)}$. This gives a significant computational advantage. Thus, the above algorithm translates to,

$$y^{(k)} = \frac{z^{(k)}}{\left\| z^{(k)} \right\|_2}; \quad Az^{(k+1)} = y^{(k)} \tag{2.142}$$

Once, the $y^{(k)}$ converges, one can calculate the corresponding eigenvalue using the Rayleigh quotient described in Eq. (2.140). Do not forget that it is the maximum eigenvalue of A^{-1}. In order to get the minimum eigenvalue of A, you have to take the reciprocal of the computed value.

Example 2.15 Compute the smallest eigenvalue of the matrix in Example 2.14 with an approximate relative error of 0.1% or less.

Solution We use the computation logic given by Eq. (2.142). So, at every step, we solve for z_{k+1} for a known y_k. At this point we observe that the matrix in Example 2.14 is a tri-diagonal. Thus, the Thomas algorithm described in Section 2.2.1.4 would be appropriate. Since, the coefficient matrix remain unaltered, the vector α can be calculated a priori using Eq. (2.86), before the start of the power method iterations. At every step of the power method, right-hand side vector β needs to be recalculated using Eq. (2.86) and then the solutions can be easily computed using Eq. (2.88).

Since, the matrix of Example 2.9 is the same as the present problem, we use the values of vector α computed there as $\alpha = [2 \quad 3.5 \quad 3.714286 \quad 1.730769]$. We tabulate the iterations of inverse power method.

Table 2.8

Iteration no.	y	β	z	λ	ε (%)
1	1	1	0.8	2.8	
	1	1.5	0.6		
	1	1.428571	0.6		
	1	1.384615	0.8		
2	0.565685	0.565685	0.424264	0.72	288.8889
	0.424264	0.707107	0.282843		
	0.424264	0.626295	0.282843		
	0.565685	0.734303	0.424264		

(contd.)

(contd.)

3	0.588348 0.392232 0.392232 0.588348	0.588348 0.686406 0.588348 0.74675	0.431455 0.274563 0.274563 0.431455	0.723077	0.425532
4	0.596559 0.379628 0.379628 0.596559	0.596559 0.677908 0.573316 0.750913	0.433861 0.271163 0.271163 0.433861	0.723529	0.062539

We stop the iteration as soon as the relative error in eigenvalue falls below the specified 0.1%. Thus, the minimum eigenvalue is given by $(1 / 0.723529) = 1.382114$.

It is important to note here that if the matrix is not a tri-diagonal, it will be computationally beneficial to perform a *LU* decomposition of the coefficient matrix in the beginning. At every iteration, the solution can be easily computed using the new right-hand side vector.

2.3.3 Inverse Power Method with Shift

By extending the above principle, one can easily determine an eigenvalue closest to any given number (θ) for a matrix A. Eigenvalues of matrix $(A - \theta I)$ are $(\lambda_i - \theta)$. Thus, the smallest absolute value of $(\lambda_i - \theta)$ will be given by the eigenvalue closest to θ. If it is the *p*th (say) eigenvalue which is closest to θ in magnitude, $(\lambda_p - \theta)$ is the minimum eigenvalue of absolute magnitude for the matrix $(A - \theta I)$. Applying the inverse power method on matrix $(A - \theta I)$, one can estimate $(\lambda_p - \theta)$ and in turn λ_p. This is important in many engineering application, because one sometimes is interested in finding the nearest eigenvalue to a predetermined number. As the eigenvalue is related to some physical quantity, this predetermined value may come from some guideline or design code provision.

2.3.4 Faddeev Leverrier Method

First method we are going to illustrate for estimating all the eigenvalues of a matrix is based on the roots of polynomial $\det(A - \lambda I) = 0$. For a square matrix of size n, this will lead to a *characteristic polynomial* of order n. The polynomial equation can be written as

$$(-1)^n (\lambda^n - a_{n-1}\lambda^{n-1} - \cdots - a_2\lambda^2 - a_1\lambda - a_0) = 0 \qquad (2.143)$$

The *Fadeev Leverrier* method is essentially a way to evaluate the coefficients (a_i's) of the *characteristic polynomial* without evaluating the determinant explicitly. We initialize a sequence of matrices as

$$A_{n-1} = A$$

Then, the coefficient a_{n-1} is given by

$$a_{n-1} = \text{trace}(A_{n-1}) \tag{2.144}$$

The other matrices in the sequence and the coefficients in the characteristic polynomial are computed as follows:

$$A_i = A(A_{i+1} - a_{i+1}I) \text{ and } a_i = \frac{\text{trace}(A_i)}{n-i}, \text{ where } i = (n-2), (n-3), \ldots, 2, 1, 0 \tag{2.145}$$

Proof of the fact that the coefficients of the polynomial are indeed given by the traces of the matrices generated recursively by Eq. (2.145) is beyond the scope of this book. The original Faddeev's proof is available in Faddeeva (1959) and Faddeeva and Faddeev (1977). A proof based on Laplace transform is available in Hou (1998).

Once the characteristic polynomial is formulated, any numerical method for finding the roots of a polynomial such as *Bairstow method* described in Section 3.4 may be used to compute the eigenvalues.

Example 2.16 Formulate the characteristic polynomial using Faddeev Leverrier method for the matrix in Example 2.14.

Solution For $n = 4$, we can express the characteristic polynomial as Eq. (2.143). Thus, $A_3 = A$ and $a_3 = \text{trace}(A) = 12$. We tabulate the computation of other a_i's below:

Table 2.9

i	$(A_{i+1} - a_{i+1}I)$	$A_i = A(A_{i+1} - a_{i+1}I)$	a_i
2	$\begin{bmatrix} -10 & -1 & 0 & 0 \\ -1 & -8 & -1 & 0 \\ 0 & -1 & -8 & -1 \\ 0 & 0 & -1 & -10 \end{bmatrix}$	$\begin{bmatrix} -19 & 6 & 1 & 0 \\ 6 & -30 & 4 & 1 \\ 1 & 3 & -30 & 6 \\ 0 & 1 & 6 & -19 \end{bmatrix}$	-49
1	$\begin{bmatrix} 30 & 6 & 1 & 0 \\ 6 & 19 & 4 & 1 \\ 1 & 4 & 19 & 6 \\ 0 & 1 & 6 & 30 \end{bmatrix}$	$\begin{bmatrix} 54 & -7 & -2 & -1 \\ -7 & 66 & -4 & -2 \\ -2 & -4 & 66 & -7 \\ -1 & -2 & -7 & 54 \end{bmatrix}$	80

(contd.)

(*contd.*)

$$
0 \quad
\begin{bmatrix}
-26 & -7 & -2 & -1 \\
-7 & -14 & -4 & -2 \\
-2 & -4 & -14 & -7 \\
-1 & -2 & -7 & -26
\end{bmatrix}
\quad
\begin{bmatrix}
-45 & 0 & 0 & 0 \\
0 & -45 & 0 & 0 \\
0 & 0 & -45 & 0 \\
0 & 0 & 0 & -45
\end{bmatrix}
\quad -45
$$

Therefore, the characteristic polynomial is given by

$$\lambda^4 - 12\lambda^3 + 49\lambda^2 - 80\lambda + 45 = 0$$

2.3.5 Similarity Transformation

In this section we will present a method that is frequently used to compute eigenvalues of a matrix. It is based on similarity transformation. A sequence of matrices is generated by repeated application of similarity transformations. The eigenvalues of the original matrix remain unchanged under similarity transformations. The original matrix in the first step and the transformed matrices thereafter are decomposed into an orthogonal matrix and an upper triangular matrix. Properties of orthogonal matrix help simplify the process of similarity transformation. Under repeated similarity transformation, the matrix transforms into a diagonal or an upper triangular matrix. Since, determinant of a diagonal or triangular matrix is the product of the diagonal elements, the diagonal elements are the eigenvalues.

But, before we can describe the method, we need some more discussion of matrix algebra. We will first present the concept of similar matrices and similarity transformation. We will proceed to establish the condition under which a matrix can be diagonalized in order to estimate the eigenvalues using this method. We will then present a method for decomposition of a matrix into an orthogonal and an upper triangular matrix. Finally, we will present the algorithm based on this decomposition for estimation of the eigenvalues. At the end, we will also present the theorem and an outline of its proof which guarantees that the method eventually converges to an upper triangular matrix.

Two $n \times n$ matrices A and B are similar if there exists another $n \times n$ invertible matrix S such that $A = SBS^{-1}$ (recall Example 2.1). In terms of linear transformation, their relationship is given by Theorem 2.5.

Notice that in order to generate a similar matrix B from original matrix A, we need to apply the following transformation: $B = S^{-1}AS$. The operation $S^{-1}AS$ will be termed as *similarity transformation* of A. Eigenvalues of two similar matrices are same. This can be easily seen from the definition of similar matrices as, $(A - \lambda I) = S(B - \lambda I)S^{-1}$. Since, determinants follow $\det(AB) = \det(A)\det(B)$, we find that A and B have same characteristic equations.

Theorem 2.5

Two matrices A and B are similar if and only if they represent the same linear transformation $T : X \mapsto X$ with respect to two different bases.

Proof: Let A represent T with respect to the basis $\{x_1, x_2, \ldots, x_n\}$ and B represent T with respect to the basis $\{y_1, y_2, \ldots, y_n\}$, an arbitrary vector x in X and its transformation Tx can be expressed as a linear combination of both the bases as follows:

$$x = \sum_i \alpha_i x_i \quad \Rightarrow \quad Tx = \sum_i (A\alpha)_i x_i \tag{2.146}$$

$$x = \sum_j \beta_j y_j \quad \Rightarrow \quad Tx = \sum_j (B\beta)_j y_j \tag{2.147}$$

Since both $\{x_1, x_2, \ldots, x_n\}$ and $\{y_1, y_2, \ldots, y_n\}$ are bases for subspace X, each y_j can also be expressed as a unique linear combination of $\{x_1, x_2, \ldots, x_n\}$ as follows:

$$y_j = \sum_i s_{ij} x_i, \quad \text{for } j = 1, 2, \ldots, n \tag{2.148}$$

Now for an arbitrary vector $x = \sum_j \beta_j y_j$ in X, we may write the following:

$$x = \sum_j \beta_j y_j \quad \Rightarrow \quad Tx = \sum_j (B\beta)_j y_j = \sum_j (B\beta)_j \sum_i s_{ij} x_i = \sum_i (SB\beta)_i x_i \tag{2.149}$$

$$x = \sum_j \beta_j y_j = \sum_j \beta_j \sum_i s_{ij} x_i = \sum_i (S\beta)_i x_i \quad \Rightarrow \quad Tx = \sum_i (AS\beta)_i x_i \tag{2.150}$$

This gives $AS\beta = SB\beta$. Since x was chosen arbitrarily, the relationship is valid for all β in $\Re^n$. So, $AS = SB$. Since, y_j's are linearly independent, so will be the columns of S. This means the matrix S is invertible. This leads to the required relation $A = SBS^{-1}$.

We leave it to the readers to work out the proof of the "only if".

Some matrices can be diagonalized by similarity transform. That is to say, matrix A can be diagonalized if it is similar to a diagonal matrix, i.e., $A = X\Lambda X^{-1}$. Then the eigenvalues of A will be the diagonal elements of Λ. Since, this also means $AX = X\Lambda$, every column of X is an eigenvector of A and together, they span the invariant subspace of A. All non-defective matrices defined in Section 2.1.2 are diagonalizable. Specifically, one will get more insight into diagonalizability by taking a closer look at Theorems 2.6 and 2.7.

Theorem 2.6

If a $n \times n$ square matrix A has n linearly independent eigenvectors as $\{x_1, x_2, \ldots, x_n\}$ then $A = X\Lambda X^{-1}$, where Λ is a diagonal matrix with eigenvalues $\{\lambda_1, \lambda_2, \ldots, \lambda_n\}$ on its diagonal and X is an $n \times n$ square matrix with eigenvectors $\{x_1, x_2, \ldots, x_n\}$ on its columns.

Proof: By definition of eigenvalues,
$Ax_i = \lambda_i x_i$ for all $i = 1, 2, \ldots, n$.
If we construct a matrix X whose columns are eigenvectors x_i's, on the right-hand side each column of X has to be multiplied by the corresponding λ_i. In Section 2.2.3 (scaling matrix), we have seen that this is equivalent to post multiplying X with a diagonal matrix whose elements are λ_i's. This essentially means, $AX = X\Lambda$. Since, eigenvectors in the columns of X are linearly independent, X is invertible. This concludes the proof $A = X\Lambda X^{-1}$.

Theorem 2.7

If a $n \times n$ square matrix A has n distinct eigenvalues, then it is diagonalizable.

Proof: In order to prove this, it will be sufficient to show that the matrix A has n linearly independent eigenvectors. Then by Theorem 2.6, the matrix will be diagonalizable.

Let us assume the distinct eigenvalues of $n \times n$ square matrix A as $\{\lambda_1, \lambda_2, \ldots, \lambda_n\}$ and the corresponding eigenvectors as $\{x_1, x_2, \ldots, x_n\}$. Fix an index k for the eigenvalues and the corresponding eigenvectors. For $k = 1$, the independence of the lone eigenvector is evident. Let us assume that the eigenvectors are independent up to k. We will examine for $k + 1$. Suppose the following is true:

$$\alpha_1 x_1 + \alpha_2 x_2 + \cdots + \alpha_k x_k + \alpha_{k+1} x_{k+1} = 0 \qquad (2.151)$$

Multiplying both the sides by matrix A and using the relation $Ax_i = \lambda_i x_i$, we get

$$\alpha_1 \lambda_1 x_1 + \alpha_2 \lambda_2 x_2 + \cdots + \alpha_k \lambda_k x_k + \alpha_{k+1} \lambda_{k+1} x_{k+1} = 0 \qquad (2.152)$$

Multiplying Eq. (2.151) with λ_{k+1} and deducting from Eq. (2.152), we obtain

$$\alpha_1 (\lambda_1 - \lambda_{k+1}) x_1 + \alpha_2 (\lambda_2 - \lambda_{k+1}) x_2 + \cdots + \alpha_k (\lambda_k - \lambda_{k+1}) x_k = 0 \qquad (2.153)$$

Since, eigenvectors (x_i's) are independent up to k, the above means all the coefficients are zero. Since, eigenvalues are also distinct, $(\lambda_i - \lambda_{k+1}) \neq 0$ for $i = 1, \ldots, k$. This means $a_1 = a_2 = \ldots = a_k = 0$. Putting this in Eq. (2.151) shows, $a_{k+1} = 0$. This concludes the proof.

Theorem 2.6 requires n independent eigenvectors and the eigenvalues need not be distinct. The second Theorem (2.7) is more binding that way. Theorems 2.6 and 2.7 can be combined to take a fresh look at the *non-defective* matrix defined in Section 2.1.2. This is done in Theorem 2.8.

Theorem 2.8

A $n \times n$ square matrix A is non-defective if and only if there exists a non-singular X such that $A = X \Lambda X^{-1}$.

Proof: Readers should be able to prove this using Theorems 2.6, 2.7, and definition of non-defective marices (Section 2.1.2). As a hint, we shall state that a *non-defective* $n \times n$ square matrix has n independent eigenvectors even if some eigenvalues may be repeated.

In an n-dimensional vector space spanned by a set of n linearly independent basis vectors, it is always possible to construct a set of n orthonormal vectors which span the same subspace. This is done by a technique known as *Gram Schmidt orthogonalization*. We will soon see that it plays an important role in computation of eigenvalues. We will demonstrate the procedure for orthogonalization by construction, which will eventually serve as the algorithm for generating an orthogonal basis for the column space of any matrix.

The Gram-Schmidt procedure provides a way to replace the independent set of columns of any matrix A with an orthonormal set of vectors spanning the same *column space*. We will show that any $n \times n$ square matrix with n linearly independent columns can be expressed as multiplication of an orthogonal matrix Q with orthonormal vectors as its columns and an upper triangular matrix R. Once again, we will show this by construction which will also serve as the algorithm for carrying out such decomposition.

2.3.5.1 Gram Schmidt *Orthogonalization*

For a set of linearly independent vectors $\{x_1, x_2, \ldots, x_n\}$ there exists an orthonormal set of vectors $\{y_1, y_2, \ldots, y_n\}$ such that $\{x_1, x_2, \ldots, x_k\}$ and $\{y_1, y_2, \ldots, y_k\}$ span the same subspace for every $k \in \{1, 2, \ldots, n\}$. We will first construct an orthonormal set of basis and then show that they span the same subspace.

Initialize y_1 as $y_1 = \dfrac{x_1}{\|x_1\|}$ and construct y_{k+1} as

$$z_{k+1} = x_{k+1} - \sum_{i=1}^{k}\left(x_{k+1}^T y_i\right) y_i \quad \text{and} \quad y_{k+1} = \dfrac{z_{k+1}}{\|z_{k+1}\|} \tag{2.154}$$

Notice that all y_i's are unit vectors. We need to test whether they are orthogonal, i.e., $y_i^T y_j = \delta_{ij}$. The δ_{ij} is the *Kronecker* delta defined as:

$$\delta_{ij} = \begin{cases} 0, & \text{for } i \neq j \\ 1, & \text{for } i = j \end{cases} \tag{2.155}$$

For $k = 1$, there is nothing to prove. Let us assume that the vectors are orthogonal up to k, i.e., $\{y_1, y_2, \ldots, y_k\}$ are orthogonal. Let us test for $k + 1$,

$$y_{k+1}^T y_j = \frac{1}{\|z_{k+1}\|}\left(x_{k+1}^T y_j - \sum_{i=1}^{k}\left(x_{k+1}^T y_i\right) y_i^T y_j\right) = \frac{1}{\|z_{k+1}\|}\left(x_{k+1}^T y_j - x_{k+1}^T y_j\right) = 0$$
$$\tag{2.156}$$

This means, $y_{k+1}^T y_j = 0$ for $j = 1, 2, \ldots, k$.

Only thing left to show is that the space (X_k) spanned by $\{x_1, x_2, \ldots, x_k\}$ and space (Y_k) spanned by $\{y_1, y_2, \ldots, y_k\}$ are same for every $k \in \{1, 2, \ldots, n\}$. By definition of y_1, it is true for $k = 1$. Assume that it is true for k, i.e., $X_k = Y_k$ and let us examine for $k + 1$.

Equation (2.154) shows that x_{k+1} is a linear combination of $\{y_1, y_2, \ldots, y_{k+1}\}$. This means $x_{k+1} \in Y_{k+1}$ and thus, $X_{k+1} \subset Y_{k+1}$.

The above relation also means $\dim X_{k+1} \leq \dim Y_{k+1}$. Dimension of the vector space X_{k+1}, $\dim X_{k+1} = k + 1$. Since, there are only $k + 1$ linearly independent vectors in Y_{k+1}, $\dim Y_{k+1} \leq k + 1$. Thus, $\dim X_{k+1} = \dim Y_{k+1}$. This essentially means $X_k = Y_k$ in light of the following proposition which we shall not prove here. Interested readers may look at any book on vector spaces for the proof.

Proposition

Let X be a subspace of a n-dimensional vector space Y, i.e., $X \subset Y$, then $\dim X \leq n$. If $\dim X = \dim Y = n$, then $X = Y$.

2.3.5.2 QR Decomposition

If a $n \times n$ matrix A has linearly independent columns then it can be decomposed as $A = QR$, where Q is a $(n \times n)$ matrix with orthonormal columns and R is a $(n \times n)$ upper triangular matrix. Let us denote the columns of A as a linearly independent set of vectors $\{x_1, x_2, \ldots, x_n\}$. Using Gram-Schmidt orthogonalization, an orthonormal set of vectors $\{y_1, y_2, \ldots, y_n\}$ can be constructed to span the same subspace as that spanned by the columns of A. Then, the composition of matrices A and Q can be written as the collection of column vectors as follows:

$$A = \begin{bmatrix} x_1, & x_2, & \ldots, & x_n \end{bmatrix} \text{ and } Q = \begin{bmatrix} y_1, & y_2, & \ldots, & y_n \end{bmatrix} \qquad (2.157)$$

We need to now express x_i's as linear combination of the y_i's and show that the coefficients (r_{ij}) lead to an upper triangular matrix R. The relation $A = QR$ leads to the following relation between the columns of A and Q,

$$x_j = \sum_{i=1}^{n} r_{ij} y_j \qquad (2.158)$$

Multiplying both sides of the above equation by y_i^T and using the relation $y_i^T y_j = \delta_{ij}$, we obtain, $r_{ij} = y_i^T x_j$. Since, the vectors $\{y_1, y_2, \ldots, y_n\}$ could be computed from the columns of matrix A, $\{x_1, x_2, \ldots, x_n\}$ using the Gram Schmidt procedure, r_{ij}'s can be easily computed using vector products. However, we still need to show that r_{ij}'s lead to an upper triangular matrix in order to save computational effort by avoiding calculation of zero elements.

In order to visualize the upper triangular matrix, let us rearrange Eq. (2.154) and change a few indices as follows:

$$x_j = z_j + \sum_{k=1}^{j-1} \left(x_j^T y_k \right) y_k \qquad (2.159)$$

Now, r_{ij}'s can be written as

$$r_{ij} = y_i^T x_j = y_i^T z_j + \sum_{k=1}^{j-1} \left(x_j^T y_k \right) y_i^T y_k = y_i^T y_j \| z_j \| + \sum_{k=1}^{j-1} \left(x_j^T y_k \right) y_i^T y_k \qquad (2.160)$$

We notice in the above equation,
- the first term is non-zero only for $i = j$
- the second term is non-zero for all $i = k$. Since, $k = 1, 2, \ldots, (j-1)$, the second term is non-zero for $i = 1, 2, \ldots, (j-1)$.

Combining the above two statements, we see that r_{ij} is non-zero for $i = 1, 2, \ldots, j$ or $i \leq j$. This constitutes an upper triangular matrix. We illustrate the relations as follows:

$$
\begin{aligned}
x_1 &= r_{11} y_1 \\
x_2 &= r_{12} y_1 + r_{22} y_2 \\
x_3 &= r_{13} y_1 + r_{23} y_2 + r_{33} y_3 \\
&\vdots \\
x_n &= r_{1n} y_1 + r_{2n} y_2 + r_{3n} y_3 + \cdots + r_{nn} y_n
\end{aligned}
\qquad (2.161)
$$

Example 2.17 illustrates the application of Gram Schmidt orthogonalization and QR decomposition.

Example 2.17 Perform a QR decomposition of the matrix given below:

$$A = \begin{bmatrix} 3 & 4 & 1 \\ 3 & 5 & 1 \\ 2 & 2 & 1 \end{bmatrix}$$

Solution Denoting the columns of A as vectors x_1, x_2, and x_3 respectively, we obtain

$$x_1 = \begin{bmatrix} 3 \\ 3 \\ 2 \end{bmatrix}, \quad x_2 = \begin{bmatrix} 4 \\ 5 \\ 2 \end{bmatrix}, \text{ and } x_3 = \begin{bmatrix} 1 \\ 1 \\ 1 \end{bmatrix}$$

We will use Gram Schmidt orthogonalization Eq. (2.154) to generate a set of orthonormal vectors.

$$\|x_1\| = \sqrt{3^2 + 3^2 + 2^2} = 4.6904$$

$$y_1 = \frac{x_1}{\|x_1\|} = \begin{bmatrix} 0.6396 \\ 0.6396 \\ 0.4264 \end{bmatrix}$$

$$z_2 = x_2 - \left(x_2^T y_1\right) y_1 = \begin{bmatrix} -0.2273 \\ 0.7727 \\ -0.8182 \end{bmatrix}$$

$$\|z_2\| = 1.1481, \quad y_2 = \frac{z_2}{\|z_2\|} = \begin{bmatrix} -0.1980 \\ 0.6730 \\ -0.7126 \end{bmatrix}$$

$$z_3 = x_3 - \left(x_3^T y_1\right) y_1 - \left(x_3^T y_2\right) y_2 = \begin{bmatrix} -0.1379 \\ 0.0690 \\ 0.1034 \end{bmatrix}$$

$$\|z_3\| = 0.1857, \quad y_3 = \frac{z_3}{\|z_3\|} = \begin{bmatrix} -0.7428 \\ 0.3714 \\ 0.5571 \end{bmatrix}$$

Now, we can constitute the Q matrix with y_1, y_2, and y_3 as the columns,

$$Q = \begin{bmatrix} 0.6396 & -0.1980 & -0.7428 \\ 0.6396 & 0.6730 & 0.3714 \\ 0.4264 & -0.7126 & 0.5571 \end{bmatrix}$$

Elements of matrix R can be computed using the relation $r_{ij} = y_i^T x_j$ as follows:

$$r_{11} = y_1^T x_1 = 4.6904, \ r_{12} = y_1^T x_2 = 6.6092, \ r_{13} = y_1^T x_3 = 1.7056,$$
$$r_{22} = y_2^T x_2 = 1.1481, \ r_{23} = y_2^T x_3 = -0.2375, \ \text{and} \ r_{33} = y_3^T x_3 = 0.1857$$

The readers can check that the entries below the diagonal are indeed zero, i.e.,

$$r_{21} = y_2^T x_1 = 0, \ r_{31} = y_3^T x_1 = 0, \ \text{and} \ r_{32} = y_3^T x_2 = 0$$

Therefore, the matrix R is

$$R = \begin{bmatrix} 4.6904 & 6.6092 & 1.7056 \\ 0 & 1.1481 & -0.2375 \\ 0 & 0 & 0.1857 \end{bmatrix}$$

2.3.5.3 *Computation of Eigenvalues*

In this section, we shall use Gram Schmidt orthogonalization and QR decomposition to carryout similarity transformation on a matrix and compute the eigenvalues.

By definition of orthogonal matrices, $Q^T = Q^{-1}$. Thus, $A = QBQ^T$ means that matrices A and B are similar and share the same eigenvalues. Orthogonal matrices are important in the sense that every real matrix can be transformed to an upper triangular matrix by similarity transformation with an orthogonal matrix. *Schur* decomposition stated in Theorem 2.9 ascertains this. A detailed proof of the theorem is beyond the scope of this book. We give an outline of the proof.

Theorem 2.9

For every real matrix A with real eigenvalues, there is an orthogonal matrix Q such that $A = QBQ^T$ where B is an upper triangular matrix. If A is complex, correspondingly there exist a unitary Q such that $A = QBQ^H$.

Proof Outline in $\Re^n$: Denote a $(n \times n)$ matrix as A_n, a $(n-1) \times (n-1)$ matrix as A_{n-1} and so on.
Determine one eigenvalue λ_1 and corresponding eigenvector x_1 for $A = A_n$. Choose

other $(n-1)$ vectors $\{y_2, \ldots, y_n\}$ such that $\{x_1, y_2, \ldots, y_n\}$ forms an orthonormal basis for the $\mathfrak{R}^n$. Now, construct a matrix X_n with $\{x_1, y_2, \ldots, y_n\}$ as columns. Then, one can write,

$$X_n^T A X_n = \begin{bmatrix} \lambda_1 & a \\ 0 & A_{n-1} \end{bmatrix} \tag{2.162}$$

where a is some real number. One can repeat that with A_{n-1} and obtain

$$X_{n-1}^T A_{n-1} X_{n-1} = \begin{bmatrix} \lambda_2 & b \\ 0 & A_{n-2} \end{bmatrix} \tag{2.163}$$

where b is some real number.

Now, if we define $Q_{n-1} = X_n \hat{X}_{n-1} = X_n \begin{bmatrix} 1 & 0 \\ 0 & X_{n-1} \end{bmatrix}$, we can write,

$$Q_{n-1}^T A Q_{n-1} = \begin{bmatrix} \lambda_1 & a_1 & a_2 \\ 0 & \lambda_2 & a_3 \\ 0 & 0 & A_{n-2} \end{bmatrix} \tag{2.164}$$

Proceeding similarly, we shall arive at $Q^T A Q = B$ where B is an upper triangular matrix with eigenvalues on the diagonal and $Q = X_n \hat{X}_{n-1} \hat{X}_{n-2} \ldots \hat{X}_2 \hat{X}_1$, a matrix with orthonormal vectors as columns.

In simple terms, Theorem 2.9 says that it is possible to reduce a real matrix A to an upper triangular matrix B by similarity transformation $Q^T A Q$ with orthogonal matrix Q and thereby determining the eigenvalues of A as the diagonal elements of B. Moreover, if the columns of matrix Q are eigenvectors of matrix A, it will be transformed to a diagonal matrix (Theorem 2.6). In summary, every real matrix through similarity transformation with an orthogonal matrix can be reduced to either a diagonal or an upper triangular matrix with eigenvalues on the diagonal. This forms the basis for the numerical method described in this section.

Finding the particular orthogonal matrix that reduces an arbitrary matrix A to an upper triangular matrix B or diagonal matrix Λ is not easy. As with other numerical methods, we rely on generating a sequence of matrices through similarity transformation that eventually converges to B or Λ.

Starting from the original square matrix, a sequence of similar matrices is generated by the application of similarity transformation with orthogonal matrices.

The orthogonal matrix (Q) obtained by decomposing the matrix A into QR at each step is used for the transformation. The sequence converges to either an upper triangular matrix or a diagonal matrix with the eigenvalues on the diagonal. Whether a matrix becomes a diagonal or an upper triangular depends on the properties of the original matrix. The algorithm is summarized below:

$$A_0 = A$$
$$A_k = Q_k R_k \tag{2.165}$$
$$A_{k+1} = Q_k^T A_k Q_k = Q_k^T Q_k R_k Q_k = R_k Q_k$$

The iteration can be stopped when the changes in the diagonal elements are below certain prescribed value. The change can be measured in absolute terms or as a relative percentage. A metric can be $\max_j \left| \dfrac{\lambda_j^{k+1} - \lambda_j^k}{\lambda_j^k} \right| \times 100$. We show an example for calculation of eigenvalues using the similarity transformation.

Example 2.18   Find the eigenvalues of the following matrix with a relative error of 0.1% or less.

$$\begin{bmatrix} 3 & 4 & 1 \\ 3 & 5 & 1 \\ 2 & 2 & 1 \end{bmatrix}$$

Solution   We start with the original matrix and perform a QR decomposition. Compute the matrix Q using Eq. (2.154) and R using Eq. (2.159). The new matrix is calculated by multiplying RQ. This process is repeated until convergence.

We tabulate the computation below:

Table 2.10

k	$A_k = R_k Q_k$	Q_k	R_k	ε (%)
1	$\begin{bmatrix} 3.0000 & 4.0000 & 1.0000 \\ 3.0000 & 5.0000 & 1.0000 \\ 2.0000 & 2.0000 & 1.0000 \end{bmatrix}$	$\begin{bmatrix} 0.6396 & -0.1980 & -0.7428 \\ 0.6396 & 0.6730 & 0.3714 \\ 0.4264 & -0.7126 & 0.5571 \end{bmatrix}$	$\begin{bmatrix} 4.6904 & 6.6092 & 1.7056 \\ 0.0000 & 1.1481 & -0.2375 \\ 0.0000 & 0.0000 & 0.1857 \end{bmatrix}$	
2	$\begin{bmatrix} 7.9545 & 2.3043 & -0.0792 \\ 0.6331 & 0.9420 & 0.2941 \\ 0.0792 & -0.1323 & 0.1034 \end{bmatrix}$	$\begin{bmatrix} 0.9968 & -0.0757 & -0.0257 \\ 0.0793 & 0.9765 & 0.2004 \\ 0.0099 & -0.2018 & 0.9794 \end{bmatrix}$	$\begin{bmatrix} 7.9801 & 2.3703 & -0.0546 \\ 0.0000 & 0.7721 & 0.2723 \\ 0.0000 & 0.0000 & 0.1623 \end{bmatrix}$	866
3	$\begin{bmatrix} 8.1420 & 1.7215 & 0.2166 \\ 0.0640 & 0.6990 & 0.4214 \\ 0.0016 & -0.0328 & 0.1589 \end{bmatrix}$	$\begin{bmatrix} 1.0000 & -0.0078 & -0.0006 \\ 0.0079 & 0.9988 & 0.0482 \\ 0.0002 & -0.0482 & 0.9988 \end{bmatrix}$	$\begin{bmatrix} 8.1423 & 1.7269 & 0.2199 \\ 0.0000 & 0.6863 & 0.4116 \\ 0.0000 & 0.0000 & 0.1790 \end{bmatrix}$	34.92

(*contd.*)

(*contd.*)

$$
4 \quad
\begin{bmatrix} 8.1556 & 1.6505 & 0.2983 \\ 0.0055 & 0.6656 & 0.4442 \\ 0.0000 & -0.0086 & 0.1788 \end{bmatrix}
\begin{bmatrix} 1.0000 & -0.0007 & 0.0000 \\ 0.0007 & 0.9999 & 0.0130 \\ 0.0000 & -0.0130 & 0.9999 \end{bmatrix}
\begin{bmatrix} 8.1557 & 1.6509 & 0.2986 \\ 0.0000 & 0.6645 & 0.4416 \\ 0.0000 & 0.0000 & 0.1845 \end{bmatrix}
\quad 11.08
$$

$$
5 \quad
\begin{bmatrix} 8.1568 & 1.6414 & 0.3199 \\ 0.0004 & 0.6587 & 0.4502 \\ 0.0000 & -0.0024 & 0.1845 \end{bmatrix}
\begin{bmatrix} 1.0000 & -0.0001 & 0.0000 \\ 0.0001 & 1.0000 & 0.0036 \\ 0.0000 & -0.0036 & 1.0000 \end{bmatrix}
\begin{bmatrix} 8.1568 & 1.6415 & 0.3199 \\ 0.0000 & 0.6587 & 0.4495 \\ 0.0000 & 0.0000 & 0.1861 \end{bmatrix}
\quad 3.1
$$

$$
6 \quad
\begin{bmatrix} 8.1568 & 1.6398 & 0.3259 \\ 0.0000 & 0.6570 & 0.4519 \\ 0.0000 & -0.0007 & 0.1861 \end{bmatrix}
\begin{bmatrix} 1.0000 & 0.0000 & 0.0000 \\ 0.0000 & 1.0000 & 0.0010 \\ 0.0000 & -0.0010 & 1.0000 \end{bmatrix}
\begin{bmatrix} 8.1568 & 1.6398 & 0.3259 \\ 0.0000 & 0.6570 & 0.4517 \\ 0.0000 & 0.0000 & 0.1866 \end{bmatrix}
\quad 0.88
$$

$$
7 \quad
\begin{bmatrix} 8.1569 & 1.6395 & 0.3276 \\ 0.0000 & 0.6565 & 0.4524 \\ 0.0000 & -0.0002 & 0.1866 \end{bmatrix}
\begin{bmatrix} 1.0000 & 0.0000 & 0.0000 \\ 0.0000 & 1.0000 & 0.0003 \\ 0.0000 & -0.0003 & 1.0000 \end{bmatrix}
\begin{bmatrix} 8.1569 & 1.6395 & 0.3276 \\ 0.0000 & 0.6565 & 0.4523 \\ 0.0000 & 0.0000 & 0.1867 \end{bmatrix}
\quad 0.25
$$

$$
8 \quad
\begin{bmatrix} 8.1569 & 1.6394 & 0.3281 \\ 0.0000 & 0.6564 & 0.4525 \\ 0.0000 & -0.0001 & 0.1867 \end{bmatrix}
\quad 0.07
$$

So, the eigenvalues are 8.1569, 0.6564, and 0.1867.

The reader should note that the eigenvectors cannot be determined in this way. The eigenvectors change during similarity transformation. However, once the eigenvalues are known, eigenvectors can be determined easily.

The QR iteration scheme outlined in this section is really in the elementary form. Although, this is the basic procedure that can be employed to estimate the eigenvalues for small and medium size matrices, it is computationally inefficient for large matrices. The method described above can always be applied if computational efficiency is not a factor. With mildly defective matrices, this method may lead to accumulation of round-off errors. Some matrix preprocessing is often performed before applying QR iterations to economize computation for large matrices. Mainly, the matrix is reduced to upper Hessenberg form (Fig. 2.7) prior to QR iteration. Detailed description of this reduction of a matrix to upper Hessenberg form is beyond the scope of this book. Interested readers may refer to other books, for example Young and Gregory (1988) and/or Golub and Van Loan (1996) for detail. Golub and Van Loan (1996) is also a good source for students interested in advanced methods for all aspects of matrix computations.

$$\begin{bmatrix}
a_{11} & a_{12} & a_{13} & a_{14} & \cdots & a_{1k} & a_{1k+1} & \cdots & a_{1n} \\
a_{21} & a_{22} & a_{23} & a_{24} & \cdots & a_{2k} & a_{2k+1} & \cdots & a_{2n} \\
0 & a_{32} & a_{33} & a_{34} & \cdots & a_{3k} & a_{3k+1} & \cdots & a_{3n} \\
0 & 0 & a_{43} & a_{44} & \cdots & a_{4k} & a_{4k+1} & \cdots & a_{4n} \\
\cdots & \cdots & \cdots & \cdots & \cdots & \cdots & \cdots & \cdots & \cdots \\
0 & 0 & 0 & 0 & \cdots & a_{kk} & a_{kk+1} & \cdots & a_{kn} \\
0 & 0 & 0 & 0 & \cdots & a_{k+1k} & a_{k+1k+1} & \cdots & a_{k+1n} \\
\cdots & \cdots & \cdots & \cdots & \cdots & \cdots & \cdots & \cdots & \cdots \\
0 & 0 & 0 & 0 & \cdots & 0 & 0 & a_{nn-1} & a_{nn}
\end{bmatrix}$$

Fig. 2.7 A matrix in Hessenberg form or a quasi uppertriangular form. In this form, only one subdiagonal elements below the main diagonal may contain non-zero elements.

EXERCISE 2.3

1. Consider the following matrix:

$$\begin{bmatrix} 7 & -2 & 1 \\ -2 & 10 & -2 \\ 1 & -2 & 7 \end{bmatrix}$$

   (a) Obtain the equation of the characteristic polynomial using Fadeev Leverrier method.
   (b) Solve the polynomial equation by Bairstow's method (Section 3.4.1) with an approximate relative error of 0.01% or less and thus, obtain all the eigenvalues.
   (c) Obtain all the eigenvalues by similarity transform with an approximate relative error of 0.01%.
   (d) Obtain the largest and smallest eigenvalues by power method and inverse power method, respectively.
   (e) Compare the eigenvalues obtained using these methods.

2. Consider the following matrix:

$$\begin{bmatrix} 2 & 2 & 1 \\ 1 & 3 & 1 \\ 1 & 2 & 2 \end{bmatrix}$$

   (a) Obtain all the eigenvalues using similarity transform.
   (b) Formulate the characteristic equation for the above matrix using Fadeev

Leverrier method. Solve it using an appropriate method for the roots of polynomials. Compare the eigenvalues obtained by the two methods.

3. Prove that the inverse power method with shift will converge to the eigenvector corresponding to the inverse of the eigenvalue nearest in magnitude to the shift applied.

4. For a square matrix A, show the following: (a) sum of the eigenvalues is equal to trace of A, (b) product of the eigenvalues is equal to det (A).

5. Formulate the iteration matrix S of Jacobi method Eq. (2.112) and Gauss Seidel method Eq. (2.113) for the Batman problem Eq. (2.2). Compute the spectral radii and spectral norm for both the iteration matrices. Can you explain why Jacobi method diverge and Gauss Seidel method converge?

SUMMARY

In the beginning of the chapter, system of linear equations was introduced through examples. We started with providing some background on matrix algebra. Intention was to give a visual sense of matrices to the readers. For this, we have introduced the matrices as a representation of linear transformation. The basic concepts of the vector space were summarized in a box. Various concepts such as, existence and uniqueness of the solution, eigenvalues, eigenvectors, and matrix norms were logically introduced.

Presentation of numerical methods started with direct methods. Gauss elimination was presented as the fundamental method. All other methods such as Gauss Jordon, LU decomposition, and Thomas algorithm were derived using Gauss elimination. Stability of the methods was analyzed and matrix conditioning through pivoting, scaling, and equilibration was introduced. A smooth transition to iterative methods was made through iterative improvement of solution using direct methods.

Two iterative methods namely Jacobi and Gauss Seidel were derived using a common framework which was also used for their convergence analysis. Gauss Seidel method was modified through rate of convergence analysis to SOR.

We presented eigenvalue estimation methods sequentially from single eigenvalue to all the eigenvalues. Power method was derived to estimate the largest eigenvalue. This framework was then used for the next two methods to estimate the smallest eigenvalue and the eigenvalue closest to any given value. At the end, we presented two methods for determining all the eigenvalues, one based on the solution of the characteristics polynomial and the other based on similarity transformation.

Throughout the chapter, we attempted to present the methods in a logical sequence. All the methods presented were derived from the basic principles. A visualization was provided wherever possible. To this end, each method and key concepts were followed by examples. Wherever we left out vital detail, the readers were provided with appropriate references. References were provided for the advanced readers as well. At the end of each section, an exercise is given for the students to practice the lessons learned. Some excel templates and C-programs are provided in the Online Resource Centre as well. Our goal will be fulfilled if this chapter helps to clarify basic concepts of linear equations to undergraduate students as well as serve as a launching pad for more advanced students. After linear equations, a natural progression will be to solve the non-linear equations. The following chapter deals with non-linear single equations and systems of non-linear equations. Watch out! Batman flies in to introduce it!

Solution of Nonlinear Equations

3.1 INTRODUCTION

Quite often we come across problems involving solution of an equation of the form $f(x) = 0$. For example, continuing with our *Batman* problem of Chapter 1, the time when he sees the wheel is given by the equation

$$f(t) = \frac{1}{2}gt^2 - 666 = 0$$

For this case, obtaining the value of t is a straightforward mathematical operation. If we further complicate the problem by specifying that *Joker* threw the wheel with an initial downward velocity of u m/s, the equation gets modified to

$$f(t) = ut + \frac{1}{2}gt^2 - 666 = 0$$

which can be solved if one knows the relevant formula for solution of a quadratic equation. In both cases, the function f is a nonlinear function of t because of the presence of t^2 term. A solution is, however, readily obtained. We may not be so fortunate in most of the practical problems involving nonlinear functions. For example, we may be interested in finding the time at which the distance travelled by the wheel (in m) is equal to the square root of the time (in sec). The problem is formulated as

$$f(t) = ut + \frac{1}{2}gt^2 - \sqrt{t} = 0$$

which may be manipulated to obtain a cubic equation, which can be solved analytically. However, the solution process is a little cumbersome and one may decide to use a numerical method to obtain the value of t. On the other hand, it is possible that an analytical solution is not obtainable (e.g., if $h = t^{1/4}$) and one has to use numerical methods. There are various other problems arising in different fields of engineering which require the solution of nonlinear equations.

In this chapter, we would discuss some of the numerical methods which could be used for this purpose. To start with, we would assume that the nonlinear function, f, is a function of only one variable, x. In Section 3.6 we would look at extension of some of the 'single variable' techniques to functions of two or more variables giving rise to a system of nonlinear equations. Also, for most of the discussion we would assume that only real roots are desired. A brief description of methods which could be used for finding complex roots is presented in Section 3.4.

3.1.1 Numerical Techniques

As discussed in Chapter 1, any method which we use to solve $f(x) = 0$ (in other words, finding the *zeros* or *roots* of the function) has to be judged on two important criteria: accuracy (how close is the computed root to the actual root) and efficiency (how much computing time it takes to obtain the root). A graphical method, in which the root is obtained by plotting the function and noting the point of intersection with the x-axis, may be very efficient in providing a *rough* estimate of the root but is not very accurate. Of course, the accuracy may be improved by successive *zoom-ins* in the vicinity of the root but then the efficiency suffers.

In any case, since this is a book on Numerical Methods, we would simply state that graphical methods have become obsolete with the advent of high-speed computing! We must confess, however, that many a times a plot of the function provides tremendous insight into the behaviour of various numerical methods.

One may borrow the idea from the graphical method and devise a numerical method in which the function is evaluated at various points until there is a change in the sign of the function value. Since most practical problems involve *well-behaved* (which, in the present case, means *continuous* or *piecewise continuous*) functions, it may be safe to assume that a root (or to be more precise, *at least* one root) lies between the two points at which the function has opposite signs. (Note that roots may also exist between two points where the function has the same sign.) In other words, we have *bracketed* the root. The only thing which is left to do now is to shrink this bracket to the desired accuracy, which can be done in a number of ways thus resulting in various *bracketing* methods. Typically an incremental search technique is used to bracket the root and the choice of increment is purely problem-specific. A large increment may not be able to bracket a root (e.g., when there are two roots close to each other) and a small increment will increase the computational time of bracketing but will generally result in smaller computational effort in refining (because the initial bracket is smaller). In the following discussion on the bracketing methods, we will assume that the root has already been bracketed and will concern ourselves only with the refinement of the estimate of the root.

Example 3.1   Bracket the roots of the following equations, given that one of the roots is close to −1 and the other two close to 1.

(a) $x^3 - 1.2500000x^2 - 1.5625250x + 1.9530938 = 0$

(b) $x^3 - 1.25020000x^2 - 1.56249999x + 1.95343750 = 0$

Solution   Table 3.1 shows the values of the functions at increments of 1, 0.1, and 0.01. The negative root is easily bracketed (between −2 and −1) for both the equations. With an increment of 1, the function values at $x = 1$ are small but there is no change of sign. Even with an increment of 0.1, we see that the function values decrease till $x = 1.2$ and starts increasing after that but there is again no change of sign. With an increment of 0.01, the two roots of the first function are bracketed, one between 1.24 and 1.25 and the other between 1.25 and 1.26. However, the roots of the second function are still not bracketed.

Table 3.1

x	f_a	f_b	x	f_a	f_b	x	f_a	f_b
−2	−7.9218562	−7.92236252	1	0.14056880	0.14073751	1.20	0.00606380	0.00614951
−1	1.2656188	1.26573749	1.1	0.05281630	0.05294551	1.21	0.00387455	0.00395569
0	1.9530938	1.95343750	1.2	0.00606380	0.00614951	1.22	0.00216130	0.00223783
1	0.1405688	0.14073751	1.3	0.00631130	0.00634951	1.23	0.00093005	0.00100193
2	1.8280438	1.82763752	1.4	0.05955880	0.05954551	1.24	0.00018680	0.00025399
						1.25	−0.00006245	0.00000001
						1.26	0.00018830	0.00024599

The plots of the two functions are virtually identical (note that the functions have nearly equal coefficients) and show the presence of a double root near 1.25 [Fig. 3.1(a)]. Although it is not apparent from the figure, there are two distinct roots at 1.245 and 1.255 for the first function and a double root at 1.25003 for the second function [Fig. 3.1(b), which zooms in near the root]. Thus, a very small increment is needed to bracket the roots if there are two roots very close to each other and it would not be possible to bracket a double root, or any root of even multiplicity.

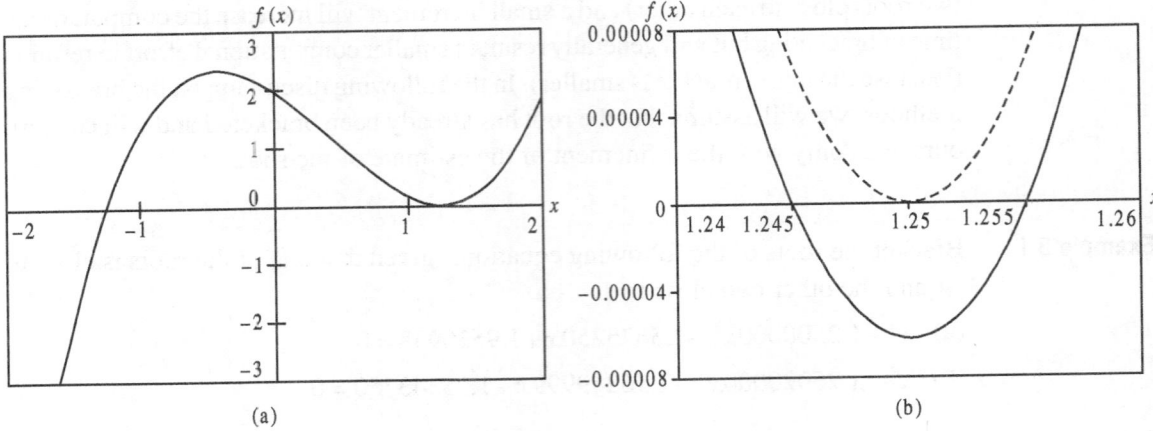

Fig. 3.1 Plot of the functions of Example 3.1: (a) General trend; (b) Near the root: (———) f_a; (– – – –) f_b.

[In general, we say that $f(x)$ has a root, ξ, of multiplicity m, if there exists a non-zero real number, c, such that

$$\lim_{x \to \xi} \frac{|f(x)|}{|x - \xi|^m} = c]$$

Sometimes one may not want to spend the effort in bracketing the root. For example, if from the physics of the problem, it is obvious that the root is near a particular value, it may be desirable to take that value as our initial estimate of the root and then try to improve the estimate using numerical methods. These methods could be called *open* methods since we have not enclosed the root and, as we will see later, the successive approximations of the root may not lie within a closed interval.

3.2 BRACKETING METHODS

Assuming that we have bracketed the root within the interval $[x_l, x_u]$ defined by the lower and upper limits, x_l and x_u, a simple way of shrinking the bracket would be to evaluate $f(x)$ at the mid-point of this interval (x_m), and then comparing its sign with $f(x_l)$ and $f(x_u)$ (for ease of notation, we denote these function values as f_m, f_l, and f_u). The bracket is then reduced to half by using x_m as x_l (if f_m and f_l have same sign) or x_u (if f_m and f_u have same sign). However, one obvious drawback of this method is that the magnitude of $f(x)$ is given no consideration in shrinking the bracket. One would expect that the root will be closer to the end where the function magnitude is smaller (and it is generally, *but not always*, true, see Fig. 3.2).

Therefore, a method may be devised which gives proper weight to the function values f_l and f_u in choosing the bracket at the next iteration. These ideas have been formalised in the subsequent subsections.

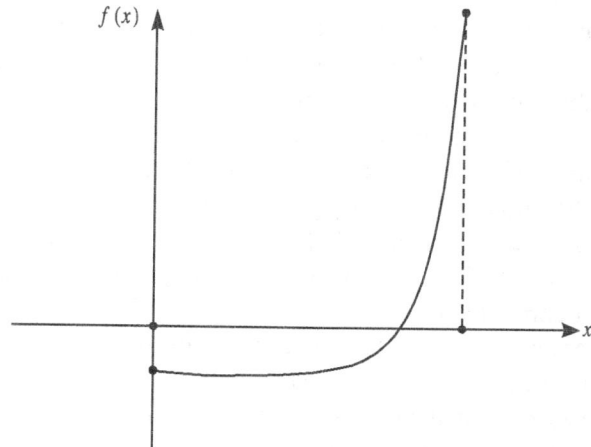

Fig. 3.2 The root is closer to the end with larger magnitude of function.

3.2.1 Bisection Method (Interval halving)

If one were given a bracket $[x_l, x_u]$ which contains the root (indicated by $f_l f_u < 0$) and asked to estimate the location of the root *without any regard to the function value at either ends* (which may work out to be a better option *in some cases*, see Problem 1, Exercise 3.2), the logical choice would be the mid-point of the bracket, $x_m = (x_l + x_u)/2$. Now if the function value at x_m is zero, we have found the root. If not, we have narrowed down the bracket to $[x_l, x_m]$ or $[x_m, x_u]$ depending on whether $f_l f_m < 0$ or $f_m f_u < 0$. Thus, at each step we reduce the bracketing *interval* to *half* by this *bisection*. The process is repeated till the desired accuracy is achieved. In absence of the true value (ξ) of the root, the estimation of accuracy may be based on the approximate error

$$\varepsilon = \text{Current approximation} - \text{Previous approximation}$$

which would be equal to half of the bracket length at any iteration (the current approximation would be the mid-point of the bracket length and the previous approximation would be either the lower or the upper end of this bracket). Since the root lies *within* the bracket, the magnitude of the approximate error will always be greater than that of the true error. This is a desirable property since the iterations are stopped when the approximate error falls below a pre-decided tolerance thus ensuring that the true error is even smaller than the desired tolerance. It should be noted that we are talking in terms of the absolute error and not relative error, which, as mentioned in Chapter 1, should be preferred. However, in situations where we do have a general idea about the magnitude of the root beforehand, the absolute error criterion will be valuable.

Throughout this book, we use the notation $x^{(i)}$ to represent the value of x at iteration i. While x^i or x_i would have been simpler, it may cause ambiguity at some places. If we denote the length of the initial bracket as $\Delta x^{(0)}$, the first estimate of the root will be at its mid-point and, no matter which half is picked as the next bracket, the length of the new bracket would be $\Delta x^{(0)}/2$ and the magnitude of the approximate error for the second estimate would be $\Delta x^{(0)}/4$. Extending this argument, the magnitude of the approximate error for the nth estimate of the root would be $\Delta x^{(0)}/2^n$ (this means that we can estimate a priori the number of iterations needed to reduce the approximate error to a specified tolerance). By using the nested interval theorem (Theorem 3.1), it is seen that the bracket will ultimately reduce to a single point, which will be the desired root. Also, the error at any iteration is equal to half of the error at the previous iteration, indicating a *linearly convergent* behaviour as described in the following box which introduces the concept of *order of convergence* of an iterative method.

For the bisection method, if we use the approximate error (which is an upper bound of the true error and approaches the true error as the iterations converge to the root), we get $p = 1$ (hence linearly convergent) and $C = 1/2$ (indicating that the error is reduced by half in every iteration). Note that for linearly convergent methods, C has to be less than 1 to ensure that the method converges to the root. For superlinearly convergent methods ($p > 1$), however, there is no such restriction. Also, it should be noted that, if the iterations are converging *quickly* towards the root, $x^{(i+1)}$ would be much closer to the root than $x^{(i)}$ and the approximate error at any iteration, $x^{(i+1)} - x^{(i)}$, is an indicator of the *true error* at the *previous iteration*, $\xi - x^{(i)}$, resulting in a conservative error estimate.

BOX 3.1 Order of Convergence of an Iterative Sequence

If an iterative sequence $x^{(0)}$, $x^{(1)}$, $x^{(2)}$ ... converges to the root ξ, the true error at iteration i is

$$e^{(i)} = \xi - x^{(i)} \quad \text{and} \quad \lim_{i \to \infty} \frac{\left| e^{(i+1)} \right|}{\left| e^{(i)} \right|^p} = C$$

then p is the *order of convergence* and C the *asymptotic error constant*. The convergence is called linear if $p = 1$, quadratic if $p = 2$, and cubic if $p = 3$. As we will see, p does not have to be an integer.

Theorem 3.1 Nested Interval Theorem

If $I_1 \supset I_2 \supset ... \supset I_n \supset ...$ is a sequence of nested, closed, bounded, non-empty intervals, then $\bigcap_{n=1}^{\infty} I_n$ is non-empty. Also, if length of the intervals approaches zero, then the intersection consists of a single point.

The theorem is proved by writing the nth interval $I_n = [l_n, u_n]$ in terms of its lower and upper ends. Since the intervals are nested, $l_1 \leq l_2 \leq ... \leq l_n \leq u_n ... \leq u_2 \leq u_1$. Therefore, l_n forms an increasing sequence bounded from above and u_n forms a decreasing sequence bounded from below. Since every monotone bounded sequence converges, l_n converges to, say, L, and u_n converges to, say, U, such that $l_n \leq L$ and $u_n \geq U \; \forall n$. If $\lim_{n \to \infty} (u_n - l_n) = 0$, we have $L = U$ and since $l_n \leq u_n \; \forall n$, $L \in \bigcap_{n=1}^{\infty} I_n$, proving that the intersection consists of a single point.

3.2.2   Linear Interpolation Method (*Regula Falsi* or *False Position*)

As discussed in the previous section, the bisection method does not account for the magnitude of the function at the end points of the bracketing interval when estimating a *better* location of the root.

A likely improvement, therefore, would be to estimate the location of the root in such a way that it is closer to the end which has a smaller function magnitude. The most common method of doing this is to use a *linear interpolation* between the two end points to estimate where the function will be zero (which indicates a *false position* of the root). Of course, one could use the function value at, say, the mid-point of the interval and perform a quadratic interpolation to locate the root. The additional function evaluation would increase the computational effort but may result in a faster convergence. In this section, we describe and analyse the linear interpolation method.

Assuming a linear variation of the function within the bracketing interval (x_l, x_u), an estimate of the location of the root could be made as $\tilde{x}$ given by

$$\tilde{x} = x_l - f_l \frac{x_u - x_l}{f_u - f_l} \tag{3.1}$$

As done earlier for the bisection method, $\tilde{x}$ will then replace x_l or x_u for the next iteration, depending on the value of $f(\tilde{x})$. If $f(x)$ is uniformly concave or convex over the interval (x_l, x_u), it can be seen (Fig. 3.3) that one end of the interval remains fixed. If we take this fixed point as $x^{(0)}$, we have

$$x^{(i+1)} = x^{(0)} - f_0 \frac{x^{(i)} - x^{(0)}}{f_i - f_0} \qquad (3.2)$$

in which f_0 and f_i represent the function values at $x^{(0)}$ and $x^{(i)}$, respectively.

To relate the error at iteration $i + 1$ with that at the ith iteration, we introduce here the concept of errors associated with interpolation. A very brief description is given here since this topic will be dealt with in detail, in Chapter 4.

Interpolation refers to approximating a given function (either continuous or in discrete form) by another function (generally polynomial) in such a way that the approximating function matches the given function at *some* selected points (assumed to be distinct). Denoting the given function by $f(x)$, the approximating polynomial by $\tilde{f}(x)$ and the grid of *match points* as $x_i \, (i = 0,1,2,...,m)$, it is clear that $\tilde{f}(x)$ will be an mth degree polynomial, the coefficients of which may be obtained by the $m + 1$ equations representing the equality of $f(x)$ and $\tilde{f}(x)$ at the grid points. However, at all other points, there would be some *remainder $R(x)$* [unless $f(x)$ happens to be a polynomial of degree m or lower]. The remainder at any point, x, may be written as (considering that the remainder has to be zero at all the grid points),

$$R(x) = f(x) - \tilde{f}(x) = A(x)(x - x_0)(x - x_1)(x - x_2)...(x - x_m) \qquad (3.3)$$

where $A(x)$ is yet to be determined function of x. To obtain the expression for $A(x)$, we now define a function

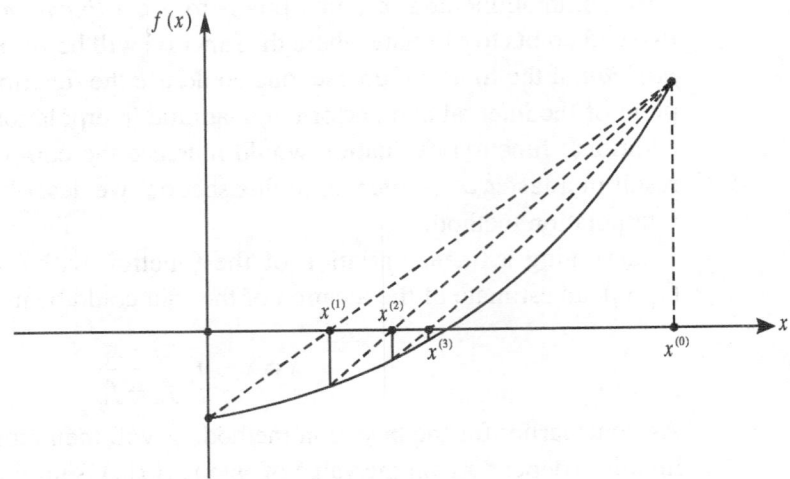

Fig. 3.3 Iterative sequence for linear interpolation for a uniformly concave function.

$$F(\chi) = f(\chi) - \tilde{f}(\chi) - A(x)(\chi - x_0)(\chi - x_1)(\chi - x_2)...(\chi - x_m) \qquad (3.4)$$

which is, clearly, zero when χ is a grid point since $f(x)$ and $\tilde{f}(x)$ are matched at these points (note that the function definition in the above equation uses the function A at the point x and NOT χ. In other words, it could be treated as a constant with respect to χ).

The important point to note is that Eq. (3.3) indicates that $F(\chi)$ will also be zero when $\chi = x$. Therefore, as the point χ moves over the interval containing $(x, x_0, x_1, x_2,...x_m)^1$, $F(\chi)$ will become zero at $(m + 2)$ points which include the $(m + 1)$ grid points and the point x, at which we want to obtain the residual, $R(x)$.

We now apply the Rolle's theorem, which states that if f is continuous on $[a, b]$ and differentiable on (a, b), and $f(a) = f(b)$, then there exists a number c in $[a, b]$ such that $f'(c) = 0$. This indicates that there are at least $(m + 1)$ points in the interval containing $(x, x_0, x_1, x_2, ..., x_m)$ at which the first derivative of the function F with respect to χ is zero. A further application of this theorem to the derivatives indicates that there are at least m points at which the second derivative of F is zero. Repeated applications of the theorem, therefore, lead to the fact that there must be a point $\hat{x} \in (x, x_0, x_1, x_2,...,x_m)$ such that the $(m + 1)$th derivative, $F^{m+1}(\hat{x}) = 0$. If we now assume that the given function, $f(x)$, has continuous derivative of at least $(m + 1)$th order and using the fact that the $(m + 1)$th derivative of $\tilde{f}(\chi)$ with respect to χ is zero (since $\tilde{f}$ is an mth degree polynomial) and that of $(\chi - x_0)(\chi - x_1)(\chi - x_2)...(\chi - x_m)$ is $(m + 1)!$, we get, from the definition of $F(\chi)$,

$$F^{m+1}(\chi) = f^{m+1}(\chi) - \tilde{f}^{m+1}(\chi) - A(x)\frac{d^{m+1}}{d(\chi)^{m+1}}(\chi - x_0)(\chi - x_1)(\chi - x_2)...(\chi - x_m) \qquad (3.5)$$

$$\Rightarrow \quad A(x) = \frac{f^{m+1}(\hat{x})}{(m+1)!} \quad \text{(on evaluating at } \chi = \hat{x})$$

resulting in

$$R(x) = \frac{f^{m+1}(\hat{x})}{(m+1)!}(x - x_0)(x - x_1)(x - x_2)...(x - x_m) \qquad (3.6)$$

The location of the point $\hat{x}$, of course, depends on x.

For linear interpolation between points $x^{(0)}$ and $x^{(i)}$, we may, therefore, write

$$f(x) = f_0 + (x - x^{(0)})\frac{f_i - f_0}{x^{(i)} - x^{(0)}} + (x - x^{(0)})(x - x^{(i)})\frac{f''(\hat{x}_1)}{2} \qquad (3.7)$$

where the first two terms on the right-hand side represent the linear interpolating polynomial, $\hat{f}(x)$, the last term represents the remainder, and $x, \hat{x}_1 \in (x^{(0)}, x^{(i)})$. Applying this equation at the root, $x = \xi$ [such that $f(\xi) = 0$], substituting

---

[1]Written compactly as $\chi \in (x, x_0, x_1, x_2, ..., x_m)$.

the value of f_0 from Eq. (3.2)

$$\left[\text{which results in } f_0 + (\xi - x^{(0)})\frac{f_i - f_0}{x^{(i)} - x^{(0)}} = (\xi - x^{(i+1)})\frac{f_i - f_0}{x^{(i)} - x^{(0)}}\right]$$

and using the mean value theorem for derivatives

[implying $\dfrac{f_i - f_0}{x^{(i)} - x^{(0)}} = f'(\hat{x}_2)$ in which $\hat{x}_2 \in (x^{(0)}, x^{(i)})$]

we get

$$0 = (\xi - x^{(i+1)})f'(\hat{x}_2) + (\xi - x^{(0)})(\xi - x^{(i)})\frac{f''(\hat{x}_1)}{2} \tag{3.8}$$

Thus

$$e^{(i+1)} = -\frac{f''(\hat{x}_1)}{2f'(\hat{x}_2)} e^{(0)} e^{(i)} \tag{3.9}$$

Assuming that the iterations converge to the root,

$$\lim_{i \to \infty} \frac{|e^{(i+1)}|}{|e^{(i)}|} = \left|\frac{f''(\hat{x}_3)e^{(0)}}{2f'(\hat{x}_4)}\right|$$

where $\hat{x}_3, \hat{x}_4, \in (x^{(0)}, \xi)$. This indicates that, similar to the bisection method, the method of false position is also linearly convergent. Since the root is always bracketed, the method will converge. However, the asymptotic error constant, which had a constant value of 1/2 for the bisection method, is now dependent on the slope and curvature of the function.

Example 3.2 Find a root of the following equation, to an accuracy of 1%, using the bisection and linear interpolation methods, given that the root has been bracketed in the interval $(-2, -1)$:

$$x^3 - 1.2500000x^2 - 1.5625250x + 1.9530938 = 0$$

Solution Table 3.2 shows the steps of computation for the bisection method.

Table 3.2

Iteration no.	x_l	x_u	f_l	f_u	$x_m = (x_l + x_u)/2$	ε_r (%)	f_m
1	−2	−1	−7.92186	1.265619	−1.5		−1.89062
2	−1.5	−1	−1.89062	1.265619	−1.25	20	5E−08
3	−1.5	−1.25	−1.89062	5E − 08	−1.375	9.090909	−0.86132
4	−1.375	−1.25	−0.86132	5E − 08	−1.3125	4.761905	−0.4104
5	−1.3125	−1.25	−0.4104	5E − 08	−1.28125	2.439024	−0.20022
6	−1.28125	−1.25	−0.20022	5E − 08	−1.26563	1.234568	−0.09888
7	−1.26563	−1.25	−0.09888	5E − 08	**−1.25781**	0.621118	Not needed

Note that the function value at the midpoint, x_m, is nearly zero at the second iteration. It may sometimes be beneficial to put a stopping criterion based on the function value, e.g., we stop the iteration when the magnitude of the function value is less than a specified limit. Here we have stopped iterating when the approximate error becomes less than 1%. The reduction in the approximate error by a factor of 2 is readily seen in the table. (Since we have computed the *relative* error, it is not exactly half of that at the previous iteration.)

The linear interpolation method converges faster since the function is almost linear in the bracket (see Fig. 3.1).

Table 3.3

Iteration no.	x_l	x_u	f_l	f_u	$\tilde{x}$ (Eq. 3.1)	ε_r (%)	$\tilde{f}$
1	−2	−1	−7.92186	1.265619	−1.13775		0.639949
2	−2	−1.13775	−7.92186	0.639949	−1.2022	5.360842	0.287417
3	−2	−1.2022	−7.92186	0.287417	−1.23013	2.270632	0.122192
4	−2	−1.23013	−7.92186	0.122192	**−1.24183**	0.941715	—

Note that one end of the bracket (in this case, the lower end) remains fixed since the function is convex upward in the entire bracket (the second derivative is negative). Thus, in Eq. (3.9), $e^{(0)}$ is equal to 0.75. The value of $(-f''/2f')$ varies from about 0.5 to 0.8 over the interval $(-2, -1.25)$, indicating that the absolute true error at any iteration should be about 0.35 to 0.6 times the true error at the previous iteration. Table 3.3 shows a factor of about 0.4 for the relative approximate error.

3.2.3 Other Possible Methods

Refinement of the bracket was done by interval halving and linear interpolation in the two methods described above. There are numerous other possibilities which could be used for this purpose. For example, instead of halving the interval, we may divide it into the golden ratio,[2] or a quadratic interpolation[3] using the function

---

[2]The golden ratio is defined as the ratio in which a line is divided so that the ratio of the smaller part to the bigger part is equal to the ratio of the bigger part to the entire line. Its value is equal to 1.618 (= .618/.382 = 1/.618). It is also called the *divine proportion* due to its frequent occurrence in nature. For refining the bracket to find the root, bisection is the optimal strategy. However, for finding a minimum of the function within a bracket, the golden ratio works out better.

[3]We fit a parabola through the three function values to get a quadratic equation, which is solved to get the value of x at which the parabola crosses the x-axis.

BOX 3.2 Ridder's Method

In this method, we modify the given function in such a way that the resulting function values at the end points of the interval and at the mid-point lie in the same straight line. The modification is done through an exponential function of the form $e^{c\frac{x-x_l}{x_u-x_l}}$ and the coefficient c is chosen so as to satisfy the linearity condition in the interval (x_l, x_u) using the two end points and the midpoint (x_m). Or, equivalently, we look for a positive number, α, such that $f_l/\alpha, f_m$, and αf_u are collinear. It is easy to show that this number is given by

$$\alpha = \frac{f_m + \text{sgn}(f_u)\sqrt{f_m^2 - f_l f_u}}{f_u}$$

and the new estimate of root, obtained by linear interpolation between f_m and αf_u, is

$$\widehat{x} = x_m - (x_u - x_m)\frac{\text{sgn}(f_u)f_m}{\sqrt{f_m^2 - f_l f_u}}$$

values at the end points along with an additional function value at the mid-point (or, for that matter, any other point in the bracket!) could be used to estimate the location of the root (Problems 3 and 4, Exercise 3.2). We may, in the linear interpolation method, devise some strategy to accelerate the rate of convergence (e.g., if one end of the bracket remains fixed for a few iterations, we use some fraction, usually ½, of the function value at that end for the interpolation at the next iteration, see Problem 5, Exercise 3.2). Or, we may manipulate the function values at the ends and at the mid-point in such a way that they fall on a straight line and then use linear interpolation (Ridder's method, see Box 3.2 and Problem 6, Exercise 3.2). Another alternative is to combine the linearly convergent bisection method and superlinearly convergent quadratic interpolation (Brent's method, see Problem 7, Exercise 3.2). These modifications generally lead to faster convergence but require either more function evaluations or more record-keeping and comparisons, and the improvement in efficiency may not be significant.

3.2.4 Remarks

- The bracketing methods have guaranteed convergence with linear convergence rate.
- Since one end of the bracket is common between two successive iterations, the function value at this point may be stored and need not be computed again.

This will reduce the total number of function evaluation making the algorithm more efficient.

- If the bracket contains more than one root, the bracketing method will find only one of them, say ξ. A common strategy to find other roots is to *factor out* $(x - \xi)$ from $f(x)$ and find the roots of the function

$$f_1(x) = \frac{f(x)}{x - \xi} = 0$$

  This strategy will not work for bracketing methods, since the bracket obtained for $f(x) = 0$ can no longer be used as a bracket for $f_1(x) = 0$, as $f_1(x)$ will have the same sign at both ends. To overcome this difficulty, *if we know that there are more than one roots in the bracket*, we may try to locate another root by expanding the initial bracket and then factoring out two roots at a time (Problem 8, Exercise 3.2). However, the open methods described in the next section would be much better at locating all roots.

- If the bracket contains a multiple root (of odd multiplicity), bisection will have no difficulty in obtaining the root. However, the linear interpolation will show a very slow convergence (Problem 9, Exercise 3.2).

- Because of linear convergence properties, typically the bracketing methods are used only as *starting* methods. As we approach the root and are relatively confident that the iterations will not diverge, we would like to use a method which has a faster convergence even if it does not guarantee convergence (as the bracketing methods do). Such *open methods* are described in the next section.

EXERCISE 3.2

(Note: For all root-finding problems, it would be a good idea to plot the function to get a feel about its nature).

1. Find a root of $f(x) = (2x)^{10} - 1.2689 = 0$ to an accuracy of 0.001 given that the root has been bracketed in [0,1]. Use the bisection method and the linear interpolation method and comment on their relative performance.

2. Solve the equation $x^7 + x^4 - 0.01 = 0$ using the linear interpolation method.

3. Solve the equation $x^3 + 3.6\,x^2 + 4.32x - 1.728 = 0$ using quadratic interpolation with three points located at the ends of the bracket and its mid-point.

4. Solve the above equation using quadratic interpolation with the third point located at a distance of one-fourth of the interval from the end having smaller function value.

5. Solve the equation $x^7 + x^4 - 0.01 = 0$ with modified linear interpolation method, reducing the function value by half if one end of the interval remains same for more than two iterations.

6. Solve the above equation using linear interpolation with mid-point using the function values multiplied by an exponential function (Ridder's method, see Box 3.2).

7. Solve the equation $x^3 - 3.6\,x^2 + 4.32\,x - 1.728 = 0$ using a quadratic inverse interpolation with three points with the provision that if the estimated root falls outside the bracket, we switch to the bisection method (Brent's method).
   [**Hint:** For the first iteration, we may take the three points as the two ends of the bracket and the estimated root obtained by linear interpolation. Using these three function values and corresponding x values, a quadratic inverse interpolation is written as $x = c_0 + c_1 f + c_2 f^2$ and the value of c_0 provides the estimate of the root. For subsequent iterations, the three points are chosen such that the root is bracketed and the most recent estimate is included.]

8. The equation $x^5 - 4.3\,x^4 + 10.04\,x^3 - 19.972\,x^2 + 24.16\,x - 11.088 = 0$ has three roots between 1 and 2. Use linear interpolation with starting bracket of $(1, 2)$ to get a root. Change the bracket to $(0, 2)$ and apply linear interpolation method to get a (possibly different) root. Factor out these two roots to deflate the given polynomial and find the third root.

9. Solve $x^3 - 4.35\,x^2 + 6.3075\,x - 3.048625 = 0$ using the linear interpolation method. Note that there is a triple root at $x = 1.45$.

10. The ratio of the speed of an object to that of light is denoted by x and is given by the following equation:

$$0.5x^3 - 4x^2 + 6x - 2 = 0$$

Find the value of x using the secant method. Start with $x_0 = 0$ and $x_1 = 1$ and iterate till the relative error is less than 1% and the answer *appears to be right*.

11. Find a root of the equation $\cos x + 1 - x = 0$ between 0 and 2π using the method of false position (with maximum approximate error of 0.1%). What is the difference between this method and the secant method?

3.3 OPEN METHODS

In the bracketing methods, to start the iterations we need to bracket the root, which may be time consuming. Once a bracket has been identified, the bracketing methods are guaranteed to converge to a root for continuous functions, but the rate of convergence is relatively slow. In this section we will look at some of the so called open methods, which need one or more *points* (not necessarily bracketing the root) to start the iterations and generally have superlinear convergence.

3.3.1 Fixed-point Iteration (Successive Substitution or One-point Iteration)

All equations of the form $f(x) = 0$ can be written as $x = \phi(x)$. For example, one of our Batman equations $f(t) = ut + \dfrac{1}{2}gt^2 - \sqrt{t} = 0$ could be written in any of the following forms:

$$t = \frac{\sqrt{t} - \dfrac{1}{2}gt^2}{u}; \quad t = \left(ut + \frac{1}{2}gt^2\right)^2;$$

$$t = \sqrt{\frac{2}{g}\left(\sqrt{t} - ut\right)}; \quad t = (u+1)t + \frac{1}{2}gt^2 - \sqrt{t} \tag{3.10}$$

If a root of $f(x) = 0$ is ξ then obviously $\xi = \phi(\xi)$ indicating that it is a *fixed point* of $\phi(x)$.[4] Now, starting from an initial guess of $x^{(0)}$, we obtain $\phi(x^{(0)})$ and compare with $x^{(0)}$. If their values are same (if we are so lucky, we may think of buying a few lottery tickets!) then $x^{(0)}$ is a root; if not, we take $x^{(1)} = \phi(x^{(0)})$ as our estimate of the root at the next iteration.

Following this process of *successive substitution* of the argument of the function $\phi(x)$, we hope to converge to the root, i.e., reach a point when two successive estimates of the root are *nearly* identical. Since the iterations require only one point to compute the next estimate (as against the use of two points in the secant method to be described later in this section), the scheme may also be called *single-point iteration* or *one-point iteration*. Sometimes the term *Picard iteration* is also used but it is more commonly used in connection with the solution of initial-value problems of ordinary differential equations.

The sequence of iteration is written as $x^{(i+1)} = \phi(x^{(i)})$ and the iterations are stopped when the approximate error $\varepsilon^{(i+1)} = |x^{(i+1)} - x^{(i)}|$ or the relative approximate error

$$\varepsilon_r^{(i+1)} = \frac{\varepsilon^{(i+1)}}{|x^{(i+1)}|}$$

falls below the desired error tolerance (sometimes $x^{(i)}$ is used in the denominator for the relative error. If the iterations are converging to the root, both these values would be almost equal). Since the root is not bracketed between two successive estimates, there is no guarantee that the method will converge. We would now look at the conditions under which this method converges and find the rate of convergence.

---

[4] The function $\phi(x)$ may be thought of as mapping the point x to a new point, say, $x^* = \phi(x)$. Generally this mapping will cause the points x and x^* to be some "distance" apart. If there is *no movement* while mapping some x, say, x_p, [i.e., $x_p = \phi(x_p)$], x_p is called a *fixed point* of the function $\phi(x)$.

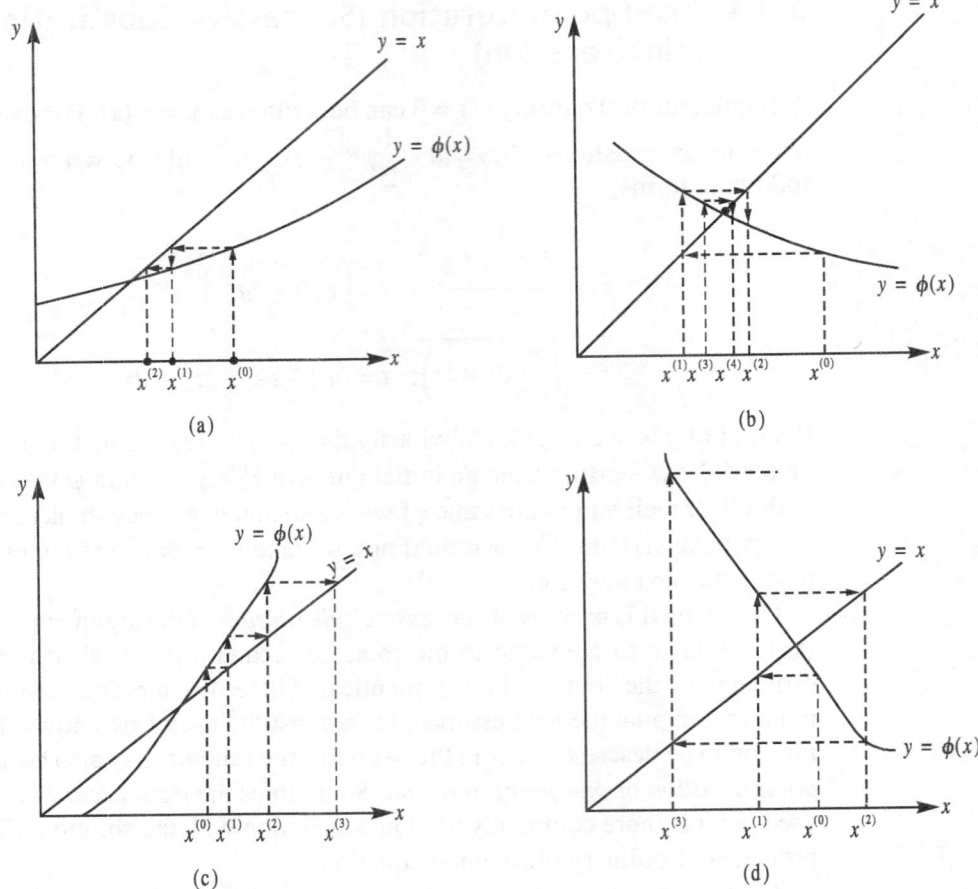

Fig. 3.4 Progress of fixed-point iteration scheme.

The true error at iteration $(i + 1)$ could be written as

$$e^{(i+1)} = \xi - x^{(i+1)} = \phi(\xi) - \phi(x^{(i)}) \tag{3.11}$$

For the analysis of the rate of convergence, we assume that the function $\phi(x)$ is *differentiable* although it is not a necessary condition for convergence (the iterations may converge even when the function is not differentiable *and is only piecewise continuous* near the root). Using the mean value theorem for derivatives, we obtain

$$e^{(i+1)} = \phi'(\tilde{x})(\xi - x^{(i)}) = \phi'(\tilde{x})e^{(i)} \tag{3.12}$$

where $\tilde{x} \in (\xi, x^{(i)})$. Thus the fixed point iteration method is linearly convergent[5]

---

[5]Assuming that $\phi'(\xi) \neq 0$. If the first $(p - 1)$ derivatives of $\phi(x)$ at $x = \xi$ are zero, and the pth derivative is not zero, the order of convergence will be p. This can be shown by expanding $\phi(x^{(i)})$ about ξ.

and, to ensure convergence, $|\phi'(\tilde{x})|$ must be less than 1. In general, $\phi'(x)$ will be different at different points and it would be logical to distinguish between a near-convergence (i.e. when $x^{(i)}$ is *near* ξ) and a far-convergence (when $x^{(i)}$ is *far* from ξ). Since we do not bracket the root in the open methods, there is a chance that the initial (or a subsequent) estimate of the root may be far away from the true root. However, for our discussions on convergence, we will concentrate on near-convergence only, i.e., we assume that the initial guess, $x^{(0)}$, is sufficiently close to the root, ξ. Following may be inferred from Eq. (3.12):

- If $|\phi'(x)| < 1$ for all x, the fixed point iteration sequence will be guaranteed to converge[6] for all $x^{(0)}$ (Problem 1, Exercise 3.3).
- If $|\phi'(x)| < 1$ for all values of x such that $|\xi - x| \leq \alpha$, the fixed-point iteration sequence will converge to ξ if $x^{(0)} \in (\xi - \alpha, \xi + \alpha)$. This is a sufficient but not a necessary condition for convergence (Problem 2, Exercise 3.3).
- If $\phi'(\tilde{x})$ is positive, the error at any iteration would have the same sign as that at the previous iteration (monotonic sequence). For negative values of the derivative, the error will oscillate between positive and negative values (Fig. 3.4).

Example 3.3   It is known that a root of the following equation lies close to $x = -1.5$

$$x^3 - 1.2500000x^2 - 1.5625250x + 1.9530938 = 0$$

Obtain this root, to an accuracy of 0.01%, using the fixed-point iteration scheme.

Solution   For the fixed-point iteration schemes, we have to decide on how to write the given equation in the functional form, $x = \phi(x)$. If we write

$$x = (x^3 - 1.2500000x^2 + 1.9530938)/1.5625250 \qquad (3.13)$$

it is easy to see that $\phi'(x) = (3x^2 - 2.5x)/1.562525$ will become larger than 1 for x values near -1, possibly leading to divergence. On the other hand, writing

$$x = (1.2500000x^2 + 1.5625250x - 1.9530938)^{1/3}$$

we have

$$\phi'(x) = (1/3)(1.2500000x^2 + 1.5625250x - 1.9530938)^{-2/3}(2.5x + 1.562525)$$

which has a smaller magnitude near the root. Table 3.4 shows the iteration using

$$x^{(i+1)} = \phi(x^{(i)}) = \left[1.2500000(x^{(i)})^2 + 1.5625250x^{(i)} - 1.9530938\right]^{1/3}$$

---

[6]We ignore the round-off errors here. There are some equations which are inherently ill-conditioned and all numerical schemes of root finding may show large errors. For example, the Wilkinson's polynomial given by $W(x) = \prod_{i=1}^{20}(x-i)$ is notoriously ill-conditioned.

Table 3.4

Iteration no.	x	ε_r (%)	$\phi(x)$
0	−1.5		−1.14073
1	−1.14073	31.49506	−1.28239
2	−1.28239	11.04697	−1.23882
3	−1.23882	3.517025	−1.25368
4	−1.25368	1.185173	−1.24877
5	−1.24877	0.393456	−1.25041
6	−1.25041	0.131321	−1.24986
7	−1.24986	0.043754	−1.25005
8	−1.25005	0.014586	−1.24998
9	**−1.24998**	0.004862	—

In Example 3.3, note the linear convergence of the iterations as the error reduces by a factor of nearly 3 at each iteration. [It is easy to see that near the root, $\phi'(x) = -1/3$.] It would be instructive to try different starting values in Table 3.4 (e.g., −2, −1, and, especially, 0 or −2.5). If we use Eq. (3.13) as the iterative scheme starting with an initial guess of −1.5, the iterations quickly diverge and if we use a starting value of −1, the iterations move towards the positive roots (in fact, it is nearly impossible to obtain the negative root of this function). In general, if we have no idea about the root, it would be difficult to decide which alternative expression for $\phi'(x)$ would be better. The fixed-point theorem may be helpful under some circumstances. It states that if $f(x)$ is continuous on $[a, b]$ and $\phi(x) \in [a,b] \; \forall \; x \in [a,b]$ then $f(x)$ has a fixed point in $[a, b]$ and if additionally $0 < |\phi'(x)| < 1 \; \forall \; x \in [a,b]$ then $\phi(x)$ has a *unique* fixed-point in $[a, b]$.

While the fixed-point iteration scheme does not require bracketing, it is still linearly convergent. It is, therefore, desirable to explore some other open methods of higher order convergence.

3.3.2 Newton-Raphson Method (or Tangent Method)

In the linear interpolation method, we approximated the function by a straight line within the bracketing interval. This line was obtained by joining the function values at either end of the interval. If we have a single point and wish to approximate the function by a straight line in the neighbourhood of this point, a natural choice would be the *tangent* at that point. Thus, starting from an initial estimate of the root, we approximate the function by the tangent at this point, and obtain a new estimate as the point where this linear approximation has a root (Fig. 3.5). The method was developed by *Newton* and, about 20 years later but independently, by *Raphson*. Heron of Alexandria was probably the earliest to use a similar method in

1st century and Al-Tusi of Iran in 12th century and Viete of France in 16th century also used the method. However, we would follow the most commonly used nomenclature of Newton's method or Newton-Raphson method. Sometimes, the iterative technique applied to a single nonlinear equation is called the Newton's method and, to a set of simultaneous nonlinear equations, is called the *Newton-Raphson method*.

The linear approximation of the function $f(x)$ near the point x_i is thus given by

$$\tilde{f}(x) = f_i + (x - x^{(i)})f'(x^{(i)}) \tag{3.14}$$

and the new estimate of the root, which is a zero of $\tilde{f}(x)$, is given by

$$x^{(i+1)} = x^{(i)} - \frac{f_i}{f'(x^{(i)})} \tag{3.15}$$

Assuming that $f(x)$ is *twice differentiable*, we may write the Taylor series expansion about $x^{(i)}$ as

$$f(x) = f(x^{(i)}) + (x - x^{(i)})f'(x^{(i)}) + \frac{(x - x^{(i)})^2}{2}f''(\hat{x}) \tag{3.16}$$

where $\hat{x} \in (x, x^{(i)})$. Letting x approach the root, ξ, and assuming that ξ is a simple root (so that $f'(x) \neq 0$ for all x close to ξ), we get

$$0 = f_i + (\xi - x^{(i)})f'(x^{(i)}) + \frac{(\xi - x^{(i)})^2}{2}f''(\hat{x}) \tag{3.17}$$

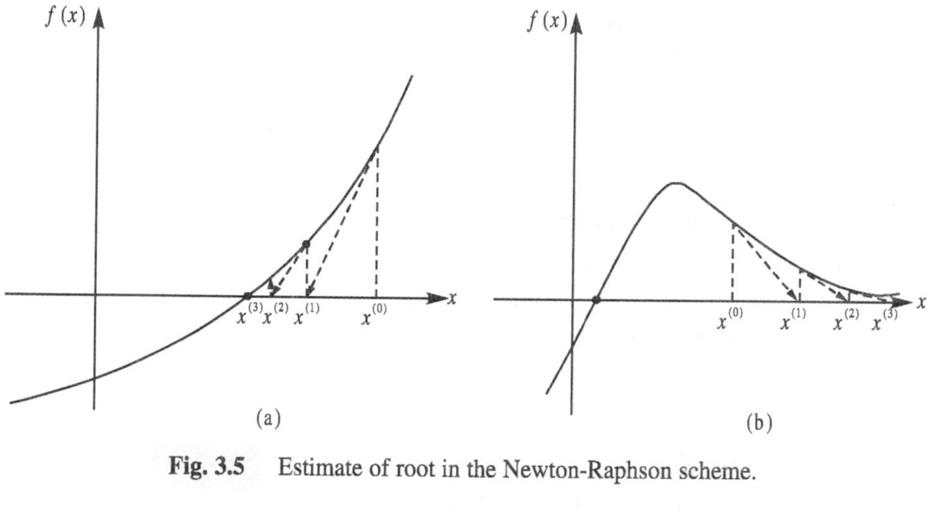

Fig. 3.5 Estimate of root in the Newton-Raphson scheme.

and

$$e^{(i+1)} = \xi - x^{(i+1)} = \xi - \left(x^{(i)} - \frac{f_i}{f'(x^{(i)})} \right) = -\frac{(\xi - x^{(i)})^2}{2f'(x^{(i)})} f''(\hat{x}) = -\frac{f''(\hat{x})}{2f'(x^{(i)})} \left(e^{(i)} \right)^2$$

(3.18)

in which $\hat{x} \in (\xi, x^{(i)})$. If the iterations converge to the root, both $x^{(i)}$ and $\hat{x}$ will approach ξ and

$$\lim_{i \to \infty} \frac{|e^{(i+1)}|}{|e^{(i)}|^2} = \left| \frac{f''(\xi)}{2f'(\xi)} \right|$$

Thus, the Newton-Raphson method is *quadratically convergent*. The quadratic convergence may also be shown by writing the Newton iteration scheme as a fixed-point scheme with

$$\phi(x^{(i)}) = x^{(i)} - \frac{f(x^{(i)})}{f'(x^{(i)})}$$

and showing that $\phi'(\xi) = 0$.

If we assume that for all points x_a and x_b *near* the root

$$\left| \frac{f''(x_a)}{2f'(x_b)} \right|$$

has an upper bound of M, we can write

$$\left| Me^{(i+1)} \right| \leq (Me^{(i)})^2$$

Therefore, if we assume that the initial guess $x^{(0)}$ is sufficiently near the root and $\left| Me^{(0)} \right| < 1$, the iterations will converge and

$$\left| e^{(i)} \right| \leq \frac{(Me^{(0)})^{2^i}}{M}$$

However, since the root is not known beforehand it is difficult to use this criterion. A more usable criterion for convergence is (Fig. 3.6):

If there is an interval $[x_l, x_u]$ such that f_l and f_u have opposite signs; $f'(x)$ is neither zero nor does it change sign in the interval; and

$$\left| \frac{f'(x)}{f(x)} \right|$$

at both x_l and x_u is less than $(x_u - x_l)$; the Newton-Raphson method will converge from any initial guess within the interval.

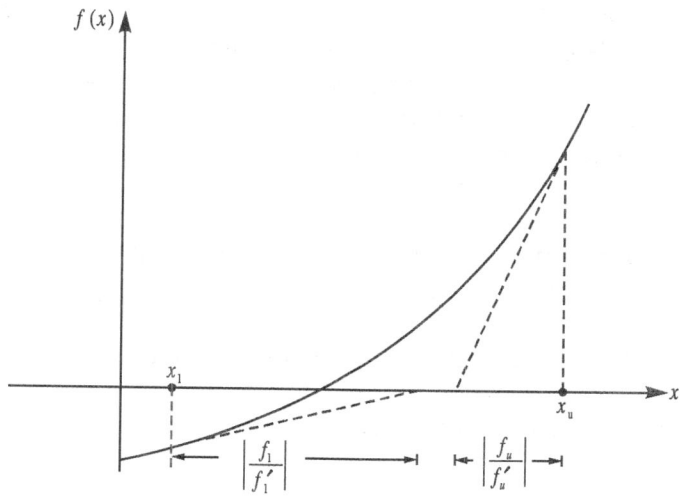

Fig. 3.6 Convergence criterion for Newton-Raphson scheme.

Example 3.4   It is known that a root of the following equation lies close to $x = 1$.

$$x^3 - 1.2500000x^2 - 1.5625250x + 1.9530938 = 0$$

Obtain this root, to an accuracy of 0.01%, using the Newton-Raphson scheme.

Solution   The function derivative is easily obtained in this case. Table 3.5 shows the iterations.

Table 3.5

Iteration no.	$x^{(i)}$	f_i	f_i'	$x^{(i+1)}$ (Eq. 3.15)	ε_r (%)
0	1	0.140569	−1.06253	1.132297	11.68394
1	1.132297	0.032945	−0.54698	1.192528	5.050672
2	1.192528	0.008007	−0.27748	1.221383	2.362551
3	1.221383	0.001962	−0.14065	1.235334	1.129262
4	1.235334	0.000473	−0.07271	1.241832	0.523306
5	1.241832	0.000104	−0.04066	1.24439	0.205515
6	1.24439	1.62E−05	−0.02798	1.244969	0.046522
7	1.244969	8.33E−07	−0.02511	**1.245002**	0.002666

Note the quadratic convergence as x approaches the root, the error being proportional to the square of the error at previous iteration. Strictly speaking, since Eq. (3.18) uses the absolute true error, the true error at any iteration should be equal to a constant times the square of the true error at the previous iteration, the constant being equal to $-f''/2f'$ at the root (e.g., this value is close to 100 at $x = 1.245$). However, since the magnitude of the root is close to 1, and the asymptotic error constant is close to 100, it turns out that the percent relative approximate error at any iteration is roughly equal to the square of that at the previous iteration, as we approach the root.

Example 3.5   It is known that a root of the following equation lies close to $x = 1$:

$$x^3 - 1.25020000x^2 - 1.56249999x + 1.95343750 = 0$$

Obtain this root, to an accuracy of 0.01%, using the Newton-Raphson scheme.

Solution   Table 3.6 shows the iterations.

Table 3.6

Iteration no.	$x^{(i)}$	f_i	f_i'	$x^{(i+1)}$ (Eq. 3.15)	ε_r (%)
0	1	0.140738	−1.0629	1.132409	11.69268
1	1.132409	0.032999	−0.54693	1.192745	5.058566
2	1.192745	0.008036	−0.27692	1.221763	2.375112
3	1.221763	0.001985	−0.13928	1.236013	1.152908
4	1.236013	0.000493	−0.06984	1.243076	0.56821
5	1.243076	0.000123	−0.03497	1.246593	0.282081
6	1.246593	3.07E–05	−0.0175	1.248347	0.140516
7	1.248347	7.67E–06	−0.00876	1.249222	0.070078
8	1.249222	1.91E–06	−0.00439	1.249658	0.034894
9	1.249658	4.75E–07	−0.00221	1.249874	0.017214
10	1.249874	1.16E–07	−0.00113	**1.249976**	0.008175

Note the *linear convergence* as x approaches the root (the error is nearly half of the error at previous iteration). The reason for reduction in the order of convergence from quadratic to linear is the presence of a double root. We will discuss more about it in Section 3.4.

The faster convergence of the Newton-Raphson method comes with the price one has to pay in terms of additional computations for obtaining the derivative of the function. The additional computational effort depends on the type of function and, for a polynomial, is almost equal to that involved in computing the function. We may define an *efficiency parameter*, η, for an iterative scheme as follows:

Let p be the order of convergence and C the asymptotic error constant for the iterative scheme, which requires a computational effort of κ times that required for evaluating the function $f(x)$. For example, for the Newton-Raphson scheme applied to a polynomial $f(x)$:

$$p = 2, \; C = \left| \frac{f''(\xi)}{2f'(\xi)} \right| \; \text{and} \; \kappa \approx 2$$

If we assume that the iterative scheme converges to the root ξ (within a specified tolerance of δ) in N iterations, and the asymptotic rate of convergence is applicable throughout the iterative sequence, we get

$$\delta \approx e^{(N)} = C \left| e^{(N-1)} \right|^P = C \left| C \left| e^{(N-2)} \right|^P \right|^P = C^{1+P} \left| e^{(N-2)} \right|^{P^2} \cdots$$

$$= C^{1+P+P^2+\dots P^{N-1}} \left| e^{(0)} \right|^{P^N} = C^{\frac{P^N-1}{P-1}} \left| e^{(0)} \right|^{P^N}$$

from which the value of N could be obtained. The computational effort would then be proportional to $N\kappa$ and the efficiency is inversely related to the effort. From the above equation, $N = N(e^{(0)}, \delta, C, p)$, i.e., a function of the initial error, desired error, asymptotic error constant and the order of convergence. To simplify the analysis, *treating the term* $C^{\frac{(P^N-1)}{(P-1)}}$ *as a constant*, for a given initial error and tolerance, we obtain $p^N = $ constant and, therefore,

$$N\kappa \propto \frac{1}{\ln p^{1/\kappa}}$$

Thus, $p^{1/\kappa}$ is a reasonably good indicator of η, the efficiency parameter of the iterative scheme.

For polynomial $f(x)$, the efficiency parameter of the Newton-Raphson method $\simeq 2^{1/2} = 1.414$. However, for some types of functions, the evaluation of derivative may require more effort than that required in the function evaluation. Therefore, sometimes a modified Newton-Raphson method is used in which the derivative is not evaluated at every iteration but is updated at every, say, fifth iteration. Also, for some functions, it may be very difficult to obtain the derivative. Therefore, it is desirable to look for other methods which do not involve the computation of the derivative but still provide a superlinear convergence.

One option is to obtain the derivative at any point numerically by evaluating the function at another neighbouring point and then using the divided difference. However, it is generally not recommended since there would be large round-off errors and, as discussed in the next section, it would be less efficient than the *secant method*.

3.3.3 Secant Method

As seen in the previous section, the Newton-Raphson method approximates the function near any point by the tangent at that point. In the secant method, the approximating straight line is taken as the *secant* (a straight line that intersects a curve at two or more points). Thus, starting from any two points (*not necessarily bracketing the root*) on the function curve, we approximate the function by the secant passing through these points, and obtain an estimate of the root as the point where this linear approximation has a root (Fig. 3.7). For the next iteration, this estimate of the root replaces *one of the previous two* points. (Generally the earlier iteration is discarded. However, an alternative would be to discard the point at which the function has a larger magnitude, see Problem 3, Exercise 3.3). Starting from two initial values, $x^{(0)}$ and $x^{(1)}$, the iterative sequence is given by

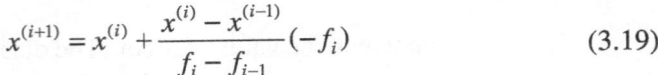

$$x^{(i+1)} = x^{(i)} + \frac{x^{(i)} - x^{(i-1)}}{f_i - f_{i-1}}(-f_i) \qquad (3.19)$$

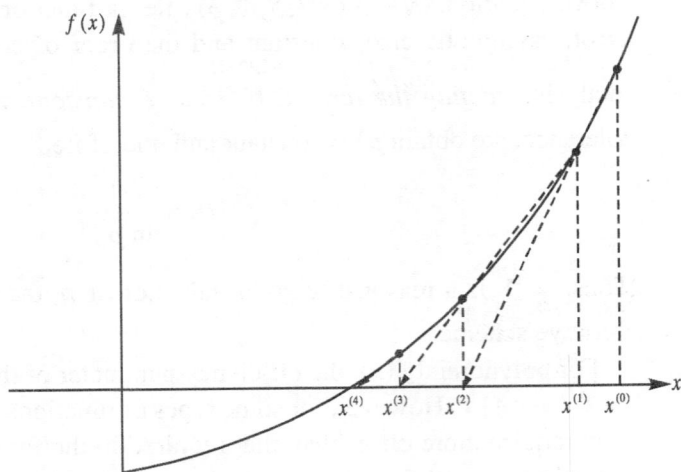

Fig. 3.7 Estimation of root in secant method.

Note that Eq. (3.19) is similar to Eq. (3.1) used in the linear interpolation method, which is not surprising since both of these are based on approximating the function by a straight line and obtaining its point of intersection with the axis. The only difference is that in linear interpolation method, the root is always bracketed between the estimates at two successive iterations. In the secant method, it is not necessary that the root lies between $x^{(i)}$ and $x^{(i+1)}$. An analysis similar to that performed for the linear interpolation method [Eq. (3.9)] results in the error of the secant method as

$$e^{(i+1)} = -\frac{f''(\hat{x}_1)}{2f'(\hat{x}_2)} e^{(i)} e^{(i-1)} \tag{3.20}$$

where $\hat{x}_1 \in (\xi, x^{(i)}, x^{(i-1)})$ and $\hat{x}_2 \in (x^{(i)}, x^{(i-1)})$. Assuming that the iterations converge to the root ξ, we get

$$\lim_{i \to \infty} \left| e^{(i+1)} \right| = \left| \frac{f''(\xi)}{2f'(\xi)} \right| \left| e^{(i)} \right| \left| e^{(i-1)} \right|$$

Thus, using the relationships

$$e^{(i+1)} = C \left| e^{(i)} \right|^p \text{ and } e^{(i-1)} = \left| \frac{e^{(i)}}{C} \right|^{1/p}$$

the order of the secant method and its asymptotic error constant are obtained by $p^2 = p + 1$ and

$$C = \left| \frac{f''(\xi)}{2f'(\xi)} \right|^{1/p}$$

implying that the order is 1.618 (better than bisection and false position but not as good as Newton-Raphson). As far as efficiency is concerned, if the computational effort in evaluating the derivative of a function is more than 0.44 times ($2^{1/1.44} = 1.618$) that required for a function evaluation, the secant method is likely to be more efficient than the Newton-Raphson method.

Example 3.6

It is known that a root of the following equation lies close to $x = 1$:

$$x^3 - 1.2500000x^2 - 1.5625250x + 1.9530938 = 0$$

Obtain this root, to an accuracy of 0.01%, using the secant method.

Solution

We start the iterations with the points $x = 0$ and $x = 1$ and use the two most recent estimates at each iteration. Table 3.7 shows the iterations.

The asymptotic error constant, C, is equal to $\left| \frac{f''(\xi)}{2f'(\xi)} \right|^{0.618}$ which is close to 17, and the order of convergence, p, is 1.618. The absolute true error (the true value of the root being 1.245002) at iterations 8, 9, and 10 is 0.000495, 0.0000717, and 0.00000334, respectively, and it is rather straightforward to see that these values are consistent with the computed values of C and p.

Table 3.7

Iteration no.	$x^{(i)}$	f_i	$x^{(i+1)}$ [Eq. (3.19)]	$\varepsilon_r (\%)$
	0	1.953094		
0	1	0.140569	1.077554	7.197238
1	1.077554	0.069158	1.152661	6.515953
2	1.152661	0.022705	1.189371	3.08653
3	1.189371	0.008906	1.213064	1.953109
4	1.213064	0.003299	1.227003	1.136029
5	1.227003	0.001248	1.235487	0.686695
6	1.235487	0.000461	1.240463	0.401159
7	1.240463	0.000164	1.243215	0.221335
8	1.243215	5.25E–05	1.244507	0.103843
9	1.244507	1.3E–05	1.24493	0.03399
10	1.24493	1.8E–06	**1.244999**	0.005494

As seen above, the secant method has a lower order of convergence than the Newton-Raphson method. In fact, it has been shown to be generally true that any iterative method which uses only one additional function evaluation at each iteration cannot have quadratic convergence. In order to improve upon the order of convergence, without using derivatives, Muller's method and Steffensen's method, as described next could be used.

3.3.4 Muller's Method

In this method, instead of a linear interpolation using the function values at $x^{(i-1)}$ and $x^{(i)}$, the function value at $x^{(i-2)}$ is also utilized to perform a quadratic interpolation (A side benefit of using a quadratic interpolation is that we may obtain complex roots also, but that will be discussed later!). Another option is to use *inverse interpolation*, i.e., expressing x as a quadratic function of f, say $g(f)$, and then obtaining the root as $g(0)$. However, this will not work in case of complex roots (see Problem 4, Exercise 3.3). Starting from three points, $x^{(0)}$, $x^{(1)}$, and $x^{(2)}$, the iterative sequence is written as

$$x^{(i+1)} = x^{(i)} + \frac{\pm\sqrt{b^2 - 4ac} - b}{2a} \qquad (3.21)$$

where a, b, and c are obtained from the quadratic interpolation using function values at $x^{(i-2)}$, $x^{(i-1)}$, and $x^{(i)}$ with the stipulation that the interpolating parabola attains a zero value at the point $x^{(i+1)}$. The analysis is simplified by shifting the origin of the coordinate system to the point $x^{(i)}$. In this new system, x_*, the quadratic polynomial is written as $f(x_*) = ax_*^2 + bx_* + c$ and matching the function values at the three grid points leads to

$$a = f\left[x^{(i)}, x^{(i-1)}, x^{(i-2)}\right]$$

$$b = f\left[x^{(i)}, x^{(i-1)}\right] + (x^{(i)} - x^{(i-1)})f\left[x^{(i)}, x^{(i-1)}, x^{(i-2)}\right]$$

$$c = f_i$$

Here, the square brackets indicate the divided differences (detailed discussion of divided difference is available in Chapter 4) defined by

$$f\left[x^{(i)}, x^{(i-1)}\right] = \frac{f_i - f_{i-1}}{x^{(i)} - x^{(i-1)}}$$

and

$$f\left[x^{(i)}, x^{(i-1)}, x^{(i-2)}\right] = \frac{f\left[x^{(i)}, x^{(i-1)}\right] - f\left[x^{(i-1)}, x^{(i-2)}\right]}{x^{(i)} - x^{(i-2)}}$$

Since the form of Eq. (3.21) is subject to severe round-off errors (Section 1.3), an alternative form is used for the iterative sequence as

$$x^{(i+1)} = x^{(i)} - \frac{2c}{b \pm \sqrt{b^2 - 4ac}} \tag{3.22}$$

The sign of the square root is chosen in such a way as to make the magnitude of the denominator large thereby choosing $x^{(i+1)}$ closer to $x^{(i)}$ (see Fig. 3.8. An alternative would be to choose the value of $x^{(i+1)}$ which gives smaller function magnitude but it would require more function computations, see Problem 5, Exercise 3.3). For $b^2 - 4ac < 0$, i.e., complex denominator, both the signs of the square root will give the same magnitude, for real values we choose the negative sign if b is negative and vice versa. As in the secant method, we again have a choice in discarding one of the three points, $x^{(i-2)}$, $x^{(i-1)}$, and $x^{(i)}$, for the next iteration. We may discard $x^{(i-2)}$, or the point farthest from $x^{(i+1)}$, or the point at which the function has the largest magnitude (see Problem 6, Exercise 3.3). Performing an analysis of interpolation error, similar to that done for the secant method, we obtain

$$\lim_{i \to \infty} e^{(i+1)} \approx -\frac{f'''(\xi)}{6f'(\xi)} e^{(i-2)} e^{(i-1)} e^{(i)} \tag{3.23}$$

If p denotes the order of the Muller's method, we get $p^3 = p^2 + p + 1$ implying that the order is 1.839 (better than secant but not as good as Newton-Raphson). The asymptotic error constant is given by $C = \left|\dfrac{f'''(\xi)}{6f'(\xi)}\right|^{(p-1)/2}$.

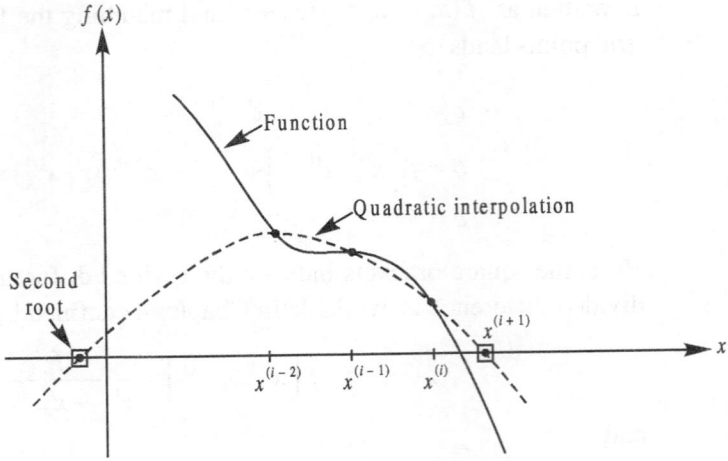

Fig. 3.8   Choosing the more appropriate root for Muller's method.

Example 3.7   It is known that a root of the following equation lies close to $x = 1$.

$$x^3 - 1.2500000x^2 - 1.5625250x + 1.9530938 = 0$$

Obtain this root, to an accuracy of 0.01%, using the Muller's method starting with the points 0, 0.5, and 1 to start the iterations.

Solution   We use the three most recent estimates at each iteration. Table 3.8 shows the iterations.

Table 3.8

Iteration no.	$x^{(i)}$	f_i	$x^{(i+1)}$ (Eq. 3.22)	ε_r (%)
	0	1.953094		
	0.5	0.984331		
0	1	0.140569	1.091296	
1	1.091296	0.058912	1.181861	7.662871
2	1.181861	0.01123	1.226125	3.61012
3	1.226125	0.00135	1.241139	1.209691
4	1.241139	0.000133	1.244831	0.296539
5	1.244831	4.35E–06	1.245002	0.013728
6	1.245002	1.25E–08	**1.245002**	4.01E–05

The order of convergence is 1.839 and the asymptotic error constant is about 4.4. Computing the absolute true errors, one can easily verify that these values hold good as the iterations approach the root. Had we taken the starting points as 0, 1, and 2, we would have encountered complex values which is desirable if we know that the function has complex roots. In this case, however, it unnecessarily complicates the computation.

3.3.5 Steffensen's Method

As discussed earlier, the Newton-Raphson method has quadratic convergence but requires computation of the derivative of the function. The secant method, on the other hand, may be thought of as approximating the derivative of the function by a secant. It does not require derivative evaluation but has sub-quadratic convergence ($p = 1.618$). In the Steffensen's method, a judicious choice of the secant is made for the approximation of the derivative, such that a quadratic convergence is achieved. The iterative scheme is written as

$$x^{(i+1)} = x^{(i)} - \frac{f_i}{\dfrac{f(x^{(i)} + f_i) - f_i}{f_i}} \tag{3.24}$$

which indicates that the function is approximated by the secant passing through $x^{(i)}$ and $x^{(i)} + f_i$. (In the next subsection, we will see another interpretation of the Steffensen's method). By expanding in a Taylor's series about the point $x^{(i)}$, and using the Taylor's series expansion of $f(x)$ about the root, ξ, it can be shown that

$$\lim_{i \to \infty} e^{(i+1)} = -\frac{f''(\xi)}{2f'(\xi)} \left[1 + f'(\xi)\right] \left[e^{(i)}\right]^2 \tag{3.25}$$

establishing the quadratic convergence of the Steffensen's method. In fact, all schemes of the form

$$x^{(i+1)} = x^{(i)} - \frac{f_i^2}{cf_i^2 + f(x^{(i)} + f_i) - f_i}$$

would have quadratic convergence. Steffensen's methods is a special case with $c = 0$.

Example 3.8   It is known that a root of the following equation lies close to $x = 1$:

$$x^3 - 1.2500000x^2 - 1.5625250x + 1.9530938 = 0$$

Obtain this root, to an accuracy of 0.01%, using the Steffensen's method.

Solution   Table 3.9 shows the iterations.

Table 3.9

Iteration no.	$x^{(i)}$	f_i	$f(x+f)$	$x^{(i+1)}$ (Eq. 3.24)	$\varepsilon_r (\%)$
0	1	0.140569	0.028568	1.176423	
1	1.176423	0.013075	0.008869	1.217067	3.33948
2	1.217067	0.002614	0.002209	1.233916	1.365463
3	1.233916	0.000581	0.000535	1.241335	0.597706
4	1.241335	0.000125	0.000119	1.24425	0.234291
5	1.24425	2.01E–05	1.96E–05	1.244954	0.056542
6	1.244954	1.2E–06	1.17E–06	**1.245002**	0.003814

The order of convergence is 2 and the asymptotic error constant is about 97, both of which may be verified by computing the absolute true errors as the iterations approach the root. Had we taken the starting point as 2, we would have reached the other root at 1.245 in 13 iterations. Starting from an initial guess of 0 requires more than 2000 iterations to reach the correct root 1.245, although the relative approximate error is very small in the initial iterations because $f(x^{(i)} + f_i)$ is of much larger magnitude than f_i. We should, however, not stop the iteration even if the approximate error is small because the function value stays very large. Table 3.10 shows an extract of these computations.

Table 3.10

Iteration no.	$x^{(i)}$	f_i	$f(x+f)$	$x^{(i+1)}$ (Eq. 3.24)	$\varepsilon_r (\%)$
0	0	1.953094	1.58334	10.31653	
1	10.31653	950.7924	8.87E+08	10.31551	0.009884
2	10.31551	950.4948	8.86E+08	10.31449	0.009888
3	10.31449	950.1971	8.85E+08	10.31347	0.009892
⋮	⋮	⋮	⋮	⋮	⋮
2475	1.25531	8.05E–06	8.27E–06	1.255007	0.024099
2476	1.255007	2.36E–07	2.42E–07	1.254998	0.00075

3.3.6 Aitken Extrapolation

Another alternative to accelerate the convergence of a linearly converging iterative sequence is to use the Aitken extrapolation. For a sequence of iterations converging to the root, ξ, if the differences between successive iterates are approximated as forming a geometric series, i.e., $x^{(i)} - x^{(i-1)} = k (x^{(i-1)} - x^{(i-2)})$, k being the multiplying factor or common ratio, the root can be obtained by extrapolation as

$$\xi = \lim_{n \to \infty} x^{(n)} = x^{(i)} + \sum_{j=1}^{\infty} (x^{(i+j)} - x^{(i+j-1)})$$

$$= x^{(i)} + (x^{(i)} - x^{(i-1)}) \sum_{j=1}^{\infty} k^j = x^{(i)} + (x^{(i)} - x^{(i-1)}) \frac{k}{1-k} \qquad (3.26)$$

Thus the extrapolated sequence, $x_*^{(i)}$, may be written as

$$x_*^{(i)} = x^{(i)} + (x^{(i)} - x^{(i-1)}) \frac{\dfrac{x^{(i)} - x^{(i-1)}}{x^{(i-1)} - x^{(i-2)}}}{1 - \dfrac{x^{(i)} - x^{(i-1)}}{x^{(i-1)} - x^{(i-2)}}} = x^{(i)} - \frac{(x^{(i)} - x^{(i-1)})^2}{x^{(i)} - 2x^{(i-1)} + x^{(i-2)}} \qquad (3.27)$$

and will generally converge faster to the root.

Example 3.9   It is known that a root of the following equation lies close to $x = -1.5$:

$$x^3 - 1.2500000x^2 - 1.5625250x + 1.9530938 = 0$$

Obtain this root, to an accuracy of 0.01%, using the fixed-point iteration scheme with Aitken extrapolation (see Box 3.3)

$$x_{i+1} = \phi(x_i) = (1.2500000x_i^2 + 1.5625250x_i - 1.9530938)^{1/3}$$

Solution   Table 3.11 shows the iteration using the Aitken extrapolation at every third step (since we need three successive iterates to apply Eq. 3.27) to refine the estimate of the root. Note that the *extrapolated* value becomes the iterate at the next step (shown by italics) and also note that at every third step $\phi(x)$ does not need to be computed (shown by a dash).

Table 3.11

Iteration no.	x	$\phi(x)$	ε_r (%)	x_* (Eq. 3.27)
0	− 1.5	− 1.14073		
1	− 1.14073	− 1.28239	31.49506	
2	− 1.28239	—	11.04697	− 1.24233
3	− 1.24233	− 1.25254	3.224835	
4	− 1.25254	− 1.24915	0.814837	
5	− 1.24915	—	0.270859	− 1.24999
6	− 1.24999	− 1.25	0.067391	
7	− 1.25	**− 1.25**	0.000553	

Comparing with Example 3.3, we note that the convergence is faster (7 iterations instead of 9) and the error is also much smaller.

BOX 3.3 Steffensen's Method from Aitken's Extrapolation

Application of Aitken's extrapolation to a fixed-point iteration scheme, written as $x_{new} = x_{old} + f(x_{old})$, would give rise to the Steffensen's method discussed in the previous subsection.

Substituting $x^{(i+1)} = x^{(i)} + f_i$ and $x^{(i+2)} = x^{(i+1)} + f(x^{(i)} + f_i) = x^{(i)} + f_i + f(x^{(i)} + f_i)$ in Eq. (3.27) written in terms of $x^{(i)}, x^{(i+1)},$ and $x^{(i+2)}$, we get

$$x_{new} = x^{(i+2)} - \frac{(x^{(i+2)} - x^{(i+1)})^2}{x^{(i+2)} - 2x^{(i+1)} + x^{(i)}}$$

$$= x^{(i)} + f_i + f(x^{(i)} + f_i) - \frac{\left[f(x^{(i)} + f_i) \right]^2}{f(x^{(i)} + f_i) - f_i}$$

$$= x^{(i)} - \frac{f_i^2}{f(x^{(i)} + f_i) - f_i}$$

which is the same as Eq. (3.24).

We may apply the extrapolation to other linearly convergent methods like bisection, false position, or the Newton-Raphson method near a multiple root (see Example 3.5), also.

EXERCISE 3.3

1. Solve the equation $x - \cos(x/2) = 0$ by using the fixed-point iteration scheme, $x = \cos(x/2)$. Since the scheme will always converge (why?), try very different initial guesses of the root, e.g., $0, -100, 100$.

2. An iterative scheme to find the square root of a number, c, may be written as $x_{i+1} = \phi(x_i) = 0.5(x_i + c/x_i)$. Find the square root of 19 with an initial guess of 4.5. It can be shown that $|\phi'(x)|$ is greater than 1 for $x < \sqrt{c/3}$. Use a starting guess of 1 and show that convergence is achieved (why?)

3. Solve the problem discussed in Example 3.6 using the secant method but instead of keeping the two most recent estimates, discard the point where the magnitude of the function is larger.

4. Find a root of $f(x) = 1.5x^2 - 5.5x - 6 = 0$ by obtaining the function values at

$x = 2, 3$, and 4, expressing x as a quadratic function $g(f)$, and then putting $f = 0$ in the inverse function, g. Use the three most recent points to repeat till convergence. Try the same technique with -6 replaced by $+6$ in the equation.

5. In Example 3.7, Muller's method was used with the sign of the square root in Eq. (3.22) chosen to make the *correction* in the estimate of the root, small. Re-work this example but choosing the sign of the radical which will give the smaller value of the function at the new estimate of the root.

6. In Example 3.7, Muller's method was used with the three most recent estimates of the root. Re-work this example but discarding the point at which the function has the largest magnitude. Try another option in which the point farthest from the most recent estimate is discarded.

7. An equation with infinite roots is $f(x) = x - \tan x = 0$. What can be said about the approximate location of the roots? Find the first three positive roots by using any of the methods described in this section.

8. Find a non-trivial root of the equation $\sin x - (x/2)^2 = 0$ using bisection method, regula-falsi method, fixed-point method, Newton-Raphson method, and secant method (for bracketing methods, use $x = 1$ and $x = 2$ as the end points). In each case, calculate true relative error (take the true root as 1.93375376) and approximate relative error at each iteration. Terminate the iterations when the approximate relative error is less than 0.01%. Plot both these errors as log (% error) vs. iteration no. for each method.

9. Find a root of $f(x) = -1 + 0.5x + 0.75x^2 - 0.25x^3$ using the Newton-Raphson method starting with $x = 0$. Comments on the results. Repeat with the starting guess of 0.5.

10. It is required to find a real zero of the function $f(x) = 6 - 18x + 14x^2 - 3x^3$. Apply the Newton-Raphson method starting with an initial guess of $x = 1$. If the method encounters some problems, explain the cause of the problem and switch to the secant method using the starting values of 1 and 2. Find the desired zero of the function with an accuracy of 0.2% in the relative approximate error.

11. Find a root of the following equation using Muller's method to an approimate error less than 0.1%:

$$x^4 \sin x - e^x = 0$$

Take three starting values as $1, 2$, and 3.

12. It is known that an angle between π and $3\pi/2$ has its tangent equal to the angle. To obtain this value, we may write $f(x) = x - \tan x = 0$ and use the Newton-Raphson method. During the iterations, it was observed that very

close to the root, two successive iterations had true errors of 0.07752 and 0.02781. Estimate the error at the next iteration. The same problem could be solved using the fixed-point iteration schemes $x = \tan x$ or $x = \tan^{-1}x$. Which of these you would prefer and why? Obtain the root using this scheme with a starting guess of $x = \pi$ and with an approximate error less than 0.01%.

13. Find the root of the polynomial, $x^4 - 2x^3 - 53x^2 + 54x + 504$, by Muller's method with a maximum approximate error of 0.01%.

3.4 COMPLEX ROOTS AND MULTIPLE ROOTS

Although most physical problems involve the determination of real roots, sometimes it is desired to obtain complex roots also. For example, eigenvalues of a real matrix (see Chapter 2) may be complex. If we use complex arithmetic in the computer program (which most scientific programming languages are capable of), the open methods described for obtaining the real roots may be easily extended to equations involving complex roots. (Bracketing methods will not work as the function values will be complex and it would not be possible to bracket the root. There is an algorithm proposed by Lehmer in 1961 that may be used to *encircle* the root, but it will not be discussed here.) Obviously, if the equation involves a polynomial with only real coefficients, we will have to start with an initial guess which is complex to reach the complex root (except the fixed-point method and Muller's method, Eq. (3.22), which may lead to a complex root even with real starting values). Sometimes, however, $f(x)$ may involve terms which lead the iterations to the complex root with real starting guess.

Example 3.10   Find one of the complex roots of the following equation:

$$x^3 - 2.850x^2 + 3.910x - 2.121 = 0$$

to an accuracy of 0.01%, using the fixed-point, Newton-Raphson, and Muller's methods.

Solution   Table 3.12 shows an extract of the iterations using the fixed-point iteration written as

$$x = \sqrt{(x^3 + 3.910x - 2.121)/2.850}$$

to create the possibility of getting a complex root. The error is based on the magnitude of the complex number.

Table 3.12

Iteration no.	x	$\varepsilon_r\,(\%)$	$f(x)$
0	1		0.98924
1	0.98924	1.087667	0.976028
2	0.976028	1.353694	0.959726
⋮	⋮	⋮	⋮
11	0.384786	54.938583	$0 + 0.443082\,i$
12	$0 + 0.443082\,i$	13.157071	$0.314402 + 0.918182\,i$
13	$0.314402 + 0.918182\,i$	54.345855	$0.569453 + 0.951449\,i$
⋮	⋮	⋮	⋮
38	$0.898444 + 1.100135\,i$	0.016361	$0.898761 + 1.099786\,i$
39	**$0.898761 + 1.099786\,i$**	0.004948	—

The iterations approach the root $0.9 + 1.1i$. The other complex root would be its conjugate.

The Newton-Raphson method must start from a complex initial guess to obtain the complex root. Table 3.13 shows the iterations.

Table 3.13

Iteration no.	$x^{(i)}$	f_i	f_i'	$x^{(i+1)}$ (Eq. 3.15)	$\varepsilon_r\,(\%)$
0	i	$0.729 + 2.910\,i$	$0.910 - 5.700\,i$	$0.478 + 0.796\,i$	7.7252
1	$0.478 + 0.796\,i$	$0.103 + 0.985\,i$	$-0.029 - 2.254\,i$	$0.631 + 0.876\,i$	14.0383
2	$0.631 + 0.876\,i$	$0.197 + 0.648\,i$	$-0.795 - 1.675\,i$	$0.730 + 0.913\,i$	7.6342
⋮	⋮	⋮	⋮	⋮	⋮
24	$0.898 + 1.101\,i$	$0.006 - 0.002\,i$	$-2.426 - 0.346\,i$	$0.898 + 1.100\,i$	0.0248
25	$0.898 + 1.100\,i$	$0.005 + 0.000\,i$	$-2.421 - 0.344\,i$	**$0.898 + 1.100\,i$**	0.0090

For Muller's method, we use the three most recent estimates at each iteration and also use the negative sign in the denominator of Eq. (3.22) (with positive sign, the iterations converge to the real root). Table 3.14 shows the iterations (the numbers are shown up to 3 decimal places only but the error computation is done with the more precise values).

Table 3.14

Iteration no.	$x^{(i)}$	f_i	$x^{(i+1)}$ (Eq. 3.22)	$\varepsilon_r (\%)$
	0.000	− 2.121		
	1.000	− 0.061		
0	2.000	2.299	− 13.761	85.466
1	−13.761	− 3201.390	2.170	534.271
2	2.170	3.159	2.806	22.694
3	2.806	8.509	3.705	24.257
4	3.705	24.108	1.768 − 0.617 i	97.908
5	1.768 − 0.617 i	0.476 − 1.744 i	1.483 − 0.766 i	12.182
6	1.483 − 0.766 i	− 0.266 − 1.124 i	1.157 − 0.863 i	15.629
7	1.157 − 0.863 i	− 0.326 − 0.506 i	0.903 − 1.018 i	6.082
8	0.903 − 1.018 i	− 0.032 − 0.176 i	0.892 − 1.098 i	3.818
9	0.892 − 1.098 i	0.017 − 0.009 i	0.900 − 1.100 i	0.476
10	0.900 − 1.100 i	0.000 − 0.000 i	**0.900 − 1.100 i**	0.007

While these methods do provide the complex roots, the use of complex arithmetic leads to much larger computation times than that for real arithmetic. A frequent occurrence of complex roots is in the computation of eigenvalues of a real matrix. Since the equation to be solved is a polynomial with all real coefficients, the complex roots will occur in conjugate pairs. Therefore, instead of finding one root at a time, it would be more practical to find a quadratic factor (which corresponds to a conjugate pair) of the polynomial. This computation would require only real arithmetic and the complex roots, if present, are readily obtained using the formula for the quadratic equation. A recursive algorithm using this idea was first suggested by Bairstow (and later modified by Lin and various other researchers).

3.4.1 Bairstow Method

Before describing the *Bairstow method*, it is helpful to look at its counterpart for obtaining a single root of a polynomial (see Box 3.4 for a quick review of characteristics of the roots of a polynomial).

We express the polynomial as

$$f(x) = \sum_{j=0}^{n} c_j x^j$$

Note that x^j represents x raised to the power j and should not be confused with $x^{(j)}$ which is the value of x at iteration j. *One way* of finding a root is to divide the polynomial by the term $(x - r)$ and obtain a remainder, R, which would be a function of r. If R is zero, r would be a root of the polynomial. If not, we could try changing

BOX 3.4 Roots of a Polynomial

Most polynomials occurring in a practical problem would have real coefficients. Hence, we describe some properties related to the roots of such polynomials.

In standard form, a polynomial is expressed as $p_n(x) = \sum_{j=0}^{n} c_{n-j} x^{n-j}$.

If an nth degree polynomial has k roots, $r_1, r_2, \ldots, r_k$, which are of multiplicity $m_1, m_2, \ldots, m_k$, respectively, then $\sum_{i=1}^{k} m_i = n$.

Considering multiple roots as separate roots, the n roots of the polynomial follow Viéte's formulae, according to which the kth symmetric sum of the roots is given by

$$\sum_{1 \le i_1 < i_2 < \ldots < i_k \le n} r_{i_1} r_{i_2} \ldots r_{i_k} = (-1)^k \frac{c_{n-k}}{c_n}, \text{ for each } k = 1, 2, \ldots, n$$

Thus the sum of all the roots ($k = 1$) is equal to $-c_{n-1}/c_n$, sum of the product of two roots at a time ($k = 2$), i.e., $r_1 r_2 + r_1 r_3 + \cdots + r_1 r_n + r_2 r_3 + r_2 r_4 + \cdots + r_{n-2} r_{n-1} + r_{n-1} r_n$, is equal to c_{n-2}/c_n, and product of all the roots ($k = n$) is equal to $(-1)^n c_0/c_n$.

If a polynomial has all real coefficients, then either all roots are real or there are even number of non-real complex roots which occur in conjugate pairs. Descartes' rule of signs states that the number of positive roots of $p_n(x) = 0$ is either equal to the number of sign changes in the coefficients of $p_n(x)$ (zero coefficients are ignored), or less than that by an *even number* (which represents the number of complex roots occurring in conjugate pairs). Similarly, the number of negative roots of $p_n(x) = 0$ is either equal to the number of sign changes in the coefficients of $p_n(-x)$ or less than that by an even number.

If we define a radius ρ as $1 + \dfrac{\max\limits_{0 \le j < n} |c_j|}{|c_n|}$, Cauchy provided the upper bound of the roots of the polynomial as ρ and the lower bound as $1/\rho_q$, where ρ_q is the radius of the polynomial $q_n(x) = x_n p_n(1/x)$, i.e., $1 + \dfrac{\max\limits_{0 < j \le n} |c_j|}{|c_0|}$. In other words, the magnitude of all roots would be

between $\dfrac{|c_0|}{|c_0| + \max\limits_{0 < j \le n} |c_j|}$ and $1 + \dfrac{\max\limits_{0 \le j < n} |c_j|}{|c_n|}$.

r by an amount Δr in such a way as to make R equal to zero. Writing the quotient as

$$\sum_{j=0}^{n-1} d_{j+1} x^j$$

and the residual as d_0 (in order to enable us to write a recursive relationship), we have

$$\sum_{j=0}^{n} c_j x^j = (x-r)\sum_{j=0}^{n-1} d_{j+1} x^j + d_0 = d_n x^n + \sum_{j=0}^{n-1}\left(d_j - r d_{j+1}\right)x^j \qquad (3.28)$$

Equating the coefficients of different powers of x, we obtain the recursive relations

$$\begin{aligned} d_n &= c_n \\ d_j &= c_j + r d_{j+1}, \quad \text{for } j = n-1 \text{ to } 0 \end{aligned} \qquad (3.29)$$

which provides us with the residual, d_0, as a function of the assumed value of the root, r. If we use the Newton-Raphson method to find Δr from

$$d_0(r) + \Delta r \frac{d}{dr}\left[d_0(r)\right] = d_0(r + \Delta r) = 0 \qquad (3.30)$$

it can be easily verified that we get the same equation (Eq. 3.15) for the iterative scheme, since $d_0(r)$ is same as $f(r)$.

In the Bairstow method, instead of the synthetic division by $(x-r)$, we carry out a division by the quadratic term $(x - r_1)(x - r_2)$, and aim to make the remainder zero. Irrespective of whether these two roots are real or complex conjugates, the quadratic term may be written in terms of real coefficients as $(x^2 - \alpha_1 x - \alpha_0)$, the quotient as

$$\sum_{j=0}^{n-2} d_{j+2} x^j$$

and the remainder as $d_0 + d_1(x - \alpha_1)$ (note that if the two roots are complex conjugates, α_0 will be negative). We then have

$$\begin{aligned} \sum_{j=0}^{n} c_j x^j &= (x^2 - \alpha_1 x - \alpha_0)\sum_{j=0}^{n-2} d_{j+2} x^j + d_1(x - \alpha_1) + d_0 \\ &= d_n x^n + (d_{n-1} - \alpha_1 d_n)x^{n-1} + \sum_{j=0}^{n-2}(d_j - \alpha_1 d_{j+1} - \alpha_0 d_{j+2})x^j \end{aligned} \qquad (3.31)$$

giving us the recursive relations

$$\begin{aligned} d_n &= c_n \\ d_{n-1} &= c_{n-1} + \alpha_1 d_n \\ d_j &= c_j + \alpha_1 d_{j+1} + \alpha_0 d_{j+2}, \quad \text{for } j = n-2 \text{ to } 0 \end{aligned} \qquad (3.32)$$

We now aim at making the remainder zero by making d_0 and d_1 equal to zero (note that both d_0 and d_1 are functions of the guess values of α_0 and α_1). A Newton-Raphson scheme could be written as (more details on solving multivariate nonlinear equations are provided in Section 3.7)

$$d_0(\alpha_0,\alpha_1) + \frac{\partial d_0(\alpha_0,\alpha_1)}{\partial \alpha_0} \Delta\alpha_0 + \frac{\partial d_0(\alpha_0,\alpha_1)}{\partial \alpha_1} \Delta\alpha_1 = d_0(\alpha_0 + \Delta\alpha_0, \alpha_1 + \Delta\alpha_1) = 0$$

$$d_1(\alpha_0,\alpha_1) + \frac{\partial d_1(\alpha_0,\alpha_1)}{\partial \alpha_0} \Delta\alpha_0 + \frac{\partial d_1(\alpha_0,\alpha_1)}{\partial \alpha_1} \Delta\alpha_1 = d_1(\alpha_0 + \Delta\alpha_0, \alpha_1 + \Delta\alpha_1) = 0$$

$$(3.33)$$

to iteratively improve the guessed values till the changes, $\Delta\alpha_0$ and $\Delta\alpha_1$, are very small. In order to compute the partial derivatives in Eq. (3.33), it is easy to see from Eq. (3.32) that

$$\frac{\partial d_n}{\partial \alpha_0} = 0$$

$$\frac{\partial d_{n-1}}{\partial \alpha_0} = 0$$

$$(3.34)$$

$$\frac{\partial d_j}{\partial \alpha_0} = d_{j+2} + \alpha_0 \frac{\partial d_{j+2}}{\partial \alpha_0} + \alpha_1 \frac{\partial d_{j+1}}{\partial \alpha_0}, \quad \text{for } j = n-2 \text{ to } 0$$

and, similarly,

$$\frac{\partial d_n}{\partial \alpha_1} = 0$$

$$\frac{\partial d_{n-1}}{\partial \alpha_1} = d_n$$

$$(3.35)$$

$$\frac{\partial d_j}{\partial \alpha_1} = d_{j+1} + \alpha_0 \frac{\partial d_{j+2}}{\partial \alpha_1} + \alpha_1 \frac{\partial d_{j+1}}{\partial \alpha_1}, \quad \text{for } j = n-2 \text{ to } 0$$

These equations could be used to obtain the derivatives required in Eq. (3.33) by a recursive algorithm applied separately to Eqs (3.34) and (3.35). However, a close look at these two equations clearly shows the equivalence of

$$\frac{\partial d_{j-1}}{\partial \alpha_0} \quad \text{and} \quad \frac{\partial d_j}{\partial \alpha_1}$$

and the *same* recursive algorithm could be applied to both the equations. Defining

$$\delta_j = \frac{\partial d_{j-1}}{\partial \alpha_0} = \frac{\partial d_j}{\partial \alpha_1}$$

we get the following algorithm for computing the derivatives required in Eq. (3.33):

$$\delta_{n-1} = d_n$$
$$\delta_{n-2} = d_{n-1} + \alpha_1 \delta_{n-1}$$
$$\delta_j = d_{j+1} + \alpha_1 \delta_{j+1} + \alpha_0 \delta_{j+2}, \quad \text{for } j = n-3 \text{ to } 0$$

Equation (3.33) can then be written as

$$\delta_1 \Delta\alpha_0 + \delta_0 \Delta\alpha_1 = -d_0$$
$$\delta_2 \Delta\alpha_0 + \delta_1 \Delta\alpha_1 = -d_1 \tag{3.36}$$

and solved to obtain the desired increments. It can be shown that the coefficient matrix for the set of linear equations in Eq. (3.36) is not singular and, therefore, a solution exists. After the iterations converge, the two roots, r_1 and r_2, are obtained from

$$r_{1,2} = 0.5\left(\alpha_1 \pm \sqrt{\alpha_1^2 + 4\alpha_0}\right) \tag{3.37}$$

Since the coefficients of the quotient (d_j) are easily obtainable using Eq. (3.32), the quadratic factorization may be repeated (till the quotient becomes quadratic or linear) to find all the roots of the polynomial, $f(x)$.

Example 3.11   Find all the roots of the equation
$$x^5 - 5.05000x^4 + 12.20000x^3 - 16.48000x^2 + 12.56440x - 4.28442 = 0$$
using the Bairstow method.

Solution   Table 3.15 shows the iterations.

Table 3.15

j	Iteration 1 $\alpha_0 = -1$ $\alpha_1 = 1$		Iteration 2 $\alpha_0 = -1.174$ $\alpha_1 = 1.467$		Iteration 3 $\alpha_0 = -1.582$ $\alpha_1 = 1.936$		Iteration 4 $\alpha_0 = -1.986$ $\alpha_1 = 2.196$		Iteration 5 $\alpha_0 = -2.018$ $\alpha_1 = 2.198$		Iteration 6 $\alpha_0 = -2.02$ $\alpha_1 = 2.2$	
	d_j	δ_j	d_j	δ_j	d_j	δ_j	d_j	δ_j	d_j	δ_j	d_j	δ_j
0	1.130	-2.096	0.484	-0.283	0.204	0.154	0.010	-0.363	0.001	-0.475	0.000	
1	0.134	0.870	0.202	0.864	0.138	0.603	0.016	0.294	0.001	0.206	0.000	
2	-5.280	3.100	-3.809	1.491	-2.669	0.728	-2.145	0.516	-2.123	0.460	-2.121	
3	7.150	-3.050	5.770	-2.116	4.589	-1.177	3.947	-0.658	3.913	-0.653	3.910	
4	-4.050	1.000	-3.583	1.000	-3.114	1.000	-2.854	1.000	-2.852	1.000	-2.850	
5	1.000		1.000		1.000		1.000		1.000		1.000	

	$\Delta\alpha_0 = -0.174$ $\Delta\alpha_1 = 0.467$		$\Delta\alpha_0 = -0.407$ $\Delta\alpha_1 = 0.470$		$\Delta\alpha_0 = -0.404$ $\Delta\alpha_1 = 0.260$		$\Delta\alpha_0 = -0.032$ $\Delta\alpha_1 = 0.002$		$\Delta\alpha_0 = -0.002$ $\Delta\alpha_1 = 0.002$			

After 5 iterations, the iterations have converged to $\alpha_0 = -2.02$ and $\alpha_1 = 2.2$. Using Eq. (3.37), the two roots are obtained as $1.1 \pm 0.9i$. The sixth iteration is performed to get the coefficients in the quotient obtained after dividing the original function, $f(x)$, by $(x^2 - 2.2x + 2.02)$. The *deflated* polynomial is thus obtained as $x^3 - 2.850$ $x^2 + 3.910\,x - 2.121$. Bairstow's method can again be applied to this polynomial, starting with an initial guess $(-2,2)$ as shown in Table 3.16.

Table 3.16

j	Iteration 1 $\alpha_0 = -2$ $\alpha_1 = 2$		Iteration 2 $\alpha_0 = -1.867$ $\alpha_1 = 1.702$		Iteration 3 $\alpha_0 = -2.016$ $\alpha_1 = 1.811$		Iteration 4 $\alpha_0 = -2.020$ $\alpha_1 = 1.800$		Iteration 5 $\alpha_0 = -2.02$ $\alpha_1 = 1.800$	
	d_j	δ_j	d_j	δ_j	d_j	δ_j	d_j	δ_j	d_j	δ_j
0	−0.001	0.510	0.174	−0.837	−0.004	−0.606	0.000	−0.671	0.000	
1	0.210	1.150	0.089	0.553	0.012	0.772	0.000	0.749	0.000	
2	−0.850	1.000	−1.148	1.000	−1.039	1.000	−1.050	1.000	−1.050	
3	1.000		1.000		1.000		1.000		1.000	

$\Delta\alpha_0 = 0.133$	$\Delta\alpha_0 = -0.150$	$\Delta\alpha_0 = -0.003$	$\Delta\alpha_0 = -0.0003$	
$\Delta\alpha_1 = -0.298$	$\Delta\alpha_1 = 0.110$	$\Delta\alpha_1 = -0.011$	$\Delta\alpha_1 = 0.0003$	

After 4 iterations, the iterations have converged to $\alpha_0 = -2.02$ and $\alpha_1 = 1.8$. Using Eq. (3.37), the two roots are obtained as $1.1 \pm 0.9i$. The fifth iteration is performed to get the coefficients in the quotient obtained after dividing the deflated function by $(x^2 - 1.8x + 2.02)$. The deflated polynomial is, obviously, a linear function given by $x - 1.05$. The fifth root is therefore 1.05.

3.4.2 Multiple Roots

A function $f(x)$ has a root, ξ, of multiplicity m, if there exists a nonzero real number, c, such that

$$\lim_{x \to \xi} \frac{|f(x)|}{|x - \xi|^m} = c$$

As we saw in Example 3.5, the otherwise quadratic convergence of Newton-Raphson method was reduced to a mere linear convergence if the desired root is a double root. Let the given function, $f(x)$, have a root (ξ) of multiplicity m (note that the function value as well as its first $(m - 1)$ derivatives are zero at $x = \xi$). We can then write

$$f(x) = A(x)(x - \xi)^m$$

where $A(x)$ is such that it does not have a zero at ξ. We can show that the Newton-Raphson scheme will be linearly convergent as follows:

$$x^{(i+1)} = x^{(i)} - \frac{f_i}{f_i'} = x^{(i)} - \frac{A(x^{(i)})(x^{(i)} - \xi)^m}{mA(x^{(i)})(x^{(i)} - \xi)^{m-1} + A'(x^{(i)})(x^{(i)} - \xi)^m}$$

from which

$$x^{(i+1)} - \xi = x^{(i)} - \xi - \frac{A(x^{(i)})(x^{(i)} - \xi)^m}{mA(x^{(i)})(x^{(i)} - \xi)^{m-1} + A'(x^{(i)})(x^{(i)} - \xi)^m}$$

$$= (x^{(i)} - \xi)\left[1 - \frac{1}{m + A'(x^{(i)})(x^{(i)} - \xi)/A(x^{(i)})}\right]$$

Therefore, in the limit, when the iterations converge to the root, we have a linear convergence

$$e^{(i+1)} = \left(1 - \frac{1}{m}\right)e^{(i)}$$

Note that for a double root, the asymptotic error constant is equal to 1/2, the same as the bisection method. In order to maintain the quadratic convergence behaviour, various modifications have been suggested. These methods aim at transforming the given function, $f(x)$, in such a way that the transformed function has a simple root at ξ.

If we know the multiplicity, m, we define a transformed function $[f(x)^{1/m}]$ and find its root using the Newton-Raphson method. We will get a quadratic convergence since the modified function has a simple root at ξ. The iterative scheme is written as

$$x^{(i+1)} = x^{(i)} - \frac{[f_i]^{1/m}}{\left[d\{f(x)\}^{1/m} / dx\right]_{x=x^{(i)}}} = x^{(i)} - m\frac{f_i}{f_i'} \tag{3.38}$$

Note that the asymptotic error constant which was earlier obtained as

$$\left|\frac{f''(\xi)}{2f'(\xi)}\right|$$

will now be based on the $(1/m)^{\text{th}}$ power of f.

Equation (3.38) will work well but assumes the fact that the multiplicity of the root is known. In some cases, we may know that there is a multiple root, but may not know its multiplicity. We can then use the fact that the derivative of the function will have a root of multiplicity one smaller than that of the function and may be written as $f'(x) = B(x)(x - \xi)^{m-1}$. Defining a new function, $f(x)/f'(x)$, it is obvious that it will have a simple root at ξ. The iterative scheme is then written as

$$x^{(i+1)} = x^{(i)} - \frac{f_i / f_i'}{d(f/f')/dx\big|_{x=x^{(i)}}} = x^{(i)} - \frac{f_i f_i'}{(f_i')^2 - f_i f_i''} \tag{3.39}$$

Example 3.12 In Example 3.5, we obtained a root (near $x = 1$) of the following equation:

$$x^3 - 1.25020000x^2 - 1.56249999x + 1.95343750 = 0$$

using the Newton-Raphson scheme, and saw that it was linearly convergent. Obtain the root using the modified schemes [Eqs (3.38) and (3.39)].

Solution Table 3.17 shows the iterations, with $m = 2$, since we know that there is a *double* root.

Table 3.17

Iteration no.	$x^{(i)}$	f_i	f_i'	$x^{(i+1)}$ (Eq. 3.38)	ε_r (%)
0	1	0.140738	− 1.0629	1.264818	20.93724
1	1.264818	0.000545	0.074243	1.250143	1.173841
2	1.250143	− 7.8E −09	0.000216	**1.250216**	0.005778

Note the much faster convergence compared to the unmodified Newton-Raphson method.

Had we not known that the multiplicity of the root is 2, we could use Eq. (3.39) as in Table 3.18.

Table 3.18

Iteration no.	$x^{(i)}$	f_i	f_i'	f_i''	$x^{(i+1)}$ (Eq. 3.39)	ε_r (%)
0	1	0.140738	− 1.0629	3.4996	1.23475	19.01193
1	1.23475	0.000585	− 0.07605	4.908098	1.250052	1.224103
2	1.250052	− 6.7E-09	− 0.00024	4.99991	**1.250034**	0.001403

In most cases, however, we will not know about the presence of a multiple root. We could, of course, use Eq. (3.39) in such cases. Though it would compute a simple root also, it would be inefficient because of the additional computation of the second derivative.

EXERCISE 3.4

1. Find all the roots of the equation $x^4 - 2x^3 - 53x^2 + 54x + 504 = 0$ by the Bairstow's method to an accuracy of 0.01%.

2. Find the multiple root of the equation $(x - 0.5236)(0.8660 \sin x - 0.5000 \cos x)$ near $x = 0.5$. (The root, clearly, is 0.5236). First use the Newton-Raphson method, then use the modified Newton-Raphson method given the fact that the multiplicity is 2. Also, assuming that the multiplicity is unknown, use the iterative scheme on the function $f(x) / f'(x)$.

3. Obtain a root of the equation $x^2 + 1 = 0$ using the Muller's method, starting with points $x_0 = 0$, $x_1 = 1$, and $x_2 = 2$.

4. Find the roots of the polynomial using Bairstow's method to an approximate error of 0.01%:

$$x^3 - 21x^2 + 129x - 234 = 0$$

3.5 DESIGN OF METHODS OF ARBITRARY ORDER

The methods discussed in this chapter were broadly divided into bracketing methods and open methods. While the basic methodology is different for these methods, all of the methods generate a sequence which (hopefully) converges to the root. In the bracketing method, the generated sequence always lies within an interval which contains the desired root and the aim is to reduce this interval to some pre-set small value. In the open methods, the aim is to generate a sequence in such a way that the distance between the root and the terms of the sequence gets progressively smaller. The rate at which the convergence is achieved is important from the point of view of computational efficiency and it would be beneficial to have methods of higher order of convergence. However, as we have seen, the Newton-Raphson method, though of higher order of convergence, may not be as efficient if the effort required in computing the derivative of the function is large. Still, a general technique to generate methods of arbitrary order of convergence would be interesting to study. There are a number of techniques which have been used to achieve a cubic or higher order of convergence. In this section we describe one such technique, proposed by Schroder.

Assuming that the function is smooth (infinitely differentiable) and writing the Taylor's series about $x^{(i)}$, we get

$$f(x^{(i)} + h) = f_i + \sum_{j=1}^{\infty} \frac{f^{[j]}(x^{(i)})}{j!} h^j \tag{3.40}$$

in which $f^{[j]}(x^{(i)})$ represents the jth derivative of the function, evaluated of $x^{(i)}$.

We assume that the first derivative of the function does not vanish near the desired root. If it does, i.e., the root is a multiple root, we may transform the function as discussed in the previous section. Since the objective is to make the function value equal to 0, we write

$$\frac{f_i}{f_i'} = -h - \frac{1}{f_i'}\sum_{j=2}^{\infty}\frac{f_i^{[j]}}{j!}h^j = -h - \sum_{j=2}^{\infty}d_j h^j \qquad (3.41)$$

in which d_j represents the jth derivative divided by the first derivative and factorial j. Using series inversion (see Box 3.5), we may now write an iterative scheme of order p as

$$x^{(i+1)} = x^{(i)} - \sum_{j=1}^{p-1}c_j\left(\frac{f_i}{f_i'}\right)^j \qquad (3.42)$$

with $c_1 = 1; c_2 = d_2; c_3 = 2d_2^2 - d_3; \dots$ Clearly, for quadratic convergence, $p = 2$, $c_1 = 1$, and we get the Newton-Raphson iteration (which was shown to have a

BOX 3.5 Inversion of Series

Given a series

$$y = -d_1 x - d_2 x^2 - \cdots$$

we can compute the coefficients of the inverse series

$$x = c_1 y + c_2 y^2 + \cdots$$

by substituting the inverse series in the forward series and equating the coefficients of various powers of y.

It is easy to show that

$$c_1 = \frac{1}{d_1};$$

$$c_2 = \frac{d_2}{d_1^3};$$

$$c_3 = \frac{2d_2^2 - d_1 d_3}{d_1^5};$$

$$\vdots$$

from which it is seen that it would simplify the computations if d_1 is taken as 1. That is the reason we have used a division by f_i' in Eq. (3.41).

quadratic convergence). To increase the order of convergence, we simply add more terms. For example, a cubic convergence is achieved by the following:

$$x^{(i+1)} = x^{(i)} - \frac{f_i}{f_i'} - \frac{f_i''}{2f_i'}\left(\frac{f_i}{f_i'}\right)^2 \tag{3.43}$$

If we truncate the series in Eq. (3.41) after one term, $j = 2$, we get a quadratic equation in h, which could be solved to obtain another form of a cubic-convergent method. It is known as the Euler scheme and is given by

$$x^{(i+1)} = x^{(i)} - \frac{2f_i}{\sqrt{(f_i')^2 - 2f_if_i'' + f_i'}}$$

Various other forms could be derived by using the fact that near the root $ff'' \ll (f')^2$ [one of these forms would be identical to Eq. (3.43)].

Higher order methods could be written but are not very common unless the higher order derivatives are easy to compute (e.g., a polynomial). It should be noted that additional computational effort is involved in evaluating the derivatives and the overall efficiency of the method may not be as good as a lower order method.

While Eq. (3.42) does produce a higher order method, it suffers from the drawback that derivatives of order higher than one are required to be computed. Various alternatives are available to attain cubic convergence using the function and its first derivative only. For example, it can be shown that we get a cubic-convergent method if in the Newton-Raphson method we replace the derivative at the point $x^{(i)}$, by an *average* derivative over the interval $(x^{(i)}, x^{(i+1)})$ in one of the following alternative forms:

$$x^{(i+1)} = x^{(i)} - \frac{f_i}{0.5\left(f_i' + f'\left(x^{(i)} - \frac{f_i}{f_i'}\right)\right)} \tag{3.44a}$$

or

$$x^{(i+1)} = x^{(i)} - \frac{f_i}{f'\left(x^{(i)} - 0.5\frac{f_i}{f_i'}\right)} \tag{3.44b}$$

Similarly, cubic convergence is achieved in the Potra-Ptak scheme

$$x^{(i+1)} = x^{(i)} - \frac{f_i + f\left(x^{(i)} - \frac{f_i}{f_i'}\right)}{f_i'} \tag{3.45}$$

Any of the methods listed here could be used to obtain a faster convergence at the expense of more function or derivative evaluations. We believe that methods with more than cubic convergence are not useful for a general problem and leave it to the reader, if interested, to derive them using Eq. (3.42).

EXERCISE 3.5

1. Show that the iteration scheme given by Eq. (3.43) has a cubic order of convergence.
2. Use the iteration scheme of Eq. (3.43) to obtain the root of the function $f(x) = x^3 - 1.2500000x^2 - 1.5625250x + 1.9530938$ near $x = -1.5$.

3.6 INTRODUCTION TO SYSTEM OF NONLINEAR EQUATIONS

A number of times we encounter a system of equations involving several unknown quantities. If these equations are all linear, we could use the techniques described in the previous chapter. However, if even one out of these equations is nonlinear, those techniques will generally not be applicable. In this section, we introduce the concept of a system of nonlinear equations and discuss the complexities as compared to a single non-linear equation. The next section describes some of the methods used for solving such systems.

Going back to the battle between Batman and Joker, let us assume that Batman is circling the Vampire State Building, keeping a distance of 2 km. We also assume that Joker is following a path such that at all times, he is at a distance of x_1 km east of the building and x_2 km north of the building with $x_1^2 - x_2^4 = 1$. Since Batman wants to catch Joker, he would like to know at which point he should wait so that he is able to complete his task. The problem could be formulated as a set of two nonlinear equations as shown below:

$$\begin{aligned}
x_1^2 + x_2^2 = 2^2 &\Rightarrow f_1(x_1, x_2) = x_1^2 + x_2^2 - 4 = 0 \\
x_1^2 - x_2^4 = 1 &\Rightarrow f_2(x_1, x_2) = x_1^2 - x_2^4 - 1 = 0
\end{aligned} \tag{3.46}$$

As we saw for a linear set of equations, these two equations may or may not have a solution. However, the nonlinearity makes it difficult to perform an analysis similar to that done for the linear system. Even if a solution exists, it may not be physically correct (e.g., a complex number in this case or a negative value for, say, mass of an object). We will, therefore, not consider analysing the system to determine the existence or otherwise of a solution and would proceed directly to the solution.

Figure 3.9 shows a contour plot of the two functions listed in Eq. (3.46). Clearly, the *four* solutions are the points of intersection of the two thick lines representing the contours of zero values. However, in most practical problems, it would not be possible to determine the total number of solutions. On the other hand, for most such problems, we would have a fairly good idea about the approximate location of *a solution* and would want to find only that particular solution. For some cases, it may be possible to reduce the system of equations to a single nonlinear equation

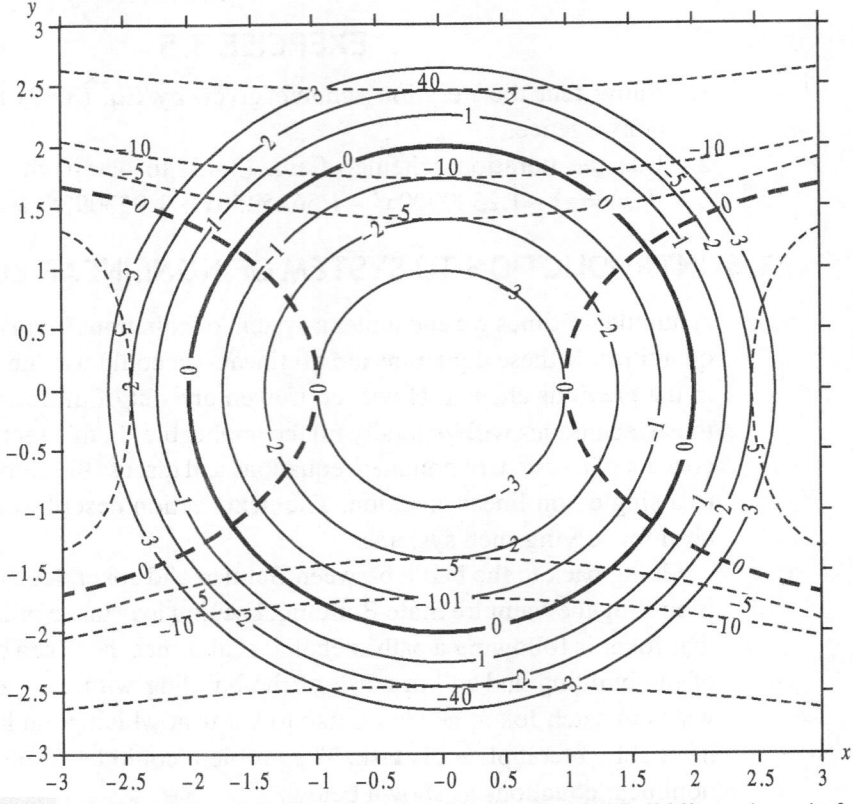

Fig. 3.9 Contour plot of the function listed in Eq. (3.46). Solid lines (———) f_1, dashed lines (- - - -) f_2.

which can then be solved using any of the techniques discussed earlier in this chapter. For example, Eq. (3.46) may be written as $3 - x_2^2 - x_2^4 = 0$, which may be analytically solved to obtain

$$x_2^2 = \frac{-1 \pm \sqrt{13}}{2} = 1.303$$

(ignoring the non-physical negative value). Thus $x_2 = \pm 1.141$ and $x_1 = \pm 1.642$, as may be seen in the figure. Again, for a general case, it is not possible to manipulate the given equations into a single equation involving one variable only. We, therefore, need to look at methods which are more generally applicable.

In the previous chapter, we looked at various methods of solving a system of linear equations and earlier in this chapter, we discussed the solution of a single nonlinear equation. It would appear that a combination of these techniques should work for the solution of a system of nonlinear equations in which n equations (at least one of them being nonlinear) have to be solved for n unknown parameters. However, not all methods which apply to a single nonlinear equation, could be extended to a system of equations. For example, taking the simplest case with two unknown quantities (x_1 and x_2) and two equations $[f_1(x_1,x_2)=0, f_2(x_1,x_2)=0]$, it is readily seen that the bracketing methods will *not* work. For a well behaved function of a single variable, there would be a root between two points at which the function has opposite values. However, if we consider the extension to two-dimensions and choose a rectangle in such a way that both the functions f_1 and f_2 show a change of sign within this rectangle, the only thing we are assured of is that zeros of both f_1 and f_2 will lie within this rectangle. However, there is no guarantee that these *contours* of zero values of f_1 and f_2 will intersect within this rectangle. For example, in Fig. 3.9, if we choose a square spanning 1 to 1.5 in both north and east directions, the function values at the four corners are computed as:

$$x_1 = 1, x_2 = 1 \implies f_1 = -2, \quad f_2 = -1$$
$$x_1 = 1.5, x_2 = 1 \implies f_1 = -0.75, \quad f_2 = 0.25$$
$$x_1 = 1.5, x_2 = 1.5 \implies f_1 = 0.5, \quad f_2 = -3.8125$$
$$x_1 = 1, x_2 = 1.5 \implies f_1 = -0.75, \quad f_2 = -5.0625$$

Since both f_1 and f_2 show a change of sign, they would have zero contours within this square. However, as we have already computed, the solution of the system of equations does not lie within this square. Thus, the bracketing methods would not work on a set of equations since the functions and their zero-value contours are all independent. The open methods, on the other hand, should work and are described in the next section.

3.7 SOLUTION METHODS

Out of the various methods described for a single nonlinear equation, the three most commonly used for a system of nonlinear equation are, the fixed-point iteration method, the Newton-Raphson method, and the secant method. Since the basic philosophy of these methods has already been described in detail, we provide a relatively brief account here. We assume that there are n equations in n unknown variables and denote these as

$$f_1(x_1, x_2, ..., x_n) = 0$$
$$f_2(x_1, x_2, ..., x_n) = 0$$
$$\vdots$$
$$f_n(x_1, x_2, ..., x_n) = 0$$
(3.47)

3.7.1 Fixed-point Iteration

Equation (3.47) may be written as

$$x_1 = \phi_1(x_1, x_2, ..., x_n)$$
$$x_2 = \phi_2(x_1, x_2, ..., x_n)$$
$$\vdots$$
$$x_n = \phi_n(x_1, x_2, ..., x_n)$$
(3.48)

Using a concise notation with vector x representing $(x_1, x_2, ..., x_n)^T$, and the vector ϕ representing $(\phi_1, \phi_2, ..., \phi_n)^T$, the sequence of iteration is written as $x^{(i+1)} = \phi(x^{(i)})$ and the iterations are stopped when some norm of the approximate error vector $\varepsilon^{(i+1)} = |x^{(i+1)} - x^{(i)}|$ or the relative approximate error falls below the desired error tolerance (generally the maximum norm is used). We would now look at the conditions under which this method converges.

The true error at iteration $(i + 1)$ could be written as (note that the true solution ξ is also a vector)

$$e^{(i+1)} = \xi - x^{(i+1)} = \phi(\xi) - \phi(x^{(i)})$$
(3.49)

As done earlier, we assume that partial derivatives of $\phi(x)$, with respect to all the variables x, exist. It is then possible to write the components of the error vector as

$$e_1^{(i+1)} = \xi_1 - x_1^{(i+1)} = \phi_1(\xi_1, \xi_2, ..., \xi_n) - \phi_1(x_1^{(i)}, x_2^{(i)}, ..., x_n^{(i)})$$

$$= \left.\frac{\partial \phi_1}{\partial x_1}\right|_{\hat{x}_1} e_1^{(i)} + \left.\frac{\partial \phi_1}{\partial x_2}\right|_{\hat{x}_2} e_2^{(i)} + \cdots + \left.\frac{\partial \phi_1}{\partial x_n}\right|_{\hat{x}_n} e_n^{(i)}$$
(3.50)

where $\hat{x}_j$ are located in the appropriate intervals, i.e., $(\xi_j, x_j^{(i)})$. Similar expressions for the other components may be easily written. Clearly, if

$$\left|\frac{\partial \phi_j}{\partial x_1}\right| + \left|\frac{\partial \phi_j}{\partial x_2}\right| + \cdots + \left|\frac{\partial \phi_j}{\partial x_n}\right| < 1, \quad \forall j \text{ from } 1 \text{ to } n$$
(3.51)

the maximum component of the error (i.e. the maximum norm) at any iteration would be smaller than the maximum norm of the error at the previous iteration and convergence is guaranteed. It is also seen that even without this condition being fulfilled, the iterations may converge if the positive and negative errors cancel out in Eq. (3.50). If we do not use the maximum norm, we could write the vector form of Eq. (3.50) as

$$e^{(i+1)} = Je^{(i)}$$

in which J is the Jacobian matrix defined by

$$J_{i,j} = \frac{\partial \phi_i}{\partial x_j}$$

Any consistent norm of the Jacobian could be used to show that for convergence, $\|J\| < 1$. Equation (3.51), clearly, is obtained when the row-sum norm is used. The following example solves the set of Eq. (3.46) to enable Batman to position himself at an appropriate location.

Example 3.13   Using the fixed-point iteration method, solve the set of Eq. (3.46) to a maximum relative error of 0.1% starting with an initial guess of (1.5,1.5).

Solution   The tables below show the iterations using the following scheme:

$$x_1 = \sqrt{4 - x_2^2}$$

$$x_2 = \left(x_1^2 - 1\right)^{1/4}$$

As described in the iterative solution of a system of linear equations (Section 2.2.2), two different methodologies could be adopted for the iterations: (i) computing all the variables at an iteration using the values of all variables at the previous iteration (similar to Jacobi), or (ii) computing all the variables at an iteration using the most recent estimates of all variables (similar to Gauss-Seidel). Table 3.19 uses scheme (i) while Table 3.20 uses scheme (ii).

Table 3.19

Iteration no.	x_1	x_2	$\phi_1(x_1,x_2)$	$\phi_2(x_1, x_2)$	ε_{1r} (%)	ε_{2r} (%)	max
0	1.5000	1.5000	1.3229	1.0574	−13.3893	−41.8612	41.8612
1	1.3229	1.0574	1.6976	0.9306	22.0754	−13.6219	22.0754
2	1.6976	0.9306	1.7703	1.1713	4.1048	20.5466	20.5466
3	1.7703	1.1713	1.6212	1.2086	−9.2000	3.0929	9.2000

(contd.)

(contd.)

4	1.6212	1.2086	1.5935	1.1296	−1.7368	−6.9975	6.9975
5	1.5935	1.1296	1.6505	1.1138	3.4521	−1.4147	3.4521
6	1.6505	1.1138	1.6611	1.1459	0.6426	2.7951	2.7951
7	1.6611	1.1459	1.6392	1.1517	−1.3378	0.5062	1.3378
8	1.6392	1.1517	1.6351	1.1397	−0.2502	−1.0557	1.0557
9	1.6351	1.1397	1.6435	1.1374	0.5116	−0.1995	0.5116
10	1.6435	1.1374	1.6451	1.1421	0.0955	0.4076	0.4076
11	1.6451	1.1421	1.6419	1.1429	−0.1966	0.0758	0.1966
12	1.6419	1.1429	1.6413	1.1411	−0.0367	−0.1561	0.1561
13	1.6413	1.1411	1.6425	1.1408	0.0754	−0.0292	0.0754

Table 3.20

Iteration no.	x_1	x_2	$\phi_1(x_1, x_2)$	$\phi_2(x_1, x_2)$*	ε_{1r} (%)	ε_{2r} (%)	max
0	1.5000	1.5000	1.3229	0.9306	−13.3893	−61.1855	61.1855
1	1.3229	0.9306	1.7703	1.2086	25.2741	23.0040	25.2741
2	1.7703	1.2086	1.5935	1.1138	−11.0965	−8.5112	11.0965
3	1.5935	1.1138	1.6611	1.1517	4.0725	3.2872	4.0725
4	1.6611	1.1517	1.6351	1.1374	−1.5913	−1.2573	1.5913
5	1.6351	1.1374	1.6451	1.1429	0.6066	0.4831	0.6066
6	1.6451	1.1429	1.6413	1.1408	−0.2334	−0.1853	0.2334
7	1.6413	1.1408	1.6427	1.1416	0.0895	0.0711	0.0895

* The updated value of x_1, i.e., $\phi_1(x_1, x_2)$, is used to compute $\phi_2(x_1, x_2)$.

The solution obtained is quite close to the exact solution but the second scheme is seen to have a faster convergence. In general, it has been observed to work better and is preferred over the first scheme. If we use a different starting point, the iterations may converge slowly or may not even converge. For example, starting with (1,1), we require 14 iterations in the first scheme and only 6 in the second; and if we start with an initial guess of (2,2), the iterations lead to complex numbers in both schemes.

Due to the similarity of this scheme with the Jacobi and Gauss-Seidel methods, and considering that the linear equations may be thought of as a special case of nonlinear equations, we could apply the convergence criteria given by Eq. (3.51) to a system of linear equations and show that a diagonally dominant coefficient matrix would result in convergence of iterations (as already established in Section 2.2.2.1).

3.7.2 Newton Method

As before, starting from an initial guess, we move along the *tangent* to estimate the location of the zeros. Expanding the functions about the estimate at the ith iteration and ignoring higher order terms, the first equation in the set of Eq. (3.47) is written as (note that the iteration counter is changed to k since we would be using i and j to represent the rows and column of matrices):

$$f_1(x^{(k+1)}) \approx f_1(x^{(k)}) + \sum_{j=1}^{n} \left. \frac{\partial f_1}{\partial x_j} \right|_{x^{(k)}} \left(x_j^{(k+1)} - x_j^{(k)} \right) = 0$$

Similar expressions for other functions are written and the following set of *linear equations* is obtained as

$$J(x^{(k)})\Delta x^{(k+1)} = -f(x^{(k)}) \qquad (3.52)$$

in which J is the Jacobian matrix defined by

$$J_{i,j} = \frac{\partial f_i(x)}{\partial x_j}$$

and Δx is the vector of desired change in the vector $x^{(k)}$ which is expected to bring the functions f closer to zero. Equation (3.52) are solved using any of the techniques described in the previous chapter. The iterations are stopped when the norm of the vector Δx becomes less than a pre-specified tolerance. Sometimes, the stopping criterion is based on the magnitude of the functions rather than the change in variables since our intent is to bring the function values close to zero. For a small system, inversion of the Jacobian may not be computationally expensive and the iterative scheme may be written as

$$x^{(k+1)} = x^{(k)} + \Delta x^{(k+1)} = x^{(k)} - J^{-1}(x^{(k)})f(x^{(k)}) \qquad (3.53)$$

For example, in the 2×2 system of equations related to the Batman problem, we have

$$\begin{Bmatrix} x_1^{(k+1)} \\ x_2^{(k+1)} \end{Bmatrix} = \begin{Bmatrix} x_1^{(k)} \\ x_2^{(k)} \end{Bmatrix} - \frac{1}{\left(\dfrac{\partial f_1}{\partial x_1}\dfrac{\partial f_2}{\partial x_2} - \dfrac{\partial f_1}{\partial x_2}\dfrac{\partial f_2}{\partial x_1} \right)_{\left(x_1^{(k)}, x_2^{(k)} \right)}} \begin{bmatrix} \dfrac{\partial f_2}{\partial x_2} & -\dfrac{\partial f_1}{\partial x_2} \\ -\dfrac{\partial f_2}{\partial x_1} & \dfrac{\partial f_1}{\partial x_1} \end{bmatrix}_{\left(x_1^{(k)}, x_2^{(k)} \right)} \begin{Bmatrix} f_1\left(x_1^{(k)}, x_2^{(k)} \right) \\ f_2\left(x_1^{(k)}, x_2^{(k)} \right) \end{Bmatrix}$$

$$(3.54)$$

However, for most practical cases, the number of variables would be larger and the inversion would not be efficient. If the evaluation of Jacobian is expensive, one may opt for *updating* the Jacobian at every, say, fifth iteration and use an LU decomposition technique to achieve economy (since the same decomposition is used for 5 iteration steps). We will not discuss these issues in this book.

Example 3.14   Using the Newton method, solve the set of Eq. (3.46) to a maximum relative error of 0.1% starting with an initial guess of (2,2).

Solution
Table 3.21 shows the iterations and the much faster convergence compared to the fixed-point method [Eq. (3.54)] has been used to compute the values of Δx_1 and Δx_2:

Table 3.21

Iteration no.	x_1	x_2	$f_1(x_1, x_2)$	$f_2(x_1, x_2)$	Δx_1	Δx_2	$\varepsilon_{1r}(\%)$	$\varepsilon_{2r}(\%)$	m a x
0	2.0000	2.0000	4.0000	– 13.0000	– 0.5278	– 0.4722	– 35.8491	– 30.9091	35.8491
1	1.4722	1.5278	0.5015	– 4.2806	0.1162	– 0.2761	7.3153	– 22.0597	22.0597
2	1.5884	1.2517	0.0897	– 0.9314	0.0495	– 0.0987	3.0230	– 8.5591	8.5591
3	1.6379	1.1530	0.0122	– 0.0844	0.0043	– 0.0114	0.2640	– 1.0027	1.0027
4	1.6423	1.1415	0.0001	– 0.0010	0.0001	– 0.0001	0.0032	– 0.0124	0.0124
5	1.6423	1.1414	2.3E-08	– 1.5E-07					

The Jacobian for the first few iterations is shown below:

$$J^{(0)} = \begin{bmatrix} 4 & 4 \\ 4 & -32 \end{bmatrix} \quad J^{(1)} = \begin{bmatrix} 2.944 & 3.056 \\ 2.944 & -14.26 \end{bmatrix}$$

$$J^{(2)} = \begin{bmatrix} 3.177 & 2.503 \\ 3.177 & -7.844 \end{bmatrix} \quad J^{(3)} = \begin{bmatrix} 3.276 & 2.306 \\ 3.276 & -6.131 \end{bmatrix}$$

The solution is obtained as (1.6423,1.1414) which is again close to the exact solution and the values of the functions at this point are very close to zero (it should be noted that the table shows truncated values only. The computations were performed in a much higher precision!). For this problem, we see that the starting guess is not very critical to the convergence. Even when starting far away from the solution at the point (10,10), the convergence is achieved in 10 iterations!

3.7.3 Secant Method

Although the Newton method works well for most problems, we do need to evaluate the partial derivatives of the given nonlinear functions. In many cases, the analytical derivatives are not available. Numerical evaluation of the derivatives using finite differences would be computationally expensive. In such cases, the secant method, discussed earlier for a single equation, may be extended to a system of equation as follows (since this method was first proposed by Broyden, it is commonly called the Broyden's method).

Given two *points* which are at a *distance* of Δx, the Jacobian may be approximated by a *Broyden* matrix, B, which satisfies $B\Delta x = \Delta f$ and an iterative scheme is written as (see Eq. 3.52)

$$B^{(k)}\Delta x^{(k+1)} = -f(x^{(k)}) \tag{3.55}$$

While for a single equation, B was uniquely determined from the approximation of the tangent by the secant, for a multidimensional case it is not possible. (There are n^2 unknowns and only n equations. For example, if

$$\Delta x = \begin{Bmatrix} 0.1 \\ 0.1 \end{Bmatrix} \text{ and } \Delta f = \begin{Bmatrix} 0.1 \\ 0.1 \end{Bmatrix}$$

it is rather easy to see that both the matrices

$$\begin{bmatrix} 1 & 0 \\ 0 & 1 \end{bmatrix} \text{ and } \begin{bmatrix} 0.5 & 0.5 \\ 0.5 & 0.5 \end{bmatrix}$$

would satisfy the Jacobian approximation.) To uniquely determine B, Broyden suggested using a constraint that the change in B from its value at the previous iteration should have a minimum norm (see Box 3.6).[7] For the first iteration, the B matrix may be taken as the analytically or numerically obtained Jacobian. However, if the Jacobian is not easy to evaluate, any approximation (e.g., an identity matrix) may be used. Convergence, however, is likely to be poor if the initial guess is not a good approximation of the Jacobian. The steps of the iterative scheme can then be written as:

1. Assume a starting point $x^{(0)}$. The physics of the problem would help in choosing an appropriate starting point.
2. Take the B matrix as the Jacobian at this point $[B^{(0)} = J(x^{(0)})]$ or an approximation to it.
3. Solve Eq. (3.55) to obtain the change in the variables, $\Delta x^{(1)}$, and obtain $x^{(1)} (= x^{(0)} + \Delta x^{(1)})$.
4. Repeat for $k = 1, 2 \ldots$ till convergence
   (a) Compute an updated B matrix as

$$B^{(k)} = B^{(k-1)} + \frac{(\Delta f^{(k)} - B^{(k-1)} \Delta x^{(k)})(\Delta x^{(k)})^T}{(\Delta x^{(k)})^T \Delta x^{(k)}} \tag{3.56}$$

   in which $\Delta x^{(k)} = x^{(k)} - x^{(k-1)}$ and $\Delta f^{(k)} = f(x^{(k)}) - f(x^{(k-1)})$.
   (b) Solve Eq. (3.55) to obtain $\Delta x^{(k+1)}$ and update the variables
   $$[\Delta x^{(k+1)} = x^{(k)} + \Delta x^{(k)}].$$

The following example illustrates the use of the Broyden's method.

---

[7]Another alternative could be to store n previous iterates and use $B \Delta x = \Delta f$ for each. However, it increases the storage requirement and is not preferred. Moreover, it is likely that the solution vectors at various iterations would be linearly dependent and an accurate solution for B may not be obtainable.

BOX 3.6 Broyden's Matrix

The iterative sequence for the Broyden matrix is obtained by stipulating that the difference between the matrices at two successive iterations have a minimum Frobenius norm. Let us assume that we have performed iterations up to the level $(k - 1)$ and need to obtain $B^{(k)}$. Denoting the difference between the two matrices as

$$\Delta B^{(k)} = B^{(k)} - B^{(k-1)}$$

and noting, from the definition of the secant, that

$$B^{(k)} \Delta x^{(k)} = \Delta f^{(k)}$$

we have,

$$\Delta B^{(k)} \Delta x^{(k)} = \Delta f^{(k)} - B^{(k-1)} \Delta x^{(k)}$$

This condition is clearly satisfied if we choose (other choices are also possible and lead to different types of methods)

$$\Delta B^{(k)} = \frac{(\Delta f^{(k)} - B^{(k-1)} \Delta x^{(k)})(\Delta x^{(k)})^T}{(\Delta x^{(k)})^T \Delta x^{(k)}}$$

To show that this choice of ΔB results in minimum Frobenius norm, let us assume that there is another matrix, say, Δ', which satisfies the secant requirement $(B^{(k-1)} + \Delta') \Delta x^{(k)} = \Delta f^{(k)}$. We may then write

$$\left\| \Delta B^{(k)} \right\| = \left\| \frac{(\Delta f^{(k)} - B^{(k-1)} \Delta x^{(k)})(\Delta x^{(k)})^T}{(\Delta x^{(k)})^T \Delta x^{(k)}} \right\|_F$$

$$= \left\| \frac{\Delta' \Delta x^{(k)} (\Delta x^{(k)})^T}{(\Delta x^{(k)})^T \Delta x^{(k)}} \right\|_F$$

$$\leq \left\| \Delta' \right\|_F \left\| \frac{\Delta x^{(k)} (\Delta x^{(k)})^T}{(\Delta x^{(k)})^T \Delta x^{(k)}} \right\|_F = \left\| \Delta' \right\|_F$$

The last step results from the fact that Frobenius norm (see Chapter 2) of the matrix formed by multiplying a vector, v, by its transpose as vv^T, is equal to $v^T v$. Thus out of all possible matrices of change in the Broyden matrix at any iteration, the one given above as ΔB would have the minimum Frobenius norm.

Example 3.15   Using the secant method, solve the set of Eq. (3.46) to a maximum relative error of 0.1% starting with an initial guess of (2,2).

Solution   Table 3.22 shows the iterations:

Table 3.22

Iteration no.	x_1	x_2	$f_1(x_1, x_2)$	$f_2(x_1, x_2)$	Δx_1	Δx_2	ε_{1r} (%)	ε_{2r} (%)	max
0	2.0000	2.0000	4.0000	−13.0000	−0.5278	−0.4722	−35.8491	−30.9091	35.8491
1	1.4722	1.5278	0.5015	−4.2806	0.0084	−0.1505	0.5701	−10.9256	10.9256
2	1.4807	1.3773	0.0893	−2.4061	0.0928	−0.1413	5.9004	−11.4292	11.4292
3	1.5735	1.2360	0.0037	−0.8582	0.0597	−0.0732	3.6550	−6.2977	6.2977
4	1.6332	1.1628	0.0195	−0.1609	0.0094	−0.0194	0.5703	−1.7004	1.7004
5	1.6426	1.1434	0.0053	−0.0109	−0.0001	−0.0019	−0.0081	−0.1661	0.1661
6	1.6424	1.1415	0.0005	−0.0001	−0.0001	−0.0001	−0.0062	−0.0061	0.0062
7	1.6423	1.1414	4.39E-05	2.27E-05					

The Broyden matrix for the first few iterations is shown below (note that for the 0th iteration it is the same as the Jacobian):

$$B^{(0)} = \begin{bmatrix} 4 & 4 \\ 4 & -32 \end{bmatrix}; \quad B^{(1)} = \begin{bmatrix} 3.472 & 3.528 \\ 8.505 & -27.97 \end{bmatrix}$$

$$B^{(2)} = \begin{bmatrix} 3.505 & 2.936 \\ 7.610 & -12.03 \end{bmatrix}; \quad B^{(3)} = \begin{bmatrix} 3.517 & 2.918 \\ 4.822 & -7.788 \end{bmatrix}$$

The solution is obtained as (1.6423,1.1414) which is again close to the exact solution. Though it takes more iterations than the Newton method, the computational effort per iteration is smaller and, depending on the type of function and number of equations, overall efficiency may be better for the secant method.

Although we did provide some discussion on convergence for the fixed-point iteration scheme, analysis of convergence of the solution methods for a set of nonlinear equations is quite complicated. Without going into the proof, we just state that the Newton's method is quadratically convergent for initial guess close to the exact solution provided the inverse of the Jacobian exists. On the other hand, the Broyden's method has superlinear convergence. (If $\left\| x^{(k+1)} - \xi \right\| \leq c_k \left\| x^{(k)} - \xi \right\|$ such that $\lim_{k \to \infty} c_k = 0$ the iteration scheme is called superlinearly convergent.) Some other methods of solving a system of nonlinear equations are also available and most of these are based on optimization. For example, in the Newton scheme, when we move along the direction of the gradient, we may not take the full Newton step but decide the step size so as to minimize the function value. These methods are, however, a little advanced for this book and are not described here.

EXERCISE 3.7

1. Two numbers, x and y, are such that $x^x + y^y = 11.72$; $x^y + y^x = 6.71$. Find the numbers by using the fixed-point iteration, Newton, and Broyden methods. Use starting guess as $x = 2$, $y = 2$ for the first two methods. For the Broyden method, use two starting points as (1,1) and (2,2).

2. In the previous problem, compare the performance of Broyden's method by using the initial estimate of the Broyden matrix as (a) Jacobian at point (1,1), (b) Jacobian at point (2,2), and (c) identity matrix.

3. Solve the following equations using the Newton method using an appropriate starting point:

$$xe^x + y\sin y + z\ln z = 6$$
$$x^2 y + y^2 z + z^2 x = 21$$
$$xyz = 7$$

4. Solve the following equations using (a) fixed-point iteration and (b) Newton-Raphson method, starting with an initial guess of $x = 1$ and $y = 1$.

$$x^2 - x + y - 0.5 = 0$$
$$x^2 - 5xy - y = 0$$

5. Use the Newton-Raphson method to solve the following equations with a maximum approximate error of 0.1% (choose $x = 1$, $y = 1$ as the starting point for the iterations):

$$x^2 + y^2 - 2 = 0$$
$$x^2 - y - 0.5x + 0.1 = 0$$

An alternative way of solving the problem is to express y in terms of x, using the second equation, and substitute in the first equation to obtain.

$$f(x) = x^2 - x^3 + 1.45x^2 + 0.1x - 1.99 = 0$$

Solve this equation using the linear interpolation (false position) method using the initial bracket $(-1, -0.5)$ and with a maximum approximate error of 2%.

SUMMARY

Various bracketing and open methods were discussed in this chapter to solve a nonlinear equation in a single variable. Analysis of these methods was performed to obtain the rate of convergence and specific cases like complex roots and multiple roots were described. *Which method to use for a particular problem* is a question which does not have a unique answer. The bracketing methods converge, but very slowly, and the open methods converge faster, if at all. The most common strategy is to start with the bracketing methods and then use one of the open methods to accelerate the convergence.

For the multivariate case, some of the methods discussed for the single variable case were extended. The fixed-point method, Newton method, and the secant method were discussed. The methods described in Chapter 2 for solving a set of linear simultaneous equations were shown to be useful here since the Newton and secant iterative schemes for the nonlinear equations involve the solution of a linear system.

The secant method has already introduced us to the approximation of the derivative of a function by using the function values at two different points and assuming the function to be linear. However, this is not the only way to approximate the function or its derivative. For example, in Muller's method, we used three function values to fit a second degree polynomial and obtained its zero to approximate the root of a given equation. There are numerous such applications where we would need to approximate a function and/or its derivatives or integrals. We describe the approximation of functions in the next chapter (Chapter 4) and the approximation of derivatives and integrals in Chapter 5.

4

Approximations of Functions

4.1 INTRODUCTION

In science and engineering, we frequently encounter problems where, based on a few measurements of a *dependent* variable (function) at corresponding values of the *independent* variable (generally distance or time), we need to estimate the value of the function for a different value of the independent variable. For example, in our *Batman* problem, suppose that the building has 200 floors and *Joker* notes down the time as the wheel passes the window of every 5th floor till it hits the ground. From this data of *time* (dependent variable) versus *distance of the floor from the roof* (independent variable), we may be required to estimate the following:

- time at which Batman (at the 13th floor) sees the wheel
- initial downward velocity, u, and the gravitational acceleration, g
- velocity and acceleration at different locations/times

On the other hand, if we were to measure the velocity at different times (say by a radar gun), we could be asked to find the distance travelled till that time. There are various methods which could be used to perform these and other similar tasks.

In this chapter, we discuss

- *interpolation* (estimating the function value at points within the data range[1] using a curve which passes through all data points), and

---

[1] We will not discuss *extrapolation*, i.e., estimating function value outside the range covered by the given data, since it may lead to large errors.

- *regression* (fitting a curve which represents the general trend of the function) and, in the next chapter, we describe
- *numerical differentiation* (estimating the function derivatives, e.g., to obtain the velocity or acceleration from distance measurements), and
- *numerical integration* (estimating the integral of the function, e.g., to obtain distance from velocity measurements)

These problems may be thought of as *approximation* problems in which we approximate the actual function, $f(x)$, by an approximate function, $\tilde{f}(x)$ over a range of interest (a, b). The need for approximation arises because of various factors. We may not know the exact nature of dependence of the function on the independent variable, *or* the functional relationship may be so complex as to preclude operations like differentiation and integration, or the measured data may have errors. The methodology we adopt, will be largely dependent on these factors. For example, if the data is exact, we would like our approximate function to pass through each data point (interpolation). On the other hand, if the data is affected by measurement errors, the approximating function *should not* pass through all the points and it would be sufficient for it to represent the general trend (regression). (Sometimes, even when the data does not contain error, we may want to do *regression* to obtain an approximate function which is smoother than the actual function.) Since various concepts involved in the analysis are simpler for a function rather than tabular data, we first discuss the *continuous* case in which we need to approximate a function $f(x)$ of a *single* independent variable x. Extensions to the *discrete* case (in which the form of the function is not specified, only the function values are given corresponding to a few values of x) and *multiple* independent variables, is conceptually similar and is described subsequently. The *continuous case*, of course, is a superset of the *discrete case* since the function values at selected points can be readily generated if the function is known.

4.2 APPROXIMATION OF FUNCTIONS

The form of the approximating function may be deduced from the knowledge of the function behaviour or from looking at the data. For example, if the function shows periodicity, it may be approximated by a combination of various sine and cosine functions. (See Section 4.7, for further discussion of periodic functions.) Similarly, if we know from the physics of the problem that the function shows an exponential decay, we could use an exponential function as the approximate function. Most of the times, however, it would not be apparent from the problem/data as to what is the exact nature of the function. Because of their simplicity, easy differentiability and integrability, and due to the fact that most functions could be expanded in terms of polynomials using the Taylor's series, polynomials have been widely used as approximating functions. Moreover, a polynomial remains a polynomial of the same degree even under a linear transformation of the dependent variable (e.g. $x^* = ax + b$). Also, there is a theorem which implies that an approximating polynomial can be brought arbitrarily close to the actual function by increasing the degree of the polynomial (Weierstrass theorem, see Theorem 4.1). Therefore, we will focus our attention mostly on polynomial approximations.

Theorem 4.1 Weierstrass Theorem

The Weierstrass approximation theorem states that "If f is a continuous real-valued function on a closed, bounded interval $[a, b]$ and if $\varepsilon > 0$, there exists a polynomial p such that $|f(x) - p(x)| < \varepsilon$ for all $x \in [a, b]$, i.e., a continuous function on a bounded interval can be uniformly approximated by polynomial functions." Many different proofs of this theorem are available and we mention here the one based on Bernstein polynomials.

Since any closed interval $[a, b]$ may be transformed to $[0, 1]$ by a linear substitution, we focus our attention on functions continuous over $[0, 1]$. Since x varies over $[0, 1]$, it may be thought of as a probability of occurrence of some event, say, E. A binomial probability, $\pi(i, n)$, representing the probability of E occurring exactly i times in n independent trials, is then written as

$$\pi(i, n) = \binom{n}{i} x^i (1 - x)^{n-i}$$

in which the binomial coefficient $\binom{n}{i}$ is equal to $\dfrac{n!}{i!(n-i)!}$. Note that it is an nth degree polynomial in x, is non-negative over $[0, 1]$, and, from elementary probability theory,

$$\sum_{i=0}^{n} \pi(i, n) = 1$$

The Bernstein polynomials of order n of a function $f(x)$ is then written as

$$B_n(x) = \sum_{i=0}^{n} \pi(i, n) f\left(\frac{i}{n}\right)$$

It is easy to see that if $f(x)$ has an upper bound, M, in the interval $[0, 1]$, then $B_n(x)$ will also have the same upper bound.

To prove the Weierstrass theorem, we write

$$\left| f(x) - B_n(x) \right| = \left| f(x) - \sum_{i=0}^{n} f\left(\frac{i}{n}\right) \pi(i,n) \right|$$

$$= \left| f(x) \sum_{i=0}^{n} \pi(i,n) - \sum_{i=0}^{n} f\left(\frac{i}{n}\right) \pi(i,n) \right|$$

$$= \left| \sum_{i=0}^{n} \left[f(x) - f\left(\frac{i}{n}\right) \right] \pi(i,n) \right|$$

$$= \sum_{\left|\frac{i}{n}-x\right|<\delta} \left| f(x) - f\left(\frac{i}{n}\right) \right| \pi(i,n) + \sum_{\left|\frac{i}{n}-x\right|\geq\delta} \left| f(x) - f\left(\frac{i}{n}\right) \right| \pi(i,n)$$

$$\leq \varepsilon + 2M \sum_{\left|\frac{i}{n}-x\right|\geq\delta} \pi(i,n)$$

In the last two lines, we have split the summation into two parts, one where the points x and i/n are within an arbitrarily close distance, δ, and the other beyond it. Since the function is continuous

$$\left| f(x) - f\left(\frac{i}{n}\right) \right| < \varepsilon \quad \text{for} \quad \left|\frac{i}{n} - x\right| < \delta$$

and the first term would be less than or equal to ε. Also, since the function has an upper bound of M, the upper bound of

$$\left| f(x) - f\left(\frac{i}{n}\right) \right|$$

will be $2M$. To evaluate $\displaystyle\sum_{\left|\frac{i}{n}-x\right|\geq\delta} \pi(i,n)$, we consider the summation

$$S = \sum_{i=0}^{n} \left(\frac{i}{n} - x\right)^2 \pi(i,n) = \frac{1}{n^2} \sum_{i=1}^{n} i^2 \pi(i,n) - \frac{2x}{n} \sum_{i=1}^{n} i\pi(i,n) + x^2 \sum_{i=0}^{n} \pi(i,n)$$

Note that in the first two terms on the RHS, the lower limit of the summation index is changed to 1 since the term corresponding to $i = 0$ will vanish. Also the summation in the third term would be unity. For the middle term, we write

$$\sum_{i=1}^{n} i\pi(i,n) = \sum_{i=1}^{n} i \frac{n!}{i!(n-i)!} x^i (1-x)^{n-i}$$

$$= \sum_{i=1}^{n} nx \frac{(n-1)!}{(i-1)!(n-i)!} x^{i-1}(1-x)^{n-i}$$

$$= nx \sum_{j=0}^{m} \frac{m!}{(j)!(m-j)!} x^j (1-x)^{m-j} \quad (\text{with } j \equiv i-1, m \equiv n-1)$$

$$= nx \sum_{j=0}^{m} \pi(j,m) = nx$$

Similarly, it can be shown that

$$\sum_{i=1}^{n} i^2 \pi(i,n) = nx(nx - x + 1)$$

The summation, S, is then obtained as

$$S = \frac{1}{n^2} nx(nx - x + 1) - \frac{2x}{n} nx + x^2 = \frac{x - x^2}{n}$$

We, therefore, have (noting that the maximum value of $x - x^2$ in the interval [0, 1] is 1/4)

$$\sum_{\left|\frac{i}{n} - x\right| \geq \delta} \pi(i,n) \leq \sum_{\left|\frac{i}{n} - x\right| \geq \delta} \frac{\left(\frac{i}{n} - x\right)^2}{\delta^2} \pi(i,n) = \frac{S}{\delta^2} = \frac{x(1-x)}{n\delta^2} \leq \frac{1}{4n\delta^2}$$

So, finally, we get

$$\left| f(x) - B_n(x) \right| \leq \varepsilon + \frac{M}{2n\delta^2}$$

which may be made arbitrarily small by taking a sufficiently large n, thus proving the Weierstrass theorem. Other proofs are also available using, for example, the Fourier series.

Once we choose the approximating function to be a polynomial, we have to decide what should be the degree of this polynomial. The answer clearly depends on how close to the actual function we want the approximation to be and how much computational effort are we willing to spend. In general, a higher-order polynomial would be closer to the function but will need more computational effort for evaluating the coefficients (for discrete case, however, a higher-order polynomial may result in a worse fit, see Fig. 4.7). We must also decide on how to quantify the nearness of the actual function, $f(x)$, and the approximating polynomial of degree m, $f_m(x)$. For example, we may use the maximum difference between $f(x)$ and $f_m(x)$ over the interval (a, b) as a measure of the error of the approximation. Or we may use the integral of the absolute value or the square of $[f(x) - f_m(x)]$. Finally, we should device an algorithm which would give us the mth-degree polynomial *nearest* to $f(x)$. We discuss these issues first, from an analytical perspective using the *Taylor's series* and the *method of least squares* and then, from a geometrical perspective in the next section.

4.2.1 Taylor's Series

Probably the simplest way of approximating a function by a polynomial is by using the Taylor's series expansion about a point:

$$f(a+h) = f(a) + hf'(a) + \frac{h^2}{2!}f''(a) + \cdots + \frac{h^i}{i!}f^{[i]}(a) + \cdots \qquad (4.1)$$

and truncating it after the desired number of terms. Similar to Chapter 3, $f^{[i]}$ denotes the ith derivative of the function. If the function and its first $(m + 1)$ derivatives are continuous over the interval $(a, a + h)$, Taylor's theorem (Theorem 4.2) states that

$$f(a+h) = f(a) + hf'(a) + \frac{h^2}{2!}f''(a) + \cdots + \frac{h^m}{m!}f^{[m]}(a) + R_m \qquad (4.2a)$$

in which the remainder, R_m, is given by

$$R_m = \int_a^{a+h} \frac{(a+h-x)^m}{m!} f^{[m+1]}(x)dx = \frac{h^{m+1}}{(m+1)!} f^{[m+1]}(\xi) \qquad (4.2b)$$

where $\xi \in (a, a+h)$. Thus, to approximate a function, $f(x)$, over the interval (a, b), by a mth degree polynomial, $f_m(x)$, we write

$$f_m(x) = f(x_0) + (x - x_0)f'(x_0) + \frac{(x - x_0)^2}{2!}f''(x_0) + \cdots + \frac{(x - x_0)^m}{m!}f^{[m]}(x_0)$$

$$(4.3)$$

where x_0 is a *judiciously chosen* point (in the absence of any other information, the most logical choice would be the mid-point of the interval. However, suppose we know that most of the times the approximation will be used to predict the function values closer to a rather than b, we may choose x_0 nearer to a.) Not only does Eq. (4.3) give us an approximation, it also provides us with an estimate of the error [Eq. (4.2b)].

Theorem 4.2 Taylor's Theorem

Writing

$$f(a+h) = f(a) + \int_a^{a+h} f'(x)\,dx$$

and integrating by parts, we obtain

$$f(a+h) = f(a) + \left[x f'(x) \right]_a^{a+h} - \int_a^{a+h} x f''(x)\,dx$$

$$= f(a) + (a+h) f'(a+h) - a f'(a) - \int_a^{a+h} x f''(x)\,dx$$

$$= f(a) + (a+h) \left[\int_a^{a+h} f''(x)\,dx + f'(a) \right] - a f'(a) - \int_a^{a+h} x f''(x)\,dx$$

$$= f(a) + h f'(a) + \int_a^{a+h} (a+h-x) f''(x)\,dx$$

A similar procedure leads to

$$f(a+h) = f(a) + h f'(a) + \frac{h^2}{2} f''(a) + \frac{1}{2} \int_a^{a+h} (a+h-x)^2 f'''(x)\,dx$$

By repeating the process (or, formally, by induction) we obtain the Taylor's theorem

$$f(a+h) = f(a) + h f'(a) + \frac{h^2}{2!} f''(a) + \cdots + \frac{h^m}{m!} f^{[m]}(a) + R_m$$

in which the remainder, R_m, is given by

$$R_m = \int_a^{a+h} \frac{(a+h-x)^m}{m!} f^{[m+1]}(x)\,dx$$

Using the second mean value theorem for integration, which states that

If $f(x)$ and $g(x)$ are continuous and integrable on $[a, b]$ and $g(x)$ has the same sign everywhere in (a, b), then there exists a number $c \in [a,b]$ such that

$$\int_a^b f(x) g(x)\,dx = f(c) \int_a^b g(x)\,dx$$

the remainder may be written as

$$R_m = \frac{h^{m+1}}{(m+1)!} f^{[m+1]}(\xi)$$

in which, $\xi \in (a, \ a+h)$.

However, since the point ξ is not known, the error estimate is generally not very useful. In most cases, though, the upper bound of the $(m + 1)$th derivative of the function over (a, b) may be obtained and will provide an upper bound for the error (the actual error is generally much smaller!). In some cases, e.g., when $f(x)$ is a $(m + 1)$th degree polynomial, $f^{[m+1]}(x)$ may be constant or nearly constant over (a, b) and the remainder may be readily computed. For example, to approximate the function $f(x) = a_0 + a_1 x + a_2 x^2$ by a first degree polynomial $f_1(x) = c_0 + c_1 x$ over the range (a, b) (*not a very realistic example, since we would hardly ever want to approximate a polynomial by another! The reason for choosing this example would be clear when we discuss* geometric interpretation of a function *later in this chapter*), we may write

$$f_1(x) = f\left(\frac{a+b}{2}\right) + \left(x - \frac{a+b}{2}\right) f'\left(\frac{a+b}{2}\right) \tag{4.4}$$

resulting in

$$c_0 = a_0 - a_2 \frac{(a+b)^2}{4} \quad \text{and} \quad c_1 = a_1 + a_2(a+b)$$

The error at any point, $f(x) - f_1(x)$, is given by the remainder term,

$$R_1 = a_2 \left(x - \frac{a+b}{2}\right)^2$$

Example 4.1 The function $f(x) = 1 + x + x^2$ has to be approximated by a linear function over the interval $(0, 1)$. Find the approximating function by using Taylor's series expansion about the points 0, 0.5, and 1, respectively.

Solution Using Eq. (4.3), the approximating linear function is written as

$$f_1(x) = f(x_0) + (x - x_0) f'(x_0) = \begin{bmatrix} 1 + x, & x_0 = 0 \\ 1.75 + 2(x - 0.5), & x_0 = 0.5 \\ 3 + 3(x - 1), & x_0 = 1 \end{bmatrix}$$

Figure 4.1 shows these three approximations along with another line D.

From this figure, the following observations are made
- Taylor's series about the mid-point is much better than that about the end-points.
- The error of this approximation grows as we move away from the mid-point (see line B and also the expression for the remainder).
- The fit does not appear to be the *best* fit since a parallel line D is much *closer* to the function. This line, however, does not represent a Taylor's series expansion of $f(x)$.

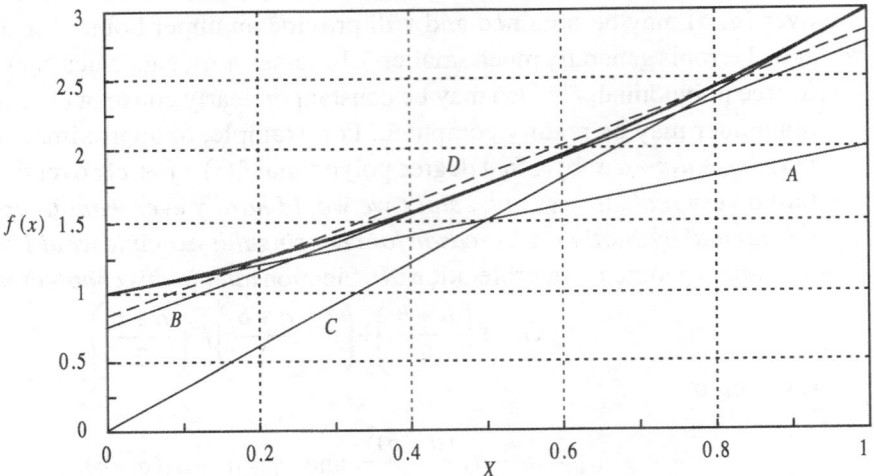

Fig. 4.1 The function $f(x) = 1 + x + x^2$ and its straight line approximation over the domain (0, 1). The function is shown by the thick line and the Taylor's series approximations about points 0 (line A), 0.5 (line B), and 1.0 (line C) are also drawn. Line D is parallel to line B.

Thus, in spite of its simplicity, Taylor's series is hardly ever used for approximating a function. However, it is very useful in performing an error analysis of other approximating schemes, as we will see on numerous occasions in this and subsequent chapters.

4.2.2 Method of Least Squares

In this method, we choose the approximating polynomial in such a way as to give us the least value of the square of the deviation $[f(x) - f_m(x)]$ integrated over the relevant domain $[a, b]$. We may write the approximating polynomial in the commonly used form,

$$f_m(x) = \sum_{j=0}^{m} c_j x^j$$

However, there are alternative ways of writing a polynomial which may be more suitable for some problems. For example, the form

$$f_m(x) = \sum_{j=0}^{m} c_j \left(x - \frac{a+b}{2} \right)^j$$

will have lower round-off errors. Other possibilities include

$$f_m(x) = \sum_{j=0}^{m} c_j p_j$$

where p_j denotes a jth degree polynomial and

$$f_m(x) = \sum_{j=0}^{m} c_j p_{m,j}$$

where $p_{m,j}$ are mth degree polynomials. We use a general notation

$$f_m(x) = \sum_{j=0}^{m} c_j \phi_j(x) \tag{4.5}$$

to represent the approximating polynomial in which the ϕ's are the (known) polynomials and the c's are the (unknown) coefficients. The problem then reduces to finding the c's which minimize

$$\int_a^b \left(f(x) - \sum_{j=0}^{m} c_j \phi_j(x) \right)^2 dx$$

We use the stationary points theorem which states that a continuous function of n variables attains an extremum only at points at which either all the n partial derivatives are zero (stationary points) or one or more of these derivatives do not exist. This leads to a set of $(m+1)$ linear simultaneous equations of the form

$$[A]\{c\} = \{b\} \tag{4.6}$$

in which

$$a_{ij} = \int_a^b \phi_i(x)\phi_j(x)dx \quad \text{and} \quad b_i = \int_a^b \phi_i(x)f(x)dx$$

Equation (4.6) are known as the *normal equations* and may be solved using any of the methods discussed in Chapter 2 to obtain the coefficients $c_0, c_1, \ldots, c_m$. For example, to approximate the function $f(x) = a_0 + a_1x + a_2x^2$ by a first degree polynomial $f_1(x) = c_0 + c_1x$ over the range (a, b), we obtain

$$\int_a^b 1.1\,dx\, c_0 + \int_a^b 1.x\,dx\, c_1 = \int_a^b 1.(a_0 + a_1x + a_2x^2)\,dx$$

$$\int_a^b x.1\,dx\, c_0 + \int_a^b x.x\,dx\, c_1 = \int_a^b x.(a_0 + a_1x + a_2x^2)\,dx$$

i.e.,

$$(b-a)c_0 + \frac{b^2 - a^2}{2}c_1 = (b-a)a_0 + \frac{b^2 - a^2}{2}a_1 + \frac{b^3 - a^3}{3}a_2$$

$$\frac{b^2 - a^2}{2}c_0 + \frac{b^3 - a^3}{3}c_1 = \frac{b^2 - a^2}{2}a_0 + \frac{b^3 - a^3}{3}a_1 + \frac{b^4 - a^4}{4}a_2 \qquad (4.7)$$

from which $c_0 = a_0 - \dfrac{b^2 + a^2 + 4ab}{6}a_2$ and $c_1 = a_1 + (b+a)a_2$.

Example 4.2   The function $f(x) = 1 + x + x^2$ has to be approximated by a linear function over the interval $[0, 1]$. Find the approximating function by using the least squares method.

Solution   Using the basis functions as $\phi_0(x) = 1$ and $\phi_0(x) = x$, and with $a = 0$ and $b = 1$, Eq. (4.7) is written as

$$c_0 + 0.5c_1 = 1 + \frac{1}{2} + \frac{1}{3} = \frac{11}{6}$$

$$\frac{1}{2}c_0 + \frac{1}{3}c_1 = \frac{1}{2} + \frac{1}{3} + \frac{1}{4} = \frac{13}{12}$$

from which, $c_0 = 5/6$ and $c_1 = 2$. The *best* approximating linear function is thus written as $f_1(x) = 5/6 + 2x$ (Line D in Fig. 4.1 shows this approximation).

This method, however, becomes quite cumbersome and the normal equations become ill-conditioned as the degree of polynomial increases. In the next section, we give a geometric interpretation of the problem of function approximation which leads to development of simpler techniques.

EXERCISE 4.2

1. Approximate the function e^x over $(-1, 1)$ by a straight line using (a) Taylor's series about $x = 0$ and (b) the method of least squares. Plot the residual for both the methods. Using these approximations, estimate the value of e^x at $x = 0$, 0.62, and 1. Compare with the exact values and comment.

2. Compare the residuals in the Taylor's series approximation at all three points in Problem 1 with the analytical expression of the remainder [Eq. (4.2b)] and verify that ξ does lie between a and $a + h$.

3. Compute the L_2 norm of the residual, i.e., $\int_{-1}^{1}\left[f(x) - f_1(x)\right]^2 dx$ for both the approximations in Problem 1. In order to reduce this norm, it is desired to approximate the function by a second degree polynomial. Obtain this polynomial and the associated L_2 norm. Is it possible that the L_2 norm of the residual is larger for a higher-degree polynomial approximation?

4. Approximate the function e^{2x-1} over $(0, 1)$ by a straight line using the method of least squares. Compare the approximating function with that obtained in Problem 1 for the domain $[-1, 1]$ and verify that these are identical (with a linear transformation of variable).

5. The function x^x has to be integrated over the interval $(0, 1)$. Since analytical integration is not possible, it is decided to approximate the function by a third degree polynomial and then perform an analytical integration of the approximating polynomial. What problem is encountered in the method of least squares? Would the Taylor's series work? If yes, estimate the value of the integral $\int_{0}^{1} x^x dx$. [Note: We will look at better methods of estimating this integral in the next chapter.]

4.3 GEOMETRIC INTERPRETATION OF FUNCTIONS AND ORTHOGONAL POLYNOMIALS

We are quite familiar with the three-dimensional Euclidean vector space in which a point is represented by its distance from an origin along three orthogonal directions (e.g., East-West, North-South, Up-Down). Sometimes a fourth dimension of time is also added in the description. Although it is difficult to visualize, higher dimensional spaces may, and probably do, exist. String theory in physics, which considers the basic building blocks to be strings rather than particles, predicts the dimensionality of the universe to be 10 (superstring theory), 11 (M-theory) or 26 (bosonic string theory). Along similar lines, we may think of an n-dimensional space (see Box 4.1) whose axes are not directions but functions and then define products of functions, angle between functions and other similar properties.

BOX 4.1 Geometric Interpretation of Functions

We may think of a space whose axes are not directions but functions. For example, the function $a_0 + a_1x$ could be represented by a point P in a two-dimensional plane having axes 1 and x, as shown in the figure below:

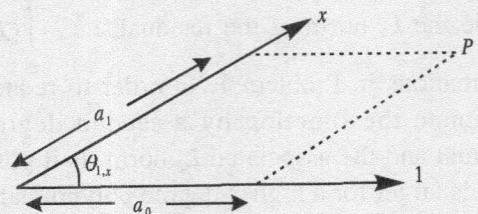

The axes need not be orthogonal but, as we will see a little later, the computations are much simpler if they are. Similar to a vector space, we may have an $(n + 1)$-dimensional function space to represent an nth degree polynomial with its axes as $1, x, x^2, \ldots, x^n$. In fact we may choose the axes as $1, 1+x, 1+x+x^2, \ldots, 1+x+\ldots+x^n$ or any other form as long as all nth degree polynomials can be represented by a linear combination of these axes. (Other functions would belong to different function spaces, e.g., periodic functions are generally represented by a space with its axes as $1, \sin ax, \cos ax, \sin 2ax, \cos 2ax, \ldots$). Analogous to the basis vectors, these axes are called *basis functions*. Once this space is defined, any function can be represented by a *point* in the corresponding space. If $f(x)$ is a polynomial of degree n it would be a point in the $(n + 1)$-dimensional function space. If $f(x)$ is not a polynomial, it may be represented in an infinite-dimensional polynomial space using the Taylor's series.

Similar to the *dot product* (scalar product) of two vectors, an *inner product* of two functions, $f(x)$ and $g(x)$ in the function space is defined as (sometimes a weight, $w(x)$, is also used in this definition, but we will introduce it later in this chapter)

$$\langle f, g \rangle = \int_a^b f(x)g(x)dx \tag{B4.1.1}$$

where (a, b) is the domain over which these functions are defined. Analogous to the vector space, the *magnitude* (norm) of a function is given as

$$\|f\| = \sqrt{\langle f, f \rangle}$$

and the *angle* between any two functions is given by

$$\theta_{f,g} = \text{arc cos} \frac{\langle f, g \rangle}{\|f\|\|g\|}$$

Using these definitions, the angle between the 1-axis and x-axis in the figure, $\theta_{1,x}$, is given by

$$\cos\theta_{1,x} = \frac{(b^2 - a^2)\big/2}{\sqrt{b-a}\sqrt{(b^3 - a^3)\big/3}} = \frac{\sqrt{3}}{2}\frac{b+a}{\sqrt{b^2 + a^2 + ba}}$$

Thus if the function is defined over the domain $(0, 1)$, the angle between the two axes would be 30°, and for the domain $(-1, 1)$, the axes would be orthogonal (and the inner product will be zero).

As described in Box 4.1, $f(x)$ is represented as a point in space whose location will depend on the nature of the function. And if we decide to approximate it by a polynomial of a lower degree, $f_m(x)$, the approximating polynomial would be another point in the $(m + 1)$-dimensional space (if $f(x)$ is a polynomial of degree n and m

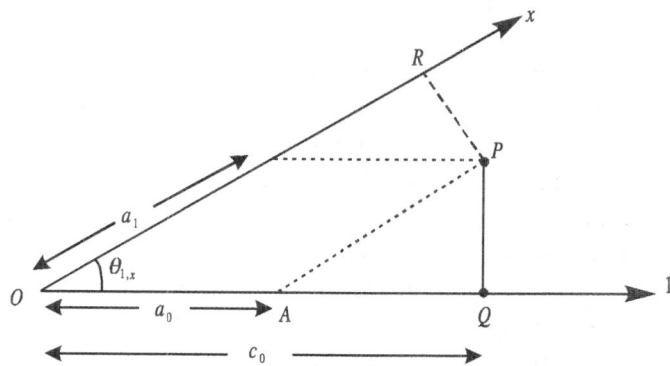

Fig. 4.2 Approximating a linear function by a constant.

is equal to or greater than n, the two points could be same). This gives us a convenient way of defining the nearness of the two functions in terms of the *distance* between these two points. Let us illustrate this with an example.

Suppose we want to approximate the function $f(x) = a_0 + a_1 x$ by a constant value $f_0(x) = c_0$ over the interval $[a, b]$. Figure 4.2 shows the point $f(x)$ in the two-dimensional function space with 1 and x as the coordinate axes (As shown in Box 4.1, the angle between the two axes depends on a and b). The *best* approximating constant function may be taken as the point on the 1-axis which is at a minimum distance from $f(x)$. For this simple problem, it is readily obtained by drawing a perpendicular from $f(x)$ on the 1-axis. On the other hand, if we want to approximate the function by a straight line passing through the origin, i.e., $f_1(x) = c_1 x$, we draw a perpendicular from $f(x)$ on the x-axis (point R in Fig. 4.2).

Similarly, Fig. 4.3 shows how to approximate the function $f(x) = a_0 + a_1 x + a_2 x^2$ over the same range by a linear function $f_1(x) = c_0 + c_1 x$. In this case, $f(x)$ is shown as a point (P) in the three-dimensional $(1, x, x^2)$ space and the best approximation is obtained as the projection (Q) of point P on the $(1, x)$ plane.

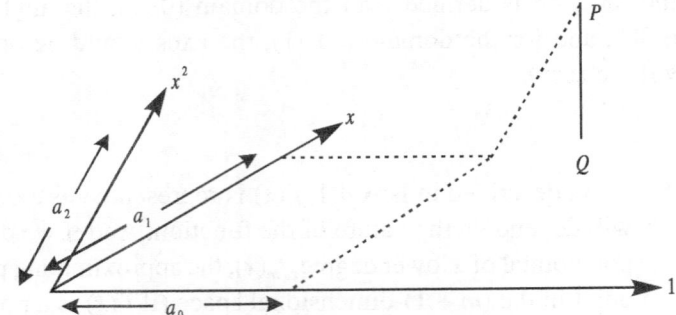

Fig. 4.3   Approximating a quadratic function by a linear function.

Extending this argument, we could say that for a function $f(x)$, the best approximating polynomial of degree m would be the projection of the function $f(x)$, on the $(m + 1)$-dimensional space. The only thing which is left to do now is to devise a method to obtain the location of this point, i.e., to choose the coefficients $c_0, c_1, \ldots, c_m$. We will do it by extending the concepts of vector space to the space representing the function.

As shown in Fig. 4.2, to obtain the best 0th order polynomial to approximate the function $f(x) = a_0 + a_1 x$ over the range (a, b), we need to find c_0 (i.e., OQ) such that PQ is perpendicular to the 1-axis. There are several ways in which it could be done. For example, from triangle APQ, we have

$$\|AQ\|\left(= \sqrt{\langle AQ, AQ \rangle} = AQ\sqrt{b-a}\right) = \|AP\|\cos\theta_{1,x}$$

$$= \sqrt{\langle a_1 x, a_1 x \rangle}\cos\theta_{1,x} = a_1 \frac{\left(b^2 - a^2\right)\big/2}{\sqrt{b-a}}$$

resulting in $c_0 = a_0 + a_1(b+a)/2$. Another method would be to obtain the projection of *OP* on the 1-axis as

$$\|OQ\| \left(= c_0 \sqrt{b-a}\right) = \frac{\langle OP, 1 \rangle}{\|1\|} = \frac{\int_a^b (a_0 + a_1 x)\,dx}{\sqrt{\int_a^b 1\,dx}} = \frac{a_0(b-a) + a_1 \dfrac{b^2 - a^2}{2}}{\sqrt{b-a}}$$

which, as it should, gives the same value of c_0. While these methods are good for illustration, their extension to higher dimensions is cumbersome. Figure 4.3 depicts the problem of approximating a three-dimensional function, $f(x) = a_0 + a_1 x + a_2 x^2$, by a two-dimensional function $f_1(x) = c_0 + c_1 x$. Finding *OQ* using geometry or inner products with the coordinate axes is possible but becomes quite involved. A much simpler technique results from the observation that the *error* term, $[f_m(x) - f(x)]$, i.e., *PQ*, should be orthogonal to the $(m+1)$-dimensional plane containing $f_m(x)$ in order for it to be minimum. This implies that *PQ* should be orthogonal to *all* the coordinate axes of $f_m(x)$. Thus, the problem of finding the best straight line approximation for a quadratic polynomial reduces to the following equations:

$$\langle f_1(x) - f(x), 1 \rangle = 0$$
$$\langle f_1(x) - f(x), x \rangle = 0$$

giving rise to the same two linear simultaneous equations in c_0 and c_1 as obtained using the method of least squares [Eq. (4.7)].

The ideas described in the preceding paragraphs could now be formalised for the general case as follows. Let $f(x)$ be a function which needs to be approximated over the interval $x = a$ to $x = b$ by an mth degree polynomial $f_m(x)$. Let the coordinate axes, or the *basis*, of the $(m+1)$-dimensional space representing the approximating polynomial be represented by ϕ_j, $j = 0$ to m. The approximating polynomial may then be written as

$$f_m(x) = \sum_{j=0}^{m} c_j \phi_j(x) \tag{4.8}$$

where c_j are coefficients which need to be determined. Using the orthogonality conditions

$$\left\langle \sum_{j=0}^{m} c_j \phi_j(x) - f(x), \phi_i(x) \right\rangle = 0, \quad \text{for } i = 0, 1, ..., m$$

we get the same set of $m+1$ linear simultaneous equations as obtained earlier [Eq. (4.7)], indicating that it is equivalent to the least squares aproximation

$$[A]\{c\}=\{b\}$$ (4.9)

in which $a_{ij}=\langle\phi_i,\phi_j\rangle$ and $b_i=\langle\phi_i,f\rangle$ (for convenience we have dropped the x showing dependence of f and ϕ on x). Equation 4.9 may be solved using any of the methods discussed in Chapter 2 to obtain the coefficients $c_0, c_1,...,c_m$. However, the computations become extremely simple if we choose an *orthogonal basis*. An orthogonal set of basis functions is one for which $\langle\phi_i,\phi_j\rangle=0$ if $i\neq j$. If, in addition, all basis functions have unit norms, i.e., $\langle\phi_i,\phi_i\rangle=1\ \forall\ i$, we get an *orthonormal* set of basis functions. For orthogonal basis, all the non-diagonal elements of the matrix A become zero and the c's are obtained directly as

$$c_i=\frac{\langle\phi_i,f\rangle}{\langle\phi_i,\phi_i\rangle}$$

And if we choose an orthonormal basis, we have $c_i=\langle\phi_i,f\rangle$. This provides us with a motivation to look closely at orthogonal functions.

From Box 4.1, we see that the functions 1 and x would be orthogonal if the domain is $(-a, a)$. The most common choice is $(-1, 1)$ and if a function $f(x^*)$ has a finite domain (a, b), we may use the transformation

$$x=\frac{2x^*-b-a}{b-a}$$

to obtain a function $f(x)$ over $(-1, 1)$. [For semi-infinite domain of the form (a, ∞) with $a > 0$, a *non-linear* transformation of the form $x = 1 - 2a/x^*$ may be used. Another possibility would be to use $x = 1 - 2\exp(x^* - a)$. However, we will not discuss these cases.] Therefore most of the subsequent discussion is based on the assumption that the function domain is $(-1, 1)$. The basis functions $\phi_0 = 1$ and $\phi_1 = x$ are thus orthogonal, i.e.

$$\int_{-1}^{1}1\cdot x\,dx=0$$

If we choose $\phi_2 = x^2$, we find that ϕ_1 and ϕ_2 are orthogonal, since

$$\int_{-1}^{1}x\cdot x^2 dx=0$$

but ϕ_0 and ϕ_2 are not, since

$$\int_{-1}^{1}1\cdot x^2 dx\neq0$$

One way of choosing ϕ_2 is to assume it to be of the form $\alpha_0 + \alpha_1 x + \alpha_2 x^2$, and using the orthogonality with ϕ_0 and ϕ_1 to show that $\alpha_2 = -3\alpha_0$ and $\alpha_1 = 0$ such that

BOX 4.2 Gram Schmidt Process for Generating Orthogonal Polynomials

Given a set of linearly independent functions $v_0, v_1, v_2,\ldots, v_m$, a set of orthogonal functions, O, generating the same subspace may be obtained by

$$O_0 = v_0;\; O_1 = v_1 - \frac{\langle v_1, O_0 \rangle}{\langle O_0, O_0 \rangle} O_0;\; O_2 = v_2 - \frac{\langle v_2, O_0 \rangle}{\langle O_0, O_0 \rangle} O_0 - \frac{\langle v_2, O_1 \rangle}{\langle O_1, O_1 \rangle} O_1$$

and

$$O_m = v_m - \frac{\langle v_m, O_0 \rangle}{\langle O_0, O_0 \rangle} O_0 - \frac{\langle v_m, O_1 \rangle}{\langle O_1, O_1 \rangle} O_1 - \cdots - \frac{\langle v_m, O_{m-1} \rangle}{\langle O_{m-1}, O_{m-1} \rangle} O_{m-1}$$

$\phi_2 = \alpha_0(1 - 3x^2)$ would form an orthogonal basis. However, the Gram Schmidt process (see Theorem 2.9 and Box 4.2) provides us a convenient algorithm for generating the orthogonal polynomials. For the second degree polynomial,

$$\phi_0 = 1,\;\; \phi_1 = x - \frac{\langle x, 1 \rangle}{\langle 1, 1 \rangle} 1 = x$$

and

$$\phi_2 = x^2 - \frac{\langle x^2, 1 \rangle}{\langle 1, 1 \rangle} 1 - \frac{\langle x^2, x \rangle}{\langle x, x \rangle} x = x^2 - \frac{1}{3}$$

Note that this ϕ_2 is the same as obtained earlier with $\alpha_0 = -1/3$. Since any arbitrary constant could be used, commonly it is chosen in such a way as to make the ϕ-value at $x = 1$ equal to 1. This set of orthogonal polynomials is known as the *Legendre polynomials*, and is described in the next subsection.

4.3.1 Legendre Polynomials

During the solution of the Laplace equation in spherical coordinates for boundary value problems, the following differential equation is obtained

$$\frac{d}{dx}\left[(1 - x^2)\frac{dp}{dx}\right] + n(n+1)p = 0 \tag{4.10}$$

in which x represents the cosine of the polar angle, and therefore has the domain $[-1, 1]$, and p represents the part of the potential which depends on the polar angle. This equation is known as the Legendre's equation and arises in many other applications also. For a physically realistic solution, it can be shown that n should be a non-negative integer and a solution of this equation is given by an nth degree polynomial. These polynomials, generally multiplied by a constant to make them equal to 1 at $x = 1$, are called the Legendre polynomials and are denoted by $P_n(x)$. They may be obtained by using the Rodrigues' formula

$$P_n(x) = \frac{1}{2^n n!} \frac{d^n}{dx^n} \left[(x^2 - 1)^n \right]$$

or using Bonnet's recursive relation

$$P_n(x) = \frac{2n-1}{n} x P_{n-1}(x) - \frac{n-1}{n} P_{n-2}(x), \qquad n = 2, 3, \dots . \qquad (4.11)$$

with $P_0(x) = 1$, $P_1(x) = x$. [Another useful recursive relation is $(1 - x^2) P_n'(x) = -nx P_n(x) + n P_{n-1}(x)$]. The legendre polynomials have the orthogonality property:

$$\left\langle P_i(x), P_j(x) \right\rangle = \int_{-1}^{1} P_i(x) P_j(x) dx = \begin{vmatrix} 0, & i \neq j \\ \dfrac{2}{2i+1}, & i = j \end{vmatrix} \qquad (4.12)$$

We may, of course, make these orthonormal by multiplying each polynomial by $\sqrt{n + 1/2}$, but it does not make much of a difference as far as the computations are concerned.

The first four polynomials are listed below:

$$P_0(x) = 1; \quad P_1(x) = x; \quad P_2(x) = \frac{1}{2}(3x^2 - 1); \quad P_3(x) = \frac{1}{2}(5x^3 - 3x)$$

From the Rodrigues' formula it is seen that the coefficient of the leading term, x^n, in $P_n(x)$ will be equal to $\dfrac{(2n)!}{2^n (n!)^2}$. In the chapter on numerical integration, we will see that it is convenient to express the orthogonal polynomials as monomials. So we could re-define the Legendre polynomials by dividing them with the coefficient of the leading term. However, we would use the generally accepted definitions.

Starting from $P_0(x)$ and $P_1(x)$ and progressing to higher-order polynomials, it is readily seen that $P_n(x)$ would be orthogonal to $1, x, x^2, \dots, x^{n-1}$. Thus, $P_n(x)$ will be orthogonal to *all* polynomials (not only Legendre polynomials) of degree equal to or less than $n - 1$.

Now, let us revisit the problem of approximating the function, $f(x^*) = a_0 + a_1 x^* + a_2 x^{*2}$, by a straight line $f_1(x^*) = c_0 + c_1 x^*$ over the domain $[a, b]$. We first change the domain to $[-1, 1]$ by using the transformation

$$x = \frac{2x^* - b - a}{b - a}$$

which results in

$$f(x) = a_0 + a_1 \frac{b+a}{2} + a_2 \left(\frac{b+a}{2}\right)^2 + \left(a_1 \frac{b-a}{2} + a_2 \frac{b^2-a^2}{2}\right) x + a_2 \left(\frac{b-a}{2}\right)^2 x^2$$

If we work with the function in domain (a, b), the polynomial orthogonal basis functions could be written as

$$1, x - \frac{a+b}{2}, \frac{3}{2}x^2 - \frac{3}{2}(a+b)x + \frac{a^2+4ab+b^2}{4}, \dots,$$

which reduces to the Legendre polynomials for $[-1, 1]$. However, it is more convenient to use the standard domain $[-1, 1]$.

Then we write $f_1(x) = \alpha_0 P_0(x) + \alpha_1 P_1(x)$ and obtain the coefficients as

$$\alpha_0 = \frac{\langle P_0(x), f(x) \rangle}{\langle P_0(x), P_0(x) \rangle} = a_0 + a_1 \frac{b+a}{2} + a_2 \frac{(b+a)^2}{6}$$

$$\alpha_1 = \frac{\langle P_1(x), f(x) \rangle}{\langle P_1(x), P_1(x) \rangle} = a_1 \frac{b-a}{2} + a_2 \frac{b^2-a^2}{2}$$

and, finally, on transforming back to the variable x^*

$$c_0 = a_0 - \frac{b^2+a^2+4ab}{6} a_2 \text{ and } c_1 = a_1 + (b+a)a_2$$

which are same as those obtained earlier.

Example 4.3
The function $f(x) = 1 + x + x^2$ has to be approximated by a linear function over the interval $[0, 1]$. Find the approximating function by using the Legendre polynomials.

Solution
Since the interval is $[0, 1]$, we would need to transform the variable such that the domain becomes $[-1, 1]$. We define $y = 2x - 1$ so that

$$f(y) = 1 + \left(\frac{y+1}{2}\right) + \left(\frac{y+1}{2}\right)^2 = 1.75 + y + 0.25y^2$$

Using the basis functions as the Legendre polynomials of order 0 and 1 as $P_0(y) = 1$ and $P_1(y) = y$, and writing the approximating polynomial as $f_1(y) = c_0 P_0(y) + c_1 P_1(y)$, we get

$$c_0 = \frac{\langle P_0(y), f(y) \rangle}{\langle P_0(y), P_0(y) \rangle} = \frac{\int_{-1}^{1} 1 \cdot (1.75 + y + 0.25y^2) dy}{2/(2 \times 0 + 1)} = \frac{11}{6}$$

$$c_1 = \frac{\langle P_1(y), f(y) \rangle}{\langle P_1(y), P_1(y) \rangle} = \frac{\int\limits_{-1}^{1} y \cdot (1.75 + y + 0.25 y^2) dy}{2/(2 \times 1 + 1)} = 1$$

Finally, transforming back to the variable x, we get

$$f_1(x) = \frac{11}{6} + 1 \times (2x - 1) = \frac{5}{6} + 2x$$

the same as that obtained in Example 4.2.

For this simple case, the advantage of using orthogonal polynomials is not apparent. However, for fitting higher order polynomials, it results in considerable saving of computations. Also, the normal equations for nonorthogonal basis become increasingly ill-conditioned as the degree of the approximating polynomial increases and may lead to large errors in the solution. Another advantage of the orthogonal polynomials is that the coefficients are independent of the degree of polynomial and addition of a higher degree term does not require re-computation of *all* coefficients. For example, after computing the coefficients of an mth degree polynomial, if we find that the approximating polynomial is not sufficiently close to the function $f(x)$, we can add the $(m + 1)$th degree basis function and find its coefficient as

$$c_{m+1} = \frac{\langle \phi_{m+1}, f \rangle}{\langle \phi_{m+1}, \phi_{m+1} \rangle}$$

All previously computed coefficients $c_0, c_1, \ldots, c_m$ will remain same. However, if we use the normal equations with non-orthogonal polynomial basis, we need to solve the entire system of $m + 2$ equations to compute the values of the coefficients $c_0, c_1, \ldots, c_{m+1}$. Note that we may write an iterative scheme (similer to those discussed in Chapter 2) which uses the previously obtained values of the coefficients as an initial guess and converges rapidly to the solution.

The fact that the *integral* of the *square* of the *error* has been minimised leads us to conclude that the approximating polynomial is the *best* in an *overall* sense. It may be argued that squaring of the error term leads to more weight being given to larger errors and using the absolute value (modulus) of the error may be more logical. Also, it may be seen (Line D in Fig. 4.1) that the error is larger near the ends of the interval compared to that in the middle. So if we have to compute the function value near the end points using the approximating polynomial, it may involve large errors. If no prior information is given about the point in (a, b) at which we want to evaluate the function, it may be a better idea to use the approximation which would minimise the maximum difference between the function and its approximation over the domain (a, b). In mathematical terms, we aim for minimising the L_2 norm (least squares), L_1 norm (least absolute deviation) or the L_∞ norm (minimax or minmax). The least squares method has already been discussed and the least absolute deviation method is not used very frequently since it is not easily amenable to analytical treatment. In the next subsection we describe the minimax method and introduce the Tchebycheff polynomials closely associated with minimax (or uniform) approximation.

4.3.2 Minimax Approximation

The least squares method provides us with an approximation of a function which is best in an overall sense since the integral of the squared error over the relevant domain is minimum. However, if we now use this approximation to estimate the function value at a particular point, we may get a larger error than that obtained from some other approximation. In fact, if we know that the function value is to be estimated close to a specified point, the Taylor's series expansion about that point is likely to be better than the least squares method. However, since the point at which the function has to be estimated is not known *a priori* but may be anywhere in the given domain, it is logical to think of an approximation which would minimize the maximum difference between the function and its approximation over the entire domain. This will ensure that no matter where the point is within the domain, the approximated value of the function will be within a certain *distance* from the actual value and this distance would be minimum for the minimax approximation. A general technique for obtaining the minimax approximation is beyond the scope of this text. However, a closely related technique based on the Tchebycheff (or Chebyshev) polynomials is described next.

4.3.3 Tchebycheff Polynomials

Consider the problem of obtaining the minimax approximation of a nth degree polynomial, $\sum_{j=0}^{n} a_j x^j$, by a $(n-1)$th degree polynomial, $\sum_{j=0}^{n-1} \alpha_j x^j$, over the domain $[-1, 1]$ (for convenience, we assume that $n > 1$ since some of the expressions derived in this section do not directly extend to $n = 1$). This is equivalent to the problem:

$$\text{minimize} \ \sup_{x \in [-1,1]} \left| \sum_{j=0}^{n} c_j x^j \right| \ \text{with} \ c_j = a_j - \alpha_j \ \text{for} \ j = 0, 1, ..., n-1 \ \text{and} \ c_n = a_n$$

(4.13)

i.e., to find the nth degree polynomial, with leading coefficient c_n, having the smallest maximum norm in $[-1, 1]$. These polynomials, with the coefficient c_n assigned a value such that the maximum norm of the polynomial becomes unity, are called *Tchebycheff polynomials* and are denoted by $T_n(x)$. Using the Tchebycheff alternation theorem (Theorem 4.3) and the fact that $T_n'(x)$ is a polynomial of degree $n - 1$,

---

Theorem 4.3 Tchebycheff Alternation Theorem

The theorem states that
An nth degree polynomial, $f_n(x)$, is the (unique) minimax approximation of the continuous function $f(x)$ over $[a, b]$, if and only if, there are *at least* $n + 2$ points $a \le x_1 < x_2 \ldots < x_{n+2} \le b$ at which the residual, $f(x) - f_n(x)$, attains its maximum magnitude with alternating signs.

Proof: We first prove the sufficient condition, i.e., show that if there are at least $n + 2$ such points, $f_n(x)$ would be minimax approximation of $f(x)$.

We denote the maximum magnitude of the residual,

$$\|f(x) - f_n(x)\|_\infty = M$$

Now, let us assume that there is another nth degree polynomial, $g_n(x)$, such that $\|f(x) - g_n(x)\|_\infty < M$.

Since $g_n(x) - f_n(x) = [f(x) - f_n(x)] - [f(x) - g_n(x)]$, and $f(x) - f_n(x) = \pm M$ at all the alternation points $(x_1, x_2, \ldots, x_{n+2})$, it is obvious that $g_n(x) - f_n(x)$ would have the same sign as $f(x) - f_n(x)$ at all these points. Therefore, $g_n(x) - f_n(x)$ would also have alternating signs at the $n + 2$ alternation points (generally called the *critical points*) and, consequently, must have at least $n + 1$ zeros. However, since $g_n(x) - f_n(x)$ is an nth degree polynomial, it must be identically zero, contradicting the assumption that $g_n(x)$ is different from $f_n(x)$. A rigorous proof of the necessary condition, i.e., to show that if there are less than $n + 2$ such points, an approximating polynomial *better than* $f_n(x)$ could be obtained, is quite involved. Here we provide a brief outline:

Let there be $m + 1$ critical points in order of increasing x as $x_0 (\ge a)$, $x_1, x_2, \ldots, x_m (\le b)$ such that the residual $f(x_i) - f_n(x_i) = (-1)^i M$ [Note that we have assumed the residual to be positive at the first critical point (x_0). The proof is similar for the case when it is negative].

We now divide the domain $[a, b]$ into $m + 1$ intervals $[a, \chi_1], [\chi_1, \chi_2], \ldots$ $[\chi_m, b]$ in such a way that each χ_i is in the open interval (x_{i-1}, x_i). This ensures that within each interval there is only one critical point.

It is then obvious that in each of these intervals there is only one point (the critical point) where the magnitude of the residual is equal to M and everywhere else it is smaller. For example, considering the first interval $[a, \chi_1]$, the residual is $+M$ at x_0 and lies in the open range $(-M, M)$ at all other points. This implies that we can find a positive δ_1 smaller than M, such that the residual in the first interval satisfies $M \geq f(x) - f_n(x) \geq -M + \delta_1$. If we have a function, say, $g(x)$, which is non-negative over the closed interval $[a, \chi_1]$ and has a maximum magnitude of N_1, it is easily shown that *in the first interval*

$$\|f(x) - f_n(x) - \alpha_1 g(x)\| < M \quad \text{where} \quad 0 \leq \alpha_1 \leq \frac{\delta_1}{N_1}$$

(to account for the case when a is a critical point, we stipulate that $g(x)$ does not vanish at a. However, since the χ_s', by definition, are not critical points, $g(x)$ may be allowed to be zero there). Similarly, for the second interval, the residual is equal to $-M$ at one point (x_1) and there exists another positive number δ_2 smaller than M, such that $M - \delta_2 \geq f(x) - f_n(x) \geq -M$.

Again, if there is a function $g(x)$ which is nonpositive over this interval and has a maximum magnitude of N_2, we have in the second interval

$$\|f(x) - f_n(x) - \alpha_2 g(x)\| < M \quad \text{where} \quad 0 \leq \alpha_2 \leq \frac{\delta_2}{N_2}$$

Extending this argument, it can be concluded that if a function $g(x)$ is alternately non-negative and non-positive in the intervals $[a, \chi_1]$, $[\chi_1, \chi_2]$, ..., $[\chi_m, b]$, and does not vanish at $x = a$ and $x = b$, then the function $f_n(x) + \alpha g(x)$ would be a better approximation [compared to $f_n(x)$] of $f(x)$ in the maximum norm over $[a, b]$, α being a positive number less than the minimum of

$$\frac{\delta_1}{N_1}, \frac{\delta_2}{N_2}, ..., \frac{\delta_{m+1}}{N_{m+1}}$$

It turns out that the function

$$g(x) = \prod_{i=1}^{m} (\chi_i - x)$$

satisfies the required conditions. Since this is a polynomial of order m, $f_n(x) + \alpha g(x)$ would be a polynomial of order n if $m \leq n$ and it would contradict the fact that $f_n(x)$ is the best approximating nth degree polynomial in the maximum norm. Therefore, m must be greater than or equal to $n + 1$ and the number of critical points (which was assumed as $m + 1$) must be greater than or equal to $n + 2$.

which can have *at most* $(n-1)$ zeros in $[-1, 1]$, we deduce that there are *exactly* $n + 1$ distinct points $-1 = x_1 < x_2 < x_3 \ldots < x_n < x_{n+1} = 1$ such that $|T_n(x_i)| = T_{max}$ for $i = 1$ to $n + 1$ (T_{max} being the maximum norm of $T_n(x)$ over $[-1, 1]$) and, applying the stationary points theorem, $T_n'(x_i) = 0$ for $i = 2$ to n (1 and $n + 1$ are excluded since these are boundary points and the derivative need not vanish even if the function has an extremum).

These properties, along with the facts listed below enable us to write expressions for the Tchebycheff polynomials as follows:

$T_{max}^2 - T_n^2(x)$ has simple zeros at $x = -1$ and $x = 1$ and double zeros at $x_2, x_3, \ldots, x_n$. [Recall that the derivative $T_n'(x_i)$ is not zero at the end points and vanishes only at points $x_2, x_3, \ldots, x_n$] therefore

$$T_{max}^2 - T_n^2(x) = A(1-x^2)\left[\prod_{i=2}^{n}(x-x_i)\right]^2$$

where A is a constant. Since the leading term in $T_n^2(x)$ is $c_n^2 x^{2n}$, it follows that

$$T_n^2(x) = T_{max}^2 - c_n^2(1-x^2)\left[\prod_{i=2}^{n}(x-x_i)\right]^2$$

Similarly, since $T_n'(x_i) = 0$ for $i = 2$ to n and the leading term in $T_n'(x)$ is $nc_n x^{n-1}$,

$$T_n'(x) = nc_n\prod_{i=2}^{n}(x-x_i) = \pm n\sqrt{\frac{T_{max}^2 - T_n^2(x)}{1-x^2}} \tag{4.14}$$

Equation (4.14) may be differentiated and the terms rearranged, using the fact that

$$\frac{d}{dx}\sqrt{T_{max}^2 - T_n^2(x)} = \frac{-T_n(x)T_n'(x)}{\sqrt{T_{max}^2 - T_n^2(x)}} = \frac{\mp nT_n(x)}{\sqrt{1-x^2}}$$

to get the following differential equation:

$$\frac{d}{dx}\left[\sqrt{1-x^2}\,\frac{dT_n(x)}{dx}\right] + \frac{n^2}{\sqrt{1-x^2}}T_n(x) = 0 \tag{4.15}$$

which is known as the Tchebycheff differential equation [note the similarity with the Legendre equation, Eq. (4.10)]. On putting $x = \cos\theta$ and using the conditions that these are polynomials with a maximum norm of unity, we get

$$T_n(x) = \cos(n\cos^{-1}x) \tag{4.16}$$

The first few Tchebycheff polynomials are listed below:

$$T_0(x) = 1 \quad T_1(x) = x, \quad T_2(x) = 2x^2 - 1, \quad T_3(x) = 4x^3 - 3x, \quad T_4(x) = 8x^4 - 8x^2 + 1$$

and it can be easily shown that the leading coefficient for $T_n(x)$ is 2^{n-1} for $n > 0$. Returning to our original problem of approximating the nth degree polynomial, with a leading coefficient a_n, by a $(n-1)$th degree polynomial using the minimax criterion, it can be seen that the residual $\sum_{j=0}^{n} c_j x^j$ should be equal to $2^{1-n} a_n T_n(x)$, from which the coefficients α could be obtained.

Example 4.4 The function $f(x) = 1 + 2x + 3x^2$ has to be approximated by a linear function over the interval $[-1, 1]$. Find the approximating function by minimizing the L_2 and L_1 norms.

Solution The minimization of L_2 norm is achieved by using the Legendre polynomials, and the approximating function is obtained as

$$f_1(x) = \frac{\langle P_0(x), f(x) \rangle}{\langle P_0(x), P_0(x) \rangle} P_0(x) + \frac{\langle P_1(x), f(x) \rangle}{\langle P_1(x), P_1(x) \rangle} P_1(x)$$

$$= \frac{\int_{-1}^{1} 1 \cdot (1 + 2x + 3x^2) dx}{2/(2 \times 0 + 1)} + \frac{\int_{-1}^{1} x \cdot (1 + 2x + 3x^2) dx}{2/(2 \times 1 + 1)} x = 2 + 2x$$

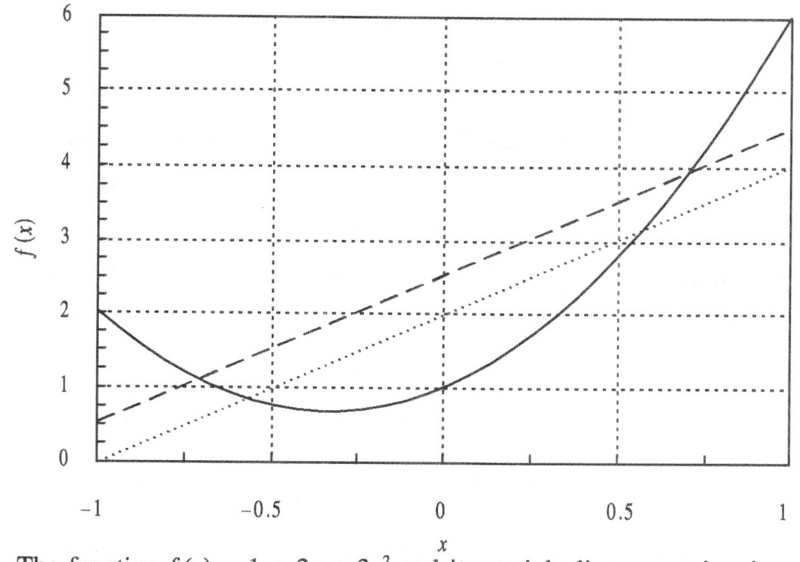

Fig. 4.4 The function $f(x) = 1 + 2x + 3x^2$ and its straight line approximation over the domain $(-1, 1)$. [Solid line: function, Dashed line: best L_1 approximation, Dotted line: best L_2 approximation.]

The minimization of the L_1 norm is obtained from the Tchebycheff polynomial of degree 2 (note that $n = 2$, and the leading coefficient $a_n = 3$), by writing the second degree polynomial with a leading coefficient of 3 and having a minimum L_1 norm as

$$c_0 + c_1 x + c_2 x^2 = 2^{-1}3(2x^2 - 1) = 3x^2 - \frac{3}{2}$$

giving $c_0 = -3/2$, $c_1 = 0$, and $c_2 = 3$.

From Eq. (4.13), $\alpha_0 = a_0 - c_0 = 5/2$ and $\alpha_1 = a_1 - c_1 = 2$ indicating that the straight line $5/2 + 2x$ is the best fit in the minimax sense. Figure 4.4 shows both of these approximations. The following may be noted from this figure:

- Minimax approximation of any function $f(x)$ by a straight line ($n = 1$) should have *at least* 3 distinct points where the residual attains its maximum magnitude with alternating signs. The residual is seen to be 1.5, -1.5, and 1.5 at $x = -1, 0$, and 1, respectively.
- If the function to be approximated is a polynomial of the next higher degree, there are *exactly* three such alternation points which include the two end points.
- The maximum magnitude of the residual using the least squares fit is 2. The least squares fit is much better in the centre but has large errors near the ends.

4.3.3.1 *Some Properties of Tchebycheff Polynomials*

(a) Similar to the Rodrigues' formula for the Legendre polynomials, we have

$$T_n(x) = \frac{(-2)^n n! \sqrt{1-x^2}}{2n!} \frac{d^n}{dx^n} \left[(1-x^2)^{n-\frac{1}{2}} \right]$$

(b) The recursive relation for the Tchebycheff polynomials is

$$T_n(x) = 2x T_{n-1}(x) - T_{n-2}(x), \quad n = 2, 3, \ldots$$
$$\text{with } T_0(x) = 1, T_1(x) = x$$

(c) In the domain $[-1, 1]$, $T_n(x)$ has n zeros, called the Tchebycheff abscissae, and $(n + 1)$ extrema, where it is equal to $+1$ or -1. These are given by

$$\text{zeros}: \quad x_i = \cos\left(\frac{2i+1}{n} \frac{\pi}{2} \right), \quad i = (n-1), (n-2), \ldots, 1, 0$$

$$\text{extrema}: \quad x_i = \cos\left(\frac{i}{n} \pi \right), \quad i = n, (n-1), \ldots, 1, 0$$

(d) $T_n(x)$ has the smallest maximum norm (equal to 1) in $[-1, 1]$ out of all nth degree polynomials with leading coefficient 2^{n-1}. In other words, for all monic polynomials, i.e., a polynomial in which the coefficient of the highest order term is unity, of nth degree, $2^{1-n} T_n(x)$ has the smallest maximum norm of 2^{1-n}.

4.3.3.2 *Orthogonality of Tchebycheff Polynomials*

Using the orthogonality of the cosine function,

$$\int_0^\pi \cos m\theta \cos n\theta \, d\theta = 0, \quad \text{for } m \neq n$$

it is easily seen that

$$\int_{-1}^1 T_m(x) T_n(x) \frac{1}{\sqrt{1-x^2}} dx = \begin{bmatrix} 0, & m \neq n \\ \pi, & m = n = 0 \\ \pi/2, & m = n \neq 0 \end{bmatrix} \tag{4.17}$$

Thus, if we modify our earlier definition of the inner product (Eq. B4.1.1) as

$$\langle f, g \rangle = \int_a^b w(x) f(x) g(x) dx \tag{4.18}$$

where $w(x)$ is a weight function, the Tchebycheff polynomials are orthogonal over the interval $(-1, 1)$ with respect to the weight function $1/\sqrt{1-x^2}$. The weight function is continuous and positive over (a, b) but may have integrable singularities at the end points. The simplest form, $w(x) = 1$, has already been discussed. Some other possibilities are: (i) $w(x) = \exp(-x^2)$ over the interval $(-\infty, \infty)$ which gives rise to orthogonal polynomials called the *Hermite polynomials* and (ii) $w(x) = \exp(-x)$ over $(0, \infty)$ which results in *Laguerre polynomials*.

Introduction of the weight allows us to manipulate the problem to account for desired objective and/or additional information. For example, if we wish to approximate a function over the interval $[-1, 1]$ but know that its argument is more likely to lie in the range $[-0.5, 0.5]$ than outside it, we will choose the weight accordingly. The Legendre and Tchebycheff polynomials are special cases of the Jacobi polynomials which are orthogonal with respect to the weight function $w(x) = (1-x)^\alpha (1+x)^\beta$. For Legendre polynomials, $\alpha = \beta = 0$ and for Tchebycheff, $\alpha = \beta = -1/2$.

Similarly, as discussed earlier, since the *unweighted* (or *ordinary*) least squares fit resulted in larger errors near the end points, having a larger weight near the ends tends to provide a fit which distributes the error more evenly over the interval. The orthogonality condition may be used to obtain the approximation of any function in terms of the Tchebycheff polynomials as described in the following example.

Example 4.5 The function $f(x) = \sqrt{1-x^2}$ has to be approximated by a second degree polynomial over the interval $[-1, 1]$. Find the approximating function by using Tchebycheff polynomials.

Solution We write the approximating polynomial as $f_2(x) = c_0 T_0(x) + c_1 T_1(x) + c_2 T_2(x)$ and obtain the coefficients by using the orthogonality property. We then have

$$f_2(x) = \frac{\langle T_0(x), f(x) \rangle}{\langle T_0(x), T_0(x) \rangle} T_0(x) + \frac{\langle T_1(x), f(x) \rangle}{\langle T_1(x), T_1(x) \rangle} T_1(x) + \frac{\langle T_2(x), f(x) \rangle}{\langle T_2(x), T_2(x) \rangle} T_2(x)$$

$$= \frac{\displaystyle\int_{-1}^{1} 1 \cdot \sqrt{1-x^2} \cdot \frac{1}{\sqrt{1-x^2}}\, dx}{\pi} + \frac{\displaystyle\int_{-1}^{1} x \cdot \sqrt{1-x^2} \cdot \frac{1}{\sqrt{1-x^2}}\, dx}{\pi/2} \cdot x$$

$$+ \frac{\displaystyle\int_{-1}^{1} (2x^2 - 1) \cdot \sqrt{1-x^2} \cdot \frac{1}{\sqrt{1-x^2}}\, dx}{\pi/2} (2x^2 - 1) = \frac{10}{3\pi} - \frac{8}{3\pi} x^2$$

Figure 4.5 shows the plot of the function and its approximation.

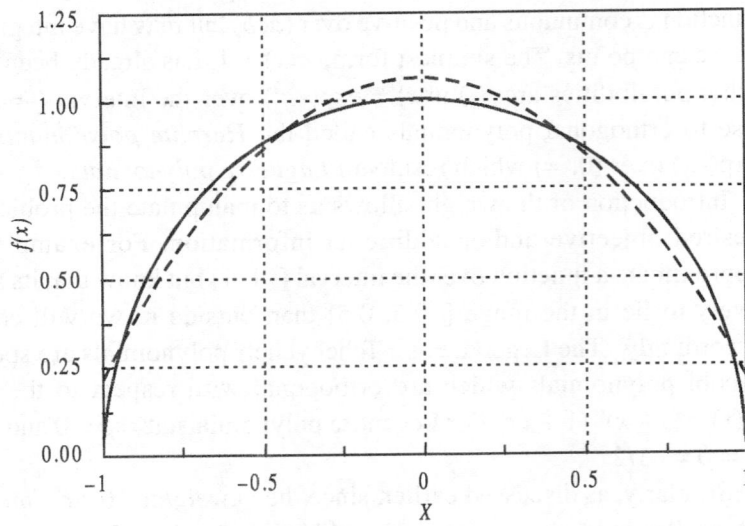

Fig. 4.5 The function $f(x) = \sqrt{1-x^2}$ and its second degree Tchebycheff polynomial approximation over the domain $[-1, 1]$. Solid line: function, Dashed line: Tchebycheff approximation.

4.3.4 Remarks on Use of Orthogonal Polynomials

- The function, $f(x)$, should be of a form which allows easy evaluation of the integrals required in the computations. If the integrals cannot be obtained analytically, e.g., $f(x) = x^x$, we may perform numerical integration (described in the next chapter) or may go for discrete data fit (described in the next section).
- For some function, e.g., $f(x) = e^x$, the Legendre polynomial fit may be easily obtained but the Tchebycheff fit requires numerical integration.

Many times the functional relationship between the dependent variable and the independent variable(s) is not known. The only information we have is the observed (or computed) value of the function, $f(x)$, at a few values of x. While numerical integration is a possibility in such cases, it is generally better to go for discrete equivalents of the methods discussed in this section.

EXERCISE 4.3

1. Approximate the function $\exp(x)$ over $[-1, 1]$ by a straight line using (a) Legendre polynomials and (b) Tchebycheff polynomials.
   For the Tchebycheff method use the following values:

$$\int_{-1}^{1} \frac{e^x dx}{\sqrt{1-x^2}} = 3.97746, \quad \int_{-1}^{1} \frac{xe^x dx}{\sqrt{1-x^2}} = 1.77550$$

2. For the previous problem, plot the residual in the Tchebycheff approximation and ascertain whether it is a minimax approximation using the alternation theorem. Comment on the result.

3. Although we have not described the methods for obtaining the best L_∞ and L_1 approximations for a general function, we list these approximations for the exponential function over the interval $[-1, 1]$:
   (a) Minimax: $f_1(x) = 1.26428 + 1.17520x$, (b) Least absolute deviation (LAD): $f_\cdot(x) = 1.12763 + 1.04219x$. The exact values of the coefficients are Minimax: $c_1 = 0.5(e - e^{-1})$, $c_0 = 0.5(e - c_1\ln c_1)$ and LAD: $c_0 = 0.5(e^{0.5} - e^{-0.5})$, $c_1 = e^{0.5} - e^{-0.5}$. Plot these two approximations along with the two obtained in Problem 1 and comment on their relative positions.

4. For all four approximations of $\exp(x)$, Legendre (minimizing L_2), Minimax (minimizing L_1), LAD (minimizing L_∞), and Tchebycheff (minimizing L_1 if the function is a polynomial of one degree higher than the approximating polynomial), obtain the L_1, L_2, and L_∞ norms of the residuals. (Note: While L_2 is straightforward to compute, other two norms cannot be readily obtained. For this problem, it would be possible to compute these norms by observing

that there are two points of intersection and the residual is negative between these points and positive elsewhere. The points of intersection are Legendre: -0.533232 and 0.620700, Minimax: -0.616401 and 0.779467, LAD: -0.5 and 0.5, and Tchebycheff: -0.665210 and 0.746749)

5. Consider the approximation of the function

$$f(\theta) = \frac{\pi^2}{2\pi^2 + \theta^2 - 2\pi\theta}$$

in the interval $[0, 2\pi]$. Approximate the function by employing a Legendre basis $\{P_j(x)\}_{j=0}^{j=3}$ after mapping the θ-domain to the x-domain in such a way that $[0, 2\pi]$ maps into $[-1, 1]$. Graphically compare the function (in the x-domain) and the approximating polynomial.

6. Approximate the function

$$f(x) = \frac{1}{1 + 25x^2}$$

over the interval $[-1, 1]$ by a second order polynomial minimizing the L_2 norm. Improve the fit by using a fourth order polynomial and plot both these approximations.

7. Consider the polynomial $\{p_n(x)\}$ given by

$$p_n(x) = \frac{\sin(n+1)\phi}{\sin\phi}$$

where $x = \cos\phi$.

Show that this polynomial satisfies the same recursion formula as the Tchebycheff polynomial. What are the first three polynomials, $p_0(x), p_1(x)$, and $p_2(x)$? Show that $\{p_n(x)\}$ forms an orthogonal system of polynomials with the weight function $w(x) = (1 - x^2)^{1/2}$, $x \in [-1, 1]$.

8. Consider the approximation of the function $f(t) = e^{t^2/24}$ in the interval $[-2\pi, 2\pi]$. First map the t-domain to the x-domain in such a way that $[-2\pi, 2\pi]$ (in t-domain) maps into $[-1, 1]$ (in x-domain). Approximate the function by employing a Legendre basis $\{P_j(x)\}_{j=0}^{j=4}$. Graphically compare the function to be approximated with the resulting approximants.

9. Approximate the function x^3 over the range $[0, 1]$ by a *straight line* using the Legendre polynomials. If we want an approximation without transformation of variable, we define functions which are orthogonal over $[0, 1]$ as opposed to Legendre polynomials which are orthogonal over $[-1, 1]$. Taking the 0th degree polynomial as 1, derive the first degree orthogonal polynomial over $[0, 1]$ and obtain the best-fit *straight line* approximation to x^3 over the range $[0, 1]$. Compare with the previously obtained result.

4.4 APPROXIMATION OF DATA

As described in the beginning of this chapter, if we have a table of data listing the values of the dependent variable, $f(x)$, corresponding to a few values of the independent variable, x, we may choose the approximating function, $\tilde{f}(x)$, either to pass through *all* data points (interpolation) or to represent the general trend of the data (regression) possibly without passing through *any* data point. We denote the data points by the set of values $\{(x_k, f(x_k)), k = 0, 1,..., n\}$. (For interpolation problems, we require all x's to be distinct.) The form of the approximating function would depend on the type of data but we again assume it to be an mth degree polynomial, $f_m(x)$, represented by Eq. (4.8), i.e.,

$$f_m(x) = \sum_{j=0}^{m} c_j \phi_j(x) \tag{4.19}$$

in which the ϕ's are the polynomial basis functions and the c's are coefficients. Obviously, for an interpolation problem, in general $m = n$ (a lower degree polynomial *may* interpolate the given data *in some cases*, e.g., three points lying on a straight line: $n = 2$, $m = 1$), and for regression, $m < n$ [when $m > n$, we will have infinite solutions, e.g., fitting a parabola ($m = 2$), through two points ($n = 1$)]. Clearly, the interpolation problems are conceptually simpler than the regression problems since we do not have to worry about the degree of approximating polynomial ($m = n$) and quantification of the residuals (all residuals are zero). Therefore, we discuss interpolation first and then move on to regression.

4.4.1 Interpolation

The problem can be stated as: given the function values at a set of $n + 1$ *distinct* points, find the polynomial (of degree *at most n*) which matches the function value at these points. Although this polynomial is unique (provided the function values are given at distinct points as shown in the next subsection), it may be expressed in many different forms. We describe below these alternative forms and look at the situations for which they are suitable.

4.4.1.1 *Conventional Form*

In the conventional representation, $\phi_j(x) = x^j$, and the coefficients c_j are obtained from the following set of linear equations representing the equality of the approximating polynomial and the function value at all data points:

$$
\begin{bmatrix}
1 & x_0 & x_0^2 & \cdots & x_0^n \\
1 & x_1 & x_1^2 & \cdots & x_1^n \\
\cdots & \cdots & \cdots & \cdots & \cdots \\
1 & x_n & x_n^2 & \cdots & x_n^n
\end{bmatrix}
\begin{Bmatrix}
c_0 \\ c_1 \\ \vdots \\ c_n
\end{Bmatrix}
=
\begin{Bmatrix}
f(x_0) \\ f(x_1) \\ \vdots \\ f(x_n)
\end{Bmatrix}
\qquad (4.20)
$$

This matrix is called the Vandermonde matrix and it is relatively straightforward to show that its determinant may be written as multiplication of terms of the form $(x_i - x_{j(\neq i)})$. This implies that a unique solution will exist for distinct points. The uniqueness may also be proved by arguing that if there is another nth degree interpolating polynomial, the difference of the two interpolating polynomials would also be an nth degree polynomial, which will vanish at all $(n+1)$ grid points (since both the interpolating polynomials reproduce the function value at the grid points). This implies that the difference is identically zero. The Vandermonde matrix is known to be ill-conditioned for large n. For example, if we take the domain to be $[-1, 1]$, increasing the number of points will (i) cause the points to come nearer to one another, and (ii) make the higher powers approach zero. Both these result in ill-conditioning of the matrix. Also, the addition of one or more data points or change in function values at an existing point necessitates the re-computation of all the coefficients. Therefore this method is not recommended.

4.4.1.2 *Lagrange Form*

The polynomials ϕ are chosen in such a way that each of them is an nth degree polynomial (denoted by L) and satisfies the following:

$$
L_i(x_j) = \delta_{ij} =
\begin{bmatrix}
1, & i = j \\
0, & i \neq j
\end{bmatrix}
$$

These are known as the Lagrange interpolating polynomials (proposed by Waring in 1779 and later by Lagrange in 1795) and it is apparent that these may be written as

$$
L_i(x) = \prod_{\substack{j=0 \\ j \neq i}}^{n} \frac{x - x_j}{x_i - x_j}
\qquad (4.21)
$$

Another useful form in which the Lagrange polynomials could be written is based on the fact that

$$\frac{d}{dx}\left[\prod_{j=0}^{n}(x-x_j)\right] = \sum_{k=0}^{n}\left[\prod_{\substack{j=0\\j\neq k}}^{n}(x-x_j)\right]$$

which, evaluated at $x = x_i$, is the denominator of Eq. (4.21), resulting in

$$L_i(x) = \frac{\displaystyle\prod_{j=0}^{n}(x-x_j)}{(x-x_i)\left[\dfrac{d}{dx}\left\{\displaystyle\prod_{j=0}^{n}(x-x_j)\right\}\right]_{x=x_i}} \qquad (4.22)$$

The equality condition at the data points then implies that the coefficients c_i are simply the function values at corresponding points, i.e., $c_i = f(x_i)$. This method is preferred when the grid points (i.e., the values of the independent variable) remain fixed but the function values keep changing, e.g., measurement of water depth at a few selected locations in a channel. If we want to predict the function value at an intermediate point, the L's have to be computed only once for that point.

The interpolated function value at that point is readily obtained for any particular set of observed values at the grid points. However, if we change the grid points, e.g., by adding or deleting an observation, the Lagrange polynomials need to be recomputed. Similarly, if the degree of the interpolating polynomial is not known *a priori* but is to be obtained from looking at the proximity of the fit to observed data for increasing number of grid points, Lagrange polynomials would not work very efficiently. For such cases Newton's (divided difference) form of the interpolating polynomials works better.

4.4.1.3 *Newton's (Divided Difference) Form*

The basis functions are chosen in such a way that ϕ_i is an ith degree polynomial defined as

$$\phi_0(x) = 1, \quad \phi_i(x) = \prod (x-x_0)(x-x_1)\ldots(x-x_{i-1}), \quad \text{for } i = 1, 2, \ldots, n+1 \qquad (4.23)$$

These are known as the Newton's divided difference polynomials and satisfy the following:

$$\phi_i(x_j) = 0, \quad \text{for } j < i$$

The coefficients, c_i, are again obtained by the equality of the interpolating polynomial and the given function at the grid points. For example, the equality at $x = x_0$ results in $c_0 = f(x_0)$ since all other basis functions (except ϕ_0) are zero at this point. At the next point ($x = x_1$), only ϕ_0 and ϕ_1 are non-zero with $\phi_0 = 1$ and

$\phi_1 = x_1 - x_0$. Therefore, the coefficient c_1 is obtained from the equality condition,

$$f(x_1) = c_0\phi_0(x_1) + c_1\phi_1(x_1) = f(x_0) + c_1(x_1 - x_0)$$

as $c_1 = \dfrac{f(x_1) - f(x_0)}{x_1 - x_0}$.

To obtain the next coefficient, c_2, we use the equality at $x = x_2$ as

$$f(x_2) = c_0\phi_0(x_2) + c_1\phi_1(x_2) + c_2\phi_2(x_2)$$
$$= f(x_0) + c_1(x_2 - x_0) + c_2(x_2 - x_0)(x_2 - x_1)$$

and obtain

$$c_2 = \frac{\dfrac{f(x_2) - f(x_1)}{x_2 - x_1} - \dfrac{f(x_1) - f(x_0)}{x_1 - x_0}}{x_2 - x_0}$$

For a more efficient recursive determination of the coefficients, we may think of the *additional* term, $c_2\phi_2(x_2)$, as the difference (at the grid point x_2) between the linear interpolating function over (x_0, x_1) and that over (x_1, x_2), see Fig. 4.6. Further, we note that the *unique* interpolating polynomial does not depend on the order of the grid points. For example, the linear interpolation over (x_0, x_1) would be same even when we use x_1 as the first point and x_0 as the second.

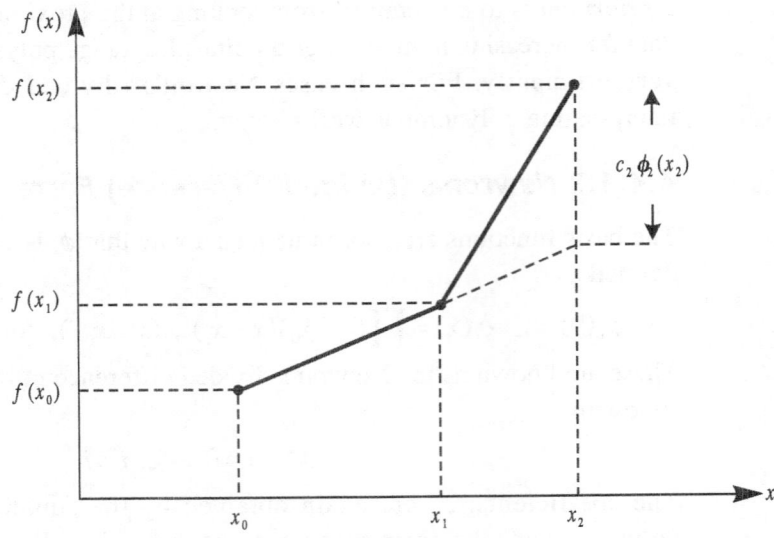

Fig. 4.6 Newton's interpolating polynomial.

The difference between these two interpolating functions at the point x_2 may then be written as

$$f(x_1) + (x_2 - x_1)\frac{f(x_2) - f(x_1)}{x_2 - x_1} - \left[f(x_1) + (x_2 - x_1)\frac{f(x_1) - f(x_0)}{x_1 - x_0} \right]$$

and, on equating it to $c_2(x_2 - x_0)(x_2 - x_1)$, we get the value of c_2.

The recursive relationship may be developed by assuming that all the coefficients up to c_i have been determined. The additional term, $c_{i+1}\phi_{i+1}(x_{i+1})$, is written as the difference (at the grid point x_{i+1}) of the ith degree interpolating polynomial passing through the points $(x_1, x_2, x_3, \ldots, x_1, x_{i+1})$ and that passing through the points $(x_1, x_2, \ldots, x_{i-1}, x_i, x_0)$ (note the change in order of points).

Clearly, since the first i grid points are same, the coefficients and basis functions up to the $(i-1)$th degree term would be identical for both these interpolating polynomials and the only difference will be in the ith term (even for this term, the basis function would be the same, see Eq. 4.23). This difference is written as

$$c_i'(x_{i+1} - x_1)(x_{i+1} - x_2)(x_{i+1} - x_3)\cdots(x_{i+1} - x_i) - c_i(x_{i+1} - x_1)(x_{i+1} - x_2)(x_{i+1} - x_3)$$

$$\cdots(x_{i+1} - x_i)$$

where c_i is the coefficient in the *original* grid point arrangement and c_i' represents the coefficient in the *modified* arrangement (removing x_0 and adding x_{i+1}).

Equating this difference to $c_{i+1}\phi_{i+1}(x_{i+1})$, we obtain

$$c_{i+1} = \frac{c_i' - c_i}{x_{i+1} - x_0}$$

We now use the standard definitions of the divided differences (i.e., the *difference* of a function *divided* by the *difference* of ordinates) as

1st divided difference: $f[x_j, x_i] = \dfrac{f(x_j) - f(x_i)}{x_j - x_i} \quad (= f[x_i, x_j])$

2nd divided difference: $f[x_k, x_j, x_i] = \dfrac{f[x_k, x_j] - f[x_j, x_i]}{x_k - x_i}$

$$(= f[x_k, x_i, x_j] = \cdots = f[x_i, x_j, x_k])$$

3rd divided difference: $f[x_l, x_k, x_j, x_i] = \dfrac{f[x_l, x_k, x_j] - f[x_k, x_j, x_i]}{x_l - x_i}$

$\ldots$

nth divided difference: $f[x_n, x_{n-1}, \ldots, x_1, x_0] = \dfrac{f[x_n, x_{n-1}, \ldots, x_1] - f[x_{n-1}, \ldots, x_1, x_0]}{x_n - x_0}$

$$(4.24)$$

BOX 4.3 Relation Between the Divided Difference and the Derivative

The function $g(x) = f(x) - f_n(x)$ will have *at least* $n + 1$ zeros (at all the grid points) in the interval spanned by the grid points $(x_0, x_1, \ldots, x_n)$. Repeated application of Rolle's theorem indicates that there would be at least one point, say ξ, in this interval at which the nth derivative of $g(x)$ will be zero.
Hence,

$$f^{[n]}(\xi) = \frac{d^n f_n(x)}{dx^n}\bigg|_{x=\xi} = \frac{d^n \sum\limits_{i=0}^{n} c_n \phi_n(x)}{dx^n}\bigg|_{x=\xi}$$

The basis functions up to the order $n - 1$ will not contribute to the nth derivative, and the only term would be

$$\frac{d^n c_n \phi_n(x)}{dx^n} = n! \, f[x_n, x_{n-1}, \ldots, x_1, x_0]$$

Therefore, a relation between the divided difference and the function derivative is obtained as

$$f[x_n, x_{n-1}, \ldots, x_1, x_0] = \frac{f^{[n]}(\xi)}{n!}$$

where $\xi \in (x_0, x_1, \ldots, x_n)$.

The coefficients c_i are then given by the divided differences as

$$\begin{aligned}
c_0 &= f[x_0] \equiv f(x_0) \\
c_1 &= f[x_1, x_0] \\
c_2 &= f[x_2, x_1, x_0] \\
&\;\;\vdots \\
c_i &= f[x_i, x_{i-1}, \ldots, x_0] \\
&\;\;\vdots \\
c_n &= f[x_n, x_{n-1}, \ldots, x_0]
\end{aligned} \tag{4.25}$$

and, for hand-computations, are efficiently obtained by using a divided difference table (see Example 4.6). The remainder, R_n, at any point x, is obtained by using an argument similar to that used in the derivation of the recursive relation, as

$$R_n(x) = f(x) - f_n(x) = \phi_{n+1}(x) f[x, x_n, x_{n-1}, \ldots, x_0] \tag{4.26}$$

Since $f(x)$ is unknown, the divided difference in Eq. (4.26) cannot be obtained.

However, if an additional point $(x_{n+1}, f(x_{n+1}))$ is available, it may be used in place of $f(x)$ to *approximate* the divided difference. This concept is similar to the one used in Chapter 1 where, in absence of the true value, an approximate error was defined as the difference between two successive iterations. Here, $R_n(x)$ is equivalent to $f_{n+1}(x) - f_n(x)$. A significant advantage of the Newton's method over the previous two forms is, therefore, the ease of estimation of error. On the other hand, if $f(x)$ is known in the functional form and its derivatives up to order $(n + 1)$ exist, we may use the relation between the divided difference and the function derivative (Box 4.3) and write (also see Eq. 3.6),

$$R_n(x) = \phi_{n+1}(x) \frac{f^{[n+1]}(\xi)}{(n+1)!} \qquad (4.27)$$

in which $\xi \in (x, x_0, x_1, ..., x_n)$. Again, since ξ is not known, we will not be able to compute the error at any point, x. We may, however, obtain an upper bound of the error from the behaviour of the $(n + 1)$th derivative of the function.

Remarks:
- The points x_j do not have to be in any particular order.
- While all the Lagrange polynomials are of the same order (n), the Newton polynomials are of increasing order $(0$ to $n)$.
- After computing the coefficients, additional effort is required in computing the interpolated value at a non-grid point. There is an algorithm (Neville's algorithm), similar to the divided difference algorithm, which provides the interpolated value at a point with almost same amount of computational effort as used in computing the coefficients. However, we will not discuss it in detail.

Example 4.6   Three weather stations, A, B, and C, are located along a straight road such that B is equidistant from A and C. There are two cities, D and E, on the same road, D being equidistant from A and B, and E from B and C. The temperatures (in °C) at a given time are recorded at A, B, and C, as 10.1, 11.3, and 11.9, respectively. Estimate the temperatures at D and E.

Solution   Although we could use any interval, we choose to have the domain AC as $(-1, 1)$. The coordinates of various points are, therefore, A: -1, B: 0, C: 1, D: -0.5, and E: 0.5. Using the measured values at $x_0 = -1$, $x_1 = 0$, and $x_2 = 1$, we obtain the second degree interpolating polynomial in the three different forms as follows:

(a) *Conventional form*
The interpolating polynomial is written as $f_2(x) = c_0 + c_1 x + c_2 x^2$ and the set of equations (see Eq. 4.20) is written as

$$\begin{bmatrix} 1 & x_0 & x_0^2 \\ 1 & x_1 & x_1^2 \\ 1 & x_2 & x_2^2 \end{bmatrix} \begin{bmatrix} c_0 \\ c_1 \\ c_2 \end{bmatrix} = \begin{Bmatrix} f(x_0) \\ f(x_1) \\ f(x_2) \end{Bmatrix} \Rightarrow \begin{bmatrix} 1 & -1 & 1 \\ 1 & 0 & 0 \\ 1 & 1 & 1 \end{bmatrix} \begin{bmatrix} c_0 \\ c_1 \\ c_2 \end{bmatrix} = \begin{Bmatrix} 10.1 \\ 11.3 \\ 11.9 \end{Bmatrix}$$

which provides the solution as $c_0 = 11.3$, $c_1 = 0.9$, and $c_2 = -0.3$. The estimated values at D ($x = -0.5$) is obtained as $10.775°C$ and at E ($x = 0.5$) as $11.675°C$.

(b) *Lagrange form*

Using Eq. (4.21), we obtain the Lagrange polynomials as

$$L_0(x) = \frac{(x - x_1)(x - x_2)}{(x_0 - x_1)(x_0 - x_2)} = \frac{x(x-1)}{2}$$

$$L_1(x) = \frac{(x - x_0)(x - x_2)}{(x_1 - x_0)(x_1 - x_2)} = \frac{(x+1)(x-1)}{-1}$$

$$L_2(x) = \frac{(x+1)x}{2}$$

The interpolating polynomial is written as

$$f_2(x) = f(x_0)L_0(x) + f(x_1)L_1(x) + f(x_2)L_2(x)$$

and it may be easily verified that it reduces to the same form as in (a).

To estimate the temperature at the points D and E, we compute the values of the Lagrange polynomials at these points as

$$L_0(-0.5) = 0.375$$
$$L_1(-0.5) = 0.75$$
$$L_2(-0.5) = -0.125$$

and

$$L_0(0.5) = -0.125$$
$$L_1(0.5) = 0.75$$
$$L_2(0.5) = 0.375$$

The interpolated temperatures are, therefore,

at *D:* $10.1 \times 0.375 + 11.3 \times 0.75 + 11.9 \times (-0.125) = 10.775 °C$

at *E:* $10.1 \times (-0.125) + 11.3 \times 0.75 + 11.9 \times 0.375 = 11.675 °C$

(c) *Newton form*

The interpolating polynomial is written as $f_2(x) = c_0 + c_1(x - x_0) + c_2 (x - x_0)(x - x_1)$

with $c_0 = f(x_0)$, $c_1 = f[x_0, x_1]$, $c_2 = f[x_0, x_1, x_2]$. Since the order of the points is not important, we would use the point B $(x = 0)$ as x_0 and A $(x = -1)$ as x_1. The divided differences are computed in a tabular form as shown below:

x	$f(x)$	$f[x_i, x_j]$	$f[x_i, x_j, x_k]$
0	11.3		
		1.2	
-1	10.1		-0.3
		0.9	
1	11.9		

The divided differences are shown in the topmost number in each column giving us

$$f[x_0, x_1] = 1.2$$

$$f[x_0, x_1, x_2] = -0.3$$

and

$$f_2(x) = 11.3 + 1.2x - 0.3x(x+1) = 11.3 + 0.9x - 0.3x^2, \text{ the same as before.}$$

- We assume that interpolation is applicable to the given data. In some cases, e.g., if C and E are located at higher ground, interpolation may not be directly applicable. We may use a *weighted* interpolation assigning higher weight to the measurement at C to estimate the temperature at E. However, this will not be discussed here.

The example above shows different ways of obtaining the *unique*[2] interpolating polynomial. The polynomial will pass through the function values at all the grid points used for its development. However, at any intermediate point, the estimated

---

[2]It is obvious that higher order polynomials would be non-unique since if $f_n(x)$ interpolates the function value at all grid points, any polynomial of the form $f_n(x) + A(x) \prod_{j=0}^{n} (x - x_j)$ would also do so since the second term vanishes at all grid points. One may try to get a higher order interpolating polynomial by omitting a few lower order terms, e.g., for three data points, x_c, x_1, and x_2, using an interpolating polynomial of the form $c_0 + c_2 x^2 + c_3 x^3$. However, a *unique* solution (or even *a solution*) is not guaranteed in such cases, see Exercise 4.4, Problem 8. Also, the computation of the interpolating polynomial is not very efficient since it cannot be expressed in Lagrange or Newton forms.

function value may be in error (unless, of course, the function itself is an nth degree polynomial or we are lucky enough to choose a point where the interpolating polynomial matches with the function value). As mentioned earlier, the interpolating polynomial is unique and, therefore, the error is also same irrespective of whether we express it in the conventional, Lagrange, or Newton form. Since the true value of the function will generally be unknown at any point other than the grid points, it is generally not possible to estimate the true error. On the other hand, for the interpolation to be useful, we must have some way of estimating the error. Equations (4.26) and (4.27) are two alternative ways of doing it.

While the three forms discussed above are applicable to any distribution of the grid points, most of the times we will have equally spaced points $x_0, x_1, \ldots, x_n$. Also, sometimes we will have complete flexibility in choosing these points, e.g., when deciding on times at which to measure the distance travelled by an object. For the first case, a set of orthonormal polynomials, closely related to Legendre polynomials and known as Gram's polynomials, are useful. For the second case, if we aim at minimising the maximum error of interpolation, discrete form of Legendre or Tchebycheff polynomials are used. These are described next.

4.4.2 Gram's Polynomials

As in Legendre polynomials, we assume that the range of the (equidistant) data is normalised such that $x_0 = -1$ and $x_n = 1$, which implies that $x_i = -1 + 2i/n$, $i = 0$ to n. The Gram's polynomials then satisfy the orthonormality condition (note that the inner product is now defined by a sum as opposed to the integral used when the function was given[3])

$$\langle G_i(x), G_j(x) \rangle = \sum_{k=0}^{n} G_i(x_k)G_j(x_k) = \begin{bmatrix} 0, & i \neq j \\ 1, & i = j \end{bmatrix} \tag{4.28}$$

The general equation for generating Gram's polynomials of order n is

$$G_{i+1}(x) = \alpha_i x G_i(x) - \frac{\alpha_i}{\alpha_{i-1}} G_{i-1}(x), \quad \text{for } i = 0, 1, 2, \ldots, n-1 \tag{4.29}$$

with $\qquad G_{-1}(x) = 0; G_0(x) = \dfrac{1}{\sqrt{n+1}}$ and $\alpha_i = \dfrac{n}{i+1}\sqrt{\dfrac{4(i+1)^2 - 1}{(n+1)^2 - (i+1)^2}}$

---

[3]The notation used here for inner product, <...>, is sometimes used to denote *only* discrete inner product with (...) used for the continuous inner product. However, we will use the same notation for both the continuous and the discrete inner products

Thus, for $n = 1$: $$G_0 = \frac{1}{\sqrt{2}}; \quad G_1 = \frac{x}{\sqrt{2}}$$

and for $n = 2$: $$G_0 = \frac{1}{\sqrt{3}}; \quad G_1 = \frac{x}{\sqrt{2}}; \quad G_2 = \sqrt{\frac{3}{2}}x^2 - \sqrt{\frac{2}{3}}$$

The interpolation formula is then given by

$$f_n(x) = \sum_{i=0}^{n} c_i G_i(x) \tag{4.30}$$

and the coefficients are obtained from the equality, $f_n(x_k) = f(x_k)$, and orthonormality property, Eq. (4.28), as

$$c_i = \sum_{k=0}^{n} f(x_k) G_i(x_k) \tag{4.31}$$

Example 4.7 Solve the problem described in Example 4.6 using the Gram's polynomials and estimate the temperature at the point E.

Solution The interpolating polynomial is written as $f_2(x) = c_0 G_0(x) + c_1 G_1(x) + c_2 G_2(x)$ with

$$G_0 = \frac{1}{\sqrt{3}}; \quad G_1 = \frac{x}{\sqrt{2}}; \quad G_2 = \sqrt{\frac{3}{2}}x^2 - \sqrt{\frac{2}{3}}$$

and the computations of the coefficients is shown in Table 4.1.

Table 4.1

k	x_k	$f(x_k)$	$G_0(x_k)$	$G_1(x_k)$	$G_2(x_k)$
0	−1	10.1	0.577350	− 0.707107	0.408248
1	0	11.3	0.577350	0	− 0.816497
2	1	11.9	0.577350	0.707107	0.408248
	$c_i = \Sigma f(x_k) G(x_k) =$		19.2258	1.27279	− 0.244949

The interpolating polynomial is again seen to be the same if we expand in powers of x. To estimate the temperatures at the point E ($x = 0.5$) we compute the Gram's polynomials at this point as $G_0 = 0.577350$; $G_1 = 0.353553$; $G_2 = -0.510310$; and obtain the temperature as

$$f(0.5) = 19.2258 \times 0.577350 + 1.27279 \times 0.353553 + 0.244949 \times 0.510310 = 11.675°C$$

While interpolation with equidistant points works well and the accuracy typically increases with a finer grid, Runge showed that, in some cases, increasing the number of grid points leads to larger errors.

For example, interpolating the function $f(x) = 1/(1 + 25x^2)$ over the interval $[-1, 1]$ using a 10th degree polynomial, we get Fig. 4.7(a).

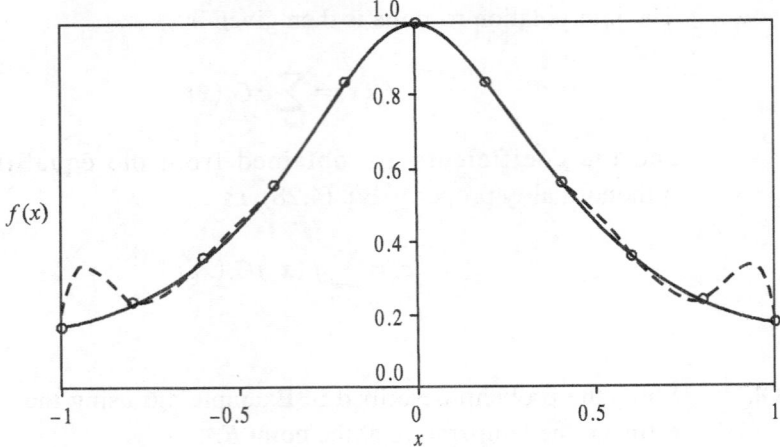

Fig. 4.7(a) Interpolation of the function $f(x) = \dfrac{1}{1+25x^2}$ using 11 equidistant points.

If we now use a 20th degree polynomial, the interpolating polynomial is as shown in Fig. 4.7(b).

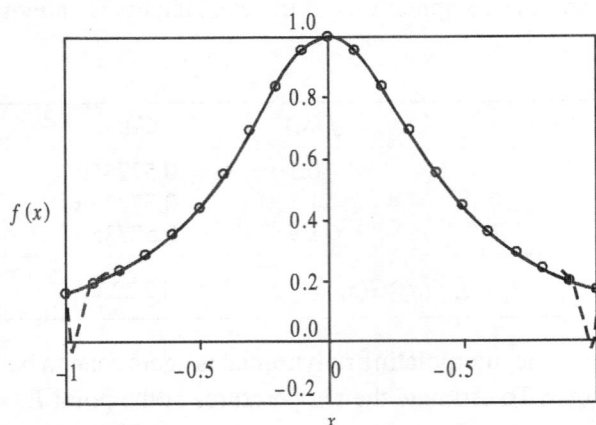

Fig. 4.7(b) Interpolation of the function $f(x) = \dfrac{1}{1+25x^2}$ using 21 equidistant points.

Therefore, in this case, equidistant interpolation may result in large errors, especially near the end points of the interval. One way to avoid this problem is to use unevenly spaced grid points (provided we have complete freedom in choosing the grid points). The pertinent question would then be: is there an arrangement of points which would be the best? In the next subsection we try to answer this question.

4.4.3 Unevenly Spaced Grid Points

From the definition of the residual, Eq. (4.27) and Eq. (4.23), it is apparent that the error of interpolation is a $(n + 1)$th degree polynomial, which vanishes at all the grid points. Location of the point ξ will depend on the location of the grid points and the point at which we have to estimate the error. However, since the function behaviour is not in our control, the best one can do is to choose the grid points in such a way as to minimise some norm of ϕ_{n+1}.

If we choose the L_2 norm, the grid points turn out to be the zeros of the Legendre polynomial, $P_{n+1}(x)$, and if we choose the L_∞ norm, the grid points should be located at the zeros of the Tchebycheff polynomial, $T_{n+1}(x)$. The L_1 norm may also be used but is not very common because of the difficulty in obtaining the grid points compared to the other two cases where the zeros of the Legendre and Tchebycheff polynomials are easily computed. The interpolating polynomial may then be obtained using the conventional form, Lagrange form, or the Newton form. However, the orthogonality property for the discrete case may be utilized to obtain the interpolating polynomial as described in the next section.

4.4.3.1 *Legendre Polynomials For Discrete Case*

If x_k are the zeros of $P_{n+1}(x)$, then the orthogonality condition is given by, for $i, j = 0, 1, 2, \ldots, n$ (compare with Eq. (4.12) for the case when the function is given and particularly note the presence of the weight):

$$\langle P_i, P_j \rangle = \sum_{k=0}^{n} P_i(x_k) P_j(x_k) w_k = 0, \qquad i \neq j$$

$$= \frac{2}{2i+1}, \quad i = j \tag{4.32}$$

where the weights are given by

$$w_k = \frac{2}{(1 - x_k^2)\left[P'_{n+1}(x_k)\right]^2} \tag{4.33}$$

The interpolating polynomial is then obtained as

$$f_n(x) = \sum_{i=0}^{n} c_i P_i(x) \qquad (4.34)$$

in which the coefficients are given by

$$c_i = \frac{\langle f, P_i \rangle}{\langle P_i, P_i \rangle} = \frac{2i+1}{2} \sum_{k=0}^{n} f(x_k) P_i(x_k) w_k \qquad (4.35)$$

Clearly, the computational complexity has increased and it may be simpler and more efficient to use the Lagrange or Newton form of the interpolating polynomials with the grid points located at the zeros of $P_{n+1}(x)$. As we see in the next subsection, the discrete version of the Tchebycheff polynomials is much simpler and is widely used.

4.4.3.2 Tchebycheff Polynomials For Discrete Case

If x_k are the zeros of $T_{n+1}(x)$, the orthogonality condition is given by, for $i, j = 0, 1, 2, \ldots, n$ (compare with Eq. (4.17) for the continuous case and note that the weight is unity):

$$\langle T_i, T_j \rangle = \sum_{k=0}^{n} T_i(x_k) T_j(x_k) = 0, \qquad i \neq j$$

$$= n+1, \qquad i = j = 0 \qquad (4.36)$$

$$= \frac{n+1}{2}, \qquad i = j \neq 0$$

The interpolating polynomial is again given as

$$f_n(x) = \sum_{i=0}^{n} c_i T_i(x) \qquad (4.37)$$

in which the coefficients are given by

$$c_i = \frac{\langle f, T_i \rangle}{\langle T_i, T_i \rangle} = \frac{2}{n+1} \sum_{k=0}^{n} f(x_k) T_i(x_k), \quad \text{for } i \geq 1$$

and

$$c_0 = \frac{1}{n+1} \sum_{k=0}^{n} f(x_k) T_0(x_k) \qquad (4.38)$$

Example 4.8   Obtain the fourth degree interpolating polynomial to $\exp(2x)$ over $[-1, 1]$ by generating five data points at zeros of the Legendre and Tchebycheff polynomials, respectively. Estimate the value of the function at the point $x = 0.85$.

Solution   For a fourth degree polynomial, we need the Legendre polynomials up to the order 5, which are listed below:

$$P_0(x) = 1, \quad P_1(x) = x, \quad P_2(x) = \frac{1}{2}(3x^2 - 1), \quad P_3(x) = \frac{1}{2}(5x^3 - 3x)$$

$$P_4(x) = \frac{1}{8}(35x^4 - 30x^2 + 3), \quad P_5(x) = \frac{1}{8}(63x^5 - 70x^3 + 15x)$$

The grid points for the data generation are located at the zeros of $P_5(x)$ which are obtained as 0, ±0.538469, ±0.906180. The weights (Eq. 4.33) are obtained as

$$w_k = \frac{2}{(1 - x_k^2) \frac{1}{64}\left[315x_k^4 - 210x_k^2 + 15\right]^2}$$

The interpolating polynomial is written as

$$f_4(x) = \sum_{i=0}^{4} c_i P_i(x)$$

and the coefficients are obtained in Table 4.2 [Note: The values of Legendre polynomials at a grid point may be directly computed by using the expressions listed above. However, it is more efficient to use the recursive relation, Eq. (4.11), to compute P_2, P_3, and P_4] :

Table 4.2

k	x_k	$F(x_k)$	w_k	$P_0(x_k)$	$P_1(x_k)$	$P_2(x_k)$	$P_3(x_k)$	$P_4(x_k)$
0	− 0.906180	0.163268	0.236927	1.00000	− 0.906180	0.731743	− 0.501031	0.245735
1	− 0.538469	0.340637	0.478629	1.00000	− 0.538469	− 0.065076	0.417382	− 0.344501
2	0.000000	1.00000	0.568889	1.00000	0.000000	− 0.500000	0.000000	0.375000
3	0.538469	2.93568	0.478629	1.00000	0.538469	− 0.065076	− 0.417382	− 0.344501
4	0.906180	6.12488	0.236927	1.00000	0.906180	0.731743	0.501031	0.245735
		$\Sigma f(x_k)P_i(x_k)w_k =$		3.62686	1.94876	0.703681	0.189276	0.039213
		(From Eq. 4.35) $c_i =$	1.81343	2.92314	1.75920	0.662465	0.176458

The value of the function at $x = 0.85$ is obtained by computing the Legendre polynomials at this point as $P_0 = 1$, $P_1 = 0.85$, $P_2 = 0.583750$, $P_3 = 0.260313$, and $P_4 = -0.050598$.

We then have

$$\begin{aligned} f_4(0.85) &= 1.81343 \times 1 + 2.92314 \times 0.85 + 1.75920 \times 0.583750 \\ &+ 0.662465 \times 0.260313 - 0.176458 \times 0.050598 = 5.48855 \end{aligned}$$

The exact value of the function is $\exp(1.7) = 5.47395$ indicating an error of only about 0.27%.

For the Tchebycheff polynomial, the required polynomials are listed below:

$$T_0(x) = 1$$
$$T_1(x) = x$$
$$T_2(x) = 2x^2 - 1$$
$$T_3(x) = 4x^3 - 3x$$
$$T_4(x) = 8x^4 - 8x^2 + 1$$
$$T_5(x) = 16x^5 - 20x^3 + 5x$$

The zeros of $T_5(x)$ are obtained as $0, \pm0.587785, \pm0.951057$. The coefficients are computed as shown in Table 4.3.

Table 4.3

k	x_k	$f(x_k)$	$T_0(x_k)$	$T_1(x_k)$	$T_2(x_k)$	$T_3(x_k)$	$T_4(x_k)$
0	−0.951057	0.149253	1.00000	− 0.951057	0.809017	− 0.587785	0.309017
1	−0.587785	0.308643	1.00000	− 0.587785	− 0.309017	0.951057	− 0.809017
2	0.000000	1.000000	1.00000	0.000000	− 1.00000	0.000000	1.00000
3	0.587785	3.239991	1.00000	0.587785	− 0.309017	− 0.951057	− 0.809017
4	0.951057	6.700037	1.00000	0.951057	0.809017	0.587785	0.309017
	$\Sigma f(x_k)T_i(x_k) =$		11.3979	7.95317	3.44460	1.06258	0.245642
	(From Eq. 4.38) $c_i =$		2.27958	3.18127	1.37784	0.425031	0.0982568

The value of the function at $x = 0.85$ is obtained by computing the Tchebycheff polynomials at this point as

$$T_0 = 1$$
$$T_1 = 0.85$$
$$T_2 = 0.445$$
$$T_3 = -0.0935$$
$$T_4 = -0.60395$$

We then have

$$f_4(0.85) = 2.27958 \times 1 + 3.18127 \times 0.85 + 1.37784 \times 0.445$$
$$-0.425031 \times 0.0935 - 0.0982568 \times 0.60395 = 5.49772$$

indicating an error of only about 0.43%. Note that the error is a little larger than that obtained using the Legendre polynomials. One of the reasons may be that the point $x = 0.85$ is closer to the grid point of Legendre polynomial (0.906180) than that of the Tchebycheff polynomial (0.951057).

As discussed earlier, if one is not at a freedom to choose the grid points, use of higher-order interpolating polynomials may result in large errors near the end points. In this case, it may be better to use piecewise interpolation, i.e., using a smaller degree polynomial to pass through a *subset* of data. [Another use of piecewise interpolation is in the solution of nonlinear equations, where a few function values in the neighbourhood of the *likely root* are utilized to perform an *inverse interpolation*, i.e., to express x as a polynomial of $f(x)$. From this inverse interpolating polynomial, the root is directly estimated by putting $f(x) = 0$.] For example, in Runge's problem (see Fig. 4.7(a)), if we connect consecutive data points by straight lines, we get a much superior interpolating polynomial than the 10th degree polynomial passing through all data points. These types of interpolating polynomials are known as *splines* and are described next.

EXERCISE 4.4

1. Estimate the value of the function at $x = 4$ from the table of data given below, using (a) Lagrange interpolating polynomial of second order using $x = 2, 3, 5$ (b) Newton's interpolating polynomial of fourth order, and (c) Gram's polynomial of second order using $x = 1, 3, 5$.

x	$f(x)$
1	1
2	12
3	54
5	375
6	756

2. The function $f(x) = 0.5 (x^3 + x^4)$ is to be approximated over the range $[1, 6]$ using a second order polynomial. Use the discrete Legendre and Tchebycheff polynomials to obtain the approximation and estimate the value of the function at $x = 4$.

3. The function x^x has to be integrated over the interval $[0, 1]$. Since analytical integration is not possible, it is decided to approximate the function by a third degree polynomial and then perform an analytical integration of the approximating polynomial.

Estimate the value of the integral $\int_0^1 x^x dx$ using both Gram's polynomials and discrete Tchebycheff polynomials by generating four data points, obtaining the approximating polynomial, and integrating it.

4. The function $f(x) = \cos x$ is to be approximated over the interval $[0, \pi/4]$ by a second order interpolating polynomial by using (i) Newton's divided difference with three points $(0, \pi/8, \pi/4)$ and (ii) Taylor's series expansion about $\pi/8$.

    Obtain the interpolating polynomials for both methods, write the error of interpolation in terms of the third derivative of the function and estimate the maximum possible error over the interval $[0, \pi/4]$. Find the true error at $x = 0.4$ and 0.75 and give reasons for the comparative performance of the two methods at these points.

5. A large number of functions have to be approximated by a second degree polynomial over the interval $[0, 1]$ using their values at three equidistant points $0, 0.5$, and 1.

    Out of Lagrange and Newton polynomials, which interpolation method would you *prefer*? why?

    Using this *preferred* method, estimate the function values at $x = 0.25$ and $x = 0.75$ for the functions $f_a(x) = x^4$ and $f_b(x) = x^3$. For the error of interpolation for $f_a(x) = x^4$, show that the value of ξ (i.e., the point at which the derivative is evaluated in the expression for the error) is linearly related to x (at which the error is to be computed).

6. A body travelling along a straight road was observed at various locations at different times:

t (s)	1.0	2.0	3.0	4.0
x (m)	4.5	23.0	80.5	213.0

    (a) Using the divided difference method, estimate the location at $t = 2.5$ s.
    (b) Another observer reported an observed location $x = 134.7$ m at $t = 3.5$ s.
        Re-esitmate the location at $t = 2.5$ s, using this additional information.

7. Estimate the value of the function at $x = 4$ from the table of data given below, using Lagrange interpolating polynomial of second order.

x	1	2	3	5	6
$f(x)$	1	12	54	375	756

8. A polynomial of the form $c_0 + c_1 x + c_3 x^3$ is used to interpolate a function whose

values are given at three distinct points. Comment on the existence and uniqueness of the interpolating polynomial for the following three sets of $\{x, f(x)\}$ values:

(i) (0, 1), (1, 2), (2, 6), (ii) (–1, 0), (0, 1), (1, 2), and (iii) (–1, 0), (0, 1), (1, 3).

4.5 SPLINE INTERPOLATION

Spline interpolation uses piecewise polynomial fitting to the set of data points. For easier presentation, we now assume that the grid points $x_0, x_1, ..., x_n$ are arranged in increasing order of x values. The grid points are also called *nodes* or *knots* (x_0 and x_n would be the *corner* nodes and the others are called *interior* nodes) and the portion between two consecutive nodes is called a *segment* or *knot span*.

There would thus be $(n + 1)$ knots and n spans. Our objective is then to obtain the splines S_i, which are polynomials of some pre-selected order, for the ith segment ($i = 0$ to $n - 1$) such that it passes through the nodes on either end of the segment (x_i and x_{i+1}). [$i = 1$ to n may also be used. In that case, the nodes at the end of the ith segment would be x_{i-1} and x_i.] The degree of these polynomials is used to classify the spline as *linear*, *quadratic*, or *cubic* spline. Generally, polynomials of higher order than cubic are not used due to their computational complexity.

Linear splines are the simplest and are uniquely defined by the data given whereas, as we will see in a short while, the quadratic and cubic splines require specification of additional constraints for a unique definition.

4.5.1 Linear Splines

For the ith segment, function value at the nodes at either end, $f(x_i)$ and $f(x_{i+1})$, uniquely define the linear spline as

$$S_i(x) = f(x_i) + (x - x_i) f[x_{i+1}, x_i], \quad i = 0, 1, 2, ..., n-1 \qquad (4.39)$$

The computations are therefore quite straightforward. Prediction of the function value at any point within the segment requires the determination of the relevant segment and the use of the corresponding spline. The drawback, as can be seen in Fig. 4.8, is that the curve is not smooth. We assume that the function which has generated the data is *smooth*. However, even if the function has a finite jump discontinuity at the knots, the spline fitting, because of its piecewise polynomial nature, could be made to work. In all subsequent discussion, though, we assume the function to be sufficiently smooth.

A number of times, we are interested in finding the first or second derivatives from the observed data (e.g., computing velocity or acceleration from distance measurements). While the first derivative of the linear spline fit would show a discontinuity at the knots, the second derivative is zero within the segment and is not defined at the knots. The linear splines are, therefore, not of much practical interest.

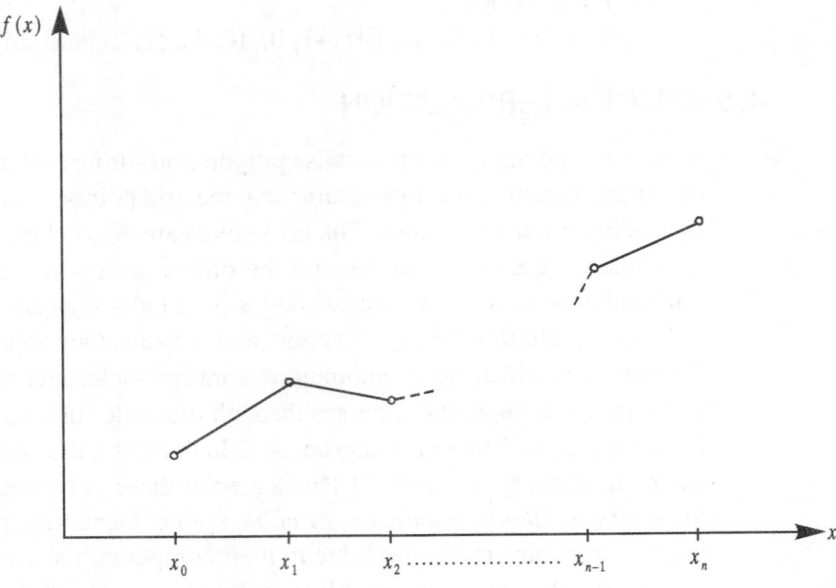

Fig. 4.8 Spline fitting using a linear spline.

4.5.2 Quadratic Splines

The discussion in the previous paragraph implies that even a quadratic polynomial would not be very practical since it would not have a continuous second derivative at the knots. However, we briefly discuss these to introduce the concept of *degree of freedom* and *constraint*.

A linear spline for the ith segment, say $c_{0,i} + c_{1,i}x$, had two undetermined constants which were uniquely defined by the function value at the nodes at either end. A quadratic spline has three coefficients while the equality of the spline interpolant and function values at the end nodes provides only two conditions. Thus, we have a *degree of freedom* of one in each segment, for a total of n, and need to provide n additional *constraints* in order to get a unique definition of the spline. One obvious constraint is the continuity of the first derivative (C^1 continuity) at the *interior* nodes, i.e., $S_i'(x_{i+1}) = S_{i+1}'(x_{i+1})$ for $i = 0, 1, 2, ..., n-2$, which provides $(n-1)$ constraints, leaving one degree of freedom.

The additional constraint required to uniquely define the quadratic spline may be chosen on the basis of any available information about the function behaviour. For a general case, some commonly used options are listed below:

- An arbitrary free parameter, say t_0, may be chosen and the quadratic spline may be expressed in terms of this parameter. By changing this parameter and looking at the resulting interpolating polynomial, one may decide the *proper* value of t_0. Considering two consecutive segments and using the C^0 continuity at the three knots and C^1 continuity at the middle knot, the quadratic spline may be written as

$$S_i(x) = f(x_i) + (x - x_i)t_i + (x - x_i)^2 \frac{t_{i+1} - t_i}{2h_i}, \quad i = 0, 1, 2, \dots, n-1 \quad (4.40)$$

in which $h_i (= x_{i+1} - x_i)$ is the length of the ith segment (note that t_i represents the first derivative of the quadratic spline at $x = x_i$). The coefficient of the quadratic term is obtained by the continuity of the first derivative $S_i'(x_{i+1}) = S_{i+1}'(x_{i+1})$, and the t's satisfy the recurrence relationship obtained from the condition, $S_i(x_{i+1}) = f(x_{i+1})$, as

$$t_{i+1} = 2f[x_{i+1}, x_i] - t_i, \quad i = 0, 1, 2, \dots, n-1 \quad (4.41)$$

If it is known that the first derivative is zero at $x = x_0$, e.g., when the function represents distance at various times and it is known that the initial velocity is zero), t_0 is equal to zero and all other t's, and therefore S_i, are readily computed.

- A zero second derivative at the first corner node (any other node could also be chosen but the computations become more complicated): $S_0''(x_0) = 0$. This implies that the 0th segment is a straight line joining $f(x_0)$ to $f(x_1)$, which is readily computed. The first segment, then, has three constraints, viz. the two function values at x_1 and x_2 and the first derivative at x_1 (which should be equal to $f[x_1, x_0]$ due to the C^1 continuity), that can be used to obtain the three coefficients in $S_1(x) = c_{0,1} + c_{1,1}x + c_{2,1}x^2$. Each successive segment is similarly computed. The same computation may be done by using the recurrence relation listed above (Eq. 4.41), with $t_0 = f[x_1, x_0]$.

- If the function is a second degree polynomial, one may expect a quadratic spline to fit it exactly. However, this does not happen if the first segment is specified as linear. The not-a-knot condition is an option which does not suffer from this drawback. It requires C^2 continuity at the first interior node: $S_0''(x_1) = S_1''(x_1)$, which implies that $S_0(x) \equiv S_1(x)$, hence the name not-a-knot.

The three function values; $f(x_0), f(x_1),$ and $f(x_2)$; are used to obtain the three coefficients $c_{0,0}(=c_{0,0}), c_{1,0}(=c_{1,1}),$ and $c_{2,0}(=c_{2,1})$. Successive segments then have three constraints as in the previous case (function values at x_i and x_{i+1} and the first derivative at x_i) and are sequentially computed. Alternatively, one could use the recurrence relationship with the continuity of the second derivative at the first interior node resulting in

$$\frac{t_1 - t_0}{h_0} = \frac{t_2 - t_1}{h_1}$$

which becomes, after using the recursion [Eq. (4.41)],

$$t_0 = f[x_1, x_0] - h_0 f[x_2, x_1, x_0]$$

Example 4.9   Obtain the quadratic spline which fits $f(x) = x^2$ sampled at points $x = 0, 1, 2$, and 3.

Solution   We first use t_0 as an arbitrary parameter and obtain the corresponding quadratic splines. Table 4.4 shows the computations:

Table 4.4

i	x_i	$f(x_i)$	$f[x_{i+1}, x_i]$	Computed t_i from Eq. (4.41) for		
				$t_0 = 0$	$t_0 = 1$	$t_0 = 2$
0	0	0	1	0	1	2
1	1	1	3	2	1	0
2	2	4	5	4	5	6
3	3	9		6	5	4

Figure 4.9 shows the plots of the three quadratic splines using Eq. (4.40) with the t values computed above.

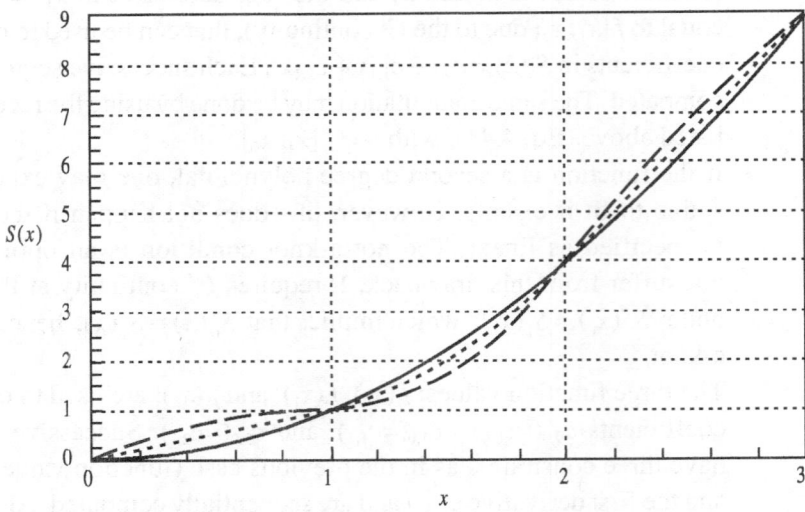

Fig. 4.9   Quadratic spline interpolation using different values of the slope at the first point, t_0. Solid line: $t_0 = 0$, dotted line: $t_0 = 1$, dashed line: $t_0 = 2$.

From the figure, it appears that the solid line ($t_0 = 0$) is a better interpolating spline as it is more smooth compared to the other curves. However, if we do not have any information about the function, there is no reason to prefer one of these curves over the others.

We now assume that the second derivative is zero at the first corner point (x_0). This results in the first segment being a straight line and $t_0 = f[x_1, x_0] = 1$. Using the recursive relations, we get the same values as shown in the table above for $t_0 = 1$.

Finally, if we apply the not-a-knot condition, we get

$$t_0 = f[x_1, x_0] - h_0 f[x_2, x_1, x_0] = 1 - 1 \times (3 - 1)/(2 - 0) = 0$$

Hence, we get the same interpolating spline as shown by the solid line in Fig. 4.9, which, incidentally, matches the exact function.

Now, we are ready to discuss the most widely used spline, the cubic spline.

4.5.3 Cubic Splines

For a cubic spline, the degree of freedom is 2 for each segment since there are four coefficients and only two constraints of the function value at either end. Again, a logical constraint would be to impose the C^2 continuity at all the *internal* nodes which provides us with $2n - 2$ constraints, one for the first derivative and the other for the second derivative. The system will then have 2 degrees of freedom and, therefore, two additional constraints have to be provided to define the unique cubic spline.

We saw earlier that a unique interpolating polynomial could be expressed in different forms. Similarly, once the two additional constraints are imposed, the unique cubic spline could also be expressed in different forms. We first discuss some ways of specifying the additional constraints and then look at a few options for expressing (and computing) the resulting cubic polynomial.

4.5.3.1 Constraints

- *Clamped*   When the function is clamped on each corner node forcing both ends to have some fixed slope, say, s_0 and s_n. This implies $S_0'(x_0) = s_0$ and $S_{n-1}'(x_n) = s_n$.
- *Natural* Curvature at the corner nodes is zero, i.e., $S_0''(x_0) = S_{n-1}''(x_n) = 0$.
- *Cyclic or periodic*   When the function is cyclic, with x_0 and x_n corresponding to the beginning and end of a cycle, respectively [necessarily, then, $f(x_0) = f(x_n)$]. The additional constraints are

$$S_0'(x_0) = S_{n-1}'(x_n) \quad \text{and} \quad S_0''(x_0) = S_{n-1}''(x_n)$$

- *Not-a-knot* When the first and last interior nodes have C^3 continuity, i.e., these do not act as a knot. Thus, $S_0(x) \equiv S_1(x)$ and $S_{n-2}(x) \equiv S_{n-1}(x)$.

4.5.3.2 Different Forms

Similar to the quadratic spline, we may write

$$S_i(x) = f(x_i) + (x - x_i)u_i + (x - x_i)^2 v_i + (x - x_i)^3 \frac{v_{i+1} - v_i}{3h_i} \quad (i = 0, 1, 2, ..., n-1)$$

$$(4.42)$$

in which the coefficient of the cubic term is obtained by using the C^2 continuity at x_{i+1}. The u's and v's (note that u_i represents the first derivative of the spline at $x = x_i$ and v_i is one-half of the second derivative) are given by the recursive relations derived from the C^1 and C^0 continuity as follows:

for $i = 0, 1, 2, ..., n-1$

$$S'_{i+1}(x_{i+1}) = S'_i(x_{i+1}) \implies u_{i+1} = u_i + 2h_i v_i + h_i(v_{i+1} - v_i)$$

$$\implies v_{i+1} = \frac{u_{i+1} - u_i}{h_i} - v_i \qquad (4.43a)$$

$$S_{i+1}(x_{i+1}) = S_i(x_{i+1}) \implies f(x_{i+1}) = f(x_i) + h_i u_i + h_i^2 v_i + h_i^3 \frac{v_{i+1} - v_i}{3h_i}$$

$$\implies f(x_{i+1}) = f(x_i) + h_i u_i + h_i^2 v_i + \frac{h_i^3}{3}\left(\frac{u_{i+1} - u_i}{h_i} - 2v_i\right)$$

$$\implies u_{i+1} = 3f[x_{i+1}, x_i] - 2u_i - h_i v_i$$

$$(4.43b)$$

Unlike the quadratic spline, however, sequential computations of the coefficients u and v will generally not be possible. In some cases both u_0 and v_0 may be specified (e.g., when the function represents distance, both the initial velocity and acceleration being zero will lead to $u_0 = v_0 = 0$) and then the spline is readily obtained using the recursive relations to compute (in the order) $u_1, v_1, u_2, v_2, ..., u_n, v_n$ [note that v_n is needed in $S_{n-1}(x)$]. However, as seen in the previous subsection, typically the constraints are specified as one at each corner.

For example, for a clamped spline, $u_0 = s_0$ and $u_n = s_n$;

for a natural spline, $v_0 = v_n = 0$;

for a cyclic spline, $u_0 = u_n$ and $v_0 = v_n$;

and for the not-a-knot condition,

$$\frac{v_1 - v_0}{h_0} = \frac{v_2 - v_1}{h_1} \quad \text{and} \quad \frac{v_{n-1} - v_{n-2}}{h_{n-2}} = \frac{v_n - v_{n-1}}{h_{n-1}}$$

In all these cases, a set of linear simultaneous equations is obtained which can be solved using any of the techniques described in Chapter 2. The recursive form listed in Eq. (4.43) gives rise to a system of $2n$ equations in $2n + 2$ unknowns (the two boundary conditions enable us to get a *unique* solution), but may be easily modified to obtain a *tridiagonal* system of $(n - 1)$ equations involving $(n + 1)$ unknowns. Two commonly used methods for achieving this are discussed below.

(a) ***Using the second derivatives as primary variables*** From the recursive relations, Eqs (4.43), we may express u in terms of v on eliminating u_{i+1}, as

$$u_i = f\left[x_{i+1}, x_i\right] - \frac{h_i(v_{i+1} + 2v_i)}{3}, \quad i = 0, 1, 2, \dots, n-1 \tag{4.44}$$

Now, writing Eq. (4.43a) as $v_i = \dfrac{u_i - u_{i-1}}{h_{i-1}} - v_{i-1}$ and substituting for u_i and u_{i-1} using Eq. (4.44), we obtain a tridiagonal system of equations as

$$h_{i-1}v_{i-1} + 2(h_{i-1} + h_i)v_i + h_iv_{i+1} = 3f\left[x_{i+1}, x_i\right] - 3f\left[x_i, x_{i-1}\right], \quad i = 1, 2, \dots, (n-1) \tag{4.45}$$

which can be solved using the Thomas algorithm utilising the two constraints. (Compare with Box 4.4 and recall that v is one-half of the second derivative).

(b) ***Using the first derivatives as primary variables*** As is clear from the above formulation, the second derivatives at the knots are obtained directly as part of the solution process. However, if one needs to find the first derivatives, differentiation of the cubic spline has to be performed. So if, for example, from the distance measurement at different times, it is desired to estimate the acceleration at these times, this formulation would work well. However, it is more likely that the first derivative at the nodes will be looked for more often than the second derivative. We describe, therefore, an alternative form of the cubic spline which will be more suitable for such cases. From the recursive relation, Eq. (4.43b), we may express v in terms of u as

$$v_i = \frac{3f\left[x_{i+1}, x_i\right] - 2u_i - u_{i+1}}{h_i}, \quad i = 0, 1, 2, \dots, n-1 \tag{4.46}$$

Now, writing Eq. (4.43a) as $v_i = \dfrac{u_i - u_{i-1}}{h_{i-1}} - v_{i-1}$ and substituting for v_i and v_{i-1} using Eq. (4.46), we obtain a tridiagonal system of equations (for $i = 1$ to $n-1$) as

$$h_i u_{i-1} + 2(h_{i-1} + h_i)u_i + h_{i-1}u_{i+1} = 3h_{i-1}f[x_{i+1}, x_i] + 3h_i f[x_i, x_{i-1}] \qquad (4.47)$$

which can again be solved using the Thomas algorithm utilising the two constraints. (Compare with Box 4.4 and recall that u is the first derivative).

BOX 4.4 Alternative Methods of Obtaining Cubic Spline Equations

Equations (4.45) could also be obtained by starting from the second derivatives as primary variables, expressing them as linear function of x (which is the linear interpolation between $S''(x_i)$ and $S''(x_{i+1})$), integrating twice, applying the C^0 continuity conditions at the two nodes to evaluate the constants of integration, and using the C^1 continuity to obtain the tridiagonal system of equations. The resulting equations are generally written as

$$(x_i - x_{i-1})S''(x_{i-1}) + 2(x_{i+1} - x_{i-1})S''(x_i) + (x_{i+1} - x_i)S''(x_{i+1})$$

$$= 6\frac{f(x_{i+1}) - f(x_i)}{x_{i+1} - x_i} - 6\frac{f(x_i) - f(x_{i-1})}{x_i - x_{i-1}},$$

and

$$S_i(x) = \frac{S''(x_{i+1})(x - x_i)^3 + S''(x_i)(x_{i+1} - x)^3}{6(x_{i+1} - x_i)} + \left[\frac{f(x_{i+1})}{x_{i+1} - x_i} - \frac{(x_{i+1} - x_i)S''(x_{i+1})}{6}\right](x - x_i)$$

$$+ \left[\frac{f(x_i)}{x_{i+1} - x_i} - \frac{(x_{i+1} - x_i)S''(x_i)}{6}\right](x_{i+1} - x)$$

We have given only an outline of the method since we believe Eq. (4.45) to be a better representation. For details, one may see Chapra and Canale (2006).

Similarly, Eq. (4.47) could be obtained by starting from the first derivatives as primary variables, expressing them as a quadratic function of x (with an undetermined coefficient, since only two conditions are available: the quadratic function should be equal to $S'(x_i)$ at $x = x_i$ and $S'(x_{i+1})$ at $x = x_{i+1}$), integrating once, applying the C^0 continuity conditions at the two nodes to evaluate the constants of integration, and using the C^1 continuity to obtain the tridiagonal system of equations. The resulting equations are usually written as

$$(x_{i+1} - x_i)S'(x_{i-1}) + 2(x_{i+1} - x_{i-1})S'(x_i) + (x_i - x_{i-1})S'(x_{i+1})$$

$$= 3(x_i - x_{i-1})\frac{f(x_{i+1}) - f(x_i)}{x_{i+1} - x_i} + 3(x_{i+1} - x_i)\frac{f(x_i) - f(x_{i-1})}{x_i - x_{i-1}}$$

and

$$S_i(x) = \frac{x_{i+1} - x}{x_{i+1} - x_i}f(x_i) + \frac{x - x_i}{x_{i+1} - x_i}f(x_{i+1}) + \frac{(x - x_i)(x_{i+1} - x)}{(x_{i+1} - x_i)^2}$$

$$\{(x_{i+1} - x)S'(x_i) - (x - x_i)S'(x_{i+1}) - (x_{i+1} - 2x + x_i)(f(x_{i+1}) - f(x_i))\}$$

The two end conditions have to be expressed in terms of the primary variables (u or v). Thus a natural spline or the not-a-knot condition (Section 4.5.3.1) does not pose any problem for the form (a) while a clamped spline is easily accounted for in the form (b). While using the natural spline with the form (b), the condition $v_0 = 0$ could be written using Eq. (4.46) as

$$2u_0 + u_1 = 3f[x_1, x_0]$$

while $v_n = 0$ translates into [using Eqs (4.43) with $i = n - 1$]

$$u_{n-1} + 2u_n = 3f[x_n, x_{n-1}]$$

Similarly, for a clamped spline with the form (a), the following could be written:

$$2v_0 + v_1 = \frac{3f[x_1, x_0] - 3s_0}{h_0}$$

$$v_{n-1} + 2v_n = \frac{3s_n - 3f[x_n, x_{n-1}]}{h_{n-1}}$$

Once the two additional constraints are specified, the cubic spline is uniquely defined. Which form of the cubic spline to use in a particular problem would depend on several factors like the ease of applying boundary conditions, requirement of first or second derivatives, and the data storage and computational efficiency requirements. For example, the recursive relations Eqs (4.43) require larger storage (both u and v) but will be more computationally efficient. The other forms require less storage (either u or v) but are likely to require more computation time. These issues are not addressed here.

Example 4.10   Obtain the natural cubic spline which interpolates $f(x) = 1/(1 + x^2)$ sampled at points $x = -2, -1, 0, 1$, and 2 by using the second derivatives as the primary variables. Estimate the function value at $x = 1.6$.

Solution   Table 4.5 shows the function values and the first divided differences:

Table 4.5

i	x_i	$f(x_i)$	$f[x_{i+1}, x_i]$
0	−2	0.2	0.3
1	−1	0.5	0.5
2	0	1.0	−0.5
3	1	0.5	−0.3
4	2	0.2	

Equation (4.45) is then written as (note that $v_0 = v_4 = 0$ for a natural spline)

$$\begin{bmatrix} 4 & 1 & 0 \\ 1 & 4 & 1 \\ 0 & 1 & 4 \end{bmatrix} \begin{Bmatrix} v_1 \\ v_2 \\ v_3 \end{Bmatrix} = \begin{Bmatrix} 0.6 \\ -3 \\ 0.6 \end{Bmatrix}$$

giving the solution as $v_1 = v_3 = 0.385714$, $v_2 = -0.942857$.

Equation (4.44) is used to obtain the coefficients u as $u_0 = 0.171429$, $u_1 = 0.557143$, $u_2 = 0$, $u_3 = -0.557143$.

Equation (4.42) is then used to write the cubic spline in the last segment (which contains the point 1.6) as

$$S_3(x) = f(x_3) + (x - x_3)u_3 + (x - x_3)^2 v_3 + (x - x_3)^3 \frac{v_4 - v_3}{3h_3}$$

$$= 0.5 - 0.557143(x-1) + 0.385714(x-1)^2 - 0.128571(x-1)^3$$

and the function value at $x = 1.6$ is estimated as 0.276800. The true value at this point is 0.280899 indicating an error of about 1.5%.

4.5.4 Hermite Interpolation

Till now, we have assumed that the *function* values are given at a set of grid points. Sometimes, however, additional information about the function may be available at these points in terms of its derivatives. For example, if we are tracking the path of an object, in addition to the distance measured at different times, we may have some velocity measurements also. This extra information will usually lead to a more accurate prediction of the position of the object at a time not coinciding with the grid points. It will enable us to use a higher degree polynomial than that possible when only the function values are given.

Thus, given the function value at a single data point we use a zero-order interpolation (a constant). But if we are given its derivative also, we will be able to use a linear interpolation with a line passing through the given function value and having a slope equal to the given derivative. Similarly, if the function and its first derivatives are given at two points, a cubic parabola may be used as the interpolating polynomial. (If we use linear interpolation, the slope of this line at the grid points may not match the values given. Similarly, a parabola passing through the two given points, will have the required slope at one of the points but it may not match the slope at the other point. If we use a fourth degree polynomial, there would be infinite possibilities which will satisfy all four given data.) This method of interpolation, using the function and its derivative(s), is known as the *Hermite* interpolation or *osculating* interpolation. Higher order derivatives may also be used, but we will only consider the first derivative.

Based on the fact that $(m + 1)$ data points can generally be interpolated with an mth degree polynomial, we may surmise that given the values of the function and its first derivative at $(n + 1)$ points, there exists a unique interpolating polynomial of degree (at most) $(2n + 1)$. For convenience, we would express it in a form similar to the Lagrange polynomials and, without loss of generality, as we did for orthogonal polynomials, we will assume that the range of data points is $[-1, 1]$, which implies that we are not considering the trivial case of a single grid point, i.e., n has to be greater than 0.

Given the set of values $\{[x_k, f(x_k), f'(x_k)], k = 0, 1, \ldots, n\}$ we write the general form of the Hermite interpolating formula as

$$f_{(d+1)(n+1)-1}(x) = \sum_{i=0}^{n} \sum_{j=0}^{d} H_{i,j} f^{[j]}(x_i) \tag{4.48}$$

in which d represents the order of derivatives specified at each point (if the first derivatives are specified, $d = 1$, if the second derivatives are *also* specified, $d = 2$), $f^{[j]}(x_i)$ denotes the jth derivative at x_i ($j = 0$ corresponds to the function value), and $H_{i,j}$ are the Hermite polynomials (of degree at most $(d + 1)(n + 1) - 1$). These are obtained from the conditions that the function value and its derivative(s) for the approximating polynomial should be equal to those specified in the problem.

In other words, the polynomial $H_{i,0}$ would be equal to 1 at $x = x_i$ and 0 at all other grid points and its derivative $H'_{i,0}$ would be zero at all grid points. Similarly, the polynomial $H_{i,1}$ would be zero at all grid points and its derivative will be 1 at $x = x_i$ and will vanish at all other grid points. Since we consider only the cases where the function and its first derivatives are given, we write

$$f_{2n+1}(x) = \sum_{i=0}^{n} \left[H_{i,0} f(x_i) + H_{i,1} f'(x_i) \right] \tag{4.49}$$

For example, for two points ($n = 1, x_0 = -1, x_1 = 1$) we have

$$f_3(x) = H_{0,0} f(x_0) + H_{1,0} f(x_1) + H_{0,1} f'(x_0) + H_{1,1} f'(x_1)$$

where all H's are cubic polynomials. As discussed in the previous paragraph, $H_{0,0}$ will satisfy

$$H_{0,0}(-1) = 1, \quad H'_{0,0}(-1) = 0, \quad H_{0,0}(1) = 0, \quad H'_{0,0}(1) = 0$$

One could write a cubic polynomial for $H_{0,0}$ in the conventional form, obtain its four coefficients using the conditions above and get

$$H_{0,0} = \frac{1}{2} - \frac{3x}{4} + \frac{x^3}{4}$$

However, it will be inefficient for large number of grid points. An alternative technique based on the Lagrange form of interpolating polynomials is described next.

Since the Lagrange polynomials, L_i (see Eq. 4.21) are nth degree polynomials which are 1 at the corresponding node (x_i) and 0 at all other nodes, if we square them we get a polynomial of degree $2n$ which will not only preserve the nodal values of 1 and 0 but the first derivative will also vanish at all nodes other than x_i. The value of the first derivative at $x = x_i$ would be equal to $2L_i(x_i)L_i'(x_i)$.

Since the Hermite polynomials are of degree $2n + 1$, and essentially have similar properties, we should be able to write ($j = 0, 1$),

$$H_{i,j} = (\alpha_{i,j} + \beta_{i,j}x)[L_i(x)]^2 \qquad (4.50)$$

with the conditions

$$\alpha_{i,0} + \beta_{i,0}x_i = 1; \qquad 2(\alpha_{i,0} + \beta_{i,0}x_i)L_i'(x_i) + \beta_{i,0} = 0$$
$$\alpha_{i,1} + \beta_{i,1}x_i = 0; \qquad 2(\alpha_{i,1} + \beta_{i,1}x_i)L_i'(x_i) + \beta_{i,1} = 1$$

Therefore, the expressions for Hermite polynomials are obtained as

$$H_{i,0} = [1 - 2(x - x_i)L_i'(x_i)][L_i(x)]^2$$
$$H_{i,1} = (x - x_i)[L_i(x)]^2 \qquad (4.51)$$

For example, with two points $(x_0 = -1, x_1 = 1)$, we have

$$L_0(x) = \frac{1-x}{2}$$

$$L_1(x) = \frac{1+x}{2}$$

and the derivatives as

$$L_0'(x) = -\frac{1}{2}$$

$$L_1'(x) = \frac{1}{2}$$

From Eq. (4.51), therefore,

$$H_{0,0} = \left[1 - 2(x+1)\left(-\frac{1}{2}\right)\right]\left(\frac{1-x}{2}\right)^2 = \frac{1}{2} - \frac{3x}{4} + \frac{x^3}{4}$$

$$H_{1,0} = \left[1 - 2(x-1)\left(\frac{1}{2}\right)\right]\left(\frac{1+x}{2}\right)^2 = \frac{1}{2} + \frac{3x}{4} - \frac{x^3}{4}$$

$$H_{0,1} = (x+1)\left(\frac{1-x}{2}\right)^2 = \frac{1}{4}(1 - x - x^2 + x^3)$$

$$H_{1,1} = (x-1)\left(\frac{1+x}{2}\right)^2 = \frac{1}{4}(-1 - x + x^2 + x^3)$$

The interpolated value at the mid-point ($x = 0$) is, therefore,

$$\tilde{f}(\text{mid-point}) = \frac{f(x_0)}{2} + \frac{f(x_1)}{2} + \frac{f'(x_0)}{4} - \frac{f'(x_1)}{4}$$

Note that the first two terms on the RHS represent the linear interpolation which would be used in absence of any data on the function derivatives. The last two terms represent a correction from the linear trend, increasing the value for a curve which tends to get flatter with increase in x and reducing it if the curve becomes steeper.

The error of interpolation could be obtained by following a methodology similar to that used for polynomial interpolation (Eq. 4.27). Here the data given is in terms of the function value and its derivative at the grid points. However, we may visualize the Hermite interpolation as the limiting case of a polynomial interpolation using only the function values at the grid points and points located at an infinitesimal distance from the grid points, i.e., the data set

$$\{(x_i, f(x_i)), (x_i + \varepsilon, f(x_i) + \varepsilon f'(x_i))\}$$

$i = 0, 1, \ldots, n$. Note that both the functions and its derivative have been incorporated in this data set. The remainder term is then obtained as

$$R = \lim_{\varepsilon \to 0} \prod_{i=0}^{n} (x - x_i)(x - x_i - \varepsilon) f[x, x_n + \varepsilon, x_n, x_{n-1} + \varepsilon, x_{n-1}, \ldots, x_0 + \varepsilon, x_0]$$

$$= \frac{f^{[2n+2]}(\xi)}{(2n+2)!} \prod_{i=0}^{n} (x - x_i)^2$$

with ξ having its usual meaning of being in the relevant interval. We could use this methodology to obtain the expressions for the Hermite polynomials also but leave it to the reader to do so, if desired.

Similar to the spline interpolation, we could use a piecewise-cubic Hermite interpolation if the function and its derivatives are given at the nodes. It will have continuous first derivatives at the nodes but the second derivative may not be continuous since each cubic polynomial may be obtained independent of neighbouring segments. We could achieve continuity of the second derivative at the knots by using higher-order interpolating polynomial, but it is not very common to do so.

One of the reasons for choosing a piecewise polynomial fitting is the presence of large oscillations in a single polynomial fit over the whole range. However, even the cubic polynomial suffers from large oscillations, though less frequently. If the data is subjected to measurement errors in the function value (or the independent variable) an interpolating polynomial would not work very well, since it will pass through the 'erroneous' points also. We would then like to go for some approximating function (again, generally polynomial) which represents the general trend of the data but does not necessarily pass through *all* the data points. This is achieved by *regression* as described next.

EXERCISE 4.5

1. Estimate the value of the function at $x = 4$ from the table of data given below, using (a) quadratic spline (assume that the derivative of the function is zero at $x = 0$) and (b) cubic spline (using the not-a-knot condition at either ends).

x	$f(x)$
1	1
2	12
3	54
5	375
6	756

2. The location of an object and its velocity were measured at different times as shown in the table below:

t (s)	Distance (m)	Velocity (m/s)
1	1	3.5
2	12	22

Use Hermite interpolation to approximate the distance as a cubic polynomial in time.

4.6 REGRESSION

The problem can be stated as: given the function values at a set of $(n + 1)$ points (*not necessarily distinct*), find the polynomial (of degree $m < n$) which is *nearest* to the function value at these points. Since we allow for the possibility of measurement errors, it is now permissible to have two different function values at the same node. Some other types of approximating functions may also be used and are described later. As we did in the continuous case, some measure of *nearness* has to be decided before attempting to obtain the *best* polynomial fit. We will use the L_2 norm (sum of squares) of the residual to quantify the error since it leads to a unique and mathematically convenient solution. While the fit based on L_1 norm (sum of absolute value) may be non-unique and difficult to obtain, that based on the L_∞ norm (maximum absolute value) is also rather tedious. However, it should be mentioned that squaring the residual tends to give more weight to the large residuals. Some other criterion may very well be used. For example, if a straight road is to be constructed near a few villages, sum of the perpendicular distance (or its square) of the villages from the road may be a good *norm* to minimize.

In general, this polynomial will depend on the selected measure of nearness. However, if the data is such that an *m*th degree polynomial is able to interpolate it, e.g., three points on a straight line ($n = 2$, $m = 1$), the regressing and interpolating polynomials will be identical (*and unique*). In the next subsection we describe the least-squares regression, which aims at minimising the L_2 norm of the residuals.

4.6.1 Least-squares Regression

Let $(x_0, f(x_0))$, $(x_1, f(x_1))$, ..., $(x_n, f(x_n))$ be the set of function values and corresponding $(n + 1)$ grid points over the interval $x = a$ to $x = b$. We need to find the *m*th degree polynomial $f_m(x)$ such that the sum of squared residuals at the grid points is a minimum.

As before, if the desired polynomial is

$$f_m(x) = \sum_{j=0}^{m} c_j \phi_j(x) \tag{4.52}$$

the c_j are to be chosen such that

$$\sum_{i=0}^{n} \left(f(x_i) - \sum_{j=0}^{m} c_j \phi_j(x_i) \right)^2 \text{ is minimized}$$

We assume that the independent variable, x, is not subject to measurement errors. Regression techniques for such cases, where both the dependent and the independent variables may have errors, are available but are not discussed here. Using the stationary point theorem, we get a set of $(m + 1)$ *linear* simultaneous equations as,

$$\sum_{i=0}^{n} \phi_k(x_i) \left(f(x_i) - \sum_{j=0}^{m} c_j \phi_j(x_i) \right) = 0, \text{ for } k = 0, 1, ..., m \tag{4.53}$$

An inherent assumption here is that the bases, ϕ, do not involve the coefficients, c. Sometimes it may not be possible to separate the bases and the coefficients which leads to *nonlinear regression* which is discussed later in this chapter. In terms of function space, Eq. (4.53) indicates that the residual is orthogonal to the basis functions and leads to

$$[A]\{c\} = \{b\} \tag{4.54}$$

in which $a_{ij} = \sum_{k=0}^{n} \phi_i(x_k)\phi_j(x_k)$ and $b_i = \sum_{k=0}^{n} \phi_i(x_k)f(x_k)$.

Equation (4.54) are the *normal equations* (see Section 4.2) and may be solved to obtain the coefficients $c_0, c_1, \ldots, c_m$.

For example, if the bases, ϕ, are chosen in the conventional form $\phi_i(x) = x^i$, we get the normal equation

$$
\begin{bmatrix}
n+1 & \sum x_i & \sum x_i^2 & \cdots & \sum x_i^m \\
\sum x_i & \sum x_i^2 & \sum x_i^3 & \cdots & \sum x_i^{m+1} \\
\cdots & \cdots & \cdots & \cdots & \cdots \\
\cdots & \cdots & \cdots & \cdots & \cdots \\
\sum x_i^m & \sum x_i^{m+1} & \sum x_i^{m+2} & \cdots & \sum x_i^{2m}
\end{bmatrix}
\begin{Bmatrix}
c_0 \\
c_1 \\
\cdots \\
\cdots \\
c_m
\end{Bmatrix}
=
\begin{Bmatrix}
\sum f(x_i) \\
\sum x_i f(x_i) \\
\cdots \\
\cdots \\
\sum x_i^m f(x_i)
\end{Bmatrix} \tag{4.55}
$$

in which the summation is over $i = 0$ to n. As before, this system of equations is ill-conditioned and it is preferable to choose an orthogonal basis, for which the coefficient matrix $[A]$ becomes diagonal and the c's are obtained directly.

For equidistant grid points, Gram's polynomial (of order m) could be used as the orthogonal basis, while if we are free to locate the points, discrete versions of Legendre polynomial of order m [with grid points at the zeros of $P_{n+1}(x)$] or Tchebycheff polynomial of order m [grid points at zeros of $T_{n+1}(x)$] could be used. For arbitrarily spaced points, the *Givens or Householder transforms* or *singular value decomposition* [Golub and van Loan (1996)] could be used to improve the condition number.

Example 4.11 The mass of a radioactive substance is measured at 2-day intervals till 8 days. Unfortunately, the reading could not be taken at 6 days due to equipment malfunction. The following table shows the other readings:

Time (day)	0	2	4	8
Mass (g)	1.000	0.7937	0.6300	0.3968

Estimate the mass at 6 days using a second-order polynomial regression.

Solution A quadratic regression is obtained by writing

$$m = c_0 + c_1 t + c_2 t^2$$

and writing the normal equations as

$$\begin{bmatrix} n+1 & \sum t_i & \sum t_i^2 \\ \sum t_i & \sum t_i^2 & \sum t_i^3 \\ \sum t_i^2 & \sum t_i^3 & \sum t_i^4 \end{bmatrix} \begin{Bmatrix} c_0 \\ c_1 \\ c_2 \end{Bmatrix} = \begin{Bmatrix} \sum m_i \\ \sum t_i m_i \\ \sum t_i^2 m_i \end{Bmatrix}$$

Table 4.6 shows the computations:

Table 4.6

i	t_i	m_i	t^2	t^3	t^4	$t\,m$	$t^2\,m$
0	0	1.0000	0	0	0	0.000	0.000
1	2	0.7937	4	8	16	1.587	3.175
2	4	0.6300	16	64	256	2.520	10.080
3	8	0.3968	64	512	4096	3.174	25.395
		Sum = 2.8205	84	584	4368	7.2818	38.65

Solution of the normal equations results in $c_0 = 0.9991$, $c_1 = -0.1102$, and $c_2 = 0.004370$. The estimated mass at $t = 6$ days is, therefore, 0.4951 g.

Since the radioactive decay follows an exponential equation, $m(t) = m(0)e^{-\lambda t}$, a linear regression between $\ln(m)$ and t would provide the value of the decay constant which could be used to estimate the mass at 6 days (see Exercise 4.6.1).

Once the least-square polynomial fit is obtained, one would like to ascertain the nearness of the fitted polynomial to the given data points. As shown in Fig. 4.10, we may qualitatively say that the fit (a) is nearer to the data than (b). To quantify it, we describe (very briefly, since we assume reader familiarity with statistical analysis) the most commonly used indicator, the coefficient of determination.

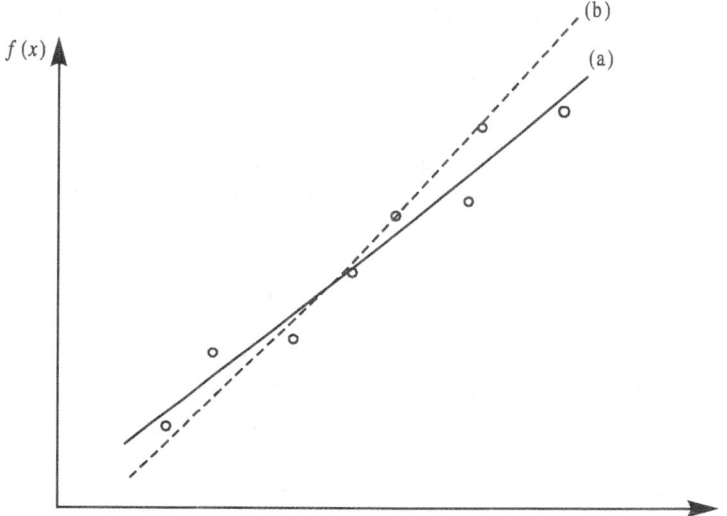

Fig. 4.10   Goodness of fit for an approximating function.

4.6.1.1 *Coefficient of Determination*

We denote the spread of data about its mean by S_t, which is the sum of the squares of the deviations from mean:

$$S_t = \sum_{i=0}^{n}\left[f(x_i)-\overline{f}\right]^2 \tag{4.56}$$

where $\overline{f}$ is the mean defined by $\overline{f} = \dfrac{\displaystyle\sum_{i=0}^{n}f(x_i)}{n+1}$.

Further, we use S_r to represent the sum of the squares of the deviations from the best-fit polynomial:

$$S_r = \sum_{i=0}^{n}\left[f(x_i)-f_m(x_i)\right]^2 \tag{4.57}$$

Obviously, since the best-fit polynomial minimizes the sum of squares of the residual, S_r will be less than S_t (or, at worst, they will be equal when the polynomial regression has only the constant term and all higher-order coefficients are zero).

The difference, $S_t - S_r$, may be thought of as the amount of variability in the data *explained* by the regression and the ratio of the explained variability to the inherent data variability, S_t, is called the coefficient of determination, r^2, given by

$$r^2 = \frac{S_t - S_r}{S_t} \tag{4.58}$$

and its square root, r, is the correlation coefficient.[4] It varies from zero (when $f_m(x) = \overline{f}$) to 1 (when the polynomial passes through all data points).

---

[4]If a linear relationship is used (i.e., $m = 1$), this correlation coefficient is identical to the Pearson correlation coefficient which measures the strength of a linear relationship between two variables and is given by

$$r = \frac{n\sum x_i y_i - \left(\sum x_i\right)\left(\sum y_i\right)}{\sqrt{n\sum x_i^2 - \left(\sum x_i\right)^2}\,\sqrt{n\sum y_i^2 - \left(\sum y_i\right)^2}}$$

which varies from -1 (for a perfectly linearly decreasing function) to $+1$ (for a perfectly linearly increasing function). Note, however, that Eq. (4.58) gives only positive r, indicating the strength of correlation but not direction. Other statistical parameters, e.g., confidence interval for the regression parameters, may also be used but are not discussed here. See Chapter 8 for further discussion.

Typically, a value of r^2 less than 0.3 indicates a poor fit while more than 0.8 indicates a good fit of the data by the regressing polynomial. Since a polynomial of a particular degree is a superset of all polynomials of a lower degree, it is obvious that the fit will improve (or, at worst, remain same) with increase in m. Thus, a possible (but not very efficient) strategy for polynomial regression would be to start with $m = 1$ and then keep on increasing m till a desirable value of r^2 is achieved. However, if there are large measurement errors or if the data does not follow a polynomial relation, a higher-order polynomial regression may produce large oscillations.

Thus, a large value of the coefficient of determination does not necessarily mean that the regression has *improved*. A plot of the data and the regression curve, therefore, must be looked at before accepting any regression.

The methodology used for a single independent variable (x) may be easily extended to the situation where the dependent variable is a function of multiple variables to obtain a *multiple regression*. For example, with two independent variables, say x and y, we may write a regression polynomial of order m_1 in x and m_2 in y as

$$f_{m_1,m_2}(x,y) = \sum_{k=0}^{m_2}\sum_{j=0}^{m_1} c_{j,k}x^j y^k \tag{4.59}$$

the $c_{j,k}$ are to be chosen such that $\displaystyle\sum_{i=0}^{n}\left(f(x_i,y_i) - \sum_{k=0}^{m_2}\sum_{j=0}^{m_1} c_{j,k}x_i^j y_i^k\right)^2$ is minimum where the total number of data points available is $n + 1$.

Using the stationary point theorem, we get

$$\sum_{i=0}^{n} x_i^{j_1} y_i^{k_1}\left(f(x_i,y_i) - \sum_{k=0}^{m_2}\sum_{j=0}^{m_1} c_{j,k}x^j y^k\right) = 0, \text{ for } j_1 = 0, 1, ..., m_1 \text{ and } k_1 = 0, 1,..., m_2$$

$$\tag{4.60}$$

a set of $(m_1 + 1)(m_2 + 1)$ *linear* simultaneous equations which may be solved to obtain the coefficients $c_{j,k}$ for $j = 0$ to m_1 and $k = 0$ to m_2. For example, if we use terms up to the first order, i.e.,

$$f_{1,1}(x,y) = c_{0,0} + c_{1,0}x + c_{0,1}y + c_{1,1}xy$$

the set of equations is written as

$$\begin{bmatrix} n+1 & \sum x_i & \sum y_i & \sum x_i y_i \\ \sum x_i & \sum x_i^2 & \sum x_i y_i & \sum x_i^2 y_i \\ \sum y_i & \sum x_i y_i & \sum y_i^2 & \sum x_i y_i^2 \\ \sum x_i y_i & \sum x_i^2 y_i & \sum x_i y_i^2 & \sum x_i^2 y_i^2 \end{bmatrix}\begin{bmatrix} c_{0,0} \\ c_{1,0} \\ c_{0,1} \\ c_{1,1} \end{bmatrix} = \begin{Bmatrix} \sum f(x_i,y_i) \\ \sum x_i f(x_i,y_i) \\ \sum y_i f(x_i,y_i) \\ \sum x_i y_i f(x_i,y_i) \end{Bmatrix} \tag{4.61}$$

in which the summation is over $i = 0$ to n. Use of this method is illustrated by the following example.

Example 4.12   The cost of fuel consumed by a truck was assumed to be linearly related to the travel distance and the load carried. Over a certain period, the following data was recorded by the driver. Obtain the underlying relationship (add the constraint that there is no cost when both the distance and the load are zero).

Distance (km)	88	210	320	88	210	320	245	65
Load factor	0.33	0.42	0.50	0.17	0.28	0.67	0.32	1.00
Cost (Rs)	140	270	400	110	250	450	280	225

Solution   We assume a relationship between the cost, C, the distance, x, and the load factor, y, as

$$C(x, y) = f_{1,1}(x, y) = c_{0,0} + c_{1,0}x + c_{0,1}y + c_{1,1}xy$$

Since the cost is zero when both the distance and the load are zero, it follows that $c_{0,0} = 0$. Also, since the relationship is linear with both variables, we may take $c_{1,1} = 0$ (see Exercise 4.6 Problem 3).

The set of equations is then written as

$$\begin{bmatrix} \sum x_i^2 & \sum x_i y_i \\ \sum x_i y_i & \sum y_i^2 \end{bmatrix} \begin{Bmatrix} c_{1,0} \\ c_{0,1} \end{Bmatrix} = \begin{Bmatrix} \sum x_i f(x_i, y_i) \\ \sum y_i f(x_i, y_i) \end{Bmatrix}$$

in which the summation is over $i = 0$ to 7.

Table 4.7 shows the computations:

Table 4.7

i	x_i	y_i	f_i	x^2	xy	y^2	xf	yf
0	88	0.33	140	7744	29.04	0.1089	12320	46.20
1	210	0.42	270	44100	88.20	0.1764	56700	113.4
2	320	0.50	400	102400	160.0	0.2500	128000	200.0
3	88	0.17	110	7744	14.96	0.0289	9680	18.70
4	210	0.28	250	44100	58.80	0.0784	52500	70.00
5	320	0.67	450	102400	214.4	0.4489	144000	301.5
6	245	0.32	280	60025	78.40	0.1024	68600	89.60
7	65	1.00	225	4225	65.00	1.000	14625	225.0
			Sum	372738	708.8	2.1939	486425	1064.4

The solution is obtained as $c_{1,0} = 0.9917$ and $c_{0,1} = 164.8$.

Thus, we have

$$\text{Cost (in Rs.)} = 0.9917 \times \text{Distance (in km)} + 164.8 \times \text{Load factor}$$

While linear regression is the preferred method, sometimes the relationship between two (or more) variables may be such that it is not possible to write it in the form of Eq. (4.52) or (4.59). For example, the exponential relationship $f(x) = c_0 e^{c_1 x}$ and the logistic function

$$f(x) = \frac{c_0}{1 + c_1 e^{c_2 x}}$$

are not of this form.

However, the exponential relationship may be readily manipulated to obtain a form suitable for linear regression as $\ln[f(x)] = \ln c_0 + c_1 x$, indicating that x should be used as the dependent variable and $\ln[f(x)]$ as the dependent variable.

Similarly, linear regression for a relationship of the type

$$f(x) = \frac{c_0 x}{1 + c_1 x}$$

may be used by regression of $1/f(x)$ on $1/x$; and for a power-law relationship $f(x) = c_0 x^{c_1}$ we may perform regression of $\ln[f(x)]$ on $\ln(x)$. However, for the logistic function no such manipulation is possible and we have to perform a *nonlinear regression*.

4.6.2 Non-linear Regression

Not all relationships between two (or more) variables could be expressed in the form of Eq. (4.52) or (4.59). For example, the logistic model of population growth

$$f(x) = \frac{c_0}{c_1 + e^{c_2 x}} \tag{4.62}$$

is not in a form suitable for linearization. In such cases, we may write the regression function as

$$f_m(x^l) \equiv f_m(x^0, x^1, ..., x^l; c_0, c_1, ..., c_m) \tag{4.63}$$

to approximate a function of $l + 1$ independent variables using $m + 1$ coefficients. The total number of data points representing the value of the dependent function corresponding to a set of values of the independent variables is assumed to be more than $m + 1$. If it is equal to $m + 1$, we are likely to get a unique answer by solving the set of nonlinear equations. If it is less than $m + 1$, a solution may not be possible. To simplify the description, in the rest of this subsection we assume that there is only one independent variable and denote it by x.

As before, we define a residual at each of the $n + 1$ data points and compute the sum of the squares of the deviations as

$$\sum_{i=0}^{n} \left[f(x_i) - f_m(x_i; c_0, c_1, ..., c_m) \right]^2$$

Again, the parameters are chosen in such a way as to minimize this sum. However, due to the nonlinear nature of the function, f_m, iterative techniques have to be used. A variety of nonlinear optimization techniques are available for this purpose but are beyond the scope of this book. Here we describe a technique based on the Newton method of linearization (see Section 3.7).

We start with an initial guess for the parameters $c_0, c_1, ..., c_m$ and assume that the nonlinear function f_m may be linearized by using a truncated Taylor series as

$$f_m\left(x_i; c_0^{(k+1)}, c_1^{(k+1)}, ..., c_m^{(k+1)}\right) = f_m\left(x_i; c_0^{(k)}, c_1^{(k)}, ..., c_m^{(k)}\right)$$

$$+ \sum_{j=0}^{m} \left. \frac{\partial f_m}{\partial c_j} \right|_{(x_i; c_0^{(k)}, c_1^{(k)}, ..., c_m^{(k)})} \left[c_j^{(k+1)} - c_j^{(k)} \right]$$

in which the superscript (k), as usual, denotes the iteration number. Starting from the known (or assumed) values at iteration level (k), the objective function to be minimized becomes linear in the unknown parameters, which are $c_0, c_1, ..., c_m$, at the iteration level $(k + 1)$. Application of the stationary point theorem then leads to the following equations [compare with Eq. (4.53)]:

$$\sum_{i=0}^{n} \left. \frac{\partial f_m}{\partial c_l} \right|_{(x_i, c^{(k)})} \left(r^{(k)}(x_i) - \sum_{j=0}^{m} \left. \frac{\partial f_m}{\partial c_j} \right|_{(x_i, c^{(k)})} \Delta c_j^{(k)} \right) = 0, \text{ for } l = 0, 1, ..., m$$

in which $r^{(k)}(x_i)$ is the residual, i.e., $f - f_m$, at the ith data point and kth iteration and $\Delta c_j^{(k)}$ is the change in the parameter value, $c_j^{(k+1)} - c_j^{(k)}$, which is expected to minimize the objective function. The approximation introduced due to the truncation of Taylor series implies that this choice of the change will generally not be optimal. The normal equations are then written in a matrix form as

$$[A]^{(k)} \{\Delta c\}^{(k)} = \{b\}^{(k)} \tag{4.64}$$

in which

$$a_{ij} = \sum_{l=0}^{n} \left. \frac{\partial f_m}{\partial c_i} \frac{\partial f_m}{\partial c_j} \right|_{(x_l, c^{(k)})} \quad \text{and} \quad b_i = \sum_{l=0}^{n} \left. \frac{\partial f_m}{\partial c_i} \right|_{(x_l, c^{(k)})} r^{(k)}(x_l)$$

If we define a Jacobian matrix, J, as (compare with Eq. (3.52) and note the difference that there we had derivatives of different functions in each row but here we have derivatives of the same function evaluated at different points)

$$
J = \begin{bmatrix}
\left.\dfrac{\partial f_m}{\partial c_0}\right|_{(x_0,c^{(k)})} & \left.\dfrac{\partial f_m}{\partial c_1}\right|_{(x_0,c^{(k)})} & \cdots & \left.\dfrac{\partial f_m}{\partial c_m}\right|_{(x_0,c^{(k)})} \\[2ex]
\left.\dfrac{\partial f_m}{\partial c_0}\right|_{(x_1,c^{(k)})} & \left.\dfrac{\partial f_m}{\partial c_1}\right|_{(x_1,c^{(k)})} & \cdots & \left.\dfrac{\partial f_m}{\partial c_m}\right|_{(x_1,c^{(k)})} \\[2ex]
\vdots & \vdots & \vdots & \vdots \\[2ex]
\left.\dfrac{\partial f_m}{\partial c_0}\right|_{(x_n,c^{(k)})} & \left.\dfrac{\partial f_m}{\partial c_1}\right|_{(x_n,c^{(k)})} & \cdots & \left.\dfrac{\partial f_m}{\partial c_m}\right|_{(x_n,c^{(k)})}
\end{bmatrix}_{(n+1)\times(m+1)}
$$

(4.65)

the normal equations may be written as

$$
[J]^T [J]\{\Delta c\} = [J]^T \{r\}
$$

(4.66)

The following example illustrates the use of the iterative procedure.

Example 4.13   The following table shows the population of a city:

Year	1951	1961	1971	1981	1991	2001
Population (in millions)	0.63	0.88	1.16	1.48	1.88	2.50

It is assumed that the growth follows a logistic model (Eq. 4.62). Estimate the parameters c_0, c_1, and c_2 by non-linear regression.

Solution   The elements of the Jacobian matrix require the derivative of the approximate function with respect to the model parameters. We have,

$$
f_2(x) \equiv f_2(x; c_0, c_1, c_2) = \frac{c_0}{c_1 + e^{c_2 x}}
$$

from which

$$
\frac{\partial f_2}{\partial c_0} = \frac{1}{c_1 + e^{c_2 x}}; \quad \frac{\partial f_2}{\partial c_1} = -\frac{c_0}{(c_1 + e^{c_2 x})^2}; \quad \frac{\partial f_2}{\partial c_2} = -\frac{c_0 x e^{c_2 x}}{(c_1 + e^{c_2 x})^2}
$$

We take 1951 as the base year ($x = 0$). If we take a different base year (e.g. 1900), the coefficients c_0 and c_1 would be different. It is obvious from the form of the logistic equation that the initial population is $c_0/(c_1 + 1)$ and the saturation population is c_0/c_1. Assuming c_1 to be small, a good guess for c_0 would be *a little more* than the *initial* population. Also, neglecting c_1, the last data point gives an estimate of c_2 as $\ln(c_0 / f_n)/x_n$. We use the starting guess as $c_0 = 0.7$ million,

$c_1 = 0.0$, and $c_2 = -0.025$ per year. The iterations are shown below:

1st iteration: $c_0^{(k-1)} = 0.7$, $c_1^{(k-1)} = 0.0$, $c_2^{(k-1)} = -0.025$

$$r = \begin{Bmatrix} -0.07000 \\ -0.01882 \\ 0.00590 \\ -0.00190 \\ -0.02280 \\ 0.05676 \end{Bmatrix}, \quad J = \begin{bmatrix} 1.000 & 0.700 & 0.000 \\ 1.284 & 1.154 & 8.988 \\ 1.649 & 1.903 & 23.082 \\ 2.117 & 3.137 & 44.457 \\ 2.718 & 5.172 & 76.112 \\ 3.490 & 8.528 & 122.162 \end{bmatrix}, \quad \Delta c = \begin{Bmatrix} -0.028721 \\ 0.014084 \\ -0.002160 \end{Bmatrix}$$

2nd iteration: $c_0^{(k-1)} = 0.6713$, $c_1^{(k-1)} = 0.01408$, $c_2^{(k-1)} = -0.02716$

$$r = \begin{Bmatrix} -0.03196 \\ -0.01522 \\ 0.03174 \\ 0.01051 \\ -0.02969 \\ 0.02531 \end{Bmatrix}, \quad J = \begin{bmatrix} 0.986 & 0.653 & 0.000 \\ 1.288 & 1.114 & 8.491 \\ 1.681 & 1.896 & 22.031 \\ 2.189 & 3.217 & 42.726 \\ 2.845 & 5.433 & 73.327 \\ 3.687 & 9.123 & 117.310 \end{bmatrix}, \quad \Delta c = \begin{Bmatrix} -0.001424 \\ 0.002638 \\ -0.000055 \end{Bmatrix}$$

3rd iteration: $c_0^{(k-1)} = 0.6699$, $c_1^{(k-1)} = 0.01144$, $c_2^{(k-1)} = -0.0271$

$$r = \begin{Bmatrix} -0.03228 \\ 0.01458 \\ 0.03035 \\ 0.00749 \\ -0.03595 \\ 0.01289 \end{Bmatrix}, \quad J = \begin{bmatrix} 0.989 & 0.655 & 0.000 \\ 1.292 & 1.118 & 8.526 \\ 1.686 & 1.905 & 22.157 \\ 2.198 & 3.327 & 43.084 \\ 2.860 & 5.480 & 74.176 \\ 3.713 & 9.235 & 119.177 \end{bmatrix}, \quad \Delta c = \begin{Bmatrix} -0.000076 \\ -0.000230 \\ -0.000015 \end{Bmatrix}$$

4th iteration: $c_0^{(k-1)} = 0.6698$, $c_1^{(k-1)} = 0.01121$, $c_2^{(k-1)} = -0.02709$

$$r = \begin{Bmatrix} -0.03235 \\ 0.01455 \\ 0.03038 \\ 0.00758 \\ -0.03585 \\ 0.01288 \end{Bmatrix}, \quad J = \begin{bmatrix} 0.989 & 0.655 & 0.000 \\ 1.292 & 1.118 & 8.529 \\ 1.687 & 1.905 & 22.165 \\ 2.198 & 3.237 & 43.084 \\ 2.860 & 5.480 & 74.176 \\ 3.713 & 9.235 & 119.177 \end{bmatrix}, \quad \Delta c = \begin{Bmatrix} -0.000003 \\ -0.000006 \\ -0.000000 \end{Bmatrix}$$

We stop the iterations here and estimate the values as

$$c_0 = 0.6698, \quad c_1 = 0.01121, \quad c_2 = -0.02709$$

Figure 4.11 shows the observed data and fitted logistic curve. Although the fit appears to be reasonable, it should be kept in mind that the saturation population of about 60 million (c_0 / c_1) appears to be unrealistic. In fact, a slight change in data (e.g., changing the population in 1951 from 0.63 million to 0.65 million) causes c_1 to be negative, which is unrealistic. Hence, we may say that the

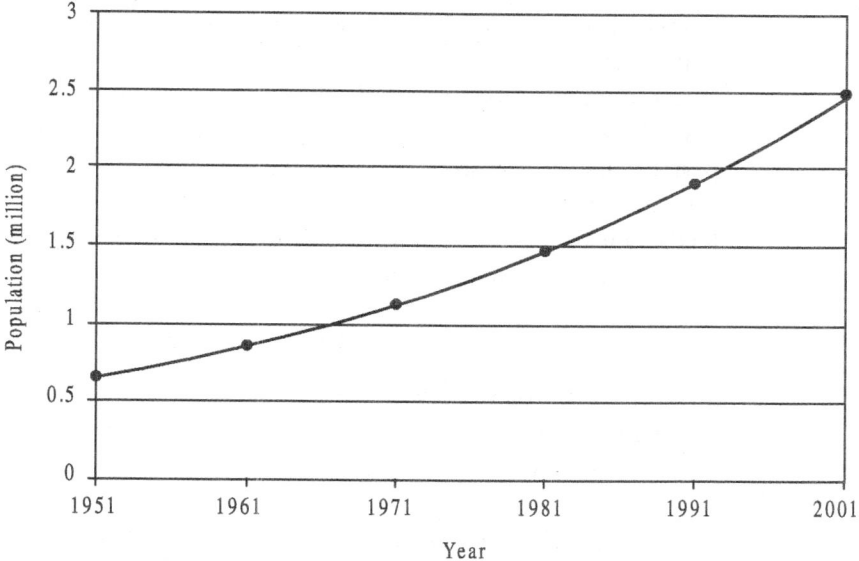

Fig. 4.11 Logistic model fit of the population data.

logistic model is not a very good model for the population growth of the city under consideration.

Another point to be noted is the sensitivity of the results to the starting guess. If we change $c_2^{(0)}$ to -0.01, the iterations do not convervge! It should also be noted that the derivatives of a nonlinear function are required in computation of the **J** matrix. If the function is such that derivatives are not easily obtainable analytically, we may have to use finite differences to estimate their values (as described in details in the next chapter). Further, to reduce the computational burden, the Jacobian may be computed at every, say, second or third iteration and LU decomposition may be used efficiently, for the iterations which have the same Jacobian.

In the analysis so far, we have assumed all data points to be equally important. However, sometimes we have more confidence in a particular measurement as compared to other data points. For example, the newer census data (year 2001) may be thought of as more reliable than the older one (1951) due to improvements in methodologies. We may then assign a higher *weight* to some observations to reflect a higher confidence. Another scenario is when we want to minimise the sum of squares of the relative errors rather than the absolute errors. In this case, the weights may be thought of as the inverse of the squares of the function values. This leads to a *weighted regression* which is not discussed in this book as it is conceptually similar to the ordinary regression.

Most of the times, we have used polynomial basis functions to perform the interpolation or regression. Sometimes, the function to be interpolated (or regressed) may be periodic in nature (e.g., the temporal variation of temperature is typically periodic with a time period of 24 hours). One option in such cases could be, to consider a single period and fit a polynomial to the function within that particular period only. With a shift of origin, this polynomial could be used to describe the function over subsequent periods also. However, use of periodic basis functions leads to a fitting function which is applicable over the entire range. In the next section we will discuss some aspects of approximations of periodic functions.

EXERCISE 4.6

1. For the problem solved in Example 4.11, use the linear regression between $\ln m$ and t to estimate the half-life of the substance and estimate the mass at 6 days. Also find the coefficient of determination and comment on the goodness of fit.

2. The enzyme reaction is assumed to follow a depletion model:

$$\frac{ds}{dt} = -\frac{\mu_{max} s}{s_{half} + s}$$

in which s is the substrate concentration at any time, t; μ_{max} is the maximum substrate consumption rate; and s_{half} is the substrate concentration corresponding to half the maximum consumption rate. This model can be used to predict the substrate concentration at any time, starting from an initial concentration of s_0, from the nonlinear equation $s_0 - s + s_{half} \ln \frac{s_0}{s} = \mu_{max} t$. The following data was obtained starting from an initial concentration, s_0, of 100 ppm:

t (min)	10	20	30	40	50
s (mM/l)	91	82	73	64	56

Estimate the parameters s_{half} and m_{max} by performing a multiple linear regression of t on the parameters $s_0 - s$ and $\ln \frac{s_0}{s}$, i.e., $t = c_{1,0}(s_0 - s) + c_{0,1} \ln \frac{s_0}{s}$. (Note that based on physics of the problem, we have assumed $c_{0,0}$ to be zero. If we include this parameter, the regression may result in a non-zero value.)

3. Re-solve Example 4.12 including the coefficient $c_{1,1}$ in the approximate relation. Compare with the solution of Example 4.12 and comment on the relative performance of the two models. Is it possible that including $c_{1,1}$ would lead to deterioration in the model performance?

4. In Artificial Neural Network (ANN) modelling, a sigmoid function is used to effect the transformation of input to a neuron into its output. The relationship between the input, I, and the scaled output, O, is written as

$$O = \frac{1}{1 + e^{-c_0 I}}$$

The following table shows a few measurements of the input and output:

I	−10	−5	0	5	10	15
O	0.10	0.25	0.49	0.75	0.89	0.96

Estimate the value of c_0 by using nonlinear regression of O on I. Then perform a linear regression of $\ln\left(\frac{1}{O} - 1\right)$ on I and estimate c_0. Compare these two values and explain the result.

5. The water level in the North Sea is mainly determined by the so called M2-tide, whose period is about 12 hours and thus has the form

$$H(t) = h_0 + a_1 \sin\frac{2\pi t}{12} + a_2 \cos\frac{2\pi t}{12}$$

where t is in hours. One has made the following measurement:

t (h)	0	2	4	6	8	10
$H(t)$ (m)	1.0	1.6	1.4	0.6	0.2	0.8

Fit $H(t)$ to the series of measurement using the method of least squares and determine h_0, a_1, and a_2.

6. An object starts from the origin at $t = 0$ and travels along the x-axis. Its position is measured at different times as given below:

Time (s)	1	2	3	4	5
Distance (m)	4.07	12.75	26.48	46.50	69.35

It is known that the object travels with a constant acceleration, i.e., $x = ut + 1/2at^2$. Estimate the initial velocity and acceleration by linearlizing the equation and then using *first order* regression.

7. For the data given in the previous problem, a quadratic spline interpolation is to be used using the six data points (including $t = 0$) and it is assumed that the distance varies linearly with time in the first interval ($t = 0$ to $t = 1$). Estimate the velocity of the object at 3 s.

8. It is known that bacteria swim against the concentration gradient of the food. The following data were measured for the concentration gradient of the food and the corresponding speed of swim:

Conc. Grad. x (μmol/cm)	0.1	0.2	0.5	1.3	4.2	9.5
Speed y (μm/min)		2.85	4.00	6.00	10.00	15.00	20.00

It was decided to fit the equation, $y = \dfrac{a\sqrt{x}}{1+b\sqrt{x}}$ to the data. Obtain least square estimates of a and b; calculate r^2 for the non-linear fit; and graphically compare the data values with the fitted curve.

9. A trignomial of the form $\hat{f}(x) = C_1 + C_2 x + C_3 \sin x$ is to be fitted to some observed values. From the normal equations, comment on when the fit will be non-unique. Obtain the least squares fit for the following data:

x	0.0	0.5	1.0	1.5	2.0
$f(x)$	1.02	1.40	1.70	2.00	2.20

10. Fit a second order polynomial to the following data and estimate y at $x = 2.5$:

x	0	1	2	3	4	5
y	2.1	7.7	13.6	27.2	40.9	61.1

11. For the data of the previous problem, estimate the value of y at $x = 2.5$ using third order Lagrange interpolating polynomials.

12. Estimate the value of e^{2x} at $x = -0.85$ and 0.85 by fitting a fourth order polynomial over the range -1 to $+1$ using (a) Legendre polynomial (b) Gram's polynomial and discrete Tchebycheff polynomials using five points and (c) Gram's polynomial and discrete Tchebycheff polynomials using 11 points (i.e., regression). Use the following integrals:

$$\int\limits_{-1}^{1} P_i(x)e^{2x}dx = \frac{-1+e^4}{2e^2}\,(i=0);\,\frac{3+e^4}{4e^2}\,(i=1);$$

$$\frac{-13+e^4}{8e^2}\,(i=2);\,\frac{77-e^4}{16e^2}\,(i=3);$$

$$\frac{-591+11e^4}{32e^2}\,(i=4)$$

4.7 PERIODIC FUNCTIONS

Let $f(x)$ be a periodic function with a period of 2π (if a function of t has a period of T, we simply define $x = 2\pi t/T$. Although the use of variable t and time period T is more practical, a period of 2π considerably simplifies the mathematical analysis).

As discussed before, either the function is given (continuous case) or its values at some grid points are known (discrete case) and we need to find an approximation which uses basis functions which are periodic over the distance (or time) 2π. Clearly, $\sin x$ and $\cos x$ could be taken as basis functions and correspond to the *fundamental frequency* of a Fourier series representation of the function (see Box 4.5).

BOX 4.5 Fourier Series

Although a Fourier series may be written in terms of any set of orthogonal basis functions, we use here the usual meaning of the Fourier series with sine and cosine as the basis functions. Given a periodic function, $f(t)$, which has a period T, and is piecewise continuous and square integrable (i.e., $\int\limits_{0}^{T}[f(t)]^2 dt$ is finite), it can be expanded in a Fourier series as

$$f(t) = \frac{a_0}{2} + \sum_{j=1}^{\infty}(a_j \cos j\omega t + b_j \sin j\omega t) \tag{B4.5.1}$$

in which $\omega(=2\pi/T)$ is the fundamental frequency and $j\omega$ is the jth harmonic. The coefficients are obtained by multiplying the above expression by $1, \cos k\omega t$, and $\sin k\omega t\,(k=1,2,\ldots,\infty)$, integrating over the interval $(0, T)$ and using the following identities $(\forall j, k \geq 1)$

$$\int_0^T \sin j\omega t \sin k\omega t \, dt = 0, \qquad j \neq k$$

$$= T/2, \quad j = k$$

$$\int_0^T \cos j\omega t \cos k\omega t \, dt = 0, \qquad j \neq k$$

$$= T/2, \quad j = k$$

$$\int_0^T \sin j\omega t \cos k\omega t \, dt = \int_0^T \sin j\omega t \, dt = \int_0^T \cos j\omega t \, dt = 0$$

as

$$a_j = \frac{2}{T} \int_0^T f(t) \cos j\omega t \, dt, \quad \forall j \geq 0$$

$$b_j = \frac{2}{T} \int_0^T f(t) \sin j\omega t \, dt, \quad \forall j \geq 1$$

(B4.5.2)

It is mathematically more convenient to express the Fourier series in its canonical form by using the transformation $x = 2\pi/T = \omega t$, such that the period becomes 2π. The Fourier series then becomes

$$f(x) = \frac{a_0}{2} + \sum_{j=1}^{\infty} (a_j \cos jx + b_j \sin jx)$$

with the coefficients (for $j \geq 0$, note that b_0 is equal to 0, though not needed in the series) .

$$a_j = \frac{1}{\pi} \int_{-\pi}^{\pi} f(x) \cos jx \, dx; \quad b_j = \frac{1}{\pi} \int_{-\pi}^{\pi} f(x) \sin jx \, dx$$

[We could use the integration over the interval $(0, 2\pi)$ but prefer the more commonly used interval of $(-\pi, \pi)$]. The Fourier series converges to $f(x)$ at all points when the function and its derivative are continuous. If the function has a finite number of jump discontinuities in each period, the Fourier series converges to the mean of the left and right limits. It should also be noted that for even functions all b's will be zero and for odd functions all a's would be zero, and we get the Fourier cosine and sine series, respectively.

The Fourier series can easily be expressed in a compact form in terms of complex exponentials by writing

$$f(x) = \sum_{j=-\infty}^{\infty} C_j e^{ijx} \qquad\qquad (B4.5.3)$$

in which i is the imaginary unit[5] $\left(= \sqrt{-1}\right)$, and using the identity

$$\int_{-\pi}^{\pi} e^{ijx} e^{-ikx}\, dx = 2\pi, \quad \text{for } j = k$$

$$= 0, \quad \text{otherwise}$$

integration of Eq. (B4.5.3) provides the values of the coefficients as

$$C_j = \frac{1}{2\pi} \int_{-\pi}^{\pi} f(x) e^{-ijx} dx, \quad \text{for } j = -\infty \text{ to } \infty$$

It is easy to show that these coefficients are related to those in the usual Fourier series (Eq. B4.5.1) as (for $j \geq 0$)

$$C_j = \frac{1}{2\pi} \int_{-\pi}^{\pi} f(x) e^{-ijx} dx = \frac{1}{2\pi} \int_{-\pi}^{\pi} f(x) \cos jx\, dx - i \frac{1}{2\pi} \int_{-\pi}^{\pi} f(x) \sin jx\, dx = \frac{a_j - i b_j}{2}$$

$$C_{-j} = \frac{1}{2\pi} \int_{-\pi}^{\pi} f(x) e^{ijx} dx = \frac{1}{2\pi} \int_{-\pi}^{\pi} f(x) \cos jx\, dx + i \frac{1}{2\pi} \int_{-\pi}^{\pi} f(x) \sin jx\, dx = \frac{a_j + i b_j}{2}$$

Although the complex form of the Fourier series is more compact and mathematically convenient, from a numerical perspective we would like to avoid complex computations. Therefore, we do not discuss the complex form of the Fourier series. However, the orthogonality condition is useful in writing its counterpart for a periodic function sampled at discrete points as shown below.

When the function is not known in an explicit form but is only sampled at a few discrete points *equispaced* over a period (for unequal spacing of data points, the basis functions become non-orthogonal and the analysis is quite complicated. We do not discuss these cases.), we assume that a single period $(-\pi, \pi)$ is divided into $(n + 1)$ equal segments such that the sampling locations are

$$x_l = -\pi + \frac{2\pi l}{n+1}, \quad l = 0, 1, 2, ..., n$$

---

[5]Sometimes j is used to represent the imaginary unit. Although it may be a little confusing since we use i and j as subscripts also, we decide to use this notation in keeping with the general usage

Note that the endpoint π is not included since the periodic nature of the function makes it redundant to sample both $-\pi$ and π. The orthogonality condition is obtained by noting that e^{imx_l} $l = 0,1,2,...,n$ is a geometric sequence with the first term equal to $e^{-im\pi}$, i.e., $(-1)^m$, and the common ratio equal to $e^{i\frac{2m\pi}{n+1}}$.

The sum of the series would be $e^{-im\pi} \dfrac{1-e^{i2\pi m}}{1-e^{i\frac{2\pi m}{n+1}}}$, which is, clearly, zero for all integer m, unless $m/(n+1)$ is also an integer, (which may be positive, negative, or zero), in which case the common ratio is 1 and the sum is readily obtained as $(-1)^m(n+1)$. Therefore, we have the following orthogonality condition:

$$\sum_{l=0}^{n} e^{ijx_l} e^{ikx_l} = (-1)^{j+k}(n+1), \quad \text{for integer } \frac{j+k}{n+1}$$

$$= 0, \quad\quad\quad\quad\quad \text{otherwise}$$

Sometimes the interval $(0, 2\pi)$ is used in which case the term $(-1)^m$ is removed since the first sampling location is $x = 0$. It would make the orthogonality conditions much simpler to write. However, to maintain uniformity with the continuous case, we use the interval $(-\pi, \pi)$.

To avoid complex numbers, we may write this orthogonality conditions in terms of sines and cosines as

$$\sum_{l=0}^{n} \sin jx_l = \sum_{l=0}^{n} \sin jx_l \cos kx_l = 0; \quad \sum_{l=0}^{n} \cos jx_l = (-1)^j(n+1), \quad \text{if } \frac{j}{n+1} \text{ is integer}$$

$$= 0, \quad\quad\quad\quad\quad \text{otherwise}$$

$$\sum_{l=0}^{n} \sin jx_l \sin kx_l = (-1)^{j+k+1}\frac{n+1}{2}, \quad \text{for integer } \frac{j+k}{n+1} \text{ and non-integer } \frac{j-k}{n+1}$$

$$= (-1)^{j+k}\frac{n+1}{2}, \quad \text{for integer } \frac{j-k}{n+1} \text{ and non-integer } \frac{j+k}{n+1}$$

$$= 0, \quad\quad\quad \text{otherwise}$$

$$\sum_{l=0}^{n} \cos jx_l \cos kx_l = (-1)^{j+k}\frac{n+1}{2}, \quad \text{when either } \frac{j-k}{n+1} \text{ or } \frac{j+k}{n+1} \text{ is integer}$$

$$= (-1)^{j+k}(n+1), \quad \text{when both } \frac{j-k}{n+1} \text{ and } \frac{j+k}{n+1} \text{ are integers} \quad \text{(B4.5.4)}$$

$$= 0, \quad\quad\quad \text{otherwise}$$

In fact, it is easily seen that sin (jx) and cos (jx) are periodic over an interval of 2π for all integer j (representing the *harmonics*) and could act as the basis functions for an approximating expression. We first discuss the continuous case and then the discrete case, in which both interpolation and regression are possible.

4.7.1 Continuous Case

Although any interval of 2π could be used to define the function, we consider the interval to be $[-\pi, \pi]$ instead of $[0, 2\pi]$ [if a function of t is defined over the interval $(0, T)$, we transform the variable as $x = 2\pi t / T - \pi$]. We assume that the approximating function involves harmonics up to and including m and write it as

$$f_m(x) = \frac{c_0}{2} + \sum_{j=1}^{m}(c_j \cos jx + s_j \sin jx) \qquad (4.67)$$

Note that the fundamental frequency is the first harmonic, and the coefficient of the zeroth harmonic (for which the sine term, obviously, vanishes) is written as $c_0/2$ to be consistent with the definition of coefficients of the cosine term in other harmonics (see Box 4.5). Also note that while a polynomial of degree m had $m + 1$ coefficients, the approximation using harmonics up to m has $2m + 1$ coefficients.

To minimize the L_2 norm (we could minimize some other norm, e.g., L_1 or L_∞, but it is not mathematically convenient) of the residual, i.e.,

$$\int_{-\pi}^{\pi} [f(x) - f_m(x)]^2 dx,$$

we use the linear regression as before (see Eq. 4.54) and obtain the following equation:

$$[A]\{c\} = \{b\} \qquad (4.68)$$

in which the unknown vector $\{c\}$ is $\{c_0, c_1, s_1, c_2, s_2, ..., c_m, s_m\}^T$, and the elements of the matrix $[A]$ and vector $\{b\}$ are given by $a_{ij} = \langle \phi_i, \phi_j \rangle$ and $b_i = \langle \phi_i, f \rangle$, with the sine and cosine terms as the basis functions which form an orthogonal set (see Box 4.5). The $(2m + 1)$ basis functions are (see Eq. 4.67) 1/2, cos x, sin x, cos $2x$, sin $2x$,..., cos mx, sin mx.

It follows that the coefficients of the approximating function are nothing but the Fourier series coefficients and we get the result that the *Fourier series truncated at the mth harmonic is the best approximation in L_2 norm using $(2m + 1)$ terms.*

4.7.1.1 *Error in Fourier Approximation*

The error of approximation at any x is, naturally, equal to the sum of the remaining terms of the Fourier series. We may, therefore, analyse the error in the x-domain

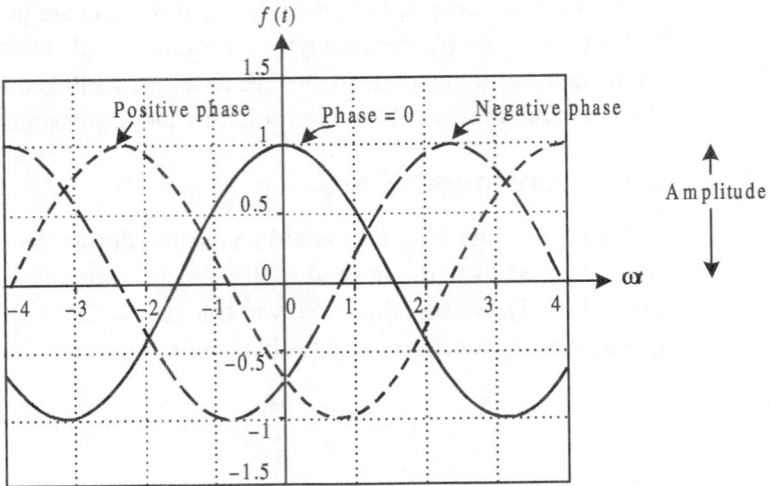

Fig. 4.12   A waveform, its amplitude and phase.

(since periodic functions are generally associated with temporal periodicity, we can call this the *time-domain* analysis). However, there is an alternative technique of analysing periodic functions which would be quite useful in a later chapter for analysing the error of numerical solution of differential equations. Here we provide a brief introduction and a detailed analysis is left for Chapter 6.

Looking at the form of the Fourier series representation of a periodic function (Eq. B4.5.1) it is readily seen that the two waveforms (sine and cosine) corresponding to a particular frequency ($j\omega$) could be combined into a single term of the form $A\cos(j\omega t + \theta)$, in which A is called the amplitude

$$\left(A = \sqrt{a_j^2 + b_j^2}\right)$$

and θ is the phase angle [$\theta = \tan^{-1}(-b_j / a_j)$] (we could also write it as a sine function with the same amplitude but a phase angle of ($\pi/2 + \theta$). The convention is to use the range [$-\pi$, π] for the phase which is defined as the angular distance of the nearest positive peak from the origin, with positive values used if this peak occurs before zero (an advanced peak) and negative if the peak occurs after zero (a delayed peak), see Fig. 4.12.

This leads to an alternative representation of the periodic function in terms of its different harmonics (i.e. in *frequency domain*) with their associated amplitudes and phases. It is known as spectral analysis and the plots of amplitude versus frequency and phase versus frequency are called the amplitude line spectra and phase line spectra, respectively. The term *line* refers to the fact that there are discrete frequencies. For a non-periodic function, we get a continuous distribution of frequencies and get the amplitude and phase spectra.

Example 4.14   For a periodic triangular function defined by $f(x) = \pi - |x|$ over a period of $[-\pi, \pi]$, obtain the Fourier series. Plot the amplitude line spectra and the phase line spectra. What would be the spectra for the approximation of this function using the terms up to the third harmonic?

Solution   The Fourier series is written as

$$f(x) = \frac{a_0}{2} + \sum_{j=1}^{\infty} (a_j \cos jx + b_j \sin jx)$$

Since the function is symmetric, all the b's would be zero. The coefficients a are obtained as

$$a_j = \frac{1}{\pi} \int_{-\pi}^{\pi} f(x) \cos jx \, dx = \frac{1}{\pi} \left[\int_{-\pi}^{0} (\pi + x) \cos jx \, dx + \int_{0}^{\pi} (\pi - x) \cos jx \, dx \right]$$

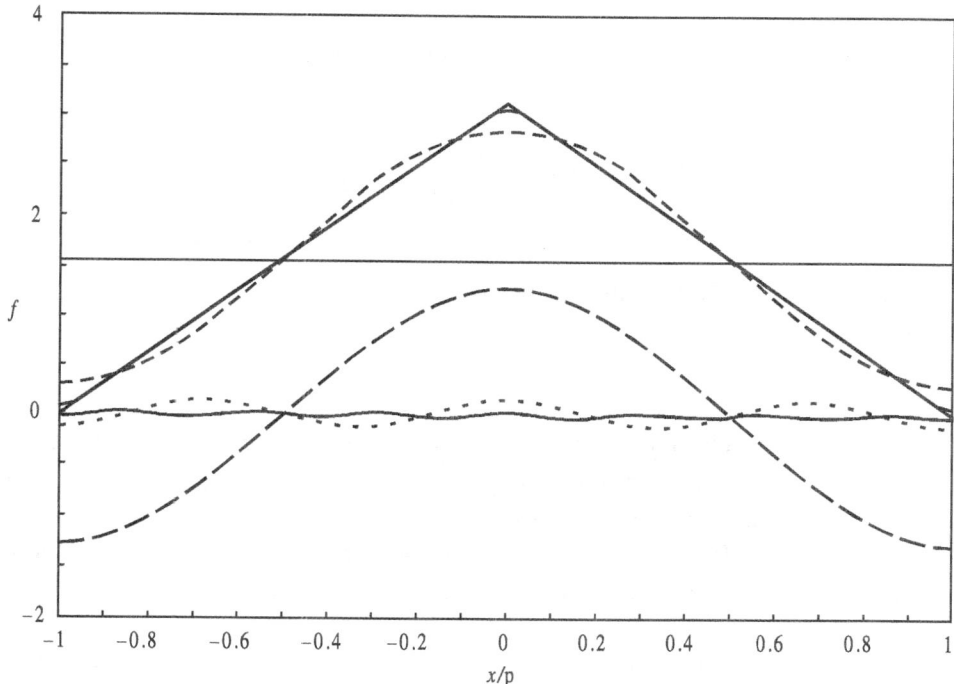

Fig. 4.13(a)   Harmonics of the triangular function.

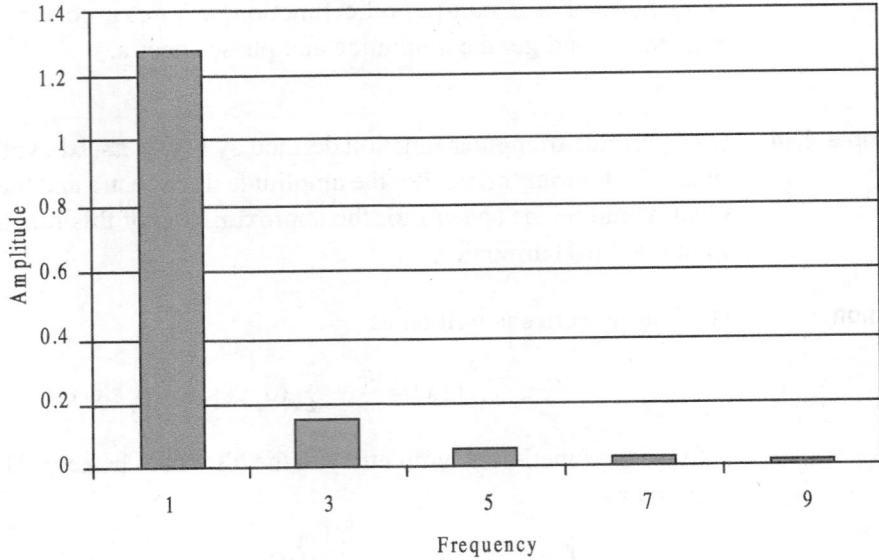

Fig. 4.13(b) Amplitude line spectra.

$$= \int_{-\pi}^{\pi} \cos jx\, dx - \frac{2}{\pi} \int_{0}^{\pi} x \cos jx\, dx$$

$$= \begin{bmatrix} \pi, & \text{for } j=0 \\[2mm] \dfrac{2}{\pi} \dfrac{1-(-1)^j}{j^2}, & \text{otherwise} \end{bmatrix}$$

resulting in the Fourier series

$$f(x) = \frac{\pi}{2} + \frac{4}{\pi} \sum_{j=1,3,5\ldots}^{\infty} \frac{\cos jx}{j^2}$$

Clearly, the phase angle is zero for all frequencies and the amplitude of the *j*th harmonic is $4/\pi j^2$. A plot of different harmonics and the amplitude line spectra is shown in Figs 4.13(a) and (b).

If we include terms up to the third harmonic only, the amplitude spectra would have zero amplitude for the fifth and higher harmonics.

Thus, we have an alternative way of defining the error of approximation in terms of the *amplitude error* and the *phase error*. If both the given periodic function and the approximation are expressed in terms of the fundamental frequency and the harmonics, each frequency may be thought of as having an amplitude error, equal to the difference between the true amplitude and the approximate amplitude, and a phase errors, equal to the difference of true and approximate phases. For example, a truncated Fourier series approximation using up to two harmonics will have zero amplitude and phase errors corresponding to the fundamental frequency and the first two harmonics, but will have amplitude errors equal to the amplitude of the true spectra for higher harmonics. Obviously, both amplitude and phase errors are important to consider and more discussion about these appears in Chapter 6.

4.7.2 Discrete Case

For a periodic function, sampled at the $(n + 1)$ points

$$x_l = -\pi + \frac{2\pi l}{n+1}, \quad l = 0,1,2,...,n$$

we would have a choice of approximating by an interpolating truncated Fourier series (which passes through all sampled points) or a regressing series (which would minimize L_2 norm of the residual at the sampling points). The discrete form of the orthogonality condition of the Fourier series (see Box 4.5) is used to derive these approximations as follows.

4.7.2.1 Interpolation

We assume that *n is even*, such that we have an odd number of grid points. In this case, we express an interpolating function which matches the sampled function values at all grid points as

$$f_m(x) = \frac{c_0}{2} + \sum_{j=1}^{m}(c_j \cos jx + s_j \sin jx) \tag{4.69}$$

where $m = n/2$. The orthogonality condition results in the values of these coefficients as

$$c_j = \frac{2}{n+1}\sum_{l=0}^{n} f(x_l)\cos jx_l \quad \text{and} \quad s_j = \frac{2}{n+1}\sum_{l=0}^{n} f(x_l)\sin jx_l \tag{4.70}$$

If n is odd, m will be equal to $(n - 1)/2$ and we add an additional term, $c_{m+1}/2\cos(m + 1)x$, in which the coefficient is given by the same formula listed above. [It is easily seen that $\sin(m + 1)x$ will vanish at all grid points and $\cos(m + 1)x_l$ will alternate between ± 1]. Note that the coefficient is divided by

2 since $j = k = m + 1$ implies that both $\dfrac{j-k}{n+1}$ and $\dfrac{j+k}{n+1}$ are integers and the relevant orthogonality condition (see Box 4.5) has to be used.

4.7.2.2 Regression

If we use fewer number of harmonics than that used for interpolation [i.e. $n/2$ for even n and $(n + 1)/2$ for odd n], we would obtain a regressing series which would not pass through all the data points. A least square regression would again lead to the conclusion that a truncated Fourier series is the best approximation. Thus, the form of the series and its coefficients would be same as that listed above for the interpolating series.

The discussion regarding Fourier series for periodic function may be extended to non-periodic functions by considering them to be periodic with an infinite period. In this case, the Fourier series becomes a Fourier integral (or Fourier transform). We do not discuss it here.

Another topic which we avoid discussing here is the efficient evaluation of the Fourier approximation for discrete data using the *Fast Fourier Transform* (FFT). We believe that this topic may be suitable for only a small section of the targeted audience of this book. We have provided the basic theory of the method here and a more detailed description is given in Chapter 8.

Example 4.15 For the periodic triangular function of the previous example, let the function be given in terms of 7 equally spaced points $(-\pi, -5\pi/7, -3\pi/7, -\pi/7, \pi/7, 3\pi/7, 5\pi/7)$. Obtain the interpolating Fourier series. Also obtain the least squares approximation of this function using the terms up to the fundamental frequency (i.e. the constant term and the first harmonic).

Solution Table 4.8 shows the sampled values of the function.

Table 4.8

i	0	1	2	3	4	5	6
x_i	−3.141593	−2.243995	−1.346397	−0.448799	0.448799	1.346397	2.243995
$f(x_i)$	0.000000	0.897598	1.795196	2.692794	2.692794	1.795196	0.897598

Since we have an odd number of grid points ($n = 6$), Eq. (4.69) is used to interpolate the function with $m = 3$. The following table shows the computations of the coefficients using Eq. (4.70) (Note that the symmetry would lead to all the coefficients s being zero. However, we show these computations in Table 4.9):

Table 4.9

l	x_l	$f(x_l)$	$f(x_l)\cos jx_l$				$f(x_l)\sin jx_l$		
			$j=0$	$j=1$	$j=2$	$j=3$	$j=1$	$j=2$	$j=3$
0	−3.141593	0.000000	0.000000	0.000000	0.000000	0.000000	0.000000	0.000000	0.000000
1	−2.243995	0.897598	0.897598	−0.559643	−0.199734	0.808708	−0.701770	0.875093	−0.389453
2	−1.346397	1.795196	1.795196	0.399469	−1.617416	−1.119286	−1.750186	−0.778906	1.403541
3	−0.448799	2.692794	2.692794	2.426123	1.678929	0.599203	−1.168359	−2.105311	−2.625280
4	0.448799	2.692794	2.692794	2.426123	1.678929	0.599203	1.168359	2.105311	2.625280
5	1.346397	1.795196	1.795196	0.399469	−1.617416	−1.119286	1.750186	0.778906	−1.403541
6	2.243995	0.897598	0.897598	−0.559643	−0.199734	0.808708	0.701770	−0.875093	0.389453
Sum			10.771175	4.531898	−0.276441	0.577249	0.000000	0.000000	0.000000
Coefficients, c_j and s_j			3.077479	1.294828	−0.078983	0.164928	0.000000	0.000000	0.000000

The interpolating function is therefore

$$f_3(x) = 1.53874 + 1.29483\cos x - 0.0789831\cos 2x + 0.164928\cos 3x.$$

As a comparison, we note that the corresponding coefficients of the Fourier series of this function (Example 4.14) are 1.57080, 1.27324, 0, and 0.141471, respectively. While the phase angle is again zero for all frequencies, the amplitude error is readily obtainable. Note particularly the presence of a small amplitude corresponding to the second harmonic although the *true amplitude* is zero. Figures 4.14(a) and (b) show a plot of this interpolating function (which, naturally, must pass through all 7 data points):

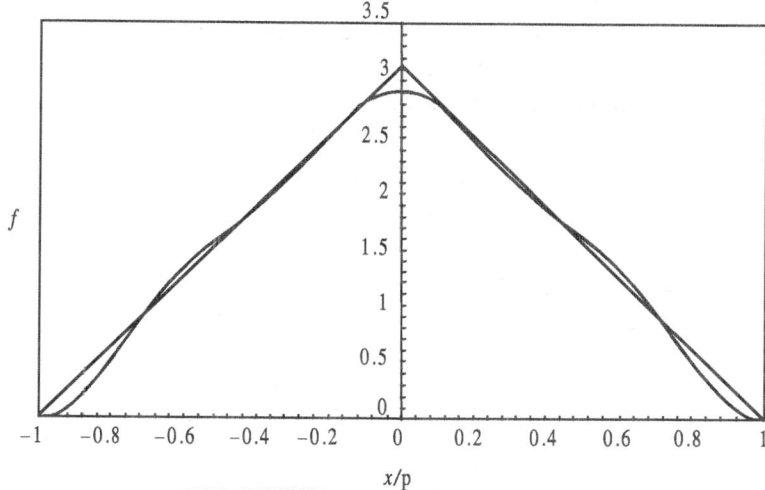

Fig. 4.14(a) Interpolating function.

As discussed, the least squares fit using the constant and the fundamental frequency is given by $f_1(x) = 1.53874 + 1.29483 \cos x$ which is shown in Fig. 4.14(b).

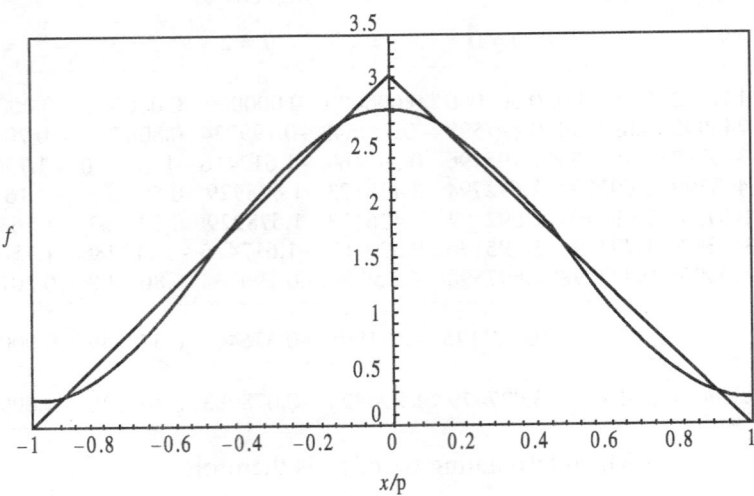

Fig. 4.14(b) Least squares fit.

EXERCISE 4.7

1. Find the Fourier series of the saw-tooth wave of period T, defined as $f(t) = t/T$ for $t = 0$ to T (you may want to define $x = \pi(2t/T - 1)$ to change the period into the standard range of $-\pi$ to π. Also note that there is a jump discontinuity at the ends of the period). Why do all cosine terms (except the constant) vanish?

2. A periodic function is defined over a period $[-\pi, \pi]$ as $f(x) = x(\pi - x)$. Find its Fourier series. Now generate a set of 8 function values starting at $x = -\pi$ at intervals of $\pi/4$ and use these to obtain the interpolating Fourier series.

3. For the function described in the question above, obtain the least square fit using terms up to the second harmonic for both the continuous function and the discrete data. Compare these approximations and obtain the phase and amplitude errors.

4. Following values of a periodic function over one period (-1 to 1) were observed:

x	-1	$-5/7$	$-3/7$	$-1/7$	$1/7$	$3/7$	$5/7$
$f(x)$	-2.00	4.87	-2.71	-0.33	3.67	0.97	2.53

Obtain the discrete Fourier series interpolation of the function.

SUMMARY

In this chapter, we discussed the motivation for approximating a given function (either as a function or as discretely sampled values) by another function (generally polynomial or trigonometric). Various methods including the Taylor's series, least squares, and orthogonal polynomials, were described for approximating a function given in a continuous form and their relative merits and drawbacks were discussed. The concept of a best-fit was discussed in terms of various error norms. For discretely sampled functions, interpolation and regression were discussed.

For interpolation, various alternative forms of the unique interpolating polynomials were described and their utility in particular conditions was emphasized. Piecewise interpolation was introduced as a method to reduce the possibility of large errors associated with equidistant interpolation. Linear and nonlinear regression was then discussed and the correlation coefficient was described. Finally, for periodic functions, both in continuous and in discrete forms, the Fourier series was used to approximate these using sine and cosine functions.

Clearly, the methods described here may be used to approximate the derivatives and the integrals of a given function (or values) by first finding the approximating function and then performing an analytical differentiation or integration. However, since the problems of estimating the derivative or integral occur quite frequently, it is desirable to obtain these approximations directly rather than finding the approximation of the function first. Although the basic philosophy is similar to what we have discussed in this chapter, there are enough differences to warrant a separate chapter. Therefore, the next chapter describes various techniques of numerical differentiation and integration.

Numerical Differentiation and Integration

5.1 INTRODUCTION

As discussed in the previous chapter, a number of times we will have to estimate the functions derivative or integral, from a set of tabulated values of the function. For example, in our *Batman* problem, if *Joker* notes down the time as the wheel passes each floor, he may be asked to estimate the velocity of the wheel at these times. On the other hand, if he measures the velocity of the wheel, he may be asked to estimate the distance between the floors. In this chapter, we discuss the methods which would enable him to do so. These methods are categorised as follows:

- *Numerical differentiation* (estimating the functions derivatives, e.g., to obtain the velocity or acceleration from distance measurements, finding the heat flux from temperature measurements), and
- *Numerical integration* (estimating the integral of the functions, e.g., to obtain distance from velocity measurements, finding the area of an object).

As before, we have two possibilities as to how the function values are given: the *continuous* case in which the function $f(x)$ is known and the *discrete* case in which the function values are given corresponding to a few values of x. We mention again that the continuous case is a superset of the discrete case since the function values at selected points can be readily generated if the function is known. Also, since differentiation is a relatively straightforward operation, we would hardly ever need to perform *numerical* differentiation of a given function. Therefore, we first discuss numerical differentiation for the discrete case and then look at numerical integration for both the discrete and continuous cases.

5.2 NUMERICAL DIFFERENTIATION

Suppose we have a table of data, listing the values of the dependent variable, $f(x)$, corresponding to a few values of the independent variable, x, and we want to estimate the function derivative at one of these points. Sometimes, though it is not very common, we may be asked to evaluate the derivative at a point which does not coincide with a grid point. A few techniques discussed here would be applicable in such cases. However, we would not address this issue further. Again, we denote the data points by the set of values $\{(x_k, f(x_k)), k = 0, 1, ..., n\}$. As we did earlier for the interpolation problem, we assume that all x_k are distinct and there is no error in data. If the data is subjected to errors of measurement, we could first obtain the regression polynomial and then obtain the derivative at any point *analytically*. We also assume that the x_k are arranged in increasing or decreasing order.

As described in the previous chapter, we should be able to find the interpolating or regression polynomial and then, the derivative at any point. However, we have also seen that a higher-order interpolating polynomial may show severe oscillations if the underlying function is not suitable for polynomial representation. Therefore, almost always, we will approximate the function by a *piecewise* polynomial and then find its derivative using the relevant polynomial. We may go for non-polynomial (e.g., rational, sinusoidal) interpolation but the procedure, and especially the error analysis, is more complicated.

The simplest continuous approximation would be a piecewise linear interpolation obtained by joining the function values at consecutive grid points, which will provide us with the *piecewise constant* first derivative. However, we will face two problems: (i) there would be two values of the first derivative at each node, one from either side, which may be significantly different from each other and (ii) the second and higher derivatives would be zero everywhere except at the grid points, where they would be undefined.

Thus, there is a need to look for alternative ways of performing the numerical differentiation. We discuss some of these techniques in the next sections. Since it is common that the given data corresponds to evenly spaced grid points, we first describe the techniques for this case and then discuss the irregularly spaced data.

5.2.1 Evenly Spaced Grid Points

In this section, we consider the case when $x_i - x_{i-1} = h$ for $i = 1, 2, ..., n$. If we assume that the function is represented by a piecewise linear polynomial, the first derivative of the function at any point, x_i, may be obtained as the

- Slope of the line in the previous segment (backward difference)

$$f_i' = \frac{f(x_i) - f(x_{i-1})}{h} \qquad (5.1)$$

We denote the estimated derivatives of the function at x_i by f_i' and the *exact* value by $f'(x_i)$.

- Slope of the line in the next segment (forward difference)

$$f_i' = \frac{f(x_{i+1}) - f(x_i)}{h} \qquad (5.2)$$

- Average of these two slopes (or the slope of the line joining the previous point and the next point) (central difference)

$$f_i' = \frac{f(x_{i+1}) - f(x_{i-1})}{2h} \qquad (5.3)$$

Clearly, the backward difference will not be valid at the first node, the forward difference will not work at the last node, and the central difference cannot be used at the first *and* last nodes. Now, if we want the second derivative at x_i, we may argue that the first derivative of the function at the mid-point of the previous segment is given by Eq. (5.1) and that at the mid-point of the next segment is given by Eq. (5.2). Since the slope is constant throughout a segment, any other point could also be taken instead of the mid-point. However, as we will see a little later, the central difference gives a better accuracy. Therefore, the second derivative may be approximated as

$$f_i'' = \frac{\dfrac{f(x_{i+1}) - f(x_i)}{h} - \dfrac{f(x_i) - f(x_{i-1})}{h}}{h} = \frac{f(x_{i+1}) - 2f(x_i) + f(x_{i-1})}{h^2} \qquad (5.4)$$

Another alternative may be to fit a second degree polynomial through the three points (x_{i-1}, x_i, x_{i+1}) and find its second derivative (which would be a constant over the interval), which results in the same expression. Of course, the same argument could be extended to arrive at expressions of the higher derivatives, which would generally involve an increasing number of grid points around (or on one side of) the point at which we want the derivative value. However, there is a more convenient way, based on the Taylor's series, of deriving these expressions, which also provides an estimate of the error in the approximation. For convenience, we first analyse the first derivative of the function and then briefly describe the higher derivatives.

Using the Taylor's series expansion (you may want to review Section 4.2 at this stage) about the point x_i, we may write

$$f(x_{i-1}) = f(x_i) - hf'(x_i) + \frac{h^2}{2!}f''(x_i) - \frac{h^3}{3!}f'''(x_i) + \cdots$$

$$f(x_{i+1}) = f(x_i) + hf'(x_i) + \frac{h^2}{2!}f''(x_i) + \frac{h^3}{3!}f'''(x_i) + \cdots \qquad (5.5)$$

from which the following can be easily obtained:

$$f'(x_i) = \frac{f(x_i) - f(x_{i-1})}{h} + h\frac{f''(x_i)}{2!} - h^2\frac{f'''(x_i)}{3!} - \cdots$$

$$= \frac{f(x_i) - f(x_{i-1})}{h} + h\frac{f''(\xi_b)}{2!}$$

$$f'(x_i) = \frac{f(x_{i+1}) - f(x_i)}{h} - h\frac{f''(x_i)}{2!} - h^2\frac{f'''(x_i)}{3!} - \cdots$$

$$= \frac{f(x_{i+1}) - f(x_i)}{h} - h\frac{f''(\xi_f)}{2!} \qquad (5.6)$$

$$f'(x_i) = \frac{f(x_{i+1}) - f(x_{i-1})}{2h} - h^2\frac{f'''(x_i)}{3!} - h^4\frac{f^v(x_i)}{5!} - \cdots$$

$$= \frac{f(x_{i+1}) - f(x_{i-1})}{2h} - h^2\frac{f'''(\xi_c)}{3!}$$

in which ξ represents a point in the appropriate interval for the backward, forward, and central difference schemes, i.e., $\xi_b \in (x_{i-1}, x_i)$; $\xi_f \in (x_i, x_{i+1})$; and $\xi_c \in (x_{i-1}, x_{i+1})$. From these relationships, the following may be noted:

- The forward and backward difference schemes for the first derivative have error of order h while the central difference scheme has $O(h^2)$ error. Note that the error, defined as (true value – approximate value), refers here to only the truncation error arising out of the chopping off of Taylor's series after finite number of terms. For small step size, h, the function values at two neighbouring points would be very close and there may be significant round-off errors in computation of the derivatives. There is, therefore, a trade-off between the truncation and round-off errors and there would be an optimum step size which would result in the minimum *total* error. However, in this chapter, we will concentrate on the truncation error assuming that use of higher precision computation will make the round-off error negligible in comparison, even for very small step sizes. The order of error implies that if the step size, h, is halved, the error will also be *roughly* half in the forward and backward schemes and one-fourth in the central difference scheme. If the second derivative of the function is constant, the error in forward and backward difference approximation of the first derivative would be *exactly* halved by halving the step size. Similarly, if the third derivative is constant, the error in central difference will be exactly one-fourth. For a general case, the error will vary linearly (or quadratically) with h only for small h.

- If the actual function is linear, i.e., the second derivative is zero, both the forward and backward schemes will be exact. Similarly, if the function is a second degree polynomial, the central difference scheme will provide the exact first derivative.

- The point ξ is not known and therefore an error estimate cannot be obtained for a general case. However, for known functions, upper and lower bounds for the second or third derivatives may be obtained and utilized to obtain the error bounds. As already discussed, though, the function is typically not known.

- Generally the error reduces as the step size, h, becomes small. However, it may sometimes happen that a reduction in step size leads to a larger error, see Exercise 5.2, Problem 1. This is due to the fact that the error term involves the derivative of the function at an *appropriate* point which changes from one step size to another. When h becomes *very small* this apparent anomaly does not exist. For an acceptable accuracy, therefore, the step size should be *small*. Of course, the acceptable step size depends on how fast the function is varying (see Exercise 5.2, Problem 2). Choosing a proper step size for sampling the function is more a matter of the *design of experiment* and will not be considered here. The issue which we address is: once the step size is chosen and function values are available at a few (evenly spaced) points, how to estimate the derivatives accurately.

Example 5.1

The location of an object at various times was measured as follows:

Time (s)	0	1	2	3	4	5	6	7	8	9
Distance (cm)	0	3	14	39	84	155	258	399	584	819

Estimate the speed of the object at 5 s using (a) backward difference, $O(h)$; (b) forward difference, $O(h)$; and (c) central difference, $O(h^2)$. First use a step size, $h = 1$ s, and then double this size ($h = 2$ s).

Solution

With a step size of 1 s, using Eqs (5.6), the speed (i.e., the first derivative of the distance) is estimated as

$$\text{Backward difference, } f_5' = \frac{f(5) - f(4)}{1} = 71 \text{ cm/s}$$

$$\text{Forward difference, } f_5' = \frac{f(6) - f(5)}{1} = 103 \text{ cm/s}$$

$$\text{Central difference, } f_5' = \frac{f(6) - f(4)}{2} = 87 \text{ cm/s}$$

for double the step size, these estimates are

$$\text{Backward difference, } f_5' = \frac{f(5) - f(3)}{2} = 58 \text{ cm/s}$$

$$\text{Forward difference, } f_5' = \frac{f(7) - f(5)}{2} = 122 \text{ cm/s}$$

$$\text{Central difference, } f_5' = \frac{f(7) - f(3)}{4} = 90 \text{ cm/s}$$

We expect the central difference estimate with the smaller step size of 1 sec to be the most accurate estimate out of those obtained here. It is apparent that the forward and backward difference estimates have larger errors and the increase in error with increase in step length is also evident.

5.2.1.1 *Improving the Accuracy*

As seen in the previous section, a simple forward or backward difference scheme estimates the first derivative with an $O(h)$ error. One way to reduce the error is to reduce the step size. However, since we do not have function measurements at closer intervals, it would generally not be feasible to do so. Using the central difference formula is likely to improve the accuracy since the error is $O(h^2)$. However, if we are at either end of the data grid, the central difference scheme is not directly applicable. This provides the motivation to devise schemes which achieve higher order of accuracy (by utilizing additional function values). Although the basic philosophy is the same, i.e., using a higher-order interpolation, there are several ways of arriving at the required formula. We discuss below some of these methods taking example of a forward difference formula which has $O(h^2)$ error:

(a) Interpolation followed by differentiation: If, instead of the linear interpolation used in Eq. (5.2), we use the point x_{i+2} to fit a parabola through the three points and then differentiate it, we obtain a more accurate approximation. Since the derivative is invariant under translation, we may, without any loss of generality, take $x_i = 0$, $x_{i+1} = h$, and $x_{i+2} = 2h$. The interpolating polynomial is then written as $f_2(x) = c_0 + c_1 x + c_2 x^2$ and the three conditions are

$$f(x_i) = c_0$$
$$f(x_{i+1}) = c_0 + hc_1 + h^2 c_2$$

$$f(x_{i+2}) = c_0 + 2hc_1 + 4h^2 c_2$$

Solving for c_0, c_1, and c_2, we obtain

$$f_2(x) = f(x_i) + \frac{-f(x_{i+2}) + 4f(x_{i+1}) - 3f(x_i)}{2h} x + \frac{f(x_{i+2}) - 2f(x_{i+1}) + f(x_i)}{2h^2} x^2$$

(5.7)

and its slope at $x = 0$ as

$$f'_i = \frac{-f(x_{i+2}) + 4f(x_{i+1}) - 3f(x_i)}{2h}$$

(5.8)

(b) Using Taylor's series: Writing the Taylor's series expansion

$$f(x_{i+2}) = f(x_i) + 2hf'(x_i) + \frac{4h^2}{2!} f''(x_i) + \frac{8h^3}{3!} f'''(x_i) + \cdots$$

(5.9)

and using Eq. (5.5) for the expansion of $f(x_{i+1})$ to eliminate the term involving the second derivative, it can be shown that

$$f'(x_i) = \frac{-f(x_{i+2}) + 4f(x_{i+1}) - 3f(x_i)}{2h} + \frac{h^2}{3} f'''(x_i) + \cdots$$

(5.10)

(c) Method of undetermined coefficients: Assuming an expression of the form

$$f'_i = c_i f(x_i) + c_{i+1} f(x_{i+1}) + c_{i+2} f(x_{i+2})$$

(5.11)

in which the c's are undetermined coefficients, we require that the first derivative be exact for all polynomials of second degree. Taking the function as $f(x) = 1$, x, and x^2, respectively, we obtain three linear equations (recall that we have taken the three points as $x = 0$, h, and $2h$) which can be solved to obtain the coefficients:

$$
\begin{aligned}
f(x) = 1 &\implies c_i + c_{i+1} + c_{i+2} = 0 \\
f(x) = x &\implies 0.c_i + hc_{i+1} + 2hc_{i+2} = 1 \\
f(x) = x^2 &\implies 0.c_i + h^2 c_{i+1} + 4h^2 c_{i+2} = 0
\end{aligned}
$$

(5.12)

which results in $c_i = -\dfrac{3}{2h}; c_{i+1} = \dfrac{2}{h}; c_{i+2} = -\dfrac{1}{2h}$.

(d) Richardson's extrapolation: From Eq. (5.6), the leading error term in the two point forward difference formula is equal to $-\dfrac{hf''(x_i)}{2}$. If we use a step size of $2h$ (in other words, we use every alternate data point), the leading error term should be roughly twice that for the step size of h. Denoting the estimate of the first derivative with step size h as $f'_i(h)$, we have

$$
\begin{aligned}
f'(x_i) &= f'_i(h) + E + O(h^2) \\
f'(x_i) &= f'_i(2h) + 2E + O(h^2)
\end{aligned}
$$

(5.13)

E being the $O(h)$ error. A new, and hopefully more accurate, estimate of $f'(x_i)$ is obtained from the above as

$$
\begin{aligned}
f'(x_i) &= 2f_i'(h) - f_i'(2h) + O(h^2) \\
&= 2\frac{f(x_{i+1}) - f(x_i)}{h} - \frac{f(x_{i+2}) - f(x_i)}{2h} + O(h^2) \\
&= \frac{-f(x_{i+2}) + 4f(x_{i+1}) - 3f(x_i)}{2h} + O(h^2)
\end{aligned}
\tag{5.14}
$$

In fact, we may carry out another step of extrapolation by using two $O(h^2)$ estimates to obtain an $O(h^3)$ estimate [we assume that the next term in the error is $O(h^3)$. When using the central difference, the next higher-order term would be $O(h^4)$] as

$$
f'(x_i) = \frac{4f_i'(h) - f_i'(2h)}{3} + O(h^3)
$$

$$
f'(x_i) = \frac{4}{3}\frac{-f(x_{i+2}) + 4f(x_{i+1}) - 3f(x_i)}{2h} - \frac{1}{3}\frac{-f(x_{i+4}) + 4f(x_{i+2}) - 3f(x_i)}{4h} + O(h^3)
$$

$$
= \frac{f(x_{i+4}) - 12f(x_{i+2}) + 32f(x_{i+1}) - 21f(x_i)}{12h} + O(h^3)
\tag{5.15}
$$

In practice, we will be writing out the expression for the *more accurate* derivative in terms of *less accurate* derivatives and not in terms of the function values. Thus, we typically stop at the first line in Eqs (5.14) and (5.15) and do not need to obtain the third line.

Thus, all the methods result in the same expression for the $O(h^2)$ estimate of the first derivative using the forward difference method. Method (a) of direct interpolation is quite cumbersome. Moreover, it does not provide an estimate of the error and is not recommended for use. An advantage of this method is that the second derivative is also easily obtained from Eq. (5.7). Method (b) is the most commonly used technique for performing the numerical differentiation and is discussed in detail in the next section. Note that it provides an estimate of the error in approximating the derivative.

Method (c) is quite straightforward but does not provide an error estimate. However, an *approximate* estimate of the error may be obtained if we assume that the error is proportional to $h^2 f'''(\xi_f)$. Then, using the function $f(x) = x^3$, it can be shown from an extension of Eq. (5.15) that the error is equal to

$$
\frac{h^2 f'''(\xi_f)}{3}
$$

Method (d) is a powerful technique and may be used repeatedly to obtain more and more accurate estimates. However, the same could be accomplished by a direct use of the Taylor's series and would provide an estimate of the error also. Moreover, the higher-order estimates typically *skip* some points [e.g., x_{i+3} in Eq. (5.15)] which is not desirable. Therefore, the method of undetermined coefficients and the Richardson's extrapolation method would not be discussed for the numerical differentiation. As we will see later in this chapter, both of these techniques are quite useful in numerical integration.

Example 5.2   For the problem discussed in Example 5.1, estimate the speed using (a) forward difference, $O(h^2)$, using step sizes of 1 s and 2 s; and (b) Richardson's extrapolation, $O(h^3)$.

Solution   With a step size of 1 s, using Eq. (5.10), the speed is estimated as

$$f_5'(h) = \frac{-f(7) + 4f(6) - 3f(5)}{2} = 84 \text{ cm/s}$$

for double the step size, the estimate is

$$f_5'(2h) = \frac{-f(9) + 4f(7) - 3f(5)}{4} = 78 \text{ cm/s}$$

Using Eq. (5.15), first line, we obtain the $O(h^3)$ estimate with Richardson's extrapolation as

$$f_5' = \frac{4f_5'(h) - f_5'(2h)}{3} = 86 \text{ cm/s}$$

As it turns out, this value is equal to the true value for this problem since the fourth derivative of the function is zero (we did not mention it earlier, and it was not really needed, but the distance was assumed to vary with time as $t + t^2 + t^3$ resulting in a speed equal to $1 + 2t + 3t^2$). It should also be noted that the error in the estimate with $h = 2$ s (which is $86 - 78 = 8$ cm/s) is exactly 4 times that with $h = 1$ s (which is 2 cm/s). It is expected since the error is $O(h^2)$ and the third derivative of the function is constant [see (Eq. 5.10) for the error term]. Moreover, the error is positive (note that the third derivative is positive) and the forward difference formulae underpredict the speed of the object.

5.2.1.2 General Formulation using Taylor's Series

We write the finite difference approximation of the nth derivative at x_i, as

$$f_i^n = \frac{1}{h^n} \sum_{j=-n_b}^{n_f} c_{i+j} f(x_{i+j}) \tag{5.16}$$

where n_b and n_f denote the number of backward and forward grid points used in the expression. Thus, $n_b = 0$ for forward difference schemes, $n_f = 0$ for backward difference schemes, and $n_b = n_f$ for central difference schemes. We may, of course, use different number of points before and after x_i. However, the non-symmetrical distribution of the grid points will imply that we loose the advantage of the cancellation of errors. Even then, sometimes it may be preferable to use non-symmetric central difference (e.g., at the second point of the data grid, it may be better to use a 5-point central difference scheme with $n_b = 1$ and $n_f = 3$ rather than a 5-point forward difference scheme with $n_f = 4$). The coefficients, c, are obtained by expanding $f(x_{i+j})$ in a Taylor's series about the point x_i and equating the coefficients of like terms on both sides. Comparing the Lagrange interpolating polynomial [Eq. (4.19) with the basis functions given by Eq. (4.21)] and Eq. (5.16), it is apparent that the coefficients may be obtained by taking the nth derivative of the Lagrange polynomial. However, it is generally easier to use Taylor's series expansion. Since there are $n_b + n_f + 1$ unknown coefficients to be obtained, we would need to equate the coefficients of the function, $f(x_i)$, and its derivatives up to the order $n_b + n_f$. For example, the forward difference approximation for the first derivative using three points ($n_b = 0$, $n_f = 2$) is written as

$$
\begin{aligned}
f_i' &= \frac{c_i f(x_i) + c_{i+1} f(x_{i+1}) + c_{i+2} f(x_{i+2})}{h} \\
&= \frac{c_i + c_{i+1} + c_{i+2}}{h} f(x_i) + (c_{i+1} + 2c_{i+2}) f'(x_i) \\
&\quad + \frac{h}{2}(c_{i+1} + 4c_{i+2}) f''(x_i) + \frac{h^2}{6}(c_{i+1} + 8c_{i+2}) f'''(x_i) + \cdots
\end{aligned}
\tag{5.17}
$$

Equating the coefficients of f, f', and f'' we get equations similar to Eq. (5.12), resulting in $c_i = -3/2$; $c_{i+1} = 2$; $c_{i+2} = -1/2$.

The error, given by true value − approximate value, is obtained from the last term in Eq. (5.17) as $\dfrac{h^2 f'''(\xi_f)}{3}$, with $\xi_f \in (x_i, x_{i+2})$.

Similarly, if we want the second derivative using the same three points, we get $c_i = 1$; $c_{i+1} = -2$; $c_{i+2} = 1$ and the error as $-hf'''(\xi_f)$. For the central difference scheme using three points ($n_b = n_f = 1$), the first derivative is written as

$$
\begin{aligned}
f_i' &= \frac{c_{i-1} f(x_{i-1}) + c_i f(x_i) + c_{i+1} f(x_{i+1})}{h} \\
&= \frac{c_{i-1} + c_i + c_{i+1}}{h} f(x_i) + (-c_{i-1} + c_{i+1}) f'(x_i) \\
&\quad + \frac{h}{2}(c_{i-1} + c_{i+1}) f''(x_i) + \frac{h^2}{6}(-c_{i-1} + c_{i+1}) f'''(x_i) + \cdots
\end{aligned}
\tag{5.18}
$$

from which, we get $c_{i-1} = -1/2$; $c_i = 0$; $c_{i+1} = 1/2$; and the error as $-h^2 f'''(\xi_c)/6$ with $(\xi_c) \in (x_{i-1}, x_{i+1})$. Note that although the leading term in the error is $O(h)$, its coefficient becomes zero. Therefore, we use an additional term in the error. For the second derivative, we get the three equations as

$$c_{i-1} + c_i + c_{i+1} = 0$$
$$-c_{i-1} + c_{i+1} = 0$$
$$c_{i-1} + c_{i+1} = 2$$

and the error as

$$\frac{h}{6}(-c_{i-1} + c_{i+1})f'''(x_i) - \frac{h^2}{24}(c_{i-1} + c_{i+1})f^{iv}(\xi_c)$$

This results in $c_{i-1} = 1; c_i = -2; c_{i+1} = 1$ and the error is obtained as $-h^2 f^{iv}(\xi_c)/12$.

It is readily seen that the leading error term in the generalized form of the finite difference approximation for the nth derivative is proportional to $h^{-n}h^{n_b+n_f+1}f^{n_b+n_f+1}(x_i)$. A dimensional analysis will show that the error in nth derivative will contain terms of the form $h^{a-n}f^{[a]}(\xi)$. For forward/backward difference, therefore, if we want an expression for the nth derivative of $O(h^a)$ accuracy, we would need to use $a + n - 1$ additional points after/before x_i. For central difference, however, as we just saw, the coefficient of the leading error term may become zero. It can be shown that for symmetrically placed points, all *even* derivatives would have symmetric coefficients and the coefficient of the leading error term will be zero. Similarly, all *odd* derivatives would have antisymmetric coefficients with $c_i = 0$. So, in the central difference scheme for the nth derivative using n_c points before and same number of points after, x_i, the accuracy would be $O(h^a)$ with $a = 2n_c + 1 - n$ for odd n and $a = 2n_c + 2 - n$ for even n, implying that a is always even.

It is not desirable to skip points close to x_i and include points farther away. For example, when using Richardson's extrapolation to obtain $O(h^3)$ estimate of the first derivative using the forward difference approximation, we effectively skip x_{i+3} but include x_{i+4}. It may be better to use the generalised form and obtain $c_i = -11/6$; $c_{i+1} = 3$; $c_{i+2} = -3/2$; $c_{i+3} = 1/3$ and the error as $-h^3 f^{iv}(\xi_f)/4$. On the other hand, if we include the point x_{i+4}, and assume that $c_{i+3} = 0$, we get $c_i = 7/4$; $c_{i+1} = 8/3$; $c_{i+2} = -1$; $c_{i+4} = 1/12$ and the error as $-h^3 f^{iv}(\xi_f)/3$. Thus, we get the same expression as was obtained using Richardson's extrapolation [Eq. (5.15)] and see that the error is larger than that obtained using four *consecutive* points. Also note that the expression for the first derivative of $O(h^2)$ accuracy using central difference approximation contains points x_{i-1} and x_{i+1} but not x_i. However, this does not mean that we have skipped the point x_i. A quick look at the derivation indicates that we did consider the coefficient c_i, it just *turned out* to be zero. In contrast, for the Richardson's extrapolation, we *forced* c_{i+3} to be zero.

A listing of some expressions is given in a convenient tabular form in Tables 5.1–5.3 for the forward, backward, and central difference schemes (note that the backward difference scheme is similar to the forward difference scheme with using $-h$ in place of h and $i - j$ in place of $i + j$). An extensive table is available in a number of references (e.g., Abramowitz and Stegun, 1964).

Table 5.1 Forward difference formulae $f_i^n = \dfrac{1}{h^n} \displaystyle\sum_{j=0}^{n_f} c_{i+j} f(x_{i+j})$

Accuracy	Derivative	c_i	c_{i+1}	c_{i+2}	c_{i+3}	c_{i+4}	Error*
$O(h)$	f_i'	-1	1				$-hf''/2$
	f_i''	1	-2	1			$-hf'''$
	f_i'''	-1	3	-3	1		$-3hf^{iv}/2$
	f_i^{iv}	1	-4	6	-4	1	$-2hf^{v}$
$O(h^2)$	f_i'	$-3/2$	2	$-1/2$			$h^2 f'''/3$
	f_i''	2	-5	4	-1		$11h^2 f^{iv}/12$
	f_i'''	$-5/2$	9	-12	7	$-3/2$	$7h^2 f^{v}/4$
$O(h^3)$	f_i'	$-11/6$	3	$-3/2$	$1/3$		$-h^3 f^{iv}/4$
	f_i''	$35/12$	$-26/3$	$19/2$	$-14/3$	$11/12$	$-5h^3 f^{v}/6$

*The derivatives in the error expressions are evaluated at some point in the appropriate interval.

Table 5.2 Backward difference formulae $f_i^n = \dfrac{1}{h^n} \displaystyle\sum_{j=-n_b}^{0} c_{i+j} f(x_{i+j})$

Accuracy	Derivative	c_{i-4}	c_{i-3}	c_{i-2}	c_{i-1}	c_i	Error
$O(h)$	f_i'				-1	1	$-hf''/2$
	f_i''			1	-2	1	hf'''
	f_i'''		-1	3	-3	1	$3hf^{iv}/2$
	f_i^{iv}	1	-4	6	-4	1	$2hf^v$
$O(h^2)$	f_i'			$1/2$	-2	$3/2$	$h^2 f'''/3$
	f_i''		-1	4	-5	2	$11h^2 f^{iv}/12$
	f_i'''	$3/2$	-7	12	-9	$5/2$	$7h^2 f^v/4$
$O(h^3)$	f_i'		$-1/3$	$3/2$	-3	$11/6$	$h^3 f^{iv}/4$
	f_i''	$11/12$	$-14/3$	$19/2$	$-26/3$	$35/12$	$5h^3 f^v/6$

Table 5.3 Central difference formulae $f_i^n = \dfrac{1}{h^n} \displaystyle\sum_{j=-n_c}^{n_c} c_{i+j} f(x_{i+j})$

Accuracy	Derivative	c_{i-2}	c_{i-1}	c_i	c_{i+1}	c_{i+2}	Error
$O(h^2)$	f_i'		$-1/2$	0	$1/2$		$-h^2 f'''/6$
	f_i''		1	-2	1		$-h^2 f^{iv}/12$
	f_i'''	$-1/2$	1	0	-1	$1/2$	$-h^2 f^v/4$
	f_i^{iv}	1	-4	6	-4	1	$-h^2 f^{vi}/6$
$O(h^4)$	f_i'	$1/12$	$-2/3$	0	$2/3$	$-1/12$	$h^4 f^v/30$
	f_i''	$-1/12$	$4/3$	$-5/2$	$4/3$	$-1/12$	$h^4 f^{vi}/90$

Example 5.3   For the problem described in Example 5.1, estimate the speed using (a) backward difference, $O(h^2)$, using step sizes of 1 s and 2 s; (b) Richardson's extrapolation, $O(h^3)$, using the two estimates obtained in (a); and (c) central difference, $O(h^4)$. Also estimate the acceleration using (a) the forward difference, $O(h)$, $O(h^2)$, and $O(h^3)$; and (b) central difference $O(h^2)$.

Solution   Using Tables 5.2 and 5.3, the speed is estimated as

Backward difference, $O(h^2)$, $h = 1$ s: $f'_5 = \dfrac{\frac{1}{2}f(3) - 2f(4) + \frac{3}{2}f(5)}{1} = 84$ cm/s

Backward difference, $O(h^2)$, $h = 2$ s: $f'_5 = \dfrac{\frac{1}{2}f(1) - 2f(3) + \frac{3}{2}f(5)}{2} = 78$ cm/s

Ricardson's extrapolation: $f'_5 = \dfrac{4f'_5(h=1) - f'_5(h=2)}{3} = 86$ cm/s

Central difference, $O(h^4)$, $h = 1$ s:

$$f'_5 = \dfrac{\frac{1}{12}f(3) - \frac{2}{3}f(4) + \frac{2}{3}f(6) - \frac{1}{12}f(7)}{1} = 86 \text{ cm/s}$$

The acceleration estimates (using Tables 5.1 and 5.3) are

Forward difference, $O(h)$: $f'_5 = \dfrac{f(5) - 2f(6) + f(7)}{1^2} = 38$ cm/s

Forward difference, $O(h^2)$: $f'_5 = \dfrac{2f(5) - 5f(6) + 4f(7) - f(8)}{1^2} = 32$ cm/s

Forward difference, $O(h^3)$:

$$f'_5 = \dfrac{\frac{35}{12}f(5) - \frac{26}{3}f(6) + \frac{19}{2}f(7) - \frac{14}{3}f(8) + \frac{11}{12}f(9)}{1^2} = 32 \text{ cm/s}$$

Forward difference, $O(h^4)$: $f'_5 = \dfrac{f(4) - 2f(5) + f(6)}{1^2} = 32$ cm/s

Knowing that the distance varies as a cubic function of time, it is easy to draw conclusions about the error along similar lines as described in Example 5.2.

5.2.2 Unevenly Spaced Grid Points

Let us now consider the case when $x_i - x_{i-1} = h_i$ for $i = 1, 2, ..., n$ such that not all h_i are equal. If we assume that the function is represented by a piecewise linear polynomial, the first derivative of the function at any point, x_i, may be obtained in a similar fashion as was done for equally spaced grid points. For example, we may use the backward difference:

$$f_i' = \frac{f(x_i) - f(x_{i-1})}{h_i}$$

and the forward difference:

$$f_i' = \frac{f(x_{i+1}) - f(x_i)}{h_{i+1}}$$

However, when we come to central difference, we may not be able to use the average of these two slopes or the slope of the line joining the function values at x_{i-1} and x_{i+1} because the segments may not be of equal length. The Taylor's series can again be utilized but the formulation must account for the unequal step sizes. For example, we may write

$$f(x_{i-1}) = f(x_i) - h_i f'(x_i) + \frac{h_i^2}{2!} f''(x_i) - \frac{h_i^3}{3!} f'''(x_i) + \cdots$$

$$f(x_{i+1}) = f(x_i) + h_{i+1} f'(x_i) + \frac{h_{i+1}^2}{2!} f''(x_i) + \frac{h_{i+1}^3}{3!} f'''(x_i) + \cdots$$

(5.19)

and get an estimate of the first derivative as

$$f_i' = \frac{f(x_{i+1}) - f(x_{i-1})}{h_{i+1} + h_i}$$

with the leading error term of $(h_i^2 - h_{i+1}^2) f''(x_i) / 2$. Or we could eliminate the term involving the second derivative to obtain

$$f_i' = -\frac{h_{i+1}}{h_i(h_i + h_{i+1})} f(x_{i-1}) + \frac{h_{i+1} - h_i}{h_{i+1} h_i} f(x_i) + \frac{h_i}{h_{i+1}(h_i + h_{i+1})} f(x_{i+1})$$

(5.20)

with the leading error term being $-h_i h_{i+1} f'''/6$. (Note that for equal segments we do get the same expressions as obtained earlier.) Another alternative would be to use Lagrange interpolating polynomials and then perform analytical integration on them. Yet another option could be the use of cubic splines, which would directly give us the first/second derivatives at the nodes (see Section 4.5). However, the general formulation discussed in the previous section can be readily applied to this case also and is described briefly here.

Similar to Eq. (5.11), we write

$$f_i^n = \sum_{j=-n_b}^{n_f} c_{i+j} f(x_{i+j})$$ (5.21)

Comparing with Eq. (5.16) we note that the h^n term has been removed since there is no single value of h.

The forward difference approximation for the first derivative using three points gives rise to [cf. Eq. (5.12)]

$$c_i + c_{i+1} + c_{i+2} = 0$$

$$0c_i + h_{i+1}c_{i+1} + (h_{i+1} + h_{i+2})c_{i+2} = 1$$

$$0c_i + h_{i+1}^2 c_{i+1} + (h_{i+1} + h_{i+2})^2 c_{i+2} = 0$$

which results in

$$c_i = -\frac{2h_{i+1} + h_{i+2}}{h_{i+1}(h_{i+1} + h_{i+2})}; \quad c_{i+1} = \frac{h_{i+1} + h_{i+2}}{h_{i+1}h_{i+2}}; \quad c_{i+2} = -\frac{h_{i+1}}{h_{i+2}(h_{i+1} + h_{i+2})}$$ (5.22)

and a leading error term of $h_{i+1}(h_{i+1} + h_{i+2})f'''(\xi_f)/6$. Since the methodology is almost identical to that discussed for the evenly spaced points, we will not discuss this topic further.

Example 5.4   For the problem described in Example 5.1, assume that the measurement at $t = 7$ s is missing. Estimate the speed at 5 s using the forward difference with the three measurements at 5, 6, and 8 s.

Solution   For $i = 5$, using Eq. (5.22), with $h_{i+1} = 6 - 5 = 1$ and $h_{i+2} = 8 - 6 = 2$, we obtain the coefficients as $c_i = 4/3$; $c_{i+1} = 3/2$; $c_{i+2} = -1/6$ and the speed is estimated from Eq. (5.21) as

$$f_5' = \sum_{j=0}^{2} c_{5+j} f(x_{5+j}) = -\frac{4}{3}f(5) + \frac{3}{2}f(6) - \frac{1}{6}f(8) = 83 \text{ cm/s}$$

The error is 3 cm/s, which matches with the expression of error listed after Eq. (5.22) (the third derivative is equal to 6).

5.2.3 Amplitude and Phase Error

We have defined the error in the numerical differentiation in terms of its magnitude. However, as we have seen earlier for periodic functions (Section 4.7), we may define the error of approximation in terms of an amplitude error and a phase error. As described in Section 4.7, a periodic function may be expressed as the sum of its various harmonics. A non-periodic function may be expressed as a Fourier integral. Hence, it is instructive to look at the error in numerical differentiation of a sinusoidal function sampled at discrete points.

Let the Fourier components of a function be in the form of $f(x) = A\cos(jx + \theta)$ (recall that A is the amplitude and θ is the phase angle). We have assumed the function to have a period of 2π and j ($= 0, 1, 2, 3, \ldots$) represents the harmonic (for functions with a period T, the analysis is similar with an additional frequency term $\omega = 2\pi/T$). Let the function be sampled at an interval of h (we assume that the period is an integer multiple of h although it is not essential to our analysis). The analytical derivatives of the jth harmonic are obtained as

$$f'(x_i) = -Aj\sin(jx_i + \phi); \quad f''(x_i) = -Aj^2\cos(jx_i + \phi); \ldots$$

If we use the lowest order central difference scheme to evaluate the first and second derivatives, we get (Table 5.3)

$$f_i' = \frac{f(x_i + h) - f(x_i - h)}{2h} = \frac{A\cos(jx_i + jh + \phi) - A\cos(jx_i - jh + \phi)}{2h}$$

$$= -A\frac{\sin jh}{h}\sin(jx_i + \phi)$$

and

$$f_i'' = \frac{f(x_i + h) - 2f(x_i) + f(x_i - h)}{h^2} = -A\frac{2(1 - \cos jh)}{h^2}\cos(jx_i + \phi)$$

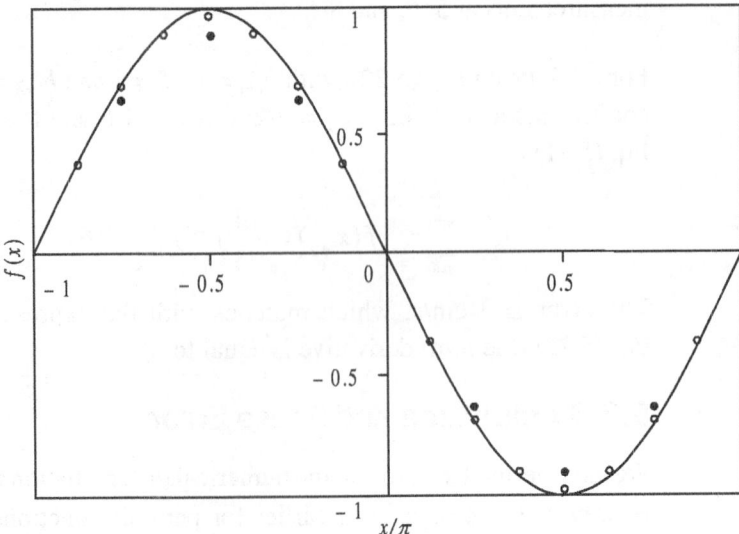

Fig. 5.1(a)   Analytical and numerical values of the first derivative for the fundamental frequency ($j = 1$). (o) $h = \pi/8$, (•) $h = \pi/4$.

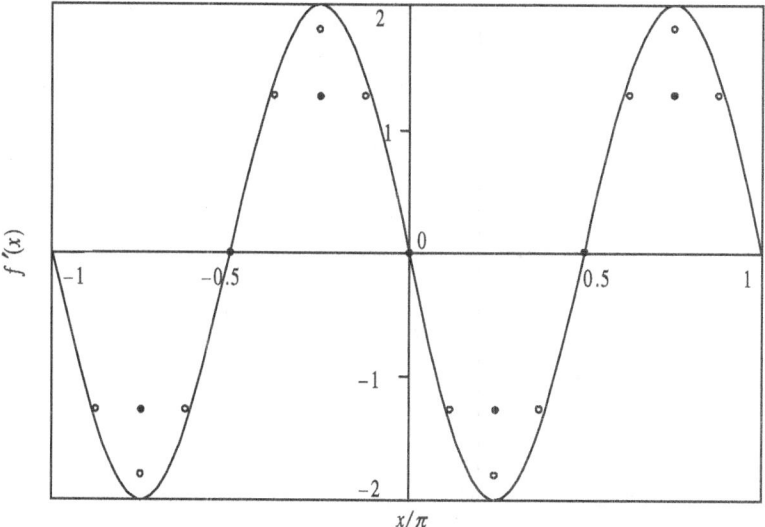

Fig. 5.1(b) Analytical and numerical values of the first derivative for the second harmonic ($j = 2$). (o) $h = \pi/8$, (•) $h = \pi/4$.

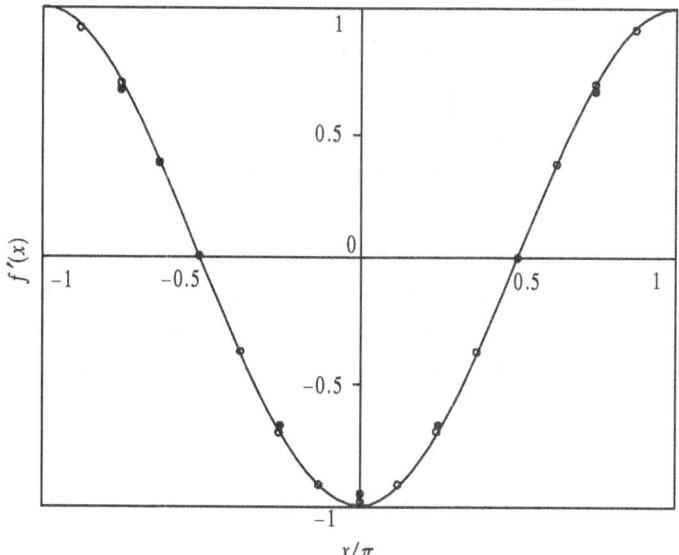

Fig. 5.2(a) Analytical and numerical values of the second derivative for the fundamental frequency ($j = 1$). (o) $h = \pi/8$, (•) $h = \pi/4$.

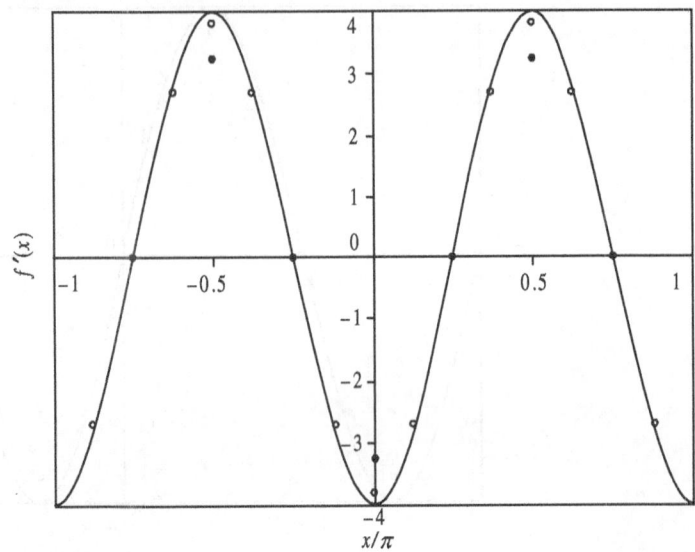

Fig. 5.2(b) Analytical and numerical values of the second derivative for the second harmonic ($j = 2$). (o) $h = \pi/8$, (•) $h = \pi/4$.

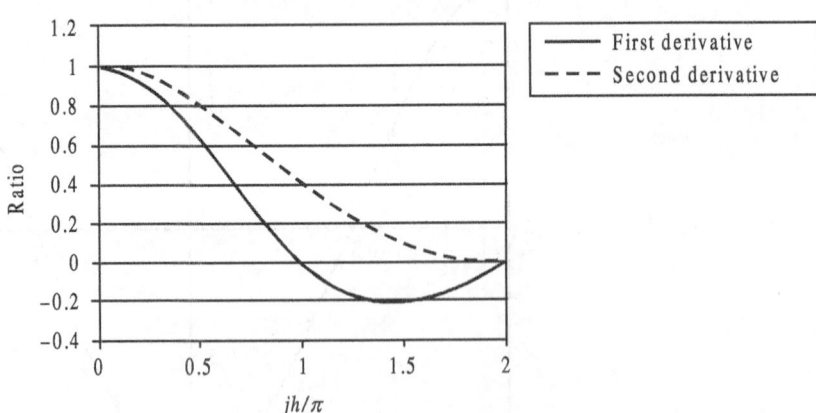

Fig. 5.3 Ratio of numerical and analytical amplitudes.

A comparison with the exact values shows that there is no phase error in the approximations. The amplitude error depends on h and is conveniently expressed in terms of the *ratio* of the approximate and exact amplitudes as $(\sin jh)/jh$ for the first derivative and $[2(1 - \cos jh)/j^2 h^2]$ for the second derivative. It is easy to verify that this ratio becomes 1 as h tends to zero. Figures 5.1 and 5.2 show the exact derivatives and their central difference approximations for the fundamental frequency and the second harmonic, respectively. The amplitude error is a function of jh and is plotted in Fig. 5.3.

From these figures, it is apparent that the numerical solution is in phase with the exact solution. The increase in the amplitude error with increase in h is also clearly seen.

If, instead of the central difference, we use the forward difference scheme to compute the derivatives (Table 5.1), we get

$$f_i' = \frac{f(x_i + h) - f(x_i)}{h} = \frac{A\cos(jx_i + jh + \phi) - A\cos(jx_i + \phi)}{h}$$

$$= -A\frac{\sin jh/2}{h/2}\sin(jx_i + jh/2 + \phi)$$

and

$$f_i'' = \frac{f(x_i + 2h) - 2f(x_i + h) + f(x_i)}{h^2} = -A\frac{2(1 - \cos jh)}{h^2}\cos(jx_i + jh + \phi)$$

Clearly, now we have a phase error also in addition to the amplitude error. Figure 5.4 shows the exact and numerical derivatives for the fundamental frequency.

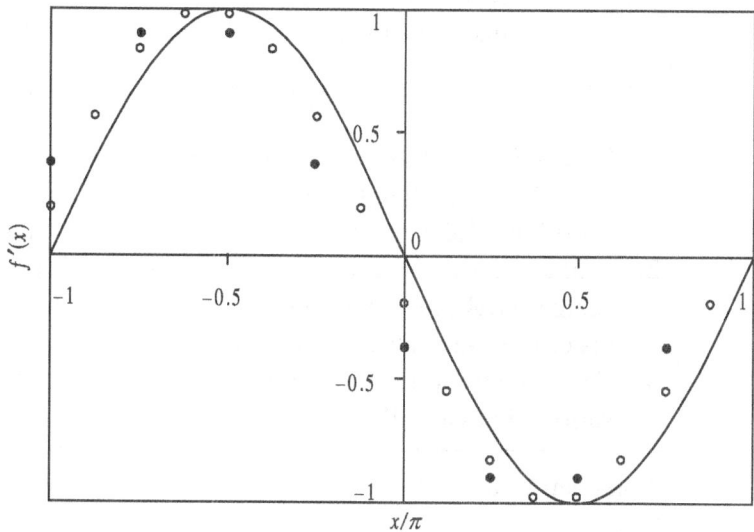

Fig. 5.4(a)   Analytical and numerical values of the first derivative for the fundamental frequency using forward difference. (o) $h = \pi/8$, (•) $h = \pi/4$.

It should be mentioned that the analysis of amplitude and phase errors is considerably simplified if we use the complex notation of the Fourier series. Here, we have represented the Fourier series in terms of the sine/cosine functions in order to avoid complex algebra. In the next chapter, complex analysis would be used to look at the behaviour of error in the numerical solution of differential equations.

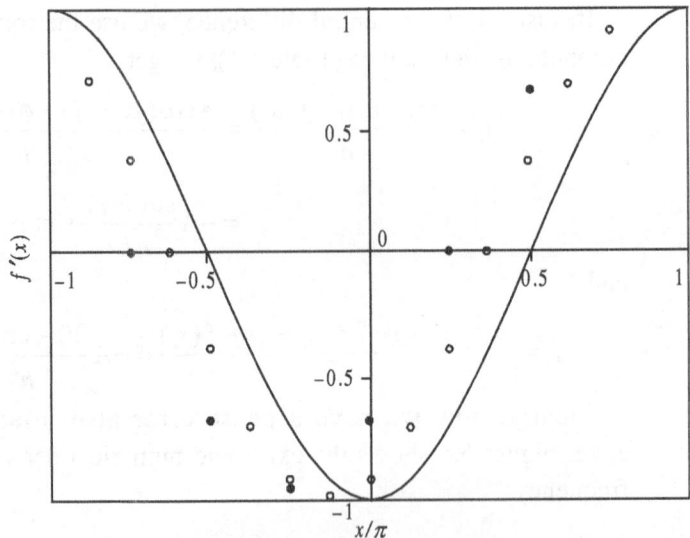

Fig. 5.4(b)   Analytical and numerical values of the second derivative for the fundamental frequency using forward difference. (o) $h = \pi/8$, (•) $h = \pi/4$.

EXERCISE 5.2

1. Estimate the derivative of the function $f(x) = 1 + x + 2x^2 - 16x^3 + 20x^4 - 3.5x^5$ at $x = 2$ using central difference approximation with two different step sizes of $h = 0.5$ and 1.0. Comment on the error vis-à-vis the step size.

2. Estimate the derivative of the functions $\exp(-0.1x)$ and $\exp(-10x)$ at $x = 0$ using central difference approximation with a step size of 0.1. Comment on the error vis-à-vis the nature of the function.

3. The velocity of an object, travelling along a straight line, was measured at various times as follows:

Time (min)	0	1	2	3	4	5	6	7	8	9	10
Velocity (cm/min)	0.00	0.65	1.72	3.48	6.39	11.18	19.09	32.12	53.60	89.02	147.41

Estimate the acceleration at 5 min using the central difference formulae of order h^4 with step sizes of 1 and 2 min. Use Richardson's extrapolation to obtain an $O(h^6)$ estimate from these values.

4. The $O(h^6)$ central difference formula for the first derivative is given by

$$f'(x_i) = -\frac{1}{60}f(x_{i-3}) + \frac{3}{20}f(x_{i-2}) - \frac{3}{4}f(x_{i-1}) + \frac{3}{4}f(x_{i+1}) - \frac{3}{20}f(x_{i+2}) + \frac{1}{60}f(x_{i+3})$$

with an error of $\dfrac{h^6}{140}f^{vii}(\xi)$. For the data given in the previous problem, estimate the acceleration at 5 min and compare with the estimate of the same order obtained using the Richardson's extrapolation. Why are these different?

5. Derive a sixth order accurate finite difference approximation for the first derivative at the point x_i using the function values at the seven points x_i, $x_{i\pm1}$, $x_{i\pm2}$, $x_{i\pm4}$ and fitting a sixth degree polynomial. Using this formula, estimate the acceleration for Problem 3.

6. For the Fourier component of the form $f(x) = A\cos(jx + \theta)$, estimate the first and second derivatives using the central difference formulae of order h^4. Obtain the amplitude and/or phase errors and compare with those of the lower order method through a plot similar to Fig. 5.3.

7. For the function

$$f(x) = \frac{\sin x}{x^3}$$

obtain finite difference approximations of f' with first order backward difference, second order central difference, and fourth order central difference at 21 equally spaced points in the interval $[\pi, 2\pi]$ and compare with the true value. For each of these methods, plot the error in the derivative at $x = 3\pi/2$ as a function of h (on a log-log scale), using four different values of h as $\pi/8$, $\pi/16$, $\pi/32$, and $\pi/256$. What are the slopes of these lines?

8. The following table is given for the values of e^x:

x	0.00	0.25	0.50	0.75	1.00	1.25	1.50	1.75	2.00
e^x	1.0000	1.2840	1.6487	2.1170	2.7183	3.4903	4.4817	5.7546	7.3891

(a) Compute

$$\left.\frac{de^x}{dx}\right|_{x=1}$$

using central difference scheme with $h = 0.25$, 0.50, and 1.00.

(b) Using the values computed in (a), obtain an estimate with maximum possible accuracy for the derivative by successive application of Richardson's extrapolation.

(c) Compute absolute values of the true relative error for each computed value of the derivative.

9. Derive a finite difference approximation for f_i'' in terms of f_i, f_{i+1}, and f_{i+2}. What is the error of this approximation?

10. The velocity of an object starting at $x = 0$ and moving along the x-axis was measured at various times as follows:

t (s)	0	10	20	30	40
v (m/s)	1.00	2.05	4.95	10.00	16.90

Estimate the acceleration at 20 s using numerical differentiation of the highest possible order. Also estimate the distance travelled in 40 s to an accuracy $O(h^6)$ applying the Romberg integration algorithm to three estimates obtained by the trapezoidal rule.

11. We are interested in fitting a piecewise Lagrange polynomial through a set of N + 1 equispaced (regular grid) discrete points by taking three points at a time. The grid points are denoted as $x_0, x_1, x_2, \ldots x_N$ and the corresponding functional values as $f_0, f_1, f_2, \ldots f_N$.

    Consider any three consecutive grid points x_i, x_{i+1}, and x_{i+2} where corresponding functional values are f_i, f_{i+1}, and f_{i+2}, respectively.

    (a) Write the expression for the Lagrange polynomial $\hat{f}(x)$ through these three points.

    (b) Using (a), obtain the expression for f_i' and $\hat{f}(x)$ at point x_i.

    (c) What is the order of truncation errors for the expressions obtained in (b)?

    (d) Compare these expressions with the well-known finite difference expressions.

12. Derive the central difference expression of accuracy $O(h^4)$ for the first derivative using two expressions of $O(h^2)$, one with step size of h, and the other with step size $2h$.

13. The displacement of an object x, is measured at different times, t. Since the velocity of the object is decreasing with time, the time interval between two successive measurements is increased by a factor of 2 such that $t_i = 2^i - 1$. Obtain the divided difference expression for the acceleration at time t in terms of the time increment $h = (t_i - t_{i-1})$ and the measured values of x_{i-1}, x_i, and x_{i+1} and show that the error is $O(h)$.

14. For the function

$$f(x) = \frac{1}{(0.5 + x)^2}$$

use Richardson's extrapolation to compute $f'(1.0)$ and $f'(5.0)$ to sixth order accuracy using the $O(h^2)$ central difference formula with initial step size of 0.5. Compute the errors at each stage.

5.3 NUMERICAL INTEGRATION

While analytical differentiation of a given function is rather straightforward, the same is not true for integration. A number of times [e.g., $\exp(-x^2)$, $\exp(-x)/x$] analytical integration is either not possible or too cumbersome. Even if the function can be analytically integrated, we may require a *weighted integration* (see Tchebycheff polynomial, Section 4.3) which may not be obtained analytically. In some cases, the function may not be known and only its values at a set of grid points are given.

In all such cases, if the integral is desired (e.g., the area enclosed by a curve, distance travelled from velocity measurements) we have to approximate it using numerical techniques (the numerical computation of an integral is also called *quadrature*). As in the case of interpolation, we will discuss both the continuous case (known function) and the discrete case (only function values known at a few points). However, we discuss the discrete case first since the continuous case involves some additional complexities. In all cases we assume that the desired integral is

$$I = \int_a^b f(x)\, dx$$

but, as we did for interpolation, in some cases we would transform the variable to have the limits cover the standard domain $(-1, 1)$.

5.3.1 Discrete Case

In this case, the function value is given in tabular form denoted by the set of values $\{(x_k, f(x_k)), k = 0, 1, \ldots, n\}$. The x's may not be equally spaced and do not have to be distinct. We assume, though, that they are *arranged in increasing order of x*. For now, we will also assume that $x_0 = a$ and $x_n = b$ so that there is no extrapolation needed for the interpolating polynomial. Sometimes, however, it may be unavoidable to use the so-called *open* formula in which the limit of integration goes beyond the data points given (e.g., $\int_0^1 x^{-1/2} dx$, the function is not defined at the lower limit and its value may be given from, say, $x = 0.01$ to 1. We discuss these improper integrals later in the chapter). As we did for differentiation, we interpolate the data with piecewise polynomials, which can then be readily integrated. If we use linear interpolation, the interpolating line and its integral are easily obtained. However, for more accurate estimate of the integral, we have to use higher-order interpolation. There are a number of ways in which it could be done, as discussed in the previous chapter. For example, we may use

- An nth degree polynomial passing through all data points (not a very good idea for large n)
- A quadratic, cubic, or higher degree spline (computationally intensive)
- A locally smooth higher-order interpolating polynomial typically imposing only C^0 continuity at the common points (preferred option since it requires much smaller computation time). We did not discuss this option for interpolation or numerical differentiation since in interpolation we wanted continuity of higher derivatives and numerical differentiation tends to magnify small deviations in data. On the other hand, integration tends to dampen the effect of small oscillations.

The desired integral, I, can be written as the sum of the function integral over various sub-intervals. For example, if we use linear interpolation we may write

$$I = \int_{x_0}^{x_1} f(x)\,dx + \int_{x_1}^{x_2} f(x)\,dx + \cdots + \int_{x_{i-1}}^{x_i} f(x)\,dx + \cdots + \int_{x_{n-1}}^{x_n} f(x)\,dx \qquad (5.23)$$

Defining the segment length as $h_i = x_i - x_{i-1}$ for $i = 1, 2, \ldots, n$ and using the fact that integration is invariant under a change of origin and the function is assumed to be linear in each segment, the estimated value of the integral, $\tilde{I}$, is written as

$$\tilde{I} = \sum_{i=1}^{n} \int_0^{h_i} f(x_{i-1}) + x f[x_{i-1}, x_i]\,dx = \sum_{i=1}^{n} \frac{f(x_{i-1}) + f(x_i)}{2} h_i \qquad (5.24)$$

Since the rightmost quantity in Eq. (5.24) represents the area of a trapezoid formed by joining the function values at x_{i-1} and x_i by a straight line, this method is also called the *trapezoidal rule*. Note that we could have written the interpolating polynomial in various alternative forms, but prefer to use the Newton divided difference form since it leads to easier integration and error analysis. We may also estimate the error in each segment, E_i, by (see Section 4.4)

$$E_i = I_i - \tilde{I}_i = \int_0^{h_i} x(x - h_i) f[x, x_{i-1}, x_i]\,dx = \int_0^{h_i} x(x - h_i) \frac{f''(\xi_i^*)}{2}\,dx \qquad (5.25)$$

where $\xi_i^* \in (x_{i-1}, x_i)$. Since $x(x - h_i) \leq 0$ in the entire interval, we can use the second mean value theorem for integrals to write

$$E_i = \frac{f''(\xi_i)}{2} \int_0^{h_i} x(x - h_i)\,dx = -\frac{h_i^3 f''(\xi_i)}{12} \qquad (5.26)$$

in which $\xi_i \in (x_{i-1}, x_i)$. This indicates that the error for a single segment is $O(h^3)$. Similar to the error in numerical differentiation, a dimensional analysis will show that the error in numerical integration will have terms of the form $h^r f^{[r-1]}(\xi)$. The error over the entire domain (a, b) is obtained by summation of individual segment errors. If the data is equidistant, i.e., $h_i = (b-a)/n \; \forall i$, we may write

$$E = \sum_{i=1}^{n} E_i = -\frac{(b-a)^3}{12n^2} \frac{\sum_{i=1}^{n} f''(\xi_i)}{n} \simeq -\frac{(b-a)^3}{12n^2} \overline{f}'' = -\frac{(b-a)h^2 \overline{f}''}{12} \qquad (5.27)$$

where $\overline{f}''$ represents the *mean* value of the second derivative of the function over the interval (a, b). We write the equation for error in terms of the mean value of the derivative rather than summation of the derivatives in order to directly compare the error for different schemes (the number of segments may be different for different methods and comparison of the sum would not be meaningful). If we assume that this mean value does not change significantly with change in h, we observe that the total error is $O(h^2)$. Thus halving the step size will reduce the error to *roughly* one-fourth (however, see Exercise 5.3, Problem 1). Since the mean value of the second derivative changes with h, the error will generally not be *exactly* one-fourth, unless the underlying function, $f(x)$, is a second degree (*or even third degree*) polynomial. As was the case with numerical differentiation, this error may be unacceptable and we again look at various alternatives for improving the accuracy.

Example 5.5 The speed of an object at various times was measured as follows:

Time (s)	0	1	2	3	4	5	6
Speed (cm/s)	0.0000	0.3466	0.3662	0.3466	0.3219	0.2986	0.2780

Estimate the distance travelled by the object at 6 s using the trapezoidal rule. First use a step size, $h = 1$ s, and then double this size ($h = 2$ s).

Solution With a step size of 1 s, using Eq. (5.24), the distance (i.e., the integral of the speed) is obtained as

$$\tilde{I} = \sum_{i=1}^{6} \frac{f(x_{i-1}) + f(x_i)}{2} h_i = h \left[\frac{f(x_0)}{2} + f(x_1) + f(x_2) + \cdots + f(x_5) + \frac{f(x_6)}{2} \right]$$

$$= 1.819 \text{ cm}$$

With double the step size, the estimated distance is

$$\tilde{I} = \sum_{i=1}^{3} \frac{f(x_{i-1}) + f(x_i)}{2} h_i = 2 \times \left[\frac{f(0)}{2} + f(2) + f(4) + \frac{f(6)}{2} \right] = 1.654 \text{ cm}$$

From Eq. (5.27), we may conclude that the error is roughly four times larger with a step size of 2 s as compared to that for the step size of 1 s. The true value of the distance is, therefore, likely to be a little larger than 1.819 cm.

5.3.1.1 *Improving the Accuracy*

The basic philosophy, again, is to use higher-order interpolation schemes but it can be done in a number of different ways:

(a) *Interpolation followed by integration* As we saw in the previous chapter, with the increase in the number of data points, it may not be a good idea to use a higher-order interpolating polynomial passing through *all* data points. The quadratic, cubic, or higher-order splines may be used but would typically require more computational time without a commensurate gain in accuracy. Therefore, we decide to use piecewise quadratic interpolation, we may write (we again use the fact that the integral is not affected by a translation and use $x_{i-2} = -h_{i-1}$; $x_{i-1} = 0$; $x_i = h_i$)

$$\tilde{I} = \sum_{i=2,4,6,\ldots n} \int_{-h_{i-1}}^{h_i} f(x_{i-2}) + (x+h_{i-1})f[x_{i-2},x_{i-1}] + x(x+h_{i-1})f[x_{i-2},x_{i-1},x_i]\, dx$$

$$= \sum_{i=2,4,6,\ldots,n} \frac{(h_{i-1}+h_i)(2h_{i-1}-h_i)}{6h_{i-1}} f(x_{i-2}) + \frac{(h_{i-1}+h_i)^3}{6h_{i-1}h_i} f(x_{i-1})$$

$$+ \frac{(h_{i-1}+h_i)(-h_{i-1}+2h_i)}{6h_i} f(x_i)$$

$$(5.28)$$

and the error as

$$E = \sum_{i=2,4,6,\ldots,n} \int_{-h_{i-1}}^{h_i} x(x+h_{i-1})(x-h_i)f[x,x_{i-2},x_{i-1},x_i]\, dx \qquad (5.29)$$

We will assume that n is even. If n is odd, a linear interpolation may be used in the first or last segment. However, it would lead to lower accuracy in the estimate of the integral. A better option would be to use a cubic for the first three or last three segments.

Unfortunately, we cannot apply the mean value theorem for integrals directly to Eq. (5.29) since $x(x + h_{i-1})(x - h_i)$ does not have a constant sign in the integration domain. For *equidistant points*, however, following Steffensen (1950) we use integration by parts to get

$$E_i = \int_{-h}^{h} x(x+h)(x-h) f[x,x_{i-2},x_{i-1},x_i]\, dx$$

$$= \left[f[x,x_{i-2},x_{i-1},x_i] \int_{-h}^{x} x(x+h)(x-h)\, dx \right]_{-h}^{h}$$

$$- \int_{-h}^{h} \frac{d\, f[x,x_{i-2},x_{i-1},x_i]}{dx} \int_{-h}^{x} x(x+h)(x-h)\, dx\, dx \qquad (5.30)$$

Note that we could divide the integral into two parts $(-h_{i-1}, 0)$ and $(0, h_i)$ such that the term $x(x + h_{i-1})(x - h_i)$ has the same sign over each of these intervals. The second mean value theorem for integrals could be applied to these individual segments. However, it is readily seen that the presence of opposite signs over these segments would imply that the error cannot be expressed in a usable form.

Also note the limits on the integral involving $x(x + h)(x - h)$. While the upper limit of this integral must be x, we could have used any constant lower limit of integration in this term. The value $-h$ is used for convenience as it makes the integral 0 for $x = -h$ as well as $x = h$. The derivative of the finite divided difference is obtained as

$$\frac{d}{dx} f[x, x_{i-2}, x_{i-1}, x_i] = \lim_{\varepsilon \to 0} \frac{f[x+\varepsilon, x_{i-2}, x_{i-1}, x_i] - f[x, x_{i-2}, x_{i-1}, x_i]}{x + \varepsilon - x}$$

$$= \lim_{\varepsilon \to 0} f[x+\varepsilon, x, x_{i-2}, x_{i-1}, x_i] = f[x, x, x_{i-2}, x_{i-1}, x_i] \qquad (5.31)$$

Now using the relation between the finite divided difference and the function derivative (see Box 4.3 and note that the relationship is applicable even when some of the points coincide) we get $f[x, x, x_{i-2}, x_{i-1}, x_i] = \dfrac{f^{iv}(\xi_i^*)}{4!}$ in which $\xi_i^* \in (x_{i-2}, x_i)$.

It is readily seen that $\displaystyle\int_{-h}^{h} (x-h)x(x+h) dx = 0$ and $\displaystyle\int_{-h}^{x} (x-h)x(x+h) dx$ is non-negative for all $x \in (-h, h)$ thus enabling us to use the second mean value theorem for integrals. We, therefore, have

$$E_i = -\int_{-h}^{h} \frac{f^{iv}(\xi_i^*)}{4!} \int_{-h}^{x} x(x+h)(x-h) \, dx \, dx$$

$$= -\frac{f^{iv}(\xi_i)}{4!} \int_{-h}^{h} \frac{(x^2 - h^2)^2}{4} dx = -\frac{h^5 f^{iv}(\xi_i)}{90}$$

in which $\xi_i \in (x_{i-2}, x_i)$. The estimated integral and total error are given by [Eqs (5.28) and (5.29)]

$$E = \frac{h}{3} \sum_{i=2,4,6,\ldots,n} [f(x_{i-2}) + 4f(x_{i-1}) + f(x_i)]$$

$$= \frac{h}{3}\left[f(x_0) + 4 \sum_{i=1,3,5,\ldots,n-1} f(x_i) + 2 \sum_{i=2,4,6,\ldots,n-2} f(x_i) + f(x_n) \right] \qquad (5.32)$$

and

$$E = \sum_{i=2,4,6,\ldots,n} -\frac{h^5 f^{iv}(\xi_i)}{90} = -\frac{(b-a)^5}{180 n^4} \frac{\displaystyle\sum_{i=2,4,6,\ldots,n} f^{iv}(\xi_i)}{n/2} = -\frac{(b-a)^5 \overline{f}^{iv}}{180 n^4}$$

(5.33)

where $\overline{f}^{iv}$ represents the *mean* value of the fourth derivative of the function over the interval (a, b). If we assume that this mean value does not change significantly with change in h, we observe that the total error is $O(h^4)$ (see Box 5.1 for general error expression). Sometimes, we say that the *degree of precision* of the quadrature scheme is 3 to indicate that *all* third degree polynomials would be exactly integrated but there are some fourth degree polynomial which cannot be exactly integrated. Also note that, although we derived the formula with a quadratic interpolation, the integral would be exact even if $f(x)$ is a cubic polynomial since the fourth derivative will be identically zero. This implies that once we perform a quadratic interpolation through 3 equidistant points, and then draw the cubic interpolating polynomial utilizing an additional (equidistant) point, the net area between these two curves would be zero, no matter what the function value is at the additional point! An easier to visualize simile is that of using a constant value to integrate a linear function. If we choose the constant value at the mid-point of the interval, any straight line passing through this point will result in the same area since the difference in area before and after the mid-point cancel out each other. Equation (5.32) is commonly called Simpson's one-third rule (because of the presence of the $h/3$ term) after Simpson in 1743 but proposed earlier by Cavalieri in 1639, Gregory in 1668, and Newton in 1676.

(b) *Using the finite difference formula* Combining the trapezoidal rule estimates over two consecutive segments and using the finite difference approximation of the second derivative in the error [Eq. (5.26)], we may write an improved estimate of the integral over the interval (x_{i-2}, x_i) as (for simplicity, we assume that the points are equidistant. For unequal spacing, the finite difference approximation of the second derivative becomes somewhat complicated)

$$\int_{x_{i-2}}^{x_i} f(x)dx \simeq \tilde{I}_i = h\frac{f(x_{i-2})+f(x_{i-1})}{2} + h\frac{f(x_{i-1})+f(x_i)}{2}$$

$$-2\frac{h^3}{12}\frac{f(x_{i-2})-2f(x_{i-1})+f(x_i)}{h^2}$$

(5.34)

in which the first two terms on the right-hand side are the trapezoidal rule estimates and the third term is the sum of errors in *both* segments. It is seen that this again results in the Simpson's 1/3 rule.

(c) *Method of undetermined coefficients* Assuming an expression of the form

$$\tilde{I}_i = c_{i-2} f(x_{i-2}) + c_{i-1} f(x_{i-1}) + c_i f(x_i) \tag{5.35}$$

in which the c's are undetermined coefficients, we require that the integral be exact for all polynomials of second degree. Taking the function as $f(x) = 1$, x, and x^2, respectively, we obtain three linear equations which can be solved to obtain the coefficients:

$$
\begin{aligned}
f(x) = 1 &\quad \Rightarrow \quad c_{i-2} + c_{i-1} + c_i = h_{i-1} + h_i \\
f(x) = x &\quad \Rightarrow \quad -h_{i-1}.c_{i-2} + 0.c_{i-1} + h_i.c_i = \frac{-h_{i-1}^2 + h_i^2}{2} \\
f(x) = x^2 &\quad \Rightarrow \quad h_{i-1}^2.c_{i-2} + 0.c_{i-1} + h_i^2.c_i = \frac{h_{i-1}^3 + h_i^3}{3}
\end{aligned}
\tag{5.36}
$$

which results in the same equation as Eq. (5.32) for equidistant points. In fact, if the points are equidistant, it may be better to write Eq. (5.35) as

$$\tilde{I}_i = h[c_{i-2} f(x_{i-2}) + c_{i-1} f(x_{i-1}) + c_i f(x_i)]$$

(d) *Richardson's extrapolation* From Eq. (5.27), the error in estimate of integral using the trapezoidal rule is equal to $-[(b-a)\bar{f}'' h^2]/12$. If we use a step size of $2h$, and assume that the mean value of second derivative is more or less same, the error term should be four times that for the step size of h. As before, a new, and probably more accurate, value is obtained as

$$\tilde{I}_i \simeq \frac{4\left[\frac{h}{2}\{f(x_{i-2}) + 2f(x_{i-1}) + f(x_i)\}\right] - \frac{2h}{2}[f(x_{i-2}) + f(x_i)]}{3} \tag{5.37}$$

in which the first term on the right-hand side represents the trapezoidal rule estimate for the left-hand side using step size h, and the second term represents the same with step size of $2h$. It is easy to see that we again get the Simpson's rule and the error, as shown earlier, is $O(h^4)$. One may then combine two estimates of $O(h^4)$ (e.g., one using step sizes of h and $2h$, and the other using $2h$ and $4h$) to obtain an $O(h^6)$ estimate and so on. Romberg proposed a general recursive form for this extrapolation well-suited for computer implementation which may be written as

$$\tilde{I}_{h,k+2} \simeq \frac{2^k \tilde{I}_{h,k} - \tilde{I}_{2h,k}}{2^k - 1} \tag{5.38}$$

BOX 5.1 Error in Integral Using Polynomial Approximation

Steffensen (1950) lists general expressions for the error in even and odd degree polynomial interpolation. For an even (say, $2m$) degree polynomial interpolation using $(2m + 1)$ equidistant points located at $0, \pm h, \pm 2h, ..., \pm mh$, the error in estimation of

$$\int_{-mh}^{mh} f(x)dx$$

is given by

$$\frac{2h^{2m+3} f^{[2m+2]}(\xi)}{(2m+2)!} \int_0^m x^2(x^2-1)(x^2-4)...(x^2-m^2)\,dx$$

in which $\xi \in (-mh, mh)$.

For an odd (say, $2m + 1$) degree polynomial using $(2m + 2)$ equidistant points located at $\pm\dfrac{h}{2}, \pm\dfrac{3h}{2}, ..., \pm\dfrac{(2m+1)h}{2}$, the error is expressed as

$$\frac{2h^{2m+3} f^{[2m+2]}(\xi)}{(2m+2)!} \int_0^{m+\frac{1}{2}} \left(x^2 - \frac{1}{4}\right)\left(x^2 - \frac{9}{4}\right)...\left(x^2 - \left(m + \frac{1}{2}\right)^2\right)dx$$

in which $\xi \in \left(-\left(m + \dfrac{1}{2}\right)h, \left(m + \dfrac{1}{2}\right)h\right)$.

For example, the error in the use of the odd degree polynomial with $m = 0$ (linear interpolation, trapezoidal rule) is obtained as

$$\frac{2h^3 f''(\xi)}{2} \int_0^{1/2} \left(x^2 - \frac{1}{4}\right)dx = -\frac{h^3 f''(\xi)}{12}$$

and that in even degree polynomial with $m = 1$ (quadratic interpolation, Simpson's rule) is obtained as

$$\frac{2h^5 f^{iv}(\xi)}{24} \int_0^1 x^2(x^2-1)\,dx = -\frac{h^5 f^{iv}(\xi)}{90}$$

in which $\tilde{I}_{h,k}$ represents the estimate of I of accuracy $O(h^k)$ with step size of h. Thus, starting from trapezoidal rule estimates, $O(h^2)$, for step sizes h, $2h$, $4h$, and $8h$, successive estimates could be obtained from

$$\tilde{I}_{h,4} = \frac{4\tilde{I}_{h,2} - \tilde{I}_{2h,2}}{3}; \quad \tilde{I}_{2h,4} = \frac{4\tilde{I}_{2h,2} - \tilde{I}_{4h,2}}{3}; \quad \tilde{I}_{4h,4} = \frac{4\tilde{I}_{4h,2} - \tilde{I}_{8h,2}}{3}$$

$$\tilde{I}_{h,6} = \frac{16\tilde{I}_{h,4} - \tilde{I}_{2h,4}}{15}; \quad \tilde{I}_{2h,6} = \frac{16\tilde{I}_{2h,4} - \tilde{I}_{4h,4}}{15}$$

$$\tilde{I}_{h,8} = \frac{64\tilde{I}_{h,6} - \tilde{I}_{2h,6}}{63}$$

with the final result of $O(h^8)$ accuracy! This algorithm, known as the *Romberg integration,* is thus a very powerful technique for performing numerical integration with very high accuracy using only a few lower accuracy estimates. Therefore, one really does not need to remember the higher accuracy formulae, only the trapezoidal rule will do! However, for unevenly spaced data it is not directly applicable.

Example 5.6 For the problem described in Example 5.5, estimate the distance travelled by the object at 6 s using the Simpson's 1/3 rule.

Solution Using Eq. (5.32), the distance is obtained as

$$\tilde{I} = \frac{h}{3}\left[f(t_0) + 4\sum_{i=1,3,5} f(t_i) + 2\sum_{i=2,4} f(t_i) + f(t_6)\right] = 1.874 \text{ cm}$$

Going back to Example 5.5, our assertion, that the true value of the distance should be a little larger than 1.819 cm, seems to be valid. Also note that the Richardson's extrapolation of values obtained in Example 5.5 provide us the estimate $(4 \times 1.819 - 1.654)/3$ cm, i.e., 1.874 cm, the same as the Simpson's rule.

Other techniques for obtaining a more accurate formula use higher-order interpolation and then perform the necessary integration. Two different philosophies could be applied at this time:

(i) As we did in (a) above, we use more points to perform the higher-order interpolation and then integrate this polynomial over the domain covered by *all* these points.

(ii) We use more points to obtain the higher-order interpolating polynomial and then integrate it over *a single segment.*

The first procedure is commonly known as the *Newton-Cotes* method (examples being the trapezoidal rule and the Simpson's rule) while the second is similar to *Adams* method. We describe below these techniques using the third-order interpolating polynomial.

5.3.1.2 *Newton-Cotes Method*

We assume that n is a multiple of 3. If it is not, the first (or last) 2 or 4 segments could be evaluated using the quadratic interpolation. We again use the fact that the integral is not affected by a translation and consider, for simplicity, the points to be equally spaced with $x_{i-3} = -h; x_{i-2} = 0; x_{i-1} = h; x_i = 2h$. We then have

$$
\tilde{I} = \sum_{i=3,6,9,\dots,n} \int_{-h}^{2h} f(x_{i-3}) + (x+h)f\left[x_{i-3}, x_{i-2}\right] + x(x+h)f\left[x_{i-3}, x_{i-2}, x_{i-1}\right]
$$

$$
+ x(x^2 - h^2) f\left[x_{i-3}, x_{i-2}, x_{i-1}, x_i\right] dx
$$

$$
= \frac{3h}{8} \sum_{i=3,6,9,\dots,n} \left[f(x_{i-3}) + 3f(x_{i-2}) + 3f(x_{i-1}) + f(x_i) \right] \tag{5.39}
$$

which is known as the Simpson's three-eighths rule (proposed much earlier by Newton), with the error given by (see Box 5.1):

$$
E = \sum_{i=3,6,9,\dots,n} \int_{-h}^{2h} x(x^2 - h^2)(x-2h) \frac{f^{iv}(\xi_i^*)}{4!} dx = -\frac{3}{80} h^5 \sum_{i=3,6,9,\dots,n} f^{iv}(\xi_i)
$$

$$
= -\frac{(b-a)^5 \bar{f}^{iv}}{80 n^4} \tag{5.40}
$$

A comparison with Eq. (5.33) shows that Simpson's 1/3 and 3/8 rules have the same order of accuracy (h^4) but the 1/3 rule is more accurate even though it is based on a lower degree polynomial! Thus a better interpolating polynomial may not necessarily lead to a more accurate integral. If we use different h, e.g., by dividing $(b-a)$ into two segments for the 1/3 rule and three segments for the 3/8 rule, the 3/8 rule will be found to be more accurate. However, we feel that a true comparison should be based on the same set of grid points. If we have, say, 6 segments and we use 3 applications of 1/3 rule or 2 applications of 3/8 rule, the

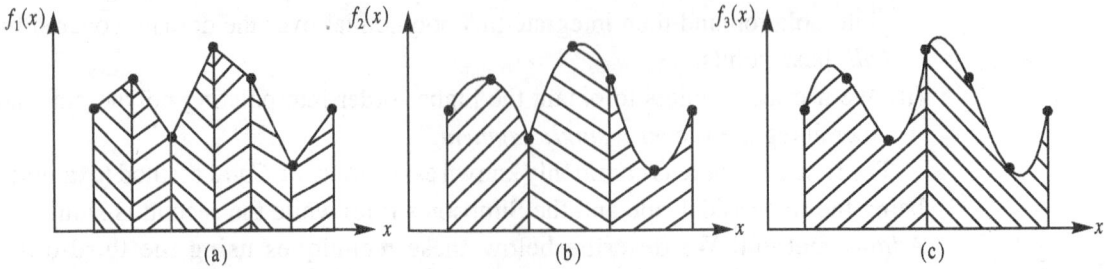

Fig. 5.5 Newton-Cotes methods; (a) trapezoidal rule, (b) Simpson's 1/3 rule, (c) Simpson's 3/8 rule.

1/3 rule will have smaller error (assuming, of course, that the fourth derivative does not change much). The 3/8 rule may be useful, however, if the number of segments is odd. In that case, the 3/8 rule may be applied to the first (or last) three segments and the 1/3 rule may be applied to the remaining (even in number) segments. Figure 5.5 shows a graphical representation of the estimates of the integral obtained using the trapezoidal rule, Simpson's 1/3 rule, and Simpson's 3/8 rule.

5.3.1.3 *Adams Method*

In the Newton-Cotes method, we use an interpolating polynomial over multiple segments and then integrate it over all those segments. As we know from our discussions on interpolation, the error of interpolation is likely to be small in the centre of the interval and large near the ends. We would, therefore, expect that in the three-segment case discussed in the previous paragraph, the interpolant would be much better over the middle segment and not-so-good over the corner segments. It would thus appear that a better accuracy may be obtained if we perform the integration only over the middle segment. As we will see, it does not lead to a more accurate integral. The reason, as before, is that a better interpolant does not necessarily mean a more accurate integral.

In case of evolving data (e.g., real-time analysis of velocity data to obtain distance travelled), as new measurements become available, we would like to have updated estimates of the integral. However, if want to apply the three-segment Newton-Cotes method, we have to wait for further measurements to get all three segments before we could apply the 3/8 rule.

It may be desirable to develop a technique in which as we add more points, the incremental integral could be easily obtained. It becomes even more desirable if the function value at any point depends on the value of integral at the previous times (we will discuss this in the chapter on differential equations). The Adams method addresses these issues as described next.

In Adams method, we write the integral as the sum of integrals over each segment, expressed as (we take $x_{i-2} = -h$; $x_{i-1} = 0$; $x_i = h$; $x_{i+1} = 2h$)

$$\tilde{I} = \sum_{i=1,2,\dots,n} \tilde{I}_i = \sum_{i=1,2,3,\dots,n} \int_0^h f(x_{i-2}) + (x+h)f[x_{i-2},x_{i-1}] + x(x+h)f[x_{i-2},x_{i-1},x_i]$$

$$+ x(x^2 - h^2)f[x_{i-2},x_{i-1},x_i,x_{i+1}]dx \quad (5.41)$$

and the error over a segment is given by

$$E_i = \int_0^h (x+h)x(x-h)(x-2h)f[x,x_{i-2},x_{i-1},x_i,x_{i+1}]dx$$

Thus, while the grid points x_{i-2}, x_{i-1}, x_i, and x_{i+1} are used to generate the third degree interpolating polynomial, the integration is performed only over one segment. We may perform the integral over the first, middle, or the last segment. Here we use the middle segment assuming that the interpolation would be more accurate over the central portion of the data points used. For an evolving data, we would typically perform the integral over the last segment. Also, note that for the first and the last segments of the entire data set the *central integral* formula will not be directly applicable since there is no data corresponding to x_{-1} and x_{n+1}. The resulting expression for the integral and the error over a segment are

$$\tilde{I}_i = \frac{h}{24}\left[-f(x_{i-2})+13f(x_{i-1})+13f(x_i)-f(x_{i+1})\right]$$

$$E_i = \frac{11h^5 f^{iv}(\xi_i)}{720} \tag{5.42}$$

The second mean value theorem for integrals is used to evaluate the error since $(x + h) \times (x-h)(x-h)$ is non-negative throughout the interval $(0, h)$ and, as before, Box 4.3 has been used to relate the derivative and the divided difference.

Disregarding the fact that Eq. (5.42) is not applicable for $i = 1$ and n, we may estimate the error in the value of the integral as

$$E = \frac{11h^5}{720}\sum_{i=1}^{n} f^{iv}(\xi_i) \simeq \frac{11(b-a)^5 \bar{f}^{iv}}{720n^4} \tag{5.43}$$

A comparison with Eq. (5.40) shows that the error is larger (and of opposite sign) than that in the Simpson's 3/8 rule. Adams methods are, therefore, generally

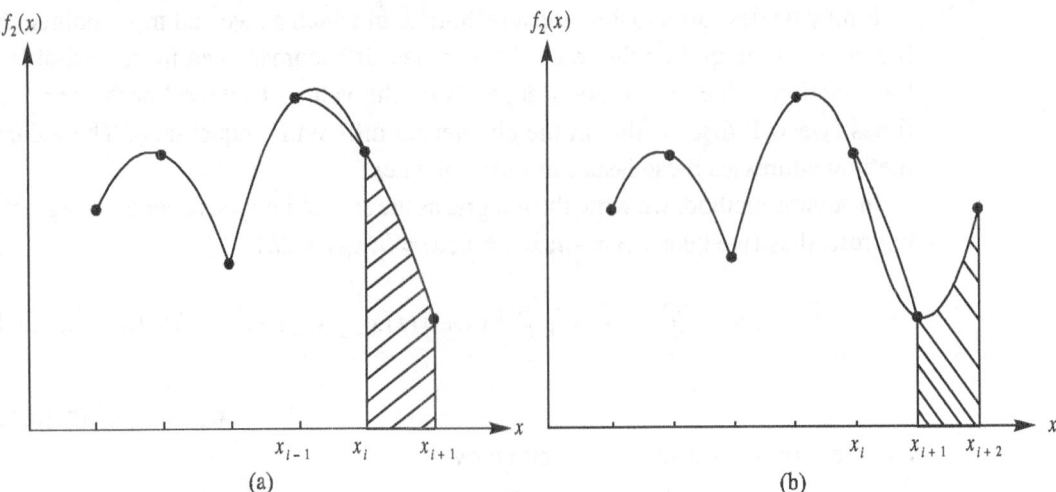

Fig. 5.6   Difference between Newton-Cotes and Adams methods.

not used for numerical integration. They are quite useful, though, for numerical solution of differential equations as discussed in Chapter 6. Figure 5.6 graphically represents the difference between the newton-Cotes and Adams methods.

Newton Cotes method uses a parabola through x_i, x_{i+1}, x_{i+2} (see Fig. 5.5b). Adams method uses a parabola through x_{i-1}, x_i, x_{i+1} (shown in Fig. 5.6(a)) and another through x_i, x_{i+1}, x_{i+2} (shown in Fig. 5.6(b)).

Table 5.4 lists some of the Newton-Cotes formulae and their error.

Table 5.4   Newton-Cotes formulae

No. of segments	Common name	Formula*	Error
1	Trapezoidal rule	$\dfrac{h}{2}[f(x_0)+f(x_1)]$	$-\dfrac{h^3 f''}{12}$
		$\dfrac{(b-a)}{2n}[f(x_0)+2\sum\limits_{i=1}^{n-1} f(x_i)+f(x_n)]$	$-\dfrac{(b-a)^3 \overline{f}''}{12n^2}$
2	Simpson's 1/3 rule	$\dfrac{h}{3}[f(x_0)+4f(x_1)+f(x_2)]$	$-\dfrac{h^5 f^{iv}}{90}$
		$\dfrac{(b-a)}{3n}[f(x_0)+4\sum\limits_{i=1,3,..}^{n-1} f(x_i)+2\sum\limits_{i=2,4,..}^{n-2} f(x_i)+f(x_n)]$	$-\dfrac{(b-a)^5 \overline{f}^{iv}}{180n^4}$
3	Simpson's 3/8 rule	$\dfrac{3h}{8}[f(x_0)+3f(x_1)+3f(x_2)+f(x_3)]$	$-\dfrac{3h^5 f^{iv}}{80}$
		$\dfrac{3(b-a)}{8n}[f(x_0)+3\sum\limits_{i=1,2,4,5,7,8,..}^{n-1} f(x_i)+2\sum\limits_{i=3,6,9,..}^{n-3} f(x_i)+f(x_n)]$	$-\dfrac{(b-a)^5 \overline{f}^{iv}}{80n^4}$
4	Boole's rule (sometimes mistyped as Bode's rule)	$\dfrac{2h}{45}[7f(x_0)+32f(x_1)+12f(x_2)+32f(x_3)+7f(x_4)]$	$-\dfrac{8h^7 f^{vi}}{945}$
		Too long to list	$-\dfrac{2(b-a)^7 \overline{f}^{vi}}{945n^6}$

*First line shows the formula and error for the first interval and the second line shows the same for the entire interval $(b-a)$. It has been assumed that n is exactly divisible by the number of segments for each formula.

Example 5.7   The speed of an object at various times was measured as follows:

Time (s)	0	1	2	3	4
Speed (cm/s)	0.0000	0.3466	0.3662	0.3466	0.3219

Estimate the distance travelled by the object at 4 s using the Boole's rule. Also estimate the distance by using (i) two applications of the Simpson's 1/3 rule, (ii) trapezoidal rule over the first segment and the 3/8 rule over next three, and (iii) 3/8 rule over the first three segments and trapezoidal rule over the last segment.

Solution   Using the Boole's rule, the distance is obtained as

$$d_4 = \frac{2}{45}[7\times0+32\times0.3466+12\times0.3662+32\times0.3466+7\times0.3219]\,\text{cm}$$

$$= 1.281\,\text{cm}$$

Using two applications of the Simpson's 1/3 rule, we get

$$d_4 = \frac{(4-0)}{3\times4}[0+4(0.3466+0.3466)+2(0.3662)+0.3219]\,\text{cm} = 1.276\,\text{cm}$$

Using trapezoidal rule over the first segment and 3/8 over the remaining three, we get

$$d_4 = \frac{1}{2}(0+0.3466)+\frac{3}{8}[0.3466+3\times0.3662+3\times0.3466+0.3219]\,\text{cm} = 1.226\,\text{cm}$$

Using 3/8 rule over the first three segments and trapezoidal rule over the last, we get

$$d_4 = \frac{3}{8}[0+3\times0.3466+3\times0.3662+0.3466]+\frac{1}{2}(0.3466+0.3219)\,\text{cm} = 1.266\,\text{cm}$$

We expect the Boole's rule to be the most accurate and the 1/3 rule to be a little less accurate. The other two methods would have a still lower accuracy because of the use of the trapezoidal rule for one segment. The results show that, *for this problem*, the application of the trapezoidal method over the last segment is "probably" more accurate than its application over the first segment.

From this discussion it is obvious that, for a given set of function values, one could obtain its integral over the range ($a = x_0$, $b = x_n$) to a desired accuracy using the Newton-Cotes *closed* (because the range of given data encloses the limits of integration) formulae. [It should be emphasized that we have *ignored round-off errors*. If these are included, a higher-order method for integration (or differentiation) may show larger *total error* than that of a smaller order method. In practice, therefore, methods of very high order are not used.] Occasionally, however, we may be required to evaluate the integral over limits extending beyond the range of data. For example, having measured the velocity of an object at time 0, 1, 2, and 3 min we want to predict where it would be at $t = 4$ min. Clearly, it involves some extrapolation and therefore should be avoided as far as possible. However, for evaluating improper integrals (discussed later in this chapter) and for obtaining a first estimate (discussed in Chapter 6 on ordinary differential equations), formulae based on extrapolation are quite useful. In such cases, *open* (both the upper and lower limits extend beyond the given data) or *semi-open* (either the lower or the upper limit extends beyond the data) formulae have to be used. These are based on obtaining the interpolating polynomial, extrapolating it to the limits of integration and then performing the integration.

5.3.1.4 *Open and Semi-open Formulae*

Suppose we want to evaluate $I = \int_a^b f(x)\,dx$ with the function values given in tabular form $\{(x_k, f(x_k)), k = 0, 1, \ldots, n\}$ in such a way that $a < x_0$ and/or $b > x_n$. We assume that the x's are equally spaced and arranged in increasing order of x. We may use the closed formulae discussed in the previous section to evaluate the integral over the range (x_0, x_n). However, since the function value is not known at a and/or b, we will have to extrapolate the function beyond the given data in order to perform the integral at either ends. For simplifying the presentation, we assume that $a = x_0$ and $b = x_n + h$ where h ($=x_n - x_0 / n$) is the spacing of the data points (in other words, we discuss the *semi-open* formulae). The extension to the case when the lower limit also extends beyond the data points or when the distance between the limits and the endmost data points is not equal to h, is straightforward, though tedious.

After using a few data points near the end for extrapolation, we could use the Newton-Cotes philosophy and perform the integration over the interval spanning all these points and the upper limit, b. Or, we could use the Adams method to perform the integration only over the interval (x_n, b). If we use the same (kth) degree polynomial to perform the *closed* integration over (x_0, x_{n-k-1}) using the appropriate Newton-Cotes formula described in the previous section and the *semi-open* integration over (x_{n-k}, b), there would be no difference in the results from the Newton-Cotes and Adams method. It is sometimes argued that the integrals over the end segments could be performed using a technique which is *one order lower* than that used for other segments since the summation of errors in the non-corner segments effectively lowers its order by one. For example, a single application of trapezoidal rule has accuracy $O(h^3)$ but the overall error over a given interval is $O(h^2)$. However, we use the same degree of polynomial interpolation for the end-segments as for the rest of the domain and describe the Adams method below.

Using linear interpolation, the trapezoidal rule could be used to obtain the integral over the range (x_0, x_n). Using the function values at x_{n-1} and x_n, we extrapolate to b and obtain

$$\int_{x_n}^{b} f(x)\, dx \simeq \frac{h}{2}[3f(x_n) - f(x_{n-1})] \tag{5.44}$$

and

$$\tilde{I} = \frac{h}{2}\left[f(x_0) + 2\sum_{i=1}^{n-2} f(x_i) + f(x_{n-1}) + 4f(x_n) \right] \tag{5.45}$$

If the lower limit of the integral is $a = x_0 - h$ (i.e., the corresponding open formula), the coefficient of $f(x_0)$ in Eq. (5.45) would also be equal to 4 and that of $f(x_1)$ will be equal to 1. However, if $n = 1$, it may be readily verified that both $f(x_0)$ and $f(x_1)$ would have a coefficient equal to 3. The error over the entire interval is equal to $3/4h^3 f''(\xi)$ where $a < \xi < b$. We believe that the semi-open formulae are more useful than the open formulae because of their use in the solution of differential equations. Hence we do not list the open formulae. The interested reader may refer to Davis and Rabinowitz (1967) for further details. The error of integration in the extrapolated segment is given by

$$E = \int_{h}^{2h} x(x-h)\frac{f''(\xi^*)}{2!}\, dx = \frac{5h^3 f''(\xi)}{12} \tag{5.46}$$

with $\xi \in (x_{n-1}, b)$. Interpolation with different order polynomials leads to similar formulae, some of which are listed in Table 5.5.

Most of the formulae we have discussed in this section have been derived for evenly spaced points. Extension to data with irregular spacing is conceptually similar but rather tedious. Fortunately, most practical problems would involve measurements at regular intervals and the Romberg's algorithm would probably be the best. Sometimes though we would not be able to avoid irregular intervals (e.g., if the function is rapidly varying in some portion, we may have to take very frequent measurements there to capture its behaviour). We hope that the material covered here would enable the reader to extend these methods to such cases and move on to the other important application of numerical integration: that for a given function.

Table 5.5   Semi-open integration formulae[*]

Extrapolating polynomial	Formula[**]	Error[+]
Constant	$hf(x_n)$	$\dfrac{h^2 f'}{2}$
	$h\left[\displaystyle\sum_{i=0}^{n} f(x_i)\right]$	
Linear	$\dfrac{h}{2}[-f(x_{n-1})+3f(x_n)]$	$\dfrac{5h^3 f''}{12}$
	$\dfrac{h}{2}\left[f(x_0)+2\displaystyle\sum_{i=1}^{n-2} f(x_i)+f(x_{n-1})+4f(x_n)\right]$	
Quadratic	$\dfrac{h}{12}[5f(x_{n-2})-16f(x_{n-1})+23f(x_n)]$	$\dfrac{9h^4 f'''}{24}$
	$\dfrac{h}{3}[f(x_0)+4\displaystyle\sum_{i=1,3,..}^{n-3} f(x_i)+2\displaystyle\sum_{i=2,4,..}^{n-4} f(x_i)+\dfrac{13}{4}f(x_{n-2})+0.f(x_{n-1})+\dfrac{27}{4}f(x_n)]$	

[*]   Recall that all the data points are equidistant and the *open* interval is beyond x_n at the same distance. Moreover, for quadratic (and higher) extrapolating polynomial, we assume that n is a multiple of the degree of the extrapolating polynomial so that the number of segments enables us to apply the closed formula from x_0 to x_n.

[**] The first line gives the integral from x_n to b and the second line from x_0 to b.

[+]   The derivatives are evaluated at some point in the appropriate interval, which is (x_{n-k}, b) for kth degree polynomial interpolation. The error is over the extrapolated segment only. The error over the entire interval is not listed as, for sufficiently large n, it would be similar to the corresponding closed formulae.

Example 5.8   The speed of an object at various times was measured as follows:

Time (s)	0	1	2	3	4
Speed (cm/s)	0.0000	0.3466	0.3662	0.3466	0.3219

Estimate the distance travelled by the object at 5 s using the constant, linear, and quadratic extrapolating polynomials.

Solution   Note that the distance is to be estimated at a time which is beyond the range of data. Hence, open formulae need to be used. From Table 5.5, the distance is estimated for various extrapolations as:

Constant: $h\left[\sum_{i=0}^{4} f(t_i)\right] = 1 \times (0.0000 + 0.3466 + 0.3662 + 0.3466 + 0.3219)$ cm

$$= 1.381 \text{ cm}$$

Linear: $\dfrac{h}{2}\left[f(t_0) + 2\sum_{i=1}^{2} f(t_i) + f(t_3) + 4f(t_4)\right]$ cm $= 1.530$ cm

Quadratic: $\dfrac{h}{3}\left[f(t_0) + 4f(t_1) + \dfrac{13}{4}f(t_2) + 0.f(t_3) + \dfrac{27}{4}f(t_4)\right]$ cm $= 1.583$ cm

We, therefore, expect the true value of the distance to be a little larger than 1.583 cm.

5.3.2 Continuous Case

If we are asked to integrate a known function, $f(x)$, over a finite interval (a, b), and the function is such that it is not possible (or very difficult) to integrate it analytically, we could use any of the methods discussed for the discrete case to approximate the integral since the function values could be generated at will at any point in the domain . [We assume that the function could be readily computed at all points in the domain. Improper integrals for which either the integration limits extend to infinity or the function becomes infinite (even indeterminate values of the form 0/0 cannot be evaluated *by the computer*) at one of the limits, are discussed in the next section.] Romberg integration would work very well since the points could be chosen to be evenly spaced [One may also think of fitting a polynomial to the given function by, say, using the Legendre polynomials and then integrating this polynomial. However, this would require the integration of the given function (see Section 4.3) which is not available!].

Since the function evaluation is generally time consuming, we would like to achieve maximum accuracy with minimum number of function evaluations. This is an additional complexity compared to the discrete case, where the function values are given at some points and the matter of function evaluation did not arise. We discuss in this section the techniques known as *Gauss quadrature*, which introduce additional degrees of freedom in the formulation by not fixing the location of the grid points *a priori*. As before, there are a number of ways in which the technique could be described and some of these are listed below:

(a) *The method of undetermined coefficients* (*and ordinates*) We assume an expression of the form

$$\tilde{I} = \sum_{i=0}^{n} c_i^* f(x_i) \tag{5.47}$$

in which the $c*$'s are undetermined coefficients, and x's are (yet to be determined) ordinates at which we would evaluate the function values. At this stage we could have changed the notation to use the range of i from 1 to m, with m indicating the number of points at which the function is evaluated. However, in keeping with the discussion till now, we prefer to follow the range 0 to n. The reader should keep in mind that the number of points are $(n + 1)$. Since there are $2n + 2$ degrees of freedom, we could specify that the integral be exact for all polynomials of degree $2n + 1$. Taking the function as $f(x) = 1, x, x^2, ..., x^{2n+1}$, respectively, we obtain $2n + 2$ nonlinear equations which can be solved to obtain the coefficients and the ordinates. For example, using two points ($n = 1$), the four equations are

$$c_0^* + c_1^* = b - a$$

$$c_0^* x_0 + c_1^* x_1 = \frac{b^2 - a^2}{2}$$

$$c_0^* x_0^2 + c_1^* x_1^2 = \frac{b^3 - a^3}{3}$$

$$c_0^* x_0^3 + c_1^* x_1^3 = \frac{b^4 - a^4}{4}$$

(5.48)

At this stage, it is convenient to transform the variable x in such a way that the range of integration becomes $[-1, 1]$ instead of $[a, b]$ (see Section 4.3). In the rest of this section, we would use the transformed variable, z, only.

Recall that $x = \left[\dfrac{b+a}{2} + \dfrac{b-a}{2} z \right]$

Also, we assume that the z's are arranged in the increasing order such that z_0 is the smallest and z_n largest. We then have

$$\int_a^b f(x) \, dx = \frac{b-a}{2} \int_{-1}^1 f(z) \, dz \simeq \frac{b-a}{2} \tilde{I}_z$$

(5.49)

where

$$\tilde{I}_z = \sum_{i=0}^n c_i f(z_i) \simeq \int_{-1}^1 f(z) \, dz$$

and the corresponding equations as (obtained by using $f(z) = 1, z, z^2, ..., z^{2n+1}$, respectively)

$$c_0 + c_1 = 2$$

$$c_0 z_0 + c_1 z_1 = 0$$

$$c_0 z_0^2 + c_1 z_1^2 = 2/3$$

$$c_0 z_0^3 + c_1 z_1^3 = 0$$

(5.50)

from which, $c_0 = c_1 = 1$; $z_0 = -\dfrac{1}{\sqrt{3}}$, $z_1 = \dfrac{1}{\sqrt{3}}$. A rough estimate of the error in $\tilde{I}_z$

may be obtained by assuming it to be proportional to $f^{2n+2}(\xi)$ in which $\xi \in (-1, 1)$.
For example, taking $f(z) = z^4$ (therefore $f^{iv}(\xi) = 24$),

$$E_z = I_z - \tilde{I}_z = \left[\frac{z^5}{5}\right]_{-1}^{1} - \sum_{i=0}^{1} c_i z_i^4 = \frac{2}{5} - \frac{2}{9} = \frac{8}{45} = \frac{f^{iv}(\xi)}{135} \tag{5.51}$$

The error in I_x will, of course, be $\dfrac{b-a}{2}$ times this error.

Equations (5.50) are nonlinear (since both c and z are unknown) and consequently
a little difficult to solve for large n. Therefore, we describe another alternative,
which leads to simpler determination of the weights and ordinates.

Since the polynomials used for obtaining the equations are arbitrary, we may use

$$f(z) = 1, (z - z_0), (z - z_0)(z - z_1), \ldots, (z - z_0)(z - z_1)\ldots(z - z_n),$$

$$z(z - z_0)(z - z_1)\ldots(z - z_n), \ldots, z^n(z - z_0)(z - z_1)\ldots(z - z_n) \tag{5.52}$$

as the $2n + 2$ polynomials and obtain the following equations:

$$\sum_{i=0}^{n} c_i = \int_{-1}^{1} 1\, dz = 2$$

$$\sum_{i=1}^{n} c_i (z_i - z_0) = \int_{-1}^{1} (z - z_0)\, dz = -2z_0$$

$$\sum_{i=2}^{n} c_i (z_i - z_0)(z_i - z_1) = \int_{-1}^{1} (z - z_0)(z - z_1)\, dz = \frac{2}{3} + 2z_0 z_1$$

$$\vdots$$

$$c_n (z_n - z_0)(z_n - z_1)\ldots(z_n - z_{n-1}) = \int_{-1}^{1} \prod_{i=0}^{n-1} (z - z_i)\, dz$$

$$0 = \int_{-1}^{1} \prod_{i=0}^{n} (z - z_i)\, dz$$

$$0 = \int_{-1}^{1} z \prod_{i=0}^{n} (z - z_i)\, dz \tag{5.53}$$

$$\vdots$$

$$0 = \int_{-1}^{1} z^n \prod_{i=0}^{n} (z - z_i)\, dz$$

It should be noted that:

- The last $n + 1$ equations do not involve the coefficients c and could be solved for the ordinates, z. The first equation involves all c's, the second does not have c_0 (since the function has a factor $z - z_0$), the third does not have c_0 and c_1, ..., and the $(n + 1)$th equation involves only c_n (and the ordinates). Therefore, once the ordinates are obtained, the coefficients can be sequentially obtained starting from the $(n + 1)$th equation.

- The last $n + 1$ equations are still nonlinear in the ordinates. However, if we define $\prod_{i=0}^{n}(z - z_i) = \sum_{i=0}^{n+1}\alpha_i z^i$, we obtain a set of $n + 1$ *linear* equations in α as it is easy to see that $\alpha_{n+1} = 1, \alpha_n = -\sum_{i=0}^{n} z_i$ and $\alpha_0 = (-1)^{n+1}\prod_{i=0}^{n} z_i$. The actual relationships are, however, not important.

$$\int_{-1}^{1}\prod_{i=0}^{n}(z - z_i)\,dz = \sum_{i=0}^{n+1}\frac{1-(-1)^{i+1}}{i+1}\alpha_i = 0$$

$$\sum_{i=0}^{n+1}\frac{1-(-1)^{i+2}}{i+2}\alpha_i = 0$$

$$\vdots \qquad\qquad (5.54)$$

$$\sum_{i=0}^{n+1}\frac{1-(-1)^{i+n+1}}{i+n+1}\alpha_i = 0$$

which could be solved to obtain the α's (recall that $\alpha_{n+1} = 1$) and the roots of the equation

$$\sum_{i=0}^{n+1}\alpha z^i = 0$$

would give the required ordinates. Another technique for linearizing uses manipulations on the original system of Eq. (5.53). On multiplying the first equation by α_0, i.e., $z_0 z_1$, the second by α_1, i.e., $-(z_1 + z_2)$, and adding them to the third equation, we get $\alpha_0 = -1/3$. Doing the same operation on the second, third, and fourth equations, we get $\alpha_1 = 0$ leading to the equation $z^2 - 1/3 = 0$, and therefore, $z_0 = -1/\sqrt{3}$, $z_1 = 1/\sqrt{3}$. For a general case, similar manipulation leads to the same set of $n + 1$ equations as Eq. (5.54).

- For each equation for α, half of the terms would be zero for odd n. [If n is even, one set of equations would have $n/2$ non-zero terms while the other set will have $n/2 + 1$ non-zero terms.] Thus, instead of solving a single set of $n + 1$ equations, we would only need to solve two sets of $(n + 1)/2$ equations. For example, for four points ($n = 3$), we obtain the following equations:

$$c_0 + c_1 + c_2 + c_3 = 2 \tag{i}$$

$$c_1(z_1 - z_0) + c_2(z_2 - z_0) + c_3(z_3 - z_0) = -2z_0 \tag{ii}$$

$$c_2(z_2 - z_0)(z_2 - z_1) + c_3(z_3 - z_0)(z_3 - z_1) = 2/3 + 2z_0 z_1 \tag{iii}$$

$$c_3(z_3 - z_0)(z_3 - z_1)(z_3 - z_2) = -\frac{2}{3}(z_0 + z_1 + z_2) - 2z_0 z_1 z_2 \tag{iv}$$

$$2\alpha_0 + \frac{2}{3}\alpha_2 + \frac{2}{5} = 0 \tag{v}$$

$$\frac{2}{3}\alpha_1 + \frac{2}{5}\alpha_3 = 0 \tag{vi}$$

$$\frac{2}{3}\alpha_0 + \frac{2}{5}\alpha_2 + \frac{2}{7} = 0 \tag{vii}$$

$$\frac{2}{5}\alpha_1 + \frac{2}{7}\alpha_3 = 0 \tag{viii}$$

(5.55)

From Eqs (vi) and (viii), $\alpha_1 = \alpha_3 = 0$ and from the (v) and (vii) equations, $\alpha_2 = -6/7$, $\alpha_3 = 3/35$.

The ordinates are therefore obtained by solving

$$z^4 - \frac{6}{7}z^2 + 3/35 = 0$$

as

$$z_{0,3} = \mp\sqrt{\frac{3}{7} + \sqrt{\frac{24}{245}}} = \mp 0.86114 \quad \text{and} \quad z_{1,2} = \mp\sqrt{\frac{3}{7} - \sqrt{\frac{24}{245}}} = \mp 0.33998$$

The coefficients are then sequentially obtained starting from Eq. (iv) as

$$c_3 = 0.3479, \quad c_2 = 0.6521, \quad c_1 = 0.6521, \quad c_0 = 0.3479$$

- An important observation could be made from the last $n + 1$ in Eq. (5.53) that the $(n + 1)$th degree polynomial

$$\prod_{i=0}^{n}(z - z_i)$$

is orthogonal to all polynomials of order n and lower. (You may want to review Section 4.3 on orthogonal polynomials). It indicates, and it will be further established in the next sub-sections, that the ordinates would be the zeros of the Legendre polynomial of order $n + 1$.

This method, though simple to use, is not as efficient as those described next. Hence, we will not discuss it further.

(b) *Based on Hermite Interpolation* In Section 4.5, we discussed the problem of interpolating when the function value *as well as* its derivative(s) is known at a few grid points. Let us assume that the function, and its first derivative, is known at $n + 1$ points: $z_0, z_1, z_2,..., z_n$ in the interval $(-1, 1)$. Using the Hermite interpolation we could write the interpolating polynomial of degree $(2n + 1)$ as (see Section 4.5)

$$f_{2n+1}(z) = \sum_{i=0}^{n} \left[H_{i,0} f(z_i) + H_{i,1} f'(z_i) \right] \tag{5.56}$$

where the H's are polynomials in z (with maximum degree $2n + 1$) and are given by

$$\begin{aligned} H_{i,0} &= \left[1 - 2(z - z_i) L_i'(z_i) \right] \left[L_i(z) \right]^2 \\ H_{i,1} &= (z - z_i) \left[L_i(z) \right]^2 \end{aligned} \tag{5.57}$$

$L(z)$ denoting the Lagrange polynomial. $\left[\text{Recall that } L_i(z) = \prod_{\substack{j=0 \\ j \neq i}} \dfrac{z - z_j}{z_i - z_j}. \right]$

Clearly, if we want a quadrature scheme of the form, Eq. (5.49), which is exact for this $(2n + 1)$ degree polynomial, we must have

$$\sum_{i=0}^{n} c_i f(z_i) = \int_{-1}^{1} \sum_{i=0}^{n} \left[H_{i,0} f(z_i) + H_{i,1} f'(z_i) \right] dz$$

This implies that we should choose the z's in such a way that

$$\int_{-1}^{1} H_{i,1}(z)dz = 0, \forall i$$

(thereby making the coefficients of the derivative terms in the integral vanish) and the coefficients would be given by

$$c_i = \int_{-1}^{1} H_{i,0}(z)dz, \forall i$$

This provides us with a methodology to find out the ordinates, z_i, and then, $H_{i,0}$, from which the coefficients of the quadrature formula are obtained.

Using the definition of the Lagrange polynomials and the fact that these are polynomials of order n, it is obvious from Eq. (5.57) that for the coefficients $H_{i,1}$ to vanish from the quadrature formula, it is sufficient (it can be shown that it is also *necessary*) that the $(n + 1)$th order polynomial

$$\prod_{i=0}^{n}(z - z_i)$$

is orthogonal to *all* polynomials of inferior order, i.e., order n or lower. As we have seen in the previous chapter, this condition is satisfied by the Legendre polynomial, $P_{n+1}(z)$, and its zeros would be the required ordinates of the quadrature scheme.

(c) *Based on Orthogonal polynomials* Let $f_{2n+1}(z)$ be a polynomial of order $(2n + 1)$ passing through the function values at all the grid points. We may then write

$$f_{2n+1}(z) = \sum_{i=0}^{n} L_i(z) f(z_i) + p_n(z) \prod_{i=0}^{n} (z - z_i) \qquad (5.58)$$

in which the first term on the right-hand side is the nth degree polynomial interpolating function in the Lagrange form, $p_n(z)$ is an arbitrary nth degree polynomial, and we have used the condition that the second term on the right-hand side must be a polynomial of degree $(2n + 1)$ and must vanish at all the grid points (since the first term, by itself, interpolates the function at all grid points). We now write

$$\int_{-1}^{1} f(z) dz \simeq \sum_{i=0}^{n} W_i f(z_i) \qquad (5.59)$$

(since the coefficients c, which we have used in the previous subsections, may be thought of as a weight assigned to each function value, we now replace them by the commonly used symbol, W) where the weights are defined by

$$W_i = \int_{-1}^{1} L_i(z) dz \qquad (5.60)$$

It is seen that the quadrature would be exact for all polynomials of order $(2n + 1)$ if

$$\int_{-1}^{1} p_n(z) \prod_{i=0}^{n} (z - z_i) dz = 0 \qquad (5.61)$$

for *all* nth degree polynomials, $p_n(z)$. Thus, the term

$$\prod_{i=0}^{n} (z - z_i)$$

should be orthogonal to all nth degree polynomials, and we again reach the conclusion that the quadrature points should be the zeros of the Legendre polynomial of order $(n + 1)$. This form of the quadrature scheme is, therefore, commonly called the *Gauss Legendre* quadrature. The weights are then obtained from Eq. (5.59) as

$$\int_{-1}^{1} L_i(z) dz$$

and turn out to be the same as

$$\int_{-1}^{1} H_{i,0} dz$$

in the previous method.

Though we had obtained a rough estimate of the error in approximating the integral using the method of undetermined coefficient with two quadrature points, it would be appropriate at this stage to provide an expression for the error in general case.

Using Eq. (4.51) for the remainder term in interpolation using the Hermite polynomials, we obtain the error in numerical integration as

$$E_z = I_z - \tilde{I}_z = \int_{-1}^{1} \frac{f^{[2n+2]}(\xi^*)}{(2n+2)!} \prod_{i=0}^{n} (z - z_i)^2 dz \qquad (5.62)$$

Applying the second mean value theorem for integrals, we get

$$E_z = \frac{f^{[2n+2]}(\xi)}{(2n+2)!} \int_{-1}^{1} \prod_{i=0}^{n} (z - z_i)^2 dz$$

Now using the facts that z_i are zeros of $P_{n+1}(z)$, and the coefficient of the leading term in $P_n(z)$ is

$$\frac{(2n)!}{2^n (n!)^2}$$

(see Section 4.3), we get

$$\prod_{i=0}^{n} (z - z_i)^2 = \left\{ \frac{2^{n+1} [(n+1)!]^2 P_{n+1}(z)}{(2n+2)!} \right\}^2$$

Using the inner product relationship

$$\int_{-1}^{1} P_n(z) P_n(z) dz = \frac{2}{2n+1}$$

(see Section 4.3), we finally obtain

$$E_z = \frac{2^{2n+3} [(n+1)!]^4}{(2n+3)[(2n+2)!]^3} f^{[2n+2]}(\xi) \qquad (5.63)$$

Note that for $n = 1$

$$E_z = \frac{f^{iv}(\xi)}{135}$$

which was obtained earlier using the method of undetermined coefficients. The weights, abscissas, and error for a few values of n are listed in Table 5.6.

| Table 5.6 | Weights, abscissas, and error for Gauss-Legendre quadrature | | |

n^*	Abscissa	Weights	Error
0	0.00000	2.0000	$\dfrac{f''(\xi)}{3}$
1	$\pm\,0.57735$	1.0000	$\dfrac{f^{iv}(\xi)}{135}$
2	0.00000	0.88889	$\dfrac{f^{vi}(\xi)}{15750}$
	$\pm\,0.77460$	0.55556	
3	$\pm\,0.33998$	0.65215	$\dfrac{f^{[8]}(\xi)}{3472875}$
	±0.86114	0.34785	
4	0.00000	0.56889	
	$\pm\,0.53847$	0.47863	$\dfrac{f^{[10]}(\xi)}{1237732650}$
	$\pm\,0.90618$	0.23693	

*Recall that the number of quadrature points is $n + 1$.

Equation (5.60) expresses the weights in a very simple form but it is quite cumbersome to evaluate. We could express the weights in a form more suitable from computational point of view by writing [see Eq. (4.22)]

$$L_i(z) = \frac{\displaystyle\prod_{j=0}^{n}(z - z_j)}{(z - z_i)\left[\dfrac{d}{dz}\left\{\displaystyle\prod_{j=0}^{n}(z - z_j)\right\}\right]_{z=z_i}} = \frac{P_{n+1}(z)}{(z - z_i)P'_{n+1}(z_i)}$$

since z_j are the zeros of $P_{n+1}(z)$ implying that

$$P_{n+1}(z) = C_{n+1}\prod_{j=0}^{n}(z - z_j)$$

with C_{n+1} denoting the coefficient of the leading term, z^{n+1}. Using the Christoffel-Darboux identity (see Box 5.2), we get

$$\int_{-1}^{1}\frac{P_{n+1}(z)}{(z - z_i)}\,dz = \frac{C_{n+1}\langle P_n, P_n\rangle}{C_n P_n(z_i)}\sum_{j=0}^{n}\frac{P_j(z_i)\displaystyle\int_{-1}^{1}P_j(z)dz}{\langle P_j, P_j\rangle} \qquad (5.64)$$

BOX 5.2

For any set of orthogonal polynomials, $\phi_i(x)$, we have

$$\sum_{j=0}^{n} \frac{\phi_j(x)\phi_j(x^*)}{\langle \phi_j, \phi_j \rangle} = \frac{\phi_{n+1}(x)\phi_n(x^*) - \phi_n(x)\phi_{n+1}(x^*)}{\frac{C_{n+1}}{C_n}\langle \phi_n, \phi_n \rangle (x - x^*)}$$

in which $\langle \cdot \rangle$ indicates the inner product. For the particular case in which x^* are the zeros of $\phi_{n+1}(x)$, represented by x_i, $i = 0, 1, \ldots, n$, we have

$$\sum_{j=0}^{n} \frac{\phi_j(x)\phi_j(x_i)}{\langle \phi_j, \phi_j \rangle} = \frac{C_n \phi_{n+1}(x)\phi_n(x_i)}{C_{n+1}\langle \phi_n, \phi_n \rangle (x - x_i)}$$

Applying it to the Legendre polynomials, we get

$$\frac{P_{n+1}(z)}{(z - z_i)} = \frac{C_{n+1}\langle P_n, P_n \rangle}{C_n P_n(z_i)} \sum_{j=0}^{n} \frac{P_j(z)P_z(z_i)}{\langle P_j, P_j \rangle}$$

From the orthogonality of Legendre polynomials, and since $P_0(z) = 1$, the rightmost integral above will vanish except for $j = 0$. From the expressions for the leading coefficient, we have

$$\frac{C_{n+1}}{C_n} = \frac{\dfrac{(2n+2)!}{2^{n+1}[(n+1)!]^2}}{\dfrac{(2n)!}{2^n(n!)^2}} = \frac{2n+1}{n+1}$$

and since $\langle P_n, P_n \rangle = \dfrac{2}{2n+1}$, we get $\displaystyle\int_{-1}^{1} \frac{P_{n+1}(z)}{(z - z_i)}\,dz = \frac{2}{(n+1)P_n(z_i)}$ and

$$W_i = \int_{-1}^{1} L_i(z)\,dz = \frac{1}{P'_{n+1}(z_i)}\int_{-1}^{1}\frac{P_{n+1}(z)}{(z - z_i)}\,dz = \frac{2}{(n+1)P_n(z_i)P'_{n+1}(z_i)} \tag{5.65}$$

or, using the relation between Legendre polynomials and derivative,[1]

---

[1] The relationship is $(1 - z^2)P'_n(z) = nP_{n-1}(z) - nzP_n(z)$. Using $(n + 1)$ in place of n, and noting that z_i are zeros of $P_{n+1}(z)$, we get $(n+1)P_n(z_i) = (1 - z_i^2)P'_{n+1}(z_i)$.

$$W_i = \frac{2}{(1-z_i^2)\left[P_{n+1}'(z_i)\right]^2} = \frac{2(1-z_i^2)}{\left[(n+1)P_n(z_i)\right]^2} \tag{5.66}$$

Note that we have treated *all* the ordinates and weights to be adjustable. Sometimes the problem may demand that some of the ordinates or weights be fixed *a priori*. Generally, by assigning one parameter we would lose an order of accuracy from the $2n + 1$ obtainable from n free points [but not always. For example, if the prescribed ordinate coincides with one of the zeros of $P_{n+1}(z)$, we still get the same accuracy.]. Therefore, it is not very common to do so. However, if it is important to use the function value at one or both end points, Radau and Lobatto quadrature schemes, respectively, could be used. We do not discuss these here.

A number of times, we may be interested in evaluating the integral using multiple applications of the quadrature scheme with increasing number of points to study the convergence properties. Table 5.6 clearly shows that there are no common points when we move from $n = 2$ to $n = 3$ or $n = 4$. Thus, we will not be able to re-use any of the previously computed function values. Since the function evaluation may be time-consuming, it would be more efficient to devise a scheme which would be able to utilise some, if not all, of the function values evaluated earlier. Kronrod (1964) proposed one such scheme which starts from, say n_1, points and adds $n_1 + 1$ points in such a way that the new abscissa include all n_1 of the old points, the free parameters being the $n_1 + 1$ new abscissas and $2n_1 + 1$ weights. These quadrature schemes, known as the Gauss Kronrod scheme, are not discussed here.

Sometimes, instead of integral of the function, we require the integral of the function multiplied by some weighting function, $w(x)$. For example, in minimax approximation of a function (see Section 4.3), we require integrals of the form

$$\int_{-1}^{1} \frac{f(x)}{\sqrt{1-x^2}} \, dx$$

One option in this case would be to treat $w(x)f(x)$ as another function, say, $g(x)$ and then apply the Gauss Legendre quadrature to this function. However, $g(x)$ may not be as well-behaved as $f(x)$ and it is preferred to use a weighted-Gauss quadrature. Apparently, Gauss had not considered the weighted schemes, which were later studied by Christoffel. The weights, W, are therefore sometimes called the Christoffel numbers. These may be written in the general form

$$\int_{-1}^{1} w(z)f(z)dz = \sum_{i=0}^{n} W_i f(z_i) + E_z \tag{5.67}$$

with

$$W_i = \int_{-1}^{1} w(z)L_i(z)dz \tag{5.68}$$

and

$$E_z = \frac{f^{[2n+2]}(\xi)}{(2n+2)!} \int_{-1}^{1} w(z)\prod_{i=0}^{n}(z-z_i)^2 dz \tag{5.69}$$

z_i being zeros of the $(n+1)$th order polynomial from the set of polynomials orthogonal over $(-1, 1)$ with respect to weight $w(z)$. The choice $w(z)=(1-z^2)^{-1/2}$ leads to the Gauss Tchebycheff quadrature and is described next. Some other weighting functions are mentioned in the section on improper integrals.

Example 5.9

For four point Gauss Legendre quadrature, obtain the location of the quadrature points and the associated weights. Use this scheme to estimate

$$\int_0^1 \exp(x-x^2)dx$$

Solution

For four quadrature points ($n=3$), the quadrature points would be located at the zeros of $P_4(z)$, and are, therefore, obtained from [see Eq. (4.11)]

$$35z^4 - 30z^2 + 3 = 0$$

Thus, we get

$$z^2 = \frac{30 \pm \sqrt{900-420}}{70} = 0.7415557, 0.1155871$$

The quadrature points are then obtained as $\pm 0.86114, \pm 0.33998$.

The weight for the point 0.86114 is obtained from Eq. (5.66) as

$$\frac{2(1-0.86114^2)}{\left[4P_3(0.86114)\right]^2} = \frac{0.51689}{\left[4\left(\frac{5}{2}0.86114^3 - \frac{3}{2}0.86114\right)\right]^2} = 0.34785$$

We may use the alternative expression to find the weight as

$$\frac{2}{(1-0.86114^2)\left[P_4'(0.86114)\right]^2} = \frac{2}{0.25844\left[\frac{35}{2}0.86114^3 - \frac{15}{2}0.86114\right]^2} = 0.34785$$

The other weights are found similarly and are as listed in Table 5.6. To estimate the integral

$$\int_{0}^{1} \exp(x-x^2)\, dx$$

we first convert it to the standard domain $(-1, 1)$ by defining $z = 2x - 1$ and the integral becomes

$$1/2 \int_{-1}^{1} \exp(z+1/2-(z+1/2)^2)\, dz$$

The estimate of this integral is obtained as

$$\tilde{I}_z = \sum_{i=0}^{3} W_i \exp\left[z_i + 1/2 - (z_i + 1/2)^2 \right] = 2.3692$$

Hence, the original integral is estimated as $2.369 / 2 = 1.1846$.

5.3.2.1 *Gauss Tchebycheff Quadrature*

We have seen (Section 4.3) that the Tchebycheff polynomials are orthogonal over $[-1, 1]$ for weight function $1/\sqrt{1-z^2}$. Therefore, for the weighted quadrature scheme with a weight of $1/\sqrt{1-z^2}$, the quadrature points would be at the zeros of $T_{n+1}(z)$ given by (see Section 4.3)

$$z_i = \cos\left[\frac{2n-2i+1}{n+1} \frac{\pi}{2} \right], \quad i = 0, 1, 2, ..., n \tag{5.70}$$

and the weights are given by [Eq. (4.22) has been used as was done in the derivation of Eq. (5.65)]:

$$W_i = \int_{-1}^{1} w(z) L_i(z) dz = \frac{1}{T'_{n+1}(z_i)} \int_{-1}^{1} \frac{w(z) T_{n+1}(z)}{(z-z_i)} dz \tag{5.71}$$

Following the same methodology as for Gauss Legendre quadrature, and using the fact that if z_i are the zeros of $T_{n+1}(z)$, then

$$T_n(z_i) = -\sqrt{1-z_i^2} \quad \text{and} \quad T'_{n+1}(z_i) = -\frac{n+1}{\sqrt{1-z_i^2}}$$

we get

$$W_i = \frac{C_{n+1} \langle T_n, T_n \rangle}{C_n T_n(z_i) T'_{n+1}(z_i)} = \frac{2^n \frac{\pi}{2}}{2^{n-1}(n+1)} = \frac{\pi}{n+1} \tag{5.72}$$

showing that all the weights are same! Similarly, the error is obtained as

$$E_z = \frac{f^{[2n+2]}(\xi)}{(2n+2)!} \int_{-1}^{1} w(z) \left[\frac{T_{n+1}(z)}{C_{n+1}} \right]^2 dz = \frac{\pi}{2^{2n+1}(2n+2)!} f^{[2n+2]}(\xi) \tag{5.73}$$

Note that all the weights turn out to be equal. However, they were not constrained to do so *a priori*. Hence there is no loss of order of precision normally associated with fixing one or more abscissas or weights.

We could compare the errors in Gauss Legendre (GL) and Gauss Tchebycheff (GT) schemes but it would not be meaningful since they integrate different functions: $f(z)$ and $f(z)/\sqrt{1-z^2}$, respectively. For evaluating

$$\int_{-1}^{1} f(z)dz$$

it would appear to be advantageous to use the GT scheme if the function involves the factor $1/\sqrt{1-z^2}$. The following example illustrates these points.

Example 5.10   Compare the results obtained from the four point ($n = 3$) Gauss Legendre and Gauss Tchebycheff quadrature schemes for estimating the integrals

(i) $\int_{-1}^{1} 1\,dz$, (ii) $\int_{-1}^{1} 1/\sqrt{1-z^2}\,dz$, (iii) $\int_{-1}^{1} \exp(z)\,dz$, and (iv) $\int_{-1}^{1} \exp(z)/\sqrt{1-z^2}\,dz$.

Solution   The quadrature points and weights for the Gauss Legendre (GL) scheme are listed in Table 5.6. The weights for the Gauss Tchebycheff (GT) scheme are all equal ($=\pi/4$) and the quadrature points are obtained from Eq. (5.70) as ± 0.92388, ± 0.38268. The estimates of the integrals are obtained below:

(i)   For $\int_{-1}^{1} 1\,dz$

GL : $f(z) = 1$, $\tilde{I} = \sum_{i=0}^{3} W_i \times 1 = 2$, which is the exact value of the integral

GT : $f(z) = \sqrt{1-z^2}$, $\tilde{I} = \sum_{i=0}^{3} W_i \times \sqrt{1-z_i^2} = 2.052$

(ii) For $\int_{-1}^{1} \dfrac{1}{\sqrt{1-z^2}}\,dz$

GL : $f(z) = \dfrac{1}{\sqrt{1-z^2}}$, $\tilde{I} = \sum_{i=0}^{3} W_i \times \dfrac{1}{\sqrt{1-z_i^2}} = 2.755$

GT : $f(z) = 1$, $\tilde{I} = \sum_{i=0}^{3} W_i \times 1 = \pi$, which is the exact value of the integral

(iii) For $\int\limits_{-1}^{1} \exp(z)\, dz$ (the exact value of the integral is 2.350)

$$GL: f(z) = \exp(z), \quad \tilde{I} = \sum_{i=0}^{3} W_i \times \exp(z_i) = 2.350$$

$$GT: f(z) = \exp(z)\sqrt{1-z^2}, \quad \tilde{I} = \sum_{i=0}^{3} W_i \times \exp(z_i)\sqrt{1-z_i^2} = 2.435$$

(iv) For $\int\limits_{-1}^{1} \dfrac{\exp(z)}{\sqrt{1-z^2}}\, dz$ (the exact value of the integral is 3.938)

$$GL: f(z) = \frac{\exp(z)}{\sqrt{1-z^2}}, \quad \tilde{I} = \sum_{i=0}^{3} W_i \times \frac{\exp(z_i)}{\sqrt{1-z_i^2}} = 3.376$$

$$GT: f(z) = 1, \quad \tilde{I} = \sum_{i=0}^{3} W_i \times \exp(z_i) = 3.977$$

5.3.3 Improper Integrals

Till now, we have made the tacit assumption that the integral to be evaluated using a numerical scheme is *proper*, i.e., it has finite limits and the integrand is defined and continuous at all points in the interval. The good thing about a proper integral is that it will always converge. For numerical integration we may be able to get around the more restrictive condition of the function being continuous at all points in the interval by suitable subdivision of the interval. As is clear from this definition, an *improper integral* would be one in which either the limit(s) of the integral is/are at $\pm\infty$ or the function to be integrated is undefined/discontinuous at any point in the interval. However, it is not clear whether an improper integral will converge or not.

As it turns out, it may converge in some cases and diverge in others. Obviously, if the integral diverges, there is no point in using a numerical method to estimate its value. Therefore, we will assume that the improper integrals to be evaluated numerically converge (see Box 5.3 for some theorems which are helpful in checking the convergence of an improper integral).

Once we know that an improper integral converges, we should look at the ways to evaluate it numerically. If the limits are finite and the integral is improper because the integrand is not defined at one of the limits (or some point within the interval), e.g., $\int_{0}^{1} x^{-1/2}\, dx$ in which the integrand is not defined at the lower limit; we may use

BOX 5.3 Convergence of Improper Integrals

Some easily verifiable results are:

The p-integral, $\int_{1}^{\infty}\dfrac{1}{x^p}\,dx$ converges for $p > 1$ (and diverges otherwise)

$\int_{0}^{1}\dfrac{1}{x^p}\,dx$ converges for $p < 1$ (and diverges otherwise)

$\int_{0}^{\infty}e^{\alpha x}\,dx$ converges for $\alpha < 0$ (and diverges otherwise)

For other improper integrals, comparison tests are used to establish their convergence. To apply these tests, it is helpful to note the absolute convergence property

(if $\int_{a}^{b}|f(x)|\,dx$ converges then so does $\int_{a}^{b}f(x)\,dx$).

Comparison test (also called Direct comparison or Standard comparison)
[2]Over the interval $[a, b)$ or $(a, b]$, if f and g are continuous and $0 \le f(x) \le g(x)$ for all x in the interval, then

$\int_{a}^{b}f(x)\,dx$ converges if

$\int_{a}^{b}g(x)\,dx$

converges (and $\int_{a}^{b}g(x)\,dx$ diverges if $\int_{a}^{b}f(x)\,dx$ diverges).

Limit Comparison test
Over the interval $[a, b)$ or $(a, b]$, if f and g are continuous and $f(x) > 0$ and $g(x) > 0$ for all x in the interval, such that

---

[2]Here a could be $-\infty$ or finite and b could be ∞ or finite.

$$\lim_{x \to b^- \text{ or } a^+} \frac{f(x)}{g(x)} = L$$

then

If $0 < L < \infty$, $\displaystyle\int_a^b f(x)\,dx$ converges *if and only if* $\displaystyle\int_a^b g(x)\,dx$ converges.

If $L = 0$, $\displaystyle\int_a^b f(x)\,dx$ converges if (note the absence of *only if*) $\displaystyle\int_a^b g(x)\,dx$ converges.

If $L = \infty$, $\displaystyle\int_a^b f(x)\,dx$ diverges if $\displaystyle\int_a^b g(x)\,dx$ diverges.

one of the semi-open methods (see Section 5.3.1.4) to estimate the integral. If we are not averse to using irregularly spaced grid points, the Gauss Legendre quadrature is likely to be the best open method for evaluating such integrals (in case the singularity lies within the interval and not at one of the ends, we may partition the integral into two parts at the singularity and separately evaluate each part).

Sometimes, a change of variable may eliminate the singularity. For example,

$$\int_0^1 \frac{f(x)}{x^p}\,dx$$

could be converted to

$$\frac{1}{p}\int_0^1 y^{\frac{1}{p}-2} f(y^{1/p})\,dy$$

by substituting $y = x^p$ (clearly, p should be equal to or smaller than 1/2 for this to work). Another option could be to truncate the interval of integration if it is ensured that the resulting error would be within permissible limits. For example, if in the integral above, $|f(x)| \le 1$ over the entire interval and assuming that p is less than 1, it can be shown that

$$\int_0^\varepsilon \frac{f(x)}{x^p}\,dx \le \int_0^\varepsilon \frac{1}{x^p}\,dx = \frac{\varepsilon^{1-p}}{1-p}.$$

This suggests that we may choose an appropriate ε and approximate the given improper integral as the proper integral

$$\int_{\varepsilon}^{1} \frac{f(x)}{x^p}\, dx.$$

The reader is referred to Davis and Rabinowitz (1967) for a detailed discussion.

Example 5.11   Evaluate $\int_{0}^{1} \frac{1}{\sqrt{1-x^2}}\, dx$ using semi-open (quadratic with $h = 0.2$) and 5-point Gauss Legendre quadrature.

Solution   It is observed that the integrand is undefined at the upper limit of the integral. The first step would, therefore, be to see whether the integral converges or not. It is readily seen that over the interval $(0, 1)$, $1/\sqrt{1-x^2} = 1/\sqrt{1+x}\sqrt{1-x} \leq 1/\sqrt{1-x}$ and both $f(x) = 1/\sqrt{1-x^2}$ and $g(x) = 1/\sqrt{1-x}$ are non-negative. Also, $\int_{0}^{1} \frac{1}{\sqrt{1-x}}\, dx$ is equivalent to $\int_{0}^{1} \frac{1}{\sqrt{x}}\, dx$, which converges (see Box 5.3, here $p = 1/2$). Hence the comparison test shows that the given improper integral converges (it is easy to see that the true value is $\pi/2$, i.e., 1.571).

Using quadratic semi-open formula (Table 5.5), with $h = 0.2$ (i.e. $n = 4$), we get

$$\tilde{I} = \frac{0.2}{3}\left[f(0) + 4f(0.2) + \frac{13}{4}f(0.4) + 0.f(0.6) + \frac{27}{4}f(0.8) \right] = 1.325$$

Using 5-point Gauss Legendre (with the transformation $z = 2x - 1$ to convert the range into the standard range), we get

$$\tilde{I}_z = 0.56889 f_z(0) + 0.47863\left[f_z(-0.53847) + f_z(0.53847) \right]$$
$$+ 0.23693\left[f_z(-0.90618) + f_z(0.90618) \right] = 2.8254$$

Hence the original integral is estimated as $2.8254/2 = 1.413$.

If either or both limits of the integral are not finite (but the value of the integral is bounded), we could use several options to first convert it into a proper integral and then numerically evaluate it. Probably the simplest option would be a substitution of variable. For example, a substitution $y = \exp(-x)$ changes the limits $(0, \infty)$ into $(0, 1)$ with the only constraint that the resulting integrand in y should be bounded over the entire interval. The truncation of the interval, as discussed for the case of integrands with singularity, may also work for some functions. The weighted Gauss quadrature [(Eq. (5.67)] for finite integration interval could be extended to infinite intervals through the Gauss Laguerre formula $(0, \infty)$ and the Gauss Hermite formula $(-\infty, \infty)$ which use the weights $\exp[-z]$ and $\exp[-z^2]$, respectively. We will, however, not discuss these here.

Example 5.12   Evaluate $\int\limits_0^\infty \dfrac{e^{-x}}{1+x^2}\, dx.$

Solution   Using the comparison test, since $\dfrac{e^{-x}}{1+x^2} \le e^{-x}$ and $\int\limits_0^\infty e^{-x}dx$ converges (Box 5.3),

$\int\limits_0^\infty \dfrac{e^{-x}}{1+x^2}\, dx$ also converges.

Now substituting $z = 2 \exp(-x) - 1$,

$$\int\limits_0^\infty \frac{e^{-x}}{1+x^2}\, dx = \int\limits_{-1}^1 \frac{dz}{2\left[1+\left(\ln \dfrac{z+1}{2}\right)^2\right]}$$

Using 4-point Gauss Legendre quadrature, we get

$$\tilde{I} = \sum_{i=0}^3 W_i \times \frac{1}{2\left[1+\left(\ln \dfrac{z_i+1}{2}\right)^2\right]} = 0.6291$$

EXERCISE 5.3

1. Estimate the integral of a fourth degree polynomial over $(0, 2)$ using the trapezoidal rule with step sizes of 1 and 2 and comment on the effect of the step size on the error. Use three different polynomials:
   (a) $1 + 2x + x^2 + 2x^3 - x^4$,
   (b) $6 + 7x + 9x^2 + 8x^3 - 5x^4$, and
   (c) $10 + 20x + 60x^2 + 41x^3 - 25x^4$.

2. A function is sampled at equidistant points as shown in the table below:

x	0	1	2	3	4	5	6	7	8	9	10	11	12
$f(x)$	1.000	1.564	2.266	3.115	4.125	5.307	6.672	8.232	10.000	11.986	14.203	16.662	19.375

Estimate the integral $\int\limits_0^{12} f(x)\, dx$ using the Simpson's 1/3 and 3/8 rules.

3. Assuming that the data in Problem 2 at $x = 12$ is not sampled, estimate the same integral using a semi-open method with quadratic extrapolation.

4. If it is known that the actual function is $f(x) = 1 + x/2 + (x/4)^2 + (x/8)^3$, obtain the errors of estimation of the methods used in Problems 2 and 3 and comment on the accuracy.

5. Use the 5-point Gauss Legendre and Gauss Tchebycheff quadratures to estimate the above integral.

6. Estimate the following integrals using the semi-open method with quadratic extrapolation, 4-point Gauss Legendre quadrature and 6-point Gauss Tchebycheff quadrature:

(i) $\int_0^1 dx/\sqrt{1-x}$

(ii) $\int_0^1 \dfrac{dx}{x^{0.1} + x^{0.9}}$ (note that the semi-open method requires extrapolation at the lower limit and the formulae given in Table 5.5 would need appropriate modification)

7. Estimate the following integrals using coordinate transformation and then applying the 5-point Gauss Legendre quadrature:

(i) $\int_0^\infty e^{-x^2} dx$ (ii) $\int_1^\infty \dfrac{dx}{x^{1.1} + x^{1.9}}$

8. The velocity of an object starting at $x = 0$ and moving along the x-axis was measured at various times as follows:

t (s)	0	10	20	30	40
v (m/s)	1.00	2.05	4.95	10.00	16.90

Estimate the distance travelled in 40 s to an accuracy $O(h^6)$ applying the Romberg integration algorithm to three estimates obtained by the trapezoidal rule.

9. The integral $\int_1^2 x^{-2} e^{-x} dx$ is to be evaluated numerically.

(a) If the integration is to be performed by using a closed-formula using function value at 6 equally spaced points, what scheme or combination of schemes will give an overall accuracy of $O(h^2)$. Find the integral using it.

(b) If the integration is to be performed by using a closed-formula using function value at 6 equally spaced points, what scheme or combination of schemes will give an overall accuracy of $O(h^4)$. Find the integral using it.

(c) If the integration is to be performed by using function values at *2 arbitrary points* what scheme will give the maximum accuracy? Find the integral using it.

10. Show that the trapezoidal rule, with $h = \dfrac{2\pi}{n+1}$ is exact for *trigonometric polynomials* of period 2π, i.e., for function of the form $\sum_{k=-n}^{n} c_k e^{ikt}$; k integer, when it is used for integration over a whole period.

11. Velocity profile in an open channel flow is shown in Fig. 5.7.

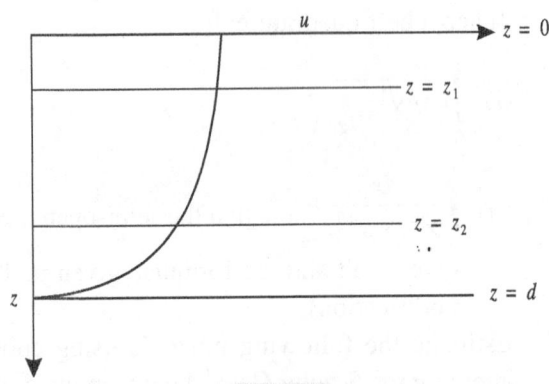

Fig. 5.7

The depth of the channel is d. The velocity at any depth is given by an arbitray function $u(z)$, graphically shown above. The mean velocity of the channel $\bar{u}$ is given by $\bar{u} = \dfrac{1}{d} \int_0^d u(z)\, dz$.

(a) Find two constants c_1 and c_2 such that $z_1 = c_1 d$, $z_2 = c_2 d$ and $\bar{u} = \dfrac{u(z_1) + u(z_2)}{2}$.

(b) The velocities u_0, u_1, u_2, and u_3 have been measured at depths of $0.2d$, $0.4d$, $0.6d$, and $0.8d$, respectively. Compute w_0, w_1, w_2, and w_3 such that the velocity at $0.5d$ can be expressed as $(w_0 u_0 + w_1 u_1 + w_2 u_2 + w_3 u_3)$.

12. Evaluate the integral of $f(x)$ from $x = 0$ to $x = 4$ using the following data:

x	0	1	2	2.5	3	4
$f(x)$	0.4	1.3	1.3	1.5	2.1	2.5

by (a) trapezoidal rule and (b) combination of Simpson's 1/3 rule and trapezoidal rule.

13. Using 3-point Gauss quadrature, find the integral of $2/(1 + 2x^2)$ from $x = -3$ to $x = 3$.

14. A waveform is approximated by the following function:
    $f(x) = 0.5$ for $x = \pi/2$ to 0
    $f(x) = 0.5 + \sin x$ for $x = 0$ to π
    $f(x) = 0.5$ for $x = \pi$ to $3\pi/2$
    Find the area under the wave using $h = \pi/2$ with (a) trapezoidal method and (b) Simpson's 1/3 rule. Comment on the relative accuracy of these results.

15. For a V-shaped channel, with bottom angle of 60°, the velocity v (in m/s) at a height y (in m) is given as $v = 2y^{1/7}$. Find the discharge (volumetric flow rate) through the channel if the flow depth is 1.5 m. Use 3-point Gauss Legendre quadrature.

16. The flow rate through a circular pipe is given by $Q = \int_0^{r_0} 2\pi r\, v\, dr$, where v is the velocity at a distance of r from the centre of pipe and r_0 is the radius of the pipe. If the velocity is approximated by $v = 2\left(1 - \dfrac{r}{r_0}\right)^{1/7}$ (in m/s), and the pipe radius is 12 cm, compute Q using (a) trapezoidal rule with $h = 2$ cm, (b) Simpson's 1/3 rule with $h = 3$ cm, (c) Simpson's 3/8 rule with $h = 4$ cm and (d) 3-point Gauss Legendre quadrature. Perform an error analysis using the true value of the flow rate as 0.0738902 m³/s.

17. Evaluate $\int_0^3 \dfrac{x^2 dx}{1+x^3}$ using step size $h = 1$; first with a combination of single applications of trapezoidal method (from $x = 0$ to 1) and Simpson's 1/3 method (from $x = 1$ to 3), and then using Simpson's 3/8 method.

18. What problems do you foresee in evaluating $\int_0^1 \dfrac{\cos x}{\sqrt{x}}\, dx$ using the Simpson's 1/3 rule. Evaluate the integral using 3-point Gauss Legendre quadrature. Manipulate the integral such that Simpson's rule could be applied and then evaluate the integral using (a) single application of Simpson's rule and (b) the 3-point Gauss Legendre quadrature. Comment on your results, given that the true value of the integral is 1.80905.

19. The integral of a function over the interval (0, 1), is approximated by the arithmetic mean of its values computed at three different points in this interval. Using a methodology similar to Gauss quadrature, obtain the location of these three points for highest order of accuracy.

20. Obtain the value of $\int_1^9 \dfrac{\ln x}{x}\, dx$ numerically using 5 points in the interval by (a) trapezoidal rule, (b) Simpson's rule, and (c) Gaussian quadrature. Compute the percentage error in each of the three cases.

SUMMARY

In this chapter, various methods for estimating the derivatives or integral of a given function (either as a function or as discretely sampled values) have been described. The use of Taylor's series to estimate the derivative from discrete data was described first and the concept of forward, backward, and central difference was introduced. Different techniques of improving the accuracy of the estimate by incorporating more data points were described.

Amplitude and phase errors in the estimation of derivatives of periodic functions were briefly described to lay the background for their more extensive use in the solution of differential equations. Numerical integration was discussed, first for a function sampled at discrete points and then for a continuous function. For the discrete case, numerical integration using a piecewise linear approximation of the function was discussed first. The improvement in accuracy by using higher-order approximations was described next. Both the Newton-Cotes and the Adams methods were described to achieve the improvement in accuracy. The Adams method was described briefly and a more detailed description would follow in the next chapter where these methods find better use in the solution of differential equations.

Open and semi-open integration formulae were discussed to account for the cases where the range of integration extends beyond the range of observed data. Numerical integration of a given function was then discussed with the emphasis placed on the use of orthogonal polynomials. Various forms of the Gauss quadrature schemes were described. Finally, numerical integration of improper integrals, which either had the range of integration going to infinity or the function value becoming infinite at some point in the range, was discussed.

The methods described in this chapter may be used to solve differential equations either by replacing the derivative by a *difference* of function values or by re-casting the equation in terms of an integral. In the next two chapters, we describe various techniques of solution of ordinary and partial differential equations which will borrow heavily from the material discussed in this (and the previous) chapter. Note that partial differential equations would need expressions for the numerical differentiation of a function of two (or more) variables, which have not been discussed in this chapter. However, the extension of the single-variable technique to multiple variables is relatively straightforward and would be described as needed in Chapter 7. Similarly, numerical integral of multiple integrals has not been discussed in this chapter but follows a philosophy similar to that of the single variable case.

6

Ordinary Differential Equations

6.1 INTRODUCTION

Batman fell off while jumping from the *Vampire State Building* to *Votre Brum Cathedral*! Yes, you read it right. He fell off because his *hook thrower* malfunctioned. Lying on the hospital bed, he was trying to piece together the events. He jumped off the *Vampire State Building* and started off the *hook thrower*. According to the design, a laser beam calculates the distance (x) between him and the target (tip of the cathedral tower in this instance) and lets out the hook at the right rate. The computer recalculates this information every 0.001 sec since he is falling continuously under the influence of gravity and the distance between him and the target is changing. The relation *between x and t* is adjustable. Since, a linear setting is most reliable, *Batman* has been using the following setting for years:

$$\frac{dx}{dt} = 1 + t + x \tag{6.1}$$

It is easy to see that his computer solves this equation every 0.001 sec with the position of batman at that instant as an initial condition. His computer has been doing it for the past 30 years! So, what went wrong this time? He sent the hook thrower to the original manufacturer for checking. The report came out to be sabotage, possibly by an accomplice of *Joker*.

The hook thrower setting has been changed to the following:

$$\frac{dx}{dt} = 1 + t + x^2 \tag{6.2}$$

You can easily see the problem now. The computer is programmed to solve Eq. (6.1) but his hook thrower is set to Eq. (6.2). Some experts have written the program for solving Eq. (6.1) some 30 years ago. In his confident self, Batman never realized that he needed to learn to solve such equations and program them in his computer, not only to avoid such disasters but also to explore the full potential (all settings linear and non-linear) of his hook thrower. So, he decided to learn it and as you can imagine, with *Batmanish* rigour. He needs to learn very accurate methods to solve them under all conditions and also learn the situations where they will fail. Here, we will try to learn with the *Batman*.

Equations (6.1) and (6.2) can be generally written as

$$\frac{dy}{dt} = f(y,t) \tag{6.3}$$

Equations of this form requires one *initial condition* in the form of $y = y_0$ at $t = t_0$ in order to obtain a solution. Initial point can be an arbitrarily chosen fixed point and for engineering problems, it is often governed by practical considerations or measurement at a known point. For *Batman*, it was readjusted every 0.001 sec. These are known as *initial value problems* (IVPs).

The independent variable does not necessarily have to be time. For example, the parabola expressed in the following form is also an IVP where the independent variable x behaves as time variable:

$$\frac{dy}{dx} = 2x, \quad y = 0 \text{ at } x = 0, \quad x \in [-1, 1] \tag{6.4}$$

We will spend much of this chapter developing methods for IVPs. Eventually, this will form the basis for solution of all other forms of differential equations including partial differential equations described in Chapter 7. Before we proceed onto mathematical rigour, let us develop a visual sense of what it means to solve an IVP. Once again remember that computers cannot carry out any operation other than $+$, $-$, $\times$, and $\div$.

In the simplest words, Eq. (6.4) is saying that, 'the *slope at any point* (x, y) *is* $2x$'. So, for Eq. (6.4), the slope in the interval $[-1, 1]$ is known a priori [Fig. 6.1(a)]. The challenge is to construct the solution y versus x [Fig. 6.1(b)] using the information of Fig. 6.1(a) and arithmetic operations.

Let us attempt to solve it graphically by using this information. Recall in numerical methods, we can only obtain a discrete approximation of a continuous function (y). We have a starting point in the form of initial condition. From the previous chapter, you are already familiar with grids. Let us choose to obtain the discrete approximation of y at x-grid size (h) of 0.2. Since, y-axis is an *axis of symmetry*, let us consider only the first quadrant. We know that the solution passes

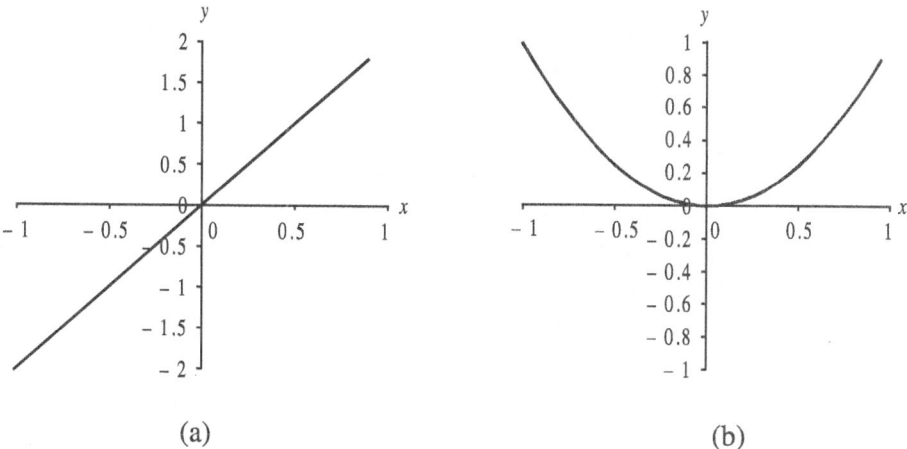

(a)   (b)

Fig. 6.1 Slope function (a) and solution (b) of Eq. (6.4).

through the point (0, 0), the initial condition [Eq. (6.4)]. We take a plane paper and mark this point. For the next point, we know the x-coordinate as $h = 0.2$, but to locate the y-coordinate, we need the direction or slope. If we have the slope, we can draw a straight line at that slope from the initial point and a vertical line from $x = 0.2$. The intersection will give us the required point.

We know that the slope in the interval (0, 0.2) is not a constant but a function of x and therefore, varies at every point. One *key assumption* in the development of numerical methods for IVP is that the *slope remains constant within the grid length*. This is also to say that function joining two points on a grid is a straight line. It is then easy to visualize that finer grid will bring us closer to reality as smaller segments of curves can be better approximated as straight lines.

Next question obviously is, *what is the value of this constant slope?* Since the slope is a continuous function, any finite interval can contain infinitely many choices of points where we can evaluate the slope and apply it over the whole interval. In fact, there could be as many choices as the number of stars in the sky. We shall see in this chapter that the primary difference between most of the numerical methods for IVPs is the way the slope is evaluated over a grid length.

For illustration, we will choose the easiest option. That is, whatever the slope is at the starting point of a grid length (an interval) remains constant over that interval. For the first interval, we then calculate the slope at the origin as zero using Eq. (6.4). Assuming the slope is constant along one grid length, we go one grid distance (0.2) along the gradient and mark the second point. At the new point ($x = 0.2$), we calculate the slope as 0.4. Once again, we travel one grid length along the gradient from this point to mark the third point. At the third point ($x = 0.4$) we pick the slope as 0.8 and proceed for one more grid. We continue this way for the entire range of x we are interested in.

Once we reach the point 1, we can join the points with *straight lines* to get an approximation of the original curve. Using this procedure, the resulting approximation is shown in Fig. 6.2(a). Since the slope is constant over each interval, it becomes a step function approximation of the continuous slope function [Fig. 6.2(b)]. As expected, refining the grid enhances the quality of approximation. We see in Fig. 6.2(a) that refinement of grid size to 0.1 has moved the approximation closer to the true solution.

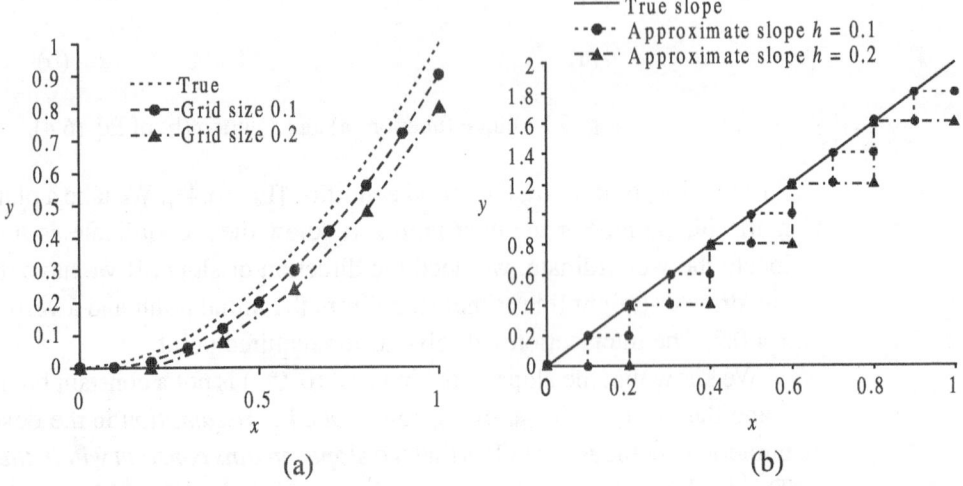

Fig. 6.2 (a) Numerical solution of Eq. (6.4) by Euler forward method (6.5) and (b) the approximate slope function.

One can easily verify that the mathematical equation for obtaining the y_{n+1} from y_n by following the above procedure is given by

$$y_{n+1} = y_n + 2x_n h \qquad (6.5)$$

The steps of obtaining y_{n+1} from y_n can now be formally put to word as follows for the generalized Problem (6.3):

- Choose a time step h and the grid $\{t_0, t_1, t_2, \ldots, t_m\}$ on t.
- Compute slopes at the present grid point as $f(y_n, t_n)$.
- Move one grid length along the slope from the present point y_n.

The process starts from the initial condition at $n = 0$ which is typically specified as $y = y_0$ at $t = t_0$.

If you put mathematical expressions for the process described above for obtaining

y_{n+1} from y_n between any two successive grid points t_n and t_{n+1}, the approximate form of the general Eq. (6.3) becomes,

$$y_{n+1} = y_n + hf(y_n, t_n) \qquad (6.6)$$

Please note, at any point of time, y_n, t_n, h and the functional form of f is known. The problem is to compute y_{n+1} at the next grid point. This can be done *explicitly* using Eq. (6.6) for any arbitrary functional form of f. A small rearrangement of (6.6) leads to,

$$\frac{y_{n+1} - y_n}{h} = f(y_n, t_n) \qquad (6.7)$$

Now compare the left-hand side of the equation with Eq. (5.2). It is easy to see that we approximated the derivative in Eq. (6.3) with a *forward difference* approximation and evaluated the functional value of f at the known point (y_n, t_n). The method for solving IVPs given by Eqs (6.6) or (6.7) is known as *Euler explicit* or *Euler forward*.

In the Euler forward method, the slope at any interval $\{t_n, t_{n+1}\}$ was approximated by the slope at the beginning of the interval or t_n. For the example of parabola, the true slope (straight line) was approximated by step functions. The true and approximate slope in the interval $\{0, 1\}$ is shown in Fig. 6.2(b) for grid sizes of 0.2 and 0.1. A natural question may arise, why not approximate the slope in the interval with, (i) the slope at the end of the interval, i.e., at t_{n+1} or (ii) average of the two slopes at t_n and t_{n+1}.

Application of these two will modify the method of Eq. (6.6) to,

Case (i):
$$y_{n+1} = y_n + hf(y_{n+1}, t_{n+1}) \qquad (6.8)$$

Case (ii):
$$y_{n+1} = y_n + h\left(\frac{f(y_n, t_n) + f(y_{n+1}, t_{n+1})}{2} \right) \qquad (6.9)$$

The method given by Eq. (6.8), although appearing very similar to Eq. (6.6), is not easy to apply. In order to illustrate this, let us apply Eq. (6.8) to approximate Eqs (6.1) and (6.2) of the Batman problem.

Eq. (6.1): $\quad y_{n+1} = y_n + h(1 + t_{n+1} + y_{n+1})$ or $y_{n+1} = \dfrac{y_n + h(1 + t_{n+1})}{(1 - h)}$

Eq. (6.2):
$$y_{n+1} = y_n + h(1 + t_{n+1} + y_{n+1}^2)$$

We observe that the right-hand side of the approximation cannot be explicitly computed anymore. In fact, if the functional form is non-linear in y as in Eq. (6.2), one needs to solve a non-linear algebraic equation at every step in order to get a solution. Such methods involving slope evaluation at the end point (t_{n+1}) of the interval are known as *implicit* methods as opposed to the *explicit* methods described earlier where the value at y_{n+1} could be computed explicitly. Thus, the method described by Eq. (6.8) for solving IVPs is known as *Euler implicit* or *Euler backward* method.

In order to have a mathematical understanding of the Case (ii) in Eq. (6.9), let us integrate the original problem (6.3) between two adjacent grid points,

$$\int_{y_n}^{y_{n+1}} dy = \int_{t_n}^{t_{n+1}} f(y,t)\,dt \tag{6.10}$$

If we numerically integrate the right-hand side using trapezoidal method, we obtain Eq. (6.9). So, this method is also known as the *trapezoidal method*. A graphical representation of the iterations of Euler forward, Euler backward, and trapezoidal method [Eqs (6.6), (6.8), and (6.9)] is shown in Fig. 6.3. You may ask again, why trapezoidal method? Why not Simpson's 1/3rd or 3/8th rule? More generally, why stop by averaging at two points? Why not consider more points to evaluate the slope?

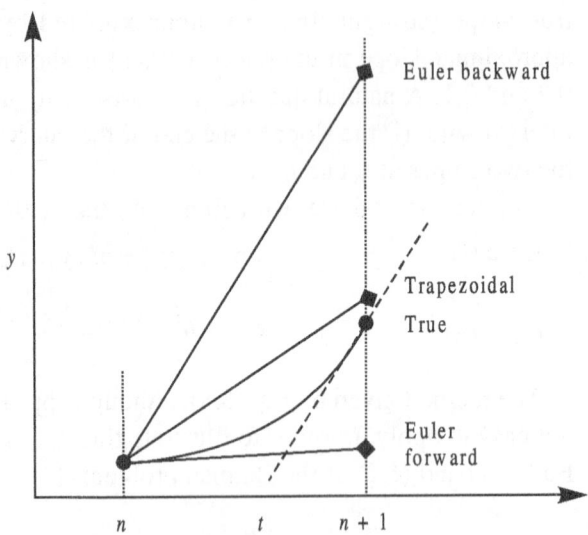

Fig. 6.3   Schematic of time stepping scheme for Euler methods and trapezoidal method. Euler forward method extends the slope at n, Euler backward extends the slope at $n+1$ and the trapezoidal method extends the average of the two slopes.

Yet another way to look at the problem will be to revisit the approximation of the derivative of the left-hand side of Eq. (6.3). So far, we only did forward difference approximation. In the previous chapter, we have learned many ways to approximate the first derivative using multiple grid points? Why not use some of those?

Now, try combining the various options of approximating the derivative with the options of single or multiple point functional evaluation of slope f, we have a mind boggling combinations in hand. So, before one goes around generating lots of such methods with random combination, let us put some system into madness.

More importantly, we have not yet asked the sensitive questions such as,

(a) Are all of these methods usable in all cases (define usability, convergence!)?

(b) Are these methods accurate?

(c) What are the advantages and disadvantages of various methods so as to enable one to choose appropriate method for a given problem (horses for the courses!)?

Euler's methods, both implicit and explicit, evaluate the functional value f at only one grid point. These are known as *single-step* methods. On the other hand, the method given by Eq. (6.9) involves evaluation of slope at *more than one* grid points. The possibilities are numerous. These will be called *multi-step* methods.

The classifications of *implicit vs. explicit* and *single-step vs. multi-step* are mutually independent. They can be combined to yield four groups of methods: (i) *single-step explicit method* (e.g., Euler explicit), (ii) *single-step implicit method* (e.g., Euler implicit), (ii) *multi-step explicit method* (will be derived in a later section), and (iv) *multi-step implicit method* (e.g., trapezoidal method of Eq. (6.9), more will be derived in a later section).

The group of methods that incorporates approximation of the derivative on the left side of Eq. (6.3) using values at more than two points is typically called *backward difference formulae* (BDFs). We shall learn about their special application in a later section.

In this chapter, we will first learn to formally derive various methods by grouping them appropriately in order to avoid future confusion. Afterwards, we will analyze some of the methods but more importantly, establish generalised protocol for analysis so that any given numerical method can be analyzed and judged on the basis of certain set standards. Lastly, we will apply the methods to solve a system of coupled IVPs, boundary-value problems (BVPs), and higher-order ordinary differential equations (ODEs).

6.2 DERIVATION OF METHODS FOR IVPS

Multi-step methods are a group which includes as their subset the single step methods. The ones derived here are more generally called *linear multi-step methods*.

They are linear because the approximations are made as a linear combination of function (and/or variable) evaluations at different grid points. In the first section, we shall approximate the slope function (multi-step methods) and in the next section we shall approximate the derivative (backward difference formulae). Combining the two methods lead to the *generalized multi-step method*. We shall use the generalized form for analysis and definitions of some convergence properties in Section 6.3.2.8.

6.2.1 Multi-Step Methods

In this section, we will derive a group of implicit and explicit methods for the general IVP given by Eq. (6.3). We rewrite the problem as follows:

$$\frac{dy}{dt} = f(y,t) \text{ where } y = y_0 \text{ at } t = t_0 \tag{6.11}$$

We shall seek valid approximations of the following forms for explicit and implicit methods.

Explicit: $$y_{n+1} = y_n + h\sum_{i=0}^{k} \alpha_i f_{n-i} \tag{6.12}$$

Implicit: $$y_{n+1} = y_n + h\sum_{i=0}^{k} \beta_i f_{n+1-i} \tag{6.13}$$

where $k = 0, 1, 2, ..., n,$ $(n + 1$ for implicit) and h is the uniform time step size, $h = t_n - t_{n-1} = t_{n+1} - t_n$.

The criteria for a valid approximation are as follows:

- Each approximation will represent the original equation with well-defined truncation error.
- The truncation error will approach zero in the limit as the time step $h \to 0$.

This is also known as the *consistency* criteria. Any numerical method derived to satisfy these criteria will be consistent with the original problem, i.e., in the limit $h \to 0$, the approximate equation will approach the original equation. We will discuss consistency, order of method, stability, and convergence in more detail in Section 6.3.

We illustrate the derivation through an explicit method with $k = 2$. Equation (6.12) then becomes

$$y_{n+1} = y_n + h(\alpha_0 f_n + \alpha_1 f_{n-1} + \alpha_2 f_{n-2}) \tag{6.14}$$

Denoting $\dfrac{dy}{dt} = y'$ and using the original equation $y' = f$, we can rewrite Eq. (6.14) as

$$y_{n+1} = y_n + h(\alpha_0 y'_n + \alpha_1 y'_{n-1} + \alpha_2 y'_{n-2}) \qquad (6.15)$$

We expand y_{n+1}, y'_{n-1}, and y'_{n-2} using Taylor's series as follows:

$$y_{n+1} = y_n + hy'_n + \frac{h^2}{2!} y''_n + \frac{h^3}{3!} y'''_n + \frac{h^4}{4!} y''''_n + o(h^5) \qquad (6.16)$$

$$y'_{n-1} = y'_n - hy''_n + \frac{h^2}{2!} y'''_n - \frac{h^3}{3!} y''''_n + o(h^4) \qquad (6.17)$$

$$y'_{n-2} = y'_n - (2h)y''_n + \frac{(2h)^2}{2!} y'''_n - \frac{(2h)^3}{3!} y''''_n + o(h^4) \qquad (6.18)$$

Putting these expressions of Eqs (6.16)–(6.18) in Eq. (6.15), we obtain

$$
\begin{aligned}
y_n + hy'_n + \frac{h^2}{2!} y''_n + \frac{h^3}{3!} y'''_n &+ \frac{h^4}{4!} y''''_n + o(h^5) = y_n + h\alpha_0 y'_n \\
&+ h\alpha_1 \left(y'_n - hy''_n + \frac{h^2}{2!} y'''_n - \frac{h^3}{3!} y''''_n + o(h^4) \right) \\
&+ h\alpha_2 \left(y'_n - (2h)y''_n + \frac{(2h)^2}{2!} y'''_n - \frac{(2h)^3}{3!} y''''_n + o(h^4) \right)
\end{aligned}
\qquad (6.19)
$$

We now perform some algebraic manipulations on Eq. (6.19) and group similar terms on the right-hand side to obtain the following:

$$
\begin{aligned}
y_n + hy'_n + \frac{h^2}{2!} y''_n + \frac{h^3}{3!} y'''_n &+ \frac{h^4}{4!} y''''_n + o(h^5) = y_n + h(\alpha_0 + \alpha_1 + \alpha_2) y'_n \\
&+ h^2 (-\alpha_1 - 2\alpha_2) y''_n + h^3 \left(\frac{\alpha_1}{2} + 2\alpha_2 \right) y'''_n + h^4 \left(-\frac{\alpha_1}{6} - \frac{4\alpha_2}{3} \right) y''''_n + o(h^5)
\end{aligned}
$$

$$(6.20)$$

Notice that the left-hand side is an exact representation of y_{n+1}, i.e., the prediction of y at the next time step. Since we have three adjustable parameters in α_i's, we can equate coefficients of three terms on the right-hand side with those on the left. So, the set of equations to determine α_i's are as follows:

$$\alpha_0 + \alpha_1 + \alpha_2 = 1$$

$$\alpha_1 + 2\alpha_2 = -\frac{1}{2}$$

$$\frac{\alpha_1}{2} + 2\alpha_2 = \frac{1}{6} \qquad (6.21)$$

The solution of the above system of equation is $\alpha_0 = 23/12, \alpha_1 = -4/3$, and $\alpha_2 = 5/12$. Putting these values in Eq. (6.14), we obtain the required numerical method as,

$$y_{n+1} = y_n + h\left(\frac{23}{12} f_n - \frac{4}{3} f_{n-1} + \frac{5}{12} f_{n-2}\right) \qquad (6.22)$$

The above analysis is general and can be used to derive an arbitrary explicit or implicit method. We can summarize the method to derive explicit or implicit method of any order in the following steps:

- Choose k and write the approximation in terms of α_i's or β_i's as in Eq. (6.14).
- Replace f with y' using the original equation.
- Expand all the terms using Taylor series in the neighbourhood of (y_n, t_n).
- Group the similar terms together on the right-hand side.
- Equate the coefficients of right-hand side terms with the left hand-side.
- Compute α_i's or β_i's.

If the derivation appears too cumbersome involving too many algebraic manipulations, we now present a more elegant way of doing the same using a table. However, we caution the students to go through the steps of derivation described above to have a thorough understanding of the process. Using the tabular form without proper understanding may lead to error.

From Eqs (6.14) and (6.15) we can construct the following table:

Table 6.1 Derivation of explicit multi-step method.

Terms	y_n	hy'_n	$h^2 y''_n$	$h^3 y'''_n$	$h^4 y''''$
y_{n+1}	1	1	½	1/6	1/24
y_n	−1	0	0	0	0
hf_n	0	$-\alpha_0$	0	0	0
hf_{n-1}	0	$-\alpha_1$	α_1	$(-1/2)\alpha_1$	$(1/6)\alpha_1$
hf_{n-2}		$-\alpha_2$	$2\alpha_2$	$-2\alpha_2$	$(4/3)\alpha_2$

Note that the entries along the rows are the coefficients of the respective Taylor series expansions. In effect, we have written down the coefficients of the expansion for the following form of Eq. (6.15):

$$y_{n+1} - y_n - h\left(\alpha_0 y'_n - \alpha_1 y'_{n-1} - \alpha_2 y'_{n-2}\right) = 0 \qquad (6.23)$$

This essentially groups the terms together in the columns. Summing the columns and setting to zero give us the required equations. For example, sum of columns 3,

4 and 5 in Table 1 gives us Eq. (6.21). Let us highlight the utility of such derivation using the tabular form with an implicit method with $k = 2$. Using Eq. (6.13), the required method can be written as:

$$y_{n+1} = y_n + h\left(\beta_0 f_{n+1} + \beta_1 f_n + \beta_2 f_{n-1}\right) \tag{6.24}$$

Now, the following table can be constructed,

Table 6.2 Derivation of a multi-step implicit method

Terms	y_n	hy'_n	$h^2 y''_n$	$h^3 y'''_n$	$h^4 y''''$
y_{n+1}	1	1	1/2	1/6	1/24
y_n	-1	0	0	0	0
hf_{n+1}	0	$-\beta_0$	$-\beta_0$	$(-1/2)\beta_0$	$(-1/6)\beta_0$
hf_n	0	$-\beta_1$	0	0	0
hf_{n-1}	0	$-\beta_2$	β_2	$(-1/2)\beta_2$	$(1/6)\beta_2$

So, the set of equations to solve for obtaining β_i's are

$$
\begin{aligned}
\beta_0 + \beta_1 + \beta_2 &= 1 \\
\beta_0 - \beta_2 &= \frac{1}{2} \\
\beta_0 + \beta_2 &= \frac{1}{3}
\end{aligned}
\tag{6.25}
$$

The solution is $\beta_0 = 5/12$, $\beta_1 = 2/3$, and $\beta_2 = -1/12$.

At this point, we are in a position to derive arbitrary order methods. The coefficients up to $k = 3$ are given in Table 6.3 for both implicit and explicit methods. The readers may wish to derive these to have some practice, especially for $k = 3$. For $k = 0$, the explicit method is the *Euler forward* and the implicit method is the *Euler backward*.

All others are generally known as *multi-step* methods. Multi-step explicit methods are also known as *Adams-Bashforth* and the implicit ones are called *Adams-Moulton* methods. Individual methods are identified by their *order*, which is related to the number of points at which the function is evaluated. For example, the method in Eq. (6.24) is a *third order Adams-Moulton* and that in Eq. (6.22) is a *third order Adams-Bashforth*. Exact nature of this *order* will be clear during the discussion of truncation error in Section 6.3.1.

Table 6.3   Single and multi-step explicit and implicit methods up to order 4.

Type	k	Method	Order	Name
Explicit	0	$y_{n+1} = y_n + hf_n$	1	Euler forward
	1	$y_{n+1} = y_n + h\left(\dfrac{3}{2} f_n - \dfrac{1}{2} f_{n-1}\right)$	2	Adams-Bashforth
	2	$y_{n+1} = y_n + h\left(\dfrac{23}{12} f_n - \dfrac{4}{3} f_{n-1} + \dfrac{5}{12} f_{n-2}\right)$	3	Adams-Bashforth
	3	$y_{n+1} = y_n + h\left(\dfrac{55}{24} f_n - \dfrac{59}{24} f_{n-1} + \dfrac{37}{24} f_{n-2} - \dfrac{3}{8} f_{n-3}\right)$	4	Adams-Bashforth
Implicit	0	$y_{n+1} = y_n + hf_{n+1}$	1	Euler backward
	1	$y_{n+1} = y_n + h\left(\dfrac{1}{2} f_{n+1} + \dfrac{1}{2} f_n\right)$	2	Trapezoidal
	2	$y_{n+1} = y_n + h\left(\dfrac{5}{12} f_{n+1} + \dfrac{2}{3} f_n - \dfrac{1}{12} f_{n-1}\right)$	3	Adams-Moulton
	3	$y_{n+1} = y_n + h\left(\dfrac{3}{8} f_{n+1} + \dfrac{19}{24} f_n - \dfrac{5}{24} f_{n-1} + \dfrac{1}{24} f_{n-2}\right)$	4	Adams-Moulton

Now that we are comfortable with the derivation of the methods, let us see application of these methods to an IVP.

In order to see the effect of various order of the methods, let us take the following example.

Example 6.1   Solve the following IVP using first to fourth order explicit and implicit methods using $h = 0.5$ and compare their accuracy by evaluating the true relative error

$$\frac{dy}{dt} = -0.6y, \quad y = 1 \text{ at } t = 0, \quad t \in 0,5$$

Solution   The true solution of the equation is $y = e^{-0.6t}$. We will now apply the methods listed in Table 6.3 to the equation. However, as soon as we try to do that, going from $t = 0$ to $t = 0.5$, we observe the following:

- Euler forward, Euler backward and trapezoidal methods only require y-values at $t = 0$, to obtain the new value of y at $t = 0.5$ (no problem here!)

- All other methods require y-values at $t = -0.5$ and/or -1.0 and/or -1.5. Since, these values are undefined, we cannot start the methods!

In general, multi-step methods are *non-self starting*. One obvious solution is to gradually go up using higher and higher-order methods. For example, use Euler forward for the first time step and second to fourth order Adams-Bashforth methods in the subsequent time steps. While this takes care of the problem of starting the higher-order methods, it increases the chance of error in the initial time steps by the use of lower-order methods. This initial error may also propagate in the future time steps (recall error propagation in Chapter 1). We will address this problem of starting of multi-step methods in Section 6.4. For the present problem, in order to make the errors of various methods comparable, we shall apply all of them independently and for equal number of time steps.

We shall evaluate the y-values at $t = 0.5$, 1.0, and 1.5 using the true solution. Then apply all the methods of Table 6.3 for rest of the time steps to reach $t = 5.0$. Since, the true solution is known, the true relative error can be calculated at each time step.

Let us first write the time stepping schemes for the explicit methods using the formulae listed in Table 6.3. We shall denote the kth order Adams-Bashforth method as ABk.

Euler forward

$$\text{(EF):} \quad y_{n+1} = y_n + h(-0.6 y_n)$$

or

$$y_{n+1} = y_n (1 - 0.6h)$$

$$\text{AB2:} \quad y_{n+1} = y_n + \frac{3}{2}h(-0.6 y_n) - \frac{1}{2}h(-0.6 y_{n-1})$$

or

$$y_{n+1} = y_n (1 - 0.9h) + 0.3h y_{n-1}$$

$$\text{AB3:} \quad y_{n+1} = y_n + \frac{23}{12}h(-0.6 y_n) - \frac{4}{3}h(-0.6 y_{n-1}) + \frac{5}{12}h(-0.6 y_{n-2})$$

$$= y_n (1 - 1.15h) + 0.8h y_{n-1} - 0.25h y_{n-2}$$

$$\text{AB4:} \quad y_{n+1} = y_n + \frac{55}{24}h(-0.6 y_n) - \frac{59}{24}h(-0.6 y_{n-1})$$

$$+ \frac{37}{24}h(-0.6 y_{n-2}) - \frac{3}{8}h(-0.6 y_{n-1})$$

$$= y_n (1 - 1.375h) + 1.475h y_{n-1} - 0.925h y_{n-2} + 0.225h y_{n-3}$$

The calculations are shown in Table 6.4

Table 6.4

t	True	EF	$\varepsilon\,(\%)$	AB2	$\varepsilon\,(\%)$	AB3	$\varepsilon\,(\%)$	AB4	$\varepsilon\,(\%)$
0	1.0000								
0.5	0.7408								
1	0.5488								
1.5	0.4066								
2	0.3012	0.2846	5.51	0.3059	1.57	0.2997	0.49	0.3017	0.16
2.5	0.2231	0.1992	10.72	0.2292	2.74	0.2214	0.77	0.2236	0.23
3	0.1653	0.1395	15.64	0.1720	4.04	0.1632	1.29	0.1661	0.47
3.5	0.1225	0.0976	20.28	0.1290	5.32	0.1204	1.65	0.1230	0.48
4	0.0907	0.0683	24.68	0.0967	6.63	0.0888	2.14	0.0914	0.79
4.5	0.0672	0.0478	28.83	0.0725	7.95	0.0655	2.52	0.0677	0.69
5	0.0498	0.0335	32.75	0.0544	9.29	0.0483	2.99	0.0504	1.14

We will now write the time stepping formulae for the implicit methods when applied to the problem. We shall denote the kth order Adams-Moulton method as AMk.

Euler backward (EB):

$$y_{n+1} = y_n + h(-0.6y_{n+1})$$

$$= \frac{y_n}{(1+0.6h)}$$

Trapezoidal (TR):

$$y_{n+1} = y_n + \frac{1}{2}h(-0.6y_n) + \frac{1}{2}h(-0.6y_{n+1})$$

$$= \frac{y_n(1-0.3h)}{(1+0.3h)}$$

AM3:

$$y_{n+1} = y_n + \frac{5}{12}h(-0.6y_{n+1}) + \frac{2}{3}h(-0.6y_n) - \frac{1}{12}h(-0.6y_{n-1})$$

$$= \frac{y_n(1-0.4h) + 0.05hy_{n-1}}{(1+0.25h)}$$

AM4:

$$y_{n+1} = y_n + \frac{3}{8}h(-0.6y_{n+1}) + \frac{19}{24}h(-0.6y_n) - \frac{5}{24}h(-0.6y_{n-1}) + \frac{1}{24}h(-0.6y_{n-2})$$
$$= \frac{y_n(1-0.475h) + 0.125hy_{n-1} - 0.025hy_{n-2}}{(1+0.225h)}$$

We tabulate the calculations in Table 6.5.

Table 6.5

t	True	EB	ε (%)	TR	ε (%)	AM3	ε (%)	AM4	ε (%)
0	1.0000								
0.5	0.7408								
1	0.5488								
1.5	0.4066								
2	0.3012	0.3127	3.84	0.3005	0.23	0.3013	0.04	0.3012	0.01
2.5	0.2231	0.2406	7.82	0.2221	0.46	0.2233	0.08	0.2231	0.02
3	0.1653	0.1851	11.95	0.1642	0.68	0.1655	0.11	0.1653	0.02
3.5	0.1225	0.1424	16.25	0.1213	0.91	0.1226	0.15	0.1224	0.03
4	0.0907	0.1095	20.70	0.0897	1.13	0.0909	0.19	0.0907	0.04
4.5	0.0672	0.0842	25.33	0.0663	1.36	0.0674	0.23	0.0672	0.05
5	0.0498	0.0648	30.14	0.0490	1.58	0.0499	0.26	0.0498	0.06

In the above example, let us make a few observations by comparing the methods:

- For the same order methods with same time step size, the implicit method gives smaller error compared to the explicit methods. This is easily seen by comparing the % error columns of EF with EB, AB2 with TR, AB3 with AM3, and AB4 with AM4.
- In the same type of method, i.e., implicit or explicit, higher the order of the method, lower is the error. This is seen by comparing the error columns of the same table. This is due to decrease in the truncation error which we shall discuss in detail at a later section.
- The error always grows with time irrespective of the order and type of the method. Therefore, if one is to calculate many time steps for a problem, it is better to use a method that gives very small error at the start so that the error in the final solution is within reasonable limit. For example, the error using AM4 grew from 0.01 to 0.06% while that for Euler backward grew from 3.84 to 30.14%. You can plot the error with time to find the rate of growth of the error and see how it varies across the methods.

We observe that Eqs (6.12) and (6.13) are actually special cases of the following general expression:

$$\sum_{i=0}^{m} \gamma_i y_{n+1-i} = h \sum_{i=0}^{k} \beta_i f_{n+1-i} \qquad (6.26)$$

In effect, all the single-step and multi-step explicit and implicit methods derived in this section are special cases of Eq. (6.26). We get Eq. (6.12) by setting $\gamma_i = 0$ for $i > 1$ and $\beta_0 = 0$ in Eq. (6.26). For Eq. (6.13), we only set $\gamma_i = 0$ for $i > 1$. Therefore, all the methods described above are essentially derived from the same concept. Equation (6.26) is the generalized form of all *linear multi-step* methods. In the next section, we derive another group of methods, which also is a special case of Eq. (6.26).

6.2.2 Backward Difference Formulae (BDF)

These methods are especially useful for their application to a specific type of problems known as *stiff equations*. The reader has to wait until Section (6.5) for a clear understanding of stiff problems and the application of BDF. We will derive the methods here for continuity of discussion in this section. The reader may find it easy to read this along with the previous sub-section because the concepts are same. These are a group of implicit methods. Generalized form of these methods are obtained from Eq. (6.26) by putting all $\beta_i = 0$ except β_0, which is set to unity. The objective is to determine the γ_i's for maximum accuracy. So, the general form becomes

$$\sum_{i=0}^{m} \gamma_i y_{n+1-i} = h f_{n+1} \qquad (6.27)$$

The derivation follows the same steps as the multi-step method. We illustrate the derivation for $m = 2$ using the tabular form. The equation for $m = 2$ is given by

$$\gamma_0 y_{n+1} + \gamma_1 y_n + \gamma_2 y_{n-1} = h f_{n+1} \qquad (6.28)$$

We tabulate the Taylor series expansions in Table 6.6.

Table 6.6   Derivation of a BDF

Terms	y_n	$h y_n'$	$h^2 y_n''$	$h^3 y_n''''$
y_{n+1}	γ_0	γ_0	$1/2\,\gamma_0$	$1/6\,\gamma_0$
y_n	γ_1	0	0	0
y_{n-1}	γ_2	$-\gamma_2$	$1/2\,\gamma_2$	$-1/6\,\gamma_2$
$h f_{n+1}$	0	-1	-1	$-1/2$

The equations can now be written as follows by summing the columns of Table 6.4:

$$\gamma_0 + \gamma_1 + \gamma_2 = 0$$
$$\gamma_0 - \gamma_2 = 1$$
$$\gamma_0 + \gamma_2 = 2$$

(6.29)

The solution of the above set of equation is $\gamma_0 = 3/2, \gamma_1 = -2$, and $\gamma_2 = 1/2$. The method is thus given by,

$$3y_{n+1} - 4y_n + y_{n-1} = 2hf_{n+1}$$

(6.30)

This is formally known as the *second order backward difference formula*. You may now compare the coefficients of this equation with the second order backward difference approximation of the first derivative in Chapter 5.

Using similar procedure, one can derive the implicit BDF of arbitrary order. In practical application however, up to sixth order BDF are useful because methods higher than sixth order are not *stiffly stable*. Full implication of this statement will be clear in Section 6.3.2.8. The BDF of up to sixth order are shown in Table 6.7.

Table 6.7 Backward difference formulae (BDF) up to order 6.

m	Method	Order
1	$y_{n+1} - y_n = hf_{n+1}$	1
2	$\frac{3}{2}y_{n+1} - 2y_n + \frac{1}{2}y_{n-1} = hf_{n+1}$	2
3	$\frac{11}{6}y_{n+1} - 3y_n + \frac{3}{2}y_{n-1} - \frac{1}{3}y_{n-2} = hf_{n+1}$	3
4	$\frac{25}{12}y_{n+1} - 4y_n + 3y_{n-1} - \frac{4}{3}y_{n-2} + \frac{1}{4}y_{n-3} = hf_{n+1}$	4
5	$\frac{137}{60}y_{n+1} - 5y_n + 5y_{n-1} - \frac{10}{3}y_{n-2} + \frac{5}{4}y_{n-3} - \frac{1}{5}y_{n-4} = hf_{n+1}$	5
6	$\frac{49}{20}y_{n+1} - 6y_n + \frac{15}{2}y_{n-1} - \frac{20}{3}y_{n-2} + \frac{15}{4}y_{n-3} - \frac{6}{5}y_{n-4} + \frac{1}{6}y_{n-5} = hf_{n+1}$	6

We will demonstrate the application of BDFs with Example 6.2.

Example 6.2   Solve the problem of Example 6.1 using BDFs of order 1 through 4 with a time step of $h = 0.5$. Evaluate the errors and compare with the multi-step methods.

Solution   In order to be able to compare the accuracies of the various methods, all methods need to be applied for the same number of time steps. Therefore, just like Example 6.1 we shall evaluate y at $t = 0.5$, 1, and 1.5 using the true solution. Thereafter, we shall apply the BDFs of up to order 4 for the rest of time steps up to $t = 5.0$. Notice, that the first order BDF is Euler backward. Applying the methods to the IVP, we write the time stepping schemes below:

BDF1 or EB:

$$y_{n+1} = y_n + h(-0.6y_{n+1})$$

$$y_{n+1} = \frac{y_n}{(1+0.6h)}$$

BDF2:

$$\frac{2}{3}y_{n+1} - 2y_n + \frac{1}{2}y_{n-1} = h(-0.6y_{n+1})$$

$$y_{n+1} = \frac{2y_n - \frac{1}{2}y_{n-1}}{\left(\frac{3}{2} + 0.6h\right)}$$

BDF3:

$$\frac{11}{6}y_{n+1} - 3y_n + \frac{3}{2}y_{n-1} - \frac{1}{3}y_{n-2} = h(-0.6y_{n+1})$$

$$y_{n+1} = \frac{3y_n - \frac{3}{2}y_{n-1} + \frac{1}{3}y_{n-2}}{\left(\frac{11}{6} + 0.6h\right)}$$

BDF4:

$$\frac{25}{12}y_{n+1} - 4y_n + 3y_{n-1} - \frac{4}{3}y_{n-2} + \frac{1}{4}y_{n-3} = h(-0.6y_{n+1})$$

$$y_{n+1} = \frac{4y_n - 3y_{n-1} + \frac{4}{3}y_{n-2} - \frac{1}{4}y_{n-3}}{\left(\frac{25}{12} + 0.6h\right)}$$

We tabulate the calculations in Table 6.8.

Table 6.8

t	True	BDF1	$\varepsilon\,(\%)$	BDF2	$\varepsilon\,(\%)$	BDF3	$\varepsilon\,(\%)$	BDF4	$\varepsilon\,(\%)$
0	1.0000								
0.5	0.7408								
1	0.5488								
1.5	0.4066								
2	0.3012	0.3127	3.84	0.2993	0.63	0.3016	0.14	0.3011	0.03
2.5	0.2231	0.2406	7.82	0.2196	1.58	0.2240	0.40	0.2229	0.11
3	0.1653	0.1851	11.95	0.1609	2.67	0.1665	0.72	0.1650	0.21
3.5	0.1225	0.1424	16.25	0.1178	3.84	0.1237	1.04	0.1221	0.30
4	0.0907	0.1095	20.70	0.0861	5.04	0.0919	1.35	0.0904	0.37
4.5	0.0672	0.0842	25.33	0.0630	6.25	0.0683	1.64	0.0669	0.44
5	0.0498	0.0648	30.14	0.0461	7.45	0.0507	1.92	0.0495	0.52

In general, we observed the following for the multi-step methods and BDFs:
- Higher-order methods are more accurate. For the same order of method applied to the same problem, errors of BDFs are in between the explicit and implicit multi-step methods with same time step.
- Multi-step *Adams-Bashforth*, *Adams-Moulton*, and *BDF*s were all *non-self starting*.
- Implicit methods may lead to non-linear algebraic equation at every time step if f is nonlinear.

There is a group of explicit methods known as *Runge-Kutta methods* that try to overcome both the difficulties mentioned in the last two points by evaluating the slope function at intermediate points within a time step. These can also be derived up to arbitrary order. We show the derivation of these methods in the next section.

6.2.3 Runge-Kutta Methods

Principle of Runge-Kutta methods is to evaluate the slope function at various intermediate points within a time step and finally use a weighted average of all the slope function evaluations to compute the y-value at the next time step. In general terms, the Runge-Kutta methods can be represented as follows:

$$y_{n+1} = y_n + h\sum_{i=0}^{p} \omega_i \phi_i \qquad (6.31)$$

where ω_i's are the weight functions and ϕ_i's are the function evaluations at the intermediate points. There are numerous ways to choose these intermediate points. Derivation involves choosing these intermediate points and the ω_i's with the goal of achieving the highest order of accuracy. In the most general form, the ϕ_i's are written as follows:

$$
\begin{aligned}
\phi_0 &= f(y_n, t_n) \\
\phi_1 &= f(y_n + h\alpha_{11}\phi_0, t_n + \beta_1 h) \\
\phi_2 &= f(y_n + h\alpha_{12}\phi_0 + h\alpha_{22}\phi_1, t_n + \beta_2 h) \\
\phi_3 &= f(y_n + h\alpha_{13}\phi_0 + h\alpha_{23}\phi_1 + h\alpha_{33}\phi_2, t_n + \beta_3 h) \\
&\vdots \\
\phi_i &= f\left(y_n + h\sum_{j=1}^{i}\alpha_{ji}\phi_{j-1}, t_n + \beta_i h\right)
\end{aligned}
\tag{6.32}
$$

At this point it will be worthwhile to compare Eq. (6.31) with the Euler forward method. You will observe that f_n in the Euler forward method have been replaced by $\sum_{i=1}^{p}\omega_i\phi_i$ where each ϕ_i's are function evaluation at different points. Locations of these points are all within one time step $[t_n, t_n + h]$, which leads to all $\beta_i \leq 1$.

Instead of one point evaluation in the Euler forward method, Runge-Kutta utilizes a weighted average of function evaluations at a number of intermediate points. Therefore, it follows that the sum of the weights (ω_i's) should be unity. Moreover, function evaluation at each ϕ_i depends on previously calculated ϕ values (ϕ_0 to ϕ_{i-1}) and can be calculated explicitly. The result of these judiciously chosen intermediate function evaluation is higher order accuracy like multi-step explicit methods but without the start-up problem.

We will demonstrate the principle of derivation of Runge-Kutta methods by taking the example of the second-order method.

Using the general form (6.31 and 6.32), the second order Runge-Kutta method may be written as:

$$
\begin{aligned}
y_{n+1} &= y_n + \omega_0 h\phi_0 + \omega_1 h\phi_1 \\
\phi_0 &= f(y_n, t_n) \\
\phi_1 &= f(y_n + \alpha_1 h\phi_0, t_n + \beta_1 h)
\end{aligned}
\tag{6.33}
$$

The objective is to estimate ω_0, ω_1, α_1, and β_1, in order to achieve maximum accuracy. Once again we will use the representation $f_n = f(y_n, t_n)$ and write ϕ_0 and ϕ_1 as follows:

$$\phi_0 = f_n$$

$$\phi_1 = f_n + \left(\alpha_1 h f_n \frac{\partial f}{\partial y}\Big|_n + \beta_1 h \frac{\partial f}{\partial t}\Big|_n \right) + \left(\alpha_1^2 h^2 f_n^2 \frac{\partial^2 f}{\partial y^2}\Big|_n \right.$$

$$\left. + 2\alpha_1\beta_1 h^2 f_n \frac{\partial f}{\partial y}\Big|_n \frac{\partial f}{\partial t}\Big|_n + \beta_1^2 h^2 \frac{\partial^2 f}{\partial t^2}\Big|_n + o(h^3) \right) \qquad (6.34)$$

Using the expansions of Eq. (6.34) in Eq. (6.33), we obtain the approximation of y_{n+1} according to second order *Runge-Kutta* method as

$$y_{n+1} = y_n + (\omega_0 + \omega_1)h f_n + \omega_1 h^2 \left(\alpha_1 f_n \frac{\partial f}{\partial y}\Big|_n + \beta_1 \frac{\partial f}{\partial t}\Big|_n \right) + \omega_1 h^3 \left(\alpha_1 f_n^2 \frac{\partial^2 f}{\partial y^2}\Big|_n \right.$$

$$\left. + 2\alpha_1\beta_1 f_n \frac{\partial f}{\partial y}\Big|_n \frac{\partial f}{\partial t}\Big|_n + \beta_1^2 \frac{\partial^2 f}{\partial t^2}\Big|_n \right) + o(h^4)$$

$$(6.35)$$

True value of y_{n+1} is given by the expansion in Eq. (6.16). Notice that, we could no longer replace the function f in Eq. (6.35) by y' as was done in Eq. (6.15). This is because of different increments given to y_n and t_n to obtain the intermediate points for function f. However, we need to use some relations to be able to relate the derivatives of y_n with f_n. This is accomplished by the use of total derivative of function f as follows:

$$\frac{d^2 y}{dt^2} = \frac{df}{dt} = \frac{\partial f}{\partial y}\frac{dy}{dt} + \frac{\partial f}{\partial t} = \left(f\frac{\partial}{\partial y} + \frac{\partial}{\partial t} \right) f \qquad (6.36)$$

Thus, using Eq. (6.36) in the Taylor series expansion shown in Eq. (6.16), we can write

$$y_{n+1} = y_n + h f_n + \frac{h^2}{2!} \left(f_n \frac{\partial f}{\partial y}\Big|_n + \frac{\partial f}{\partial t}\Big|_n \right) + \frac{h^3}{3!}\frac{d^3 y}{dt^3}\Big|_n + o(h^4) \qquad (6.37)$$

Equation (6.37) is the true expansion of y_{n+1} whereas Eq. (6.35) is the approximation given by second order *Runge-Kutta* method. We can now determine the unknown constants ω_0, ω_1, α_1, and β_1 by equating the coefficients of the similar terms in the two equations.

However, we realize immediately that we can get three equations for four unknowns by matching the terms up to $o(h^2)$. This naturally leads to multiple solutions. At the same time, if we want to simulate up to $o(h^3)$, we will land up with six equations with four unknowns which will have no solution. Thus, we settle for the former and have more than one form of second order *Runge-Kutta* method. The equations are:

$$\omega_0 + \omega_1 = 1$$

$$\omega_1 \alpha_1 = \frac{1}{2}$$

$$\omega_1 \beta_1 = \frac{1}{2}$$

(6.38)

Thus, we can find the solution in terms of one independent parameter:

$$\omega_1 = \frac{1}{2\alpha_1}, \quad \beta_1 = \alpha_1, \quad \text{and} \quad \omega_0 = 1 - \frac{1}{2\alpha_1}$$

(6.39)

By choosing various values of α_1, we can obtain different second order *Runge-Kutta* methods. Some of the commonly used forms are shown in Table 6.6. Following along the similar line, one can derive higher-order methods as well. Some examples of third and fourth order *Runge-Kutta* methods are also shown in Table 6.9. We leave it for the readers to go through the derivation or at least convince themselves that these are indeed consistent approximations of the original IVP. We show the application of these methods for the problem of Example 6.1.

Table 6.9   Commonly used Runge-Kutta methods up to order 4.

Second order *Runge-Kutta* method

$$\alpha_1 = \frac{1}{2} = \beta_1, \quad \omega_0 = 0, \quad \omega_1 = 1$$

$$y_{n+1} = y_n + h\phi_1$$

$$\phi_0 = f(y_n, t_n)$$

$$\phi_1 = f\left(y_n + \frac{1}{2}h\phi_0, t_n + \frac{1}{2}h\right)$$

Second order *Runge-Kutta* method, also known as *Heun's* method

$$\alpha_1 = 1 = \beta_1, \quad \omega_0 = \frac{1}{2}, \quad \omega_1 = \frac{1}{2}$$

$$y_{n+1} = y_n + h\left[\frac{1}{2}\phi_0 + \frac{1}{2}\phi_1\right]$$

$$\phi_0 = f(y_n, t_n)$$

$$\phi_1 = f(y_n + h\phi_0, t_n + h)$$

(contd.)

(contd.)

Second order *Runge-Kutta* method, also known as *Ralston's method*

$$\alpha_1 = \frac{3}{4} = \beta_1, \quad \omega_0 = \frac{1}{3}, \quad \omega_1 = \frac{2}{3}$$

$$y_{n+1} = y_n + h\left[\frac{1}{3}\phi_0 + \frac{2}{3}\phi_1\right]$$

$$\phi_0 = f(y_n, t_n)$$

$$\phi_1 = f\left(y_n + \frac{3}{4}h\phi_0, t_n + \frac{3}{4}h\right)$$

Third order *Runge-Kutta* method

$$y_{n+1} = y_n + h\left[\frac{1}{6}\phi_0 + \frac{2}{3}\phi_1 + \frac{1}{6}\phi_2\right]$$

$$\phi_0 = f(y_n, t_n)$$

$$\phi_1 = f\left(y_n + \frac{1}{2}h\phi_0, t_n + \frac{1}{2}h\right)$$

$$\phi_2 = f(y_n - h\phi_0 + 2h\phi_1, t_n + h)$$

Fourth order *Runge-Kutta* method

$$y_{n+1} = y_n + h\left[\frac{1}{6}\phi_0 + \frac{1}{3}(\phi_1 + \phi_2) + \frac{1}{6}\phi_3\right]$$

$$\phi_0 = f(y_n, t_n)$$

$$\phi_1 = f\left(y_n + \frac{1}{2}h\phi_0, t_n + \frac{1}{2}h\right)$$

$$\phi_2 = f\left(y_n + \frac{1}{2}h\phi_1, t_n + \frac{1}{2}h\right)$$

$$\phi_3 = f(y_n + h\phi_2, t_n + h)$$

Example 6.3 Solve the problem of Example 6.1 using the Runge-Kutta methods of order 2-4 with a time step of $h = 0.5$. Evaluate the true relative errors at each time step and compare with the multi-step and BDF methods.

Solution Although it is possible to apply all order Runge-Kutta methods from the first time step, in order to compare the errors with Examples 6.1 and 6.2, we shall apply these methods for the same number of time steps. Therefore, similar to the previous examples, we will evaluate the true solution for the first three time steps and apply the Runge-Kutta method thereafter.

Second order Runge-Kutta method

We apply only one of the three (Heun's method) shown in Table 6.6. We leave the other two for the readers to work out. The method when applied to the problem gives the following time stepping scheme:

$$\phi_0 = -0.6 y_n, \phi_1 = -0.6(y_n + h\phi_0), y_{n+1} = y_n + \frac{1}{2}h(\phi_0 + \phi_1)$$

We tabulate the calculations in Table 6.10.

Table 6.10

t	y (True)	ϕ_0	ϕ_1	y	ε (%)
0	1.0000				
0.5	0.7408				
1	0.5488				
1.5	0.4066				
2	0.3012	−0.2439	−0.1708	0.3029	0.56
2.5	0.2231	−0.1817	−0.1272	0.2257	1.13
3	0.1653	−0.1354	−0.0948	0.1681	1.70
3.5	0.1225	−0.1009	−0.0706	0.1252	2.28
4	0.0907	−0.0751	−0.0526	0.0933	2.85
4.5	0.0672	−0.0560	−0.0392	0.0695	3.44
5	0.0498	−0.0417	−0.0292	0.0518	4.02

Iteration scheme and calculation for the third order Runge-Kutta method is shown in Table 6.11.

$$\phi_0 = -0.6 y_n, \quad \phi_1 = -0.6\left(y_n + \frac{1}{2}h\phi_0\right), \quad \phi_2 = -0.6(y_n - h\phi_0 + 2h\phi_1)$$

$$y_{n+1} = y_n + h\left(\frac{1}{6}\phi_0 + \frac{2}{3}\phi_1 + \frac{1}{6}\phi_2\right)$$

Table 6.11

t	y (True)	ϕ_0	ϕ_1	ϕ_2	y	ε (%)
0	1.0000					
0.5	0.7408					
1	0.5488					
1.5	0.4066					
2	0.3012	−0.2439	−0.21	−0.1927	0.3011	0.04
2.5	0.2231	−0.1806	−0.15	−0.1427	0.2229	0.09
3	0.1653	−0.1338	−0.11	−0.1057	0.1651	0.13
3.5	0.1225	−0.0991	−0.08	−0.0783	0.1222	0.17
4	0.0907	−0.0733	−0.06	−0.0579	0.0905	0.21
4.5	0.0672	−0.0543	−0.05	−0.0429	0.0670	0.26
5	0.0498	−0.0402	−0.03	−0.0318	0.0496	0.30

Fourth order Runge-Kutta Method:

$$\phi_0 = -0.6y_n, \quad \phi_1 = -0.6\left(y_n + \frac{1}{2}h\phi_0\right), \quad \phi_2 = -0.6\left(y_n + \frac{1}{2}h\phi_1\right), \quad \phi_3 = -0.6(y_n + h\phi_2)$$

$$y_{n+1} = y_n + h\left(\frac{1}{6}\phi_0 + \frac{1}{3}\phi_1 + \frac{1}{3}\phi_2 + \frac{1}{6}\phi_3\right)$$

We tabulate the calculations in Table 6.12.

Table 6.12

t	y (True)	ϕ_0	ϕ_1	ϕ_2	ϕ_3	y	ε (%)
0	1.0000						
0.5	0.7408						
1	0.5488						
1.5	0.4066						
2	0.3012	−0.2439	−0.2074	−0.2128	−0.1801	0.3012	0.00
2.5	0.2231	−0.1807	−0.1536	−0.1577	−0.1334	0.2231	0.01
3	0.1653	−0.1339	−0.1138	−0.1168	−0.0988	0.1653	0.01
3.5	0.1225	−0.0992	−0.0843	−0.0865	−0.0732	0.1225	0.01
4	0.0907	−0.0735	−0.0625	−0.0641	−0.0542	0.0907	0.01
4.5	0.0672	−0.0544	−0.0463	−0.0475	−0.0402	0.0672	0.02
5	0.0498	−0.0403	−0.0343	−0.0352	−0.0298	0.0498	0.02

Errors in the Runge-Kutta methods are clearly comparable to those of the implicit multi-step methods. They do not have the start-up problem either. If you started to wonder why one would need the other methods, wait until we analyze all the methods in various ways in the next section. We will establish different kinds of errors associated with the methods. This will not only lead to better understanding of the methods but will also help us to determine the suitability of a method for a given problem.

A natural question that may arise is whether Runge-Kutta methods of order higher than 4 are possible. The answer is yes but there are a few drawbacks. To understand this, let us take a look at Table 6.6. We have evaluated the slope function ϕ at 2 points for the second order method, at 3 points for the third order method and at 4 points for the fourth order method. That is to say, for an nth order Runge-Kutta method with $n = 2$ to 4, we need to evaluate the slope function at n points or stages. This cannot be done with the Runge-Kutta methods of higher-order and therefore, the methods become uneconomical.

It has been shown (Butcher, 1965) that to attain orders of $n = 5$ and 6, one needs to evaluate the slope function on at least 6 and 7 points, respectively. With 8 and 9 point evaluation of slope function, maximum order possible are 6 and 7, respectively. For even higher number of points $(m \geq 10)$, the maximum attainable order of accuracy is $\leq (m-2)$. This is the reason for the popularity of the Runge-Kutta methods of up to order 4. The higher orders are almost never used.

EXERCISE 6.2

1. Solve the differential equation $\dfrac{dy}{dx} = x^2 y - 2y$ with $y(0) = 1$ and $h = 0.1$, over the interval $x \in [0,1]$ using the following methods:
    (a) Euler forward
    (b) Euler backward
    (c) Trapezoidal
    (d) Second order Runge-Kutta
    (e) Solve analytically and compare (i) the solutions graphically with the analytical solution and, (ii) numerically the true relative errors of the above methods at every time step.

2. Solve the differential equation $\dfrac{dy}{dx} = 10 \sin \pi x$ with the initial condition $y(0) = 0$ for one complete period of the sine function. Use a uniform grid with 10 intervals in a period. Use the following explicit methods for solution and compare the solutions graphically and numerically with the true solution:
    (a) 4th order Runge-Kutta method
    (b) 4th order Adams-Bashforth. Use values from (a) for startup.

3. Solve the following equation using the explicit multi-step methods of order 1 to 4 in $x \in [2,5]$ with $h = 0.2$:

$$x\frac{dy}{dx} + y = x \text{ where } y = 1 \text{ at } x = 2$$

Use the values from first order method to start the second order, values from second order methods to start the third order and the values from third order method to start the fourth order. Solve the equation analytically. Now compare the solutions graphically with the analytical solution.

4. Consider a *LR*-circuit with a resistance (R) and an inductance (L). A time varying potential of $E(t)$ is applied to the circuit. Application of Kirchoff's law leads to the following differential equation for the current (i):

$$L\frac{di}{dt} + Ri = E(t)$$

Compute the current at every 0.2 hours for 3 hours due to application of a square voltage

$$E(t) = \begin{cases} 0, & t < 0 \\ 24, & 0 \leq t \leq 1 \\ 0, & t > 1 \end{cases}$$

Given are the values of inductance $L = 12$ H and the resistance $R = 18\ \Omega$. The initial condition is $i = 0$ at $t = 0$. Use fourth order Runge-Kutta method and fourth order Adams-Moulton, fourth order Adams-Bashforth and fourth order BDF methods for solution. Compare the solutions graphically with the analytical solution. (*Hint:* You may use progressively higher-order implicit method to start the fourth order multi-step methods).

5. Discretize the *Bernoulli* equation given by $\dfrac{dy}{dx} + \phi(x)y = \psi(x)y^n$ using fourth order Runge-Kutta scheme. Use it to solve the following equations with $h = 0.2$:

   (a) $\dfrac{dy}{dx} + \dfrac{y}{x} + x^2 y^2 = 0; \quad y(1) = 1; \quad x \in (1, 2)$

   (b) $\dfrac{dy}{dx} + y - \dfrac{1}{\sqrt{y}} = 0; \quad y(1) = 1; \quad x \in (1, 2)$

6. Solve the equations in Question 5, using analytical, Euler forward, Euler backward, and trapezoidal methods. Compare graphically, all four (including the fourth order Runge-Kutta approximate solutions) with the analytical solution.

7. Dissolved oxygen deficit (y) in a stream due to pollutants is given by the *Streeter-Phelps* equation as follows:

$$u\dfrac{dy}{dx} + k_a y = k_d L_0 e^{-\frac{k_d}{u}x}$$

   where the stream velocity $u = 10$ km per day, $k_d = 0.5$ per day, $k_a = 0.2$ per day, and $L_0 = 12$ mg/l. The initial condition $y(0) = 3$ mg/l. The x is measured along the length of the river. Using fourth order Runge-Kutta method, compute y for a stretch of 10 km by taking $h = 0.2$ km. Graphically compare the approximate solution with the analytical solution.

8. Consider the following equation:

$$\dfrac{dy}{dx} + \dfrac{y}{x} = \dfrac{\sin x}{x^2}; \quad y(1) = 2$$

   (a) Solve the equation using Euler forward, Euler backward, trapezoidal, and fourth order Runge-Kutta methods in $x \in (1, 5)$. Use $h = 0.2$.

(b) Analytical solution of the equation is given by

$$y = \frac{2}{x} + \frac{1}{x} \int_{1}^{x} \frac{\sin \phi}{\phi} d\phi, \quad \forall x \geq 1$$

Compute the analytical solution at each grid point using Simpson's 1/3rd rule. Plot the time series of error for each method.

9. Solve the differential equation $\dfrac{dy}{dt} + 10y = 99e^{-t}$ with the initial condition $y(0) = 2$ using, (a) Euler's forward (explicit) method, and (b) Euler backward (implicit) method, to obtain the values of y at $t = 1.0$. Use time steps of 0.01, 0.02, 0.025, and 0.05. Find the analytical solution and graphically compare the solutions of both the methods with varying time steps. Observe the behaviour of the errors with time step of the two methods.

10. Consider the form of IVP in Eq. (6.10). Now, use Simpson's 1/3rd rule and 3/8th rule (Table 5.4) for the integral on the right-hand side to obtain two different methods. Can you categorize these methods? (explicit vs. implicit, single-step vs. multi-step, etc.)

11. Table 6.9 shows one form of third order Runge-Kutta method. Are there other forms? Can you derive a third order general Runge-Kutta method with variable coefficients (like one floating variable we found in the second order Runge-Kutta method).

12. Derive a method for the general IVP Eq. (6.3) using an approximation of the form Eq. (6.26) with $m = 2$ and $k = 2$.

6.3 ERROR ANALYSIS OF METHODS FOR IVPS

In the preceding section, we have derived many different methods for solving an initial value problem. During the process of derivation, we often truncated an infinite series such as, Taylor's series at some finite number of terms. Regardless to say that it introduces *truncation error* (Chapter 1) in the approximation. The introduced error may manifest itself in different forms in the final solution. The error may grow with time or attenuate. For a periodic solution, it may impart errors in amplitude and/or phase. In this section, we shall establish some ways to analyze the methods for various kinds of errors.

6.3.1 Truncation Error

Concept of truncation error was first introduced in Chapter 1 as the error introduced by approximating an infinite series by a finite one. We will now formalize the concept in the context of numerical methods for ordinary differential equation. We will define various types of truncation error and formalize a process by which one can calculate the truncation error or establish the order of error for any given method.

We will analyse truncation error in two ways. In all the initial value problems, value of y at the initial point (y_0) is known exactly. Numerical methods calculate the value of y_{n+1} given the value of y_n. That is, y_1 is calculated from y_0 after a time step of h, y_2 is calculated from y_1, and so on.

We will first calculate the error incurred in one time step and this will be called the *local truncation error* (*LTE*). We will then calculate the error involved in the approximation of the original differential Eq. (6.11) when we approximate it using a truncated series. This will be called the *global truncation error* (*GTE*). We will now highlight these through examples.

Notice, that at any time step, true value of y_{n+1} is given by the Taylor's series expansion shown in Eq. (6.16). We rewrite it here:

$$y_{n+1} = y_n + h y'_n + \frac{h^2}{2!} y''_n + \frac{h^3}{3!} y'''_n + \frac{h^4}{4!} y''''_n + o(h^5) \tag{6.40}$$

Equations of the form (6.26) or (6.31) calculate an approximate value of y_{n+1}. Let us denote the approximate value as $\tilde{y}_{n+1}$. If we consider the third order Adams-Bashforth method, this approximate value is given by the following truncated series:

$$\tilde{y}_{n+1} = y_n + h\left(\frac{23}{12} f_n - \frac{4}{3} f_{n-1} + \frac{5}{12} f_{n-2} \right) \tag{6.41}$$

We have already seen in Eq. (6.19) that the right-hand side can be represented as an infinite series using the exact relation of the original ODE, $y'_n = f_n$ and expanding y'_{n-1} and y'_{n-2} in Taylor's series. Putting the values $\alpha_0 = 23/12$, $\alpha_1 = -4/3$, and $\alpha_2 = 5/12$ on the right-hand side of Eq. (6.19), Eq. (6.41) can be written as

$$\tilde{y}_{n+1} = y_n + h y'_n + \frac{h^2}{2} y''_n + \frac{h^3}{6} y''' - \frac{h^4}{3} y''''_n + o(h^5) \tag{6.42}$$

Recall from Chapter 1, the definition of error as the difference between the true value and the approximate value. Therefore, an expression of the *local truncation error* (*LTE*) can be obtained by deducting Eq. (6.42) from Eq. (6.40) as follows:

$$LTE = y_{n+1} - \tilde{y}_{n+1} = \frac{3}{8} h^4 y''''_n + o(h^5) \tag{6.43}$$

Notice that all the terms of order h^4 and higher survives as their coefficients do not match between the true and approximate series. One is often interested in the first non-zero term or the leading order term of the truncation error. If the expression contains h^p as the leading order term we would term the method to have a local truncation error of order p.

In this case, we say, *this Adams-Bashforth method has a fourth order local truncation error or the method has fourth order accuracy locally.* It is important to note that we could also obtain this leading order local truncation error term by summing the terms on the last column of Table 6.1. This shows that, if we derive the method using the tabular form, we can also obtain the local truncation error during the process of derivation.

Notice in Eq. (6.40) that the true y_{n+1} was expressed in terms of Taylor's series around the previous time step values (y_n, t_n). This was then compared with the approximate expression of Eq. (6.41) which involves calculation of approximate y_{n+1} based on the previous time step value. Therefore, the error expression involved error incurred in only one time step, from n to $(n + 1)$ in this case. The *GTE* on the other hand is the accumulated error in n (say) time steps. Both true and approximate values of y_n need to be expressed in terms of the initial conditions (y_0, t_0) and the difference will result in the *global truncation error* expression. Let us first illustrate this with the help of the Euler forward method. Let us rewrite Eq. (6.40), the Taylor series expansion of y_{n+1} in the neighbourhood of (y_n, t_n) up to the second order term with remainder:

$$y_{n+1} = y_n + hy_n' + \frac{h^2}{2!} y''(\xi) \quad \text{where } \xi \in (t_n, t_{n+1}) \tag{6.44}$$

Since, $y' = f$ and $\tilde{y}_{n+1} = y_n + hf_n$ for Euler forward, the above equation can also be written as:

$$y_{n+1} = y_n + hf_n + \frac{h^2}{2!} y''(\xi) = \tilde{y}_{n+1} + \frac{h^2}{2!} y''(\xi) \quad \text{where } \xi \in (t_n, t_{n+1}) \tag{6.45}$$

The above equation is equivalent to the *LTE* expression of Eq. (6.43) for the Euler forward method which gives the error incurred for one time application of the Euler forward method starting from time n to $(n + 1)$ as

$$LTE = \frac{h^2}{2!} y''(\xi)$$

This expression assumes that the value of y_n at t_n is known exactly. For an IVP, only the initial value (y_0, t_0) is known exactly and all subsequent values are computed using the approximate numerical scheme. So, similar errors are encountered at each step. Starting from the initial value, the first three applications of the Euler forward method is as follows:

$$y_1 = y_0 + hf_0 + \frac{h^2}{2!} y''(\xi_0) \quad \text{where} \quad \xi_0 \in (t_0, t_1)$$

$$y_2 = y_1 + hf_1 + \frac{h^2}{2!} y''(\xi_1) = y_0 + h(f_0 + f_1) + \frac{h^2}{2!} [y''(\xi_0) + y''(\xi_1)]$$

where $\xi_1 \in (t_1, t_2)$

$$y_3 = y_2 + hf_2 + \frac{h^2}{2!}y''(\xi_2) = y_0 + h(f_0 + f_1 + f_2) + \frac{h^2}{2!}[y''(\xi_0) + y''(\xi_1) + y''(\xi_2)]$$

where $\xi_2 \in (t_2, t_3)$.

Therefore, repeated application of the Euler forward method for $(n + 1)$ time steps leads to the following expression:

$$y_{n+1} = \left[y_0 + h\sum_{i=0}^{n} f_i \right] + \frac{h^2}{2!}\sum_{i=0}^{n} y''(\xi_i) \quad \text{where } \xi_i \in (t_i, t_{i+1}) \tag{6.46}$$

Equation (6.46) is an exact representation of the value y_{n+1}. The term in the square bracket in Eq. (6.46) represents the approximate value of y_{n+1} by applying Euler forward method $(n + 1)$ times, i.e., $\tilde{y}_{n+1} = y_0 + h\sum_{i=0}^{n} f_i$. Therefore, the above equation can be represented as

$$y_{n+1} = \tilde{y}_{n+1} + \frac{h^2}{2!}\sum_{i=0}^{n} y''(\xi_i) \quad \text{where } \xi_i \in (t_i, t_{i+1}) \tag{6.47}$$

Therefore, the *GTE* of the Euler forward method is given by

$$GTE = \frac{h^2}{2!}\sum_{i=0}^{n} y''(\xi_i) \quad \text{where } \xi_i \in (t_i, t_{i+1}) \tag{6.48}$$

At this point, the reader may want to stop and compare the expression of *LTE* in Eq. (6.45) with that of *GTE* in Eq. (6.48). Repeated application of the Euler forward method for $(n + 1)$ times starting from a known initial point translated into summation of the residual terms for each step. For most of our practical problems of interest, the solution y is expected to be sufficiently smooth (i.e., higher derivatives exist!) and bounded. In that case, application of the second mean-value theorem for integrals yield the following result:

$$\sum_{i=0}^{n} y''(\xi_i) = (n+1)y''(\bar{\xi}) \quad \text{where } \bar{\xi} \in (t_0, t_{n+1})$$

However, for uniform time steps, $(n + 1) = (t_{n+1} - t_0)/h$. Therefore, the global truncation error is

$$GTE = \frac{h}{2!}(t_{n+1} - t_0)y''(\bar{\xi}) \quad \text{where } \bar{\xi} \in (t_0, t_{n+1}) \tag{6.49}$$

Therefore, the global truncation error of the Euler forward method is first order (of the order h) compared to the local truncation error which is second order (of the order h^2). In general, the *GTE* is always one order less than the *LTE*. This happens because of the integration (or summation) of errors which translates into one h term in the denominator. Similar analysis can also be done for a general multi-step method. Let us consider k-step multi-step methods of Eqs (6.12) and (6.13) as follows:

$$\tilde{y}_{n+1} = y_n + h \sum_{i=0 \text{ or } -1}^{k} \alpha_i f_{n-i} \tag{6.50}$$

The index i starts from 0 for explicit methods and starts from -1 for implicit methods. The *LTE* expression for such a method can be written similar to Eq. (6.43) as:

$$y_{n+1} = \tilde{y}_{n+1} + \frac{h^p}{p!} y^{(p)}(\xi) \quad \text{where } \xi \in (t_n, t_{n+1}) \tag{6.51}$$

where $p = (k + 2)$ for explicit and $(k + 3)$ for implicit methods, $y^{(p)}$ is the pth derivative and $\tilde{y}_{n+1}$ is given by Eq. (6.50).

We have seen that a multi-step method cannot start with only initial value known. For the method of Eq. (6.50), let us assume that the exact values of y for the first k time steps are known and the method is applied for further n time steps. This leads to the following approximation:

$$\tilde{y}_{n+k} = y_k + \sum_{j=0 \text{ or } -1}^{k} \alpha_j \sum_{i=k-j}^{(n+k-1)-j} f_i$$

$$y_{n+k} = \left[y_k + \sum_{j=0 \text{ or } -1}^{k} \alpha_j \sum_{i=k-j}^{(n+k-1)-j} f_i \right] + \sum_{i=k}^{n+k-1} \frac{h^p}{p!} y^{(p)}(\xi_i) = \tilde{y}_{n+k} + \sum_{i=k}^{n+k-1} \frac{h^p}{p!} y^{(p)}(\xi_i) \tag{6.52}$$

where $\xi_i \in (t_i, t_{i+1})$

Once again applying the second mean-value theorem for integrals for sufficiently smooth and bounded y, we may write,

$$\sum_{i=k}^{n+k-1} y^{(p)}(\xi_i) = n y^{(p)}(\bar{\xi}) = \frac{t_{n+k} - t_k}{h} y^{(p)}(\bar{\xi}) \quad \text{where } \bar{\xi} \in (t_k, t_{n+k})$$

Applying this relation, the global truncation error of the method becomes

$$GTE = \frac{h^{p-1}}{p!} (t_{n+k} - t_k) y^{(p)}(\bar{\xi}) \quad \text{where } \bar{\xi} \in (t_k, t_{n+k})$$

The k-step method with a *LTE* of order p has the global truncation error of order $(p - 1)$.

Recall that we called Eq. (6.48) as third order Adams-Bashforth method in

Table 6.3 although the method has fourth order *LTE*. All the methods are named by the order of the leading order term in the *GTE*.

It is important to note that the local truncation error is one order higher than the global truncation error. This is generally the case for a consistent scheme for IVP. This can also be understood from the fact that in order to obtain global truncation error, *LTE* is integrated over many intervals of length 'h' resulting in a division by h. By the same logic, local truncation error for the second order Runge-Kutta method comes out to be o(h^3). We leave it to the readers to show that the global truncation error is indeed o(h^2) and thus the name *second order*.

Since the truncation error is proportional to h^p, one can anticipate the error to decrease with h. This way one can get arbitrarily close to the true solution by decreasing h. Once we fix an error target of ε, we can always choose an h that satisfies the criterion. Therefore, h becomes a function of ε and we write it as $h(\varepsilon)$. At this point, we are ready to formally define consistency.

Definition 6.1   A numerical method for an IVP is *consistent* if the *GTE* is such that for any $\varepsilon > 0$, there exists a $h(\varepsilon) > 0$ for which $\left|GTE\right| < \varepsilon$.

This also means that the global truncation error approaches zero and the approximate equation approaches the original equation in the limit $h \to 0$. In other words,

$$\lim_{h \to 0} GTE = 0$$

If the leading order term of a numerical method is o(h^p), it is consistent if $p > 0$.

Let us now see the implication of truncation error on the accuracy of the method. We observed that the leading order term in the truncation error for all consistent numerical method derived so far could be represented as

$$GTE = Kh^p \tag{6.53}$$

where K is a constant. Since the time step h is small, higher-order terms will decrease more rapidly. Thus, the leading order term is likely to dominate the absolute magnitude of the truncation error. We can define the order of accuracy of a numerical method as follows:

Definition 6.2   A consistent numerical method is said to have an order of accuracy of p, if p is the largest positive integer such that, $\left|GTE\right| \leq Kh^p$ for $0 < h \leq H$ where K and H are constants. Therefore, one can write

$$\ln(GTE) = \ln(K) + p\ln(h) \tag{6.54}$$

Thus, for a constant h, we would expect a linear relationship between ln (GTE) and p with ln(h) as the slope. This means that the truncation error will decrease with increasing order of the methods at a rate of ln(h) or higher-order method has less error. In practice, however, the slope is often steeper than ln(*h*). Graphically, this can be shown for the methods we have discussed so far using the values calculated in Examples 6.1 to 6.3.

In computation, total error is a composite of the effect of truncation error and machine round-off error. We can only calculate the total error by comparing the numerical solution with the true solution. Let us plot the natural logarithm of errors at *t* = 5 for explicit, implicit, BDFs and Runge-Kutta methods with *p* (Fig. 6.4).

For all the methods, the variation is linear but the slope is larger than ln *h*. Error in implicit and Runge-Kutta method decreased more rapidly compared to explicit and BDF, with the increasing order of methods.

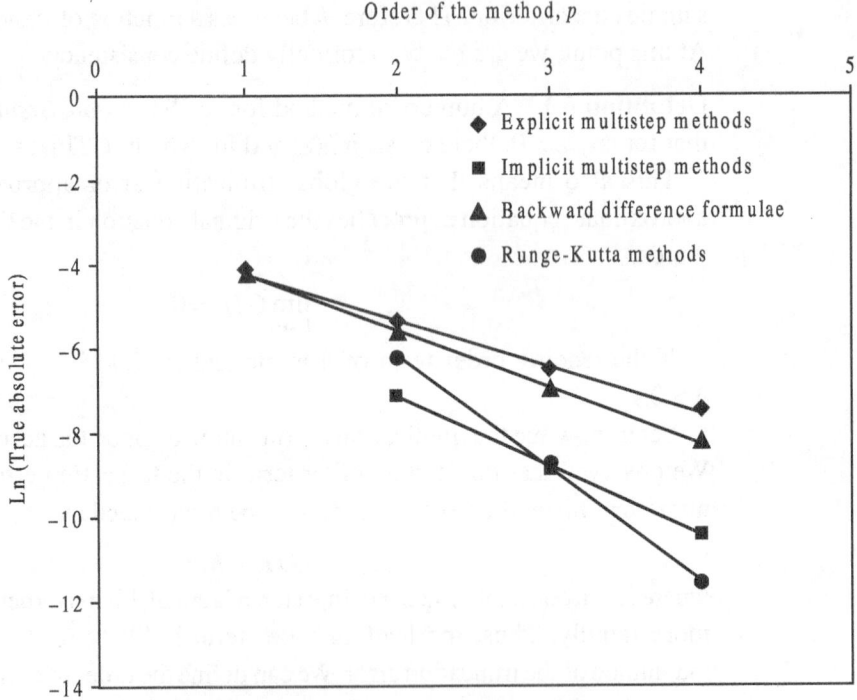

Fig. 6.4   Plot of natural logarithm of true absolute error at *t* = 5.0 in Examples 6.1–6.3 with the order of the methods (*p*). The time step *h* was constant at 0.5 but all the slopes are larger than ln(*h*).

For the same method (constant *p*), we will expect a linear relationship between ln(*GTE*) and ln(*h*) with the slope being *p*. This means that refining the grid will also decrease the truncation error at the rate of *p*. This provides a numerical way to estimate the order of a method. We demonstrate this by taking Example 6.1 with trapezoidal method.

Example 6.4   Numerically determine the order of the trapezoidal method by taking Example 6.1.

Solution   We solve the problem using trapezoidal method with time steps of 0.1, 0.25, 0.5, 0.75, 1.0, 1.5 and 2.0. Since, the chosen h values have the least common multiple as 6.0, we plot the natural logarithm of error at $t = 6.0$ with the natural logarithm of the time step h. The values are shown in Table 6.13.
True solution at $t = 6.0$ is $y = 1.0e^{-0.6 \times 6.0} = 0.027323722$.

Table 6.13

h	y at $t = 6.0$ Using trapezoidal method	True absolute error (ε)	$\ln (h)$	$\ln (\varepsilon)$
0.1	0.0273	2.95096E-05	−2.30258509	−10.4307943
0.25	0.0271	0.000184435	−1.38629436	−8.59821441
0.5	0.0266	0.000737721	−0.69314718	−7.21194488
0.75	0.0257	0.001659689	−0.28768207	−6.40112497
1	0.0244	0.002949648	0	−5.8260693
1.5	0.0207	0.006623322	0.405465108	−5.01715824
2	0.0156	0.011698722	0.693147181	−4.44827564

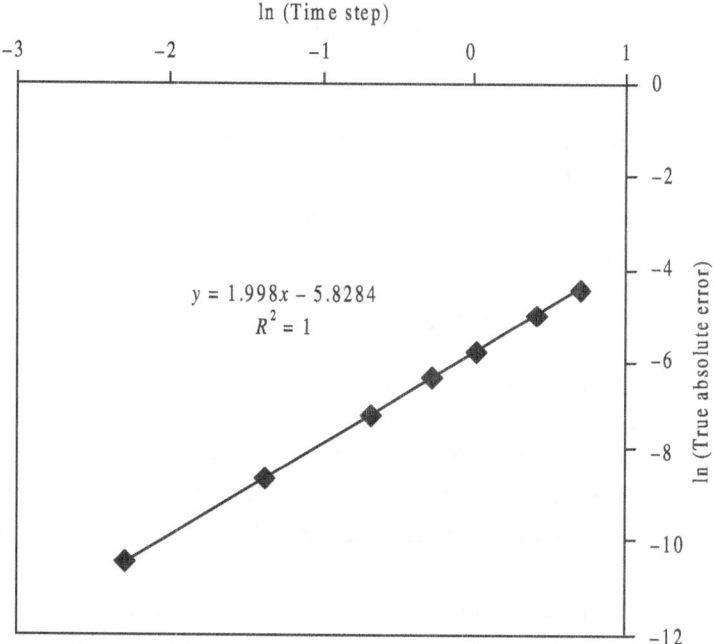

Fig. 6.5   Natural logarithm of the true absolute error at $t = 0.6$ plotted with the natural logarithm of the time step (h) for the trapezoidal method. Theoretical order of the method is 2 and the numerically obtained order is 1.998.

We now fit a straight line through the points in the plot of $\ln(h)$ vs. $\ln(\varepsilon)$ (Fig. 6.5). The slope of the fitted line is 1.998 and we know that the trapezoidal method is second order accurate globally. This is a convenient way of determining the order of the global truncation error of a method.

In this section, we were concerned with the accuracy of a numerical method with respect to the true solution. Sometimes, a *consistent* numerical method may produce oscillating or diverging result (e.g., goes to infinity) even though the true solution may be finite and monotonous.

In order to visualize this, let us consider the problem of Example 6.1. Let us solve it using Euler forward method with the time step sizes of $h = 0.25, 0.5, 1.0,$ 2.0, 3.0, and 4.0. The solutions are compared with true solution in Fig. (6.6a). The solution is well behaved up to $h = 1.0$. At $h = 2.0$, the solution oscillated for the first few time steps but gradually died down as the time progressed. The magnitude of oscillations are much larger at $h = 3.0$ although it is decreasing as the time progressed. At $h = 4.0$, the magnitude of oscillations are increasing with time.

Compare this observation with the solution using Euler backward method with the same time step sizes (Fig. 6.6b). No oscillation was observed for any of the time steps. Only the error increased with the size of the time step which is expected from the truncation error analysis.

At this point, recall that both Euler forward and Euler backward are first order method. But there is nothing in the truncation error analysis that indicated the oscillation observed in the Euler forward method. This behaviour is due to the stability of a numerical method and often a more important aspect than the truncation error. We introduce it in the next section.

6.3.2 Stability Analysis

In this section, we will attempt to understand why some numerical methods produce oscillations while some others don't for the same IVP using same time steps. Notice that in Fig. 6.6(a), the oscillations keep on increasing with time at $h = 4.0$. Eventually, it may (will) produce unbounded solution in a few time steps although the true solution is bounded. This leads to the concept of stability.

The true solution or the analytical solution is stable when it does not grow unbounded. For such a problem, if a numerical method produces solution that is *bounded* for all possible choices of time steps, the method is *unconditionally stable*. If the numerical method produces *unbounded* solution for all possible choices of time steps, the method is *unconditionally unstable*.

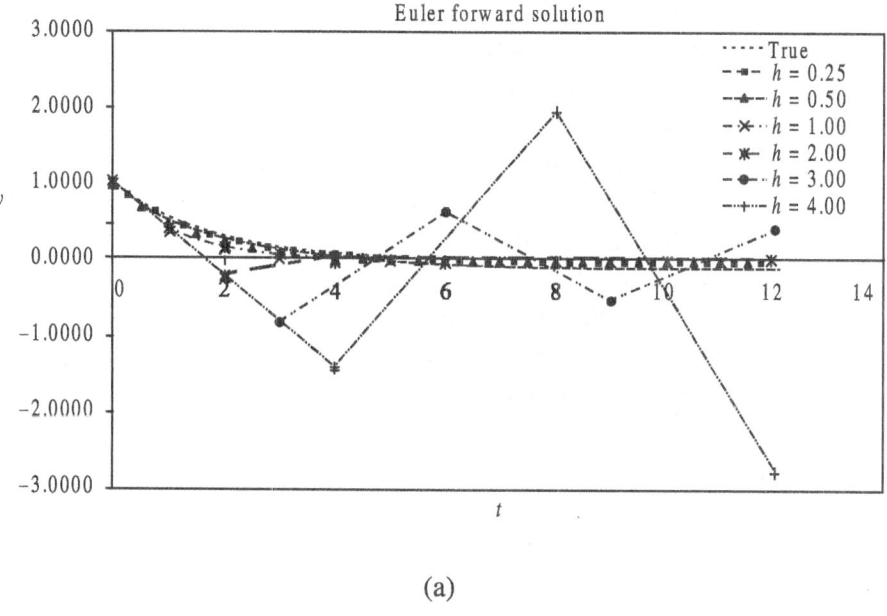

(a)

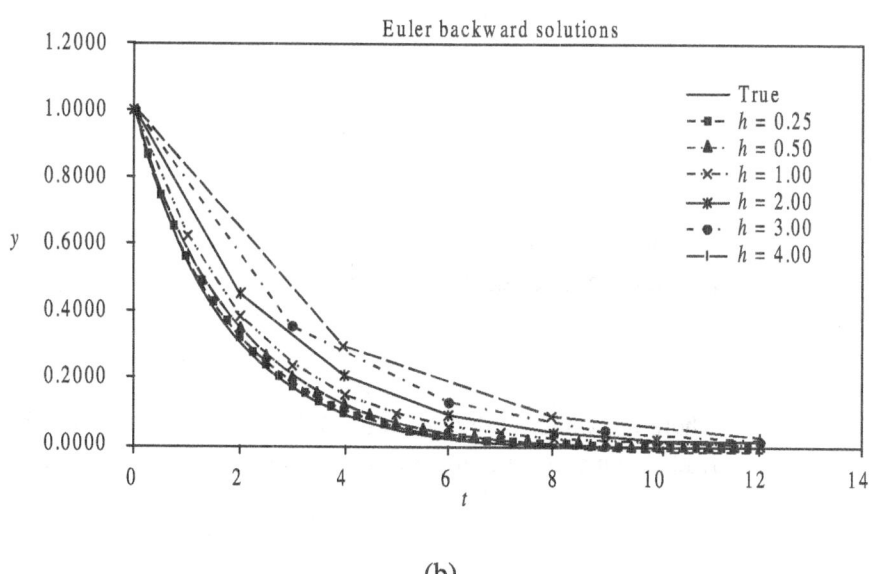

(b)

Fig. 6.6 Solution of Example Problem (6.1) with different time steps using (a) Euler forward and (b) Euler backward methods. The time step sizes above 2.0 shows oscillations in the solutions using Euler forward method.

However, majority of the numerical methods derived in Section 6.2 are *conditionally stable* which produces bounded solution only for certain time step values. At this juncture, we will proceed with these working definitions of stability. In the last subsection (6.3.2.8), we will formalize the definition of stability and absolute stability.

A given numerical method can be stable for one IVP but unstable for another with the same time step. The conditions of stability may be different for two different IVPs using the same numerical method. Therefore, we need a model problem, which we can use to compare various numerical methods. Let us introduce the origin of the model problem for the general IVP (Eq. 6.11). Since the solution starts from the fixed initial point (y_0, t_0), let us look at the behaviour of the equation in the neighbourhood of (y_0, t_0) by expanding the function $f(y, t)$ using Taylor's series,

$$
\begin{aligned}
\frac{dy}{dt} = f(y_0, t_0) + (y - y_0) \frac{\partial f}{\partial y}\bigg|_{(y_0, t_0)} + (t - t_0) \frac{\partial f}{\partial t}\bigg|_{(y_0, t_0)} \\
+ \frac{(y - y_0)^2}{2} \frac{\partial^2 f}{\partial y^2}\bigg|_{(y_0, t_0)} + (y - y_0)(t - t_0) \frac{\partial^2 f}{\partial y \partial t}\bigg|_{(y_0, t_0)} \\
+ \frac{(t - t_0)^2}{2} \frac{\partial^2 f}{\partial t^2}\bigg|_{(y_0, t_0)} + \cdots
\end{aligned}
\tag{6.55}
$$

If we only collect the linear terms and ignore the quadratic and higher-order terms, the above equation may be written as follows:

$$
\frac{dy}{dt} = \alpha + \beta t + \lambda y + \cdots
\tag{6.56}
$$

where α, β, and λ are constants. Out of the three terms on the right, the last term leads to an exponential solution. The above equation can be generalized as

$$
\frac{dy}{dt} = \lambda y + \gamma(t)
$$

where $\gamma(t)$ is any function of t. For any approximate solution ($\tilde{y}$) of the above equation computed using a numerical method, we can write

$$
\frac{d\tilde{y}}{dt} = \lambda \tilde{y} + \gamma(t)
$$

Since the error (ε) is defined as $\varepsilon = y - \tilde{y}$, by deducting the approximate equation from the true equation, we obtain

$$\frac{d\varepsilon}{dt} = \lambda\varepsilon$$

Therefore, the error follows the equation without the function $\gamma(t)$. The function $\gamma(t)$ has no effect on the error because it does not contain the dependent variable and can be evaluated exactly at every time step. Since, the major effect on the growth and decay of the error is provided by the λy term, the model problem typically used for the analysis of numerical methods is

$$\frac{dy}{dt} = \lambda y \tag{6.57}$$

In order to obtain the model equation, the full expression of Eq. (6.55) was linearized. Stability analysis performed using this model equation belongs to the group of *linear stability analysis*. In this book, we will restrict ourselves to *linear stability analysis* using model Eq. (6.57).

To generalize the applicability of stability analysis to wide array of problems, we will allow λ to assume complex values, i.e.

$$\lambda = \lambda_r + i\lambda_i \tag{6.58}$$

The analytical solution with the initial condition $y = y_0$ at $t = 0$ is

$$y = y_0 e^{\lambda t} \tag{6.59}$$

It is a common practice to plot the stability regions in the $\lambda_r h - \lambda_i h$ plane although the time step h has no meaning in the context of analytical solution. One looks at the analytical solution Eq. (6.59) on a discrete grid as,

$$y_n = y_0 e^{\lambda h \left(\frac{t_n}{h}\right)} = y_0 e^{\lambda h n} \tag{6.60}$$

Since, we will deal with a constant time step of h, it does not alter anything in terms of the analytical solution but helps in future comparison of stability regions of the numerical methods.

It is easy to see that y will grow unbounded for all positive values of real part of λ ($\lambda_r > 0$) irrespective of whether λ_i is positive or negative. Therefore, the analytical solution of the model problem is bounded for all $\lambda_r < 0$. This is true for majority of the engineering problems. This region of bounded analytical solution is shown as the shaded region (grey) in Fig. 6.7 (complete left half plane). For $\lambda_r > 0$, the analytical solution grows unbounded.

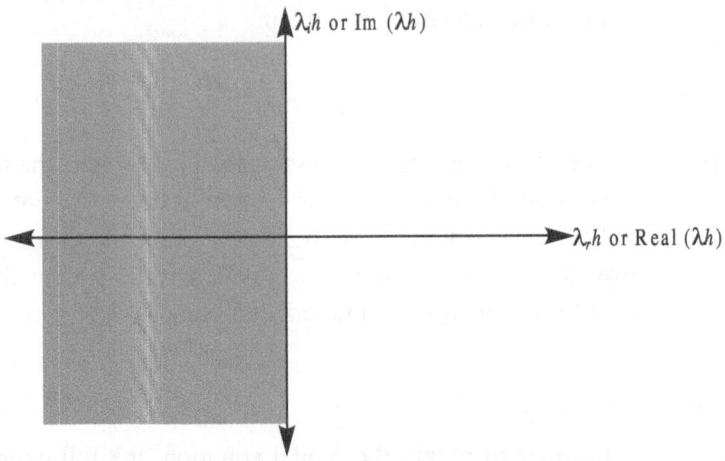

Fig. 6.7 Region of bounded analytical solution for the model problem.

We will compute the stability regions of the numerical methods and compare them with the model problem. If the region includes the complete shaded region of Fig. 6.7 for a method, we will call that method as unconditionally stable. For other methods with conditional stability, we will compute the condition or the ranges of λ_r and λ_i for which the methods are stable. Although, the analytical solution of the model problem grows unbounded for $\lambda_r > 0$, some numerical methods can still produce stable solution without oscillations and match the true solution reasonably well. Therefore, while formulating the stability regions of the numerical methods, we will not restrict ourselves to $\lambda_r \leq 0$.

We will now present the detailed stability analysis of three methods, Euler forward, Euler backward, and trapezoidal. This will be followed by a presentation of a general framework that can be used to obtain stability regions of other methods derived in Section 6.2. We will give one example formulation from each group of methods. At the end, we will formalize the definition of stability and characterize types of stability.

6.3.2.1 *Euler Forward*

The Euler forward method is given in Table 6.3 as follows:

$$y_{n+1} = y_n + hf_n \tag{6.61}$$

Applying this method to the model problem of Eq. (6.57), we obtain

$$y_{n+1} = y_n + \lambda h y_n = \left(1 + \lambda_r h + i\lambda_i h\right) y_n = \sigma y_n \tag{6.62}$$

where

$$\sigma = \frac{y_{n+1}}{y_n} = 1 + \lambda h = 1 + \lambda_r h + i\lambda_i h$$

is the amplification in the value of y in one time step, going from y_n to y_{n+1}. This σ is known as the *amplification factor*. By recursive application of Eq. (6.62), one can relate the value of y at any time step n with the initial condition y_0 as follows:

$$y_n = \sigma^n y_0 \qquad (6.63)$$

It is clear from the above equation that for any finite initial condition y_0, the solution will be bounded, if and only if, the absolute magnitude of the *amplification factor* is less than or equal to unity. That is to say,

$$|\sigma| = |(1 + \lambda_r h) + i\lambda_i h| \leq 1 \qquad (6.64)$$

The above equation can be written in the following form:

$$(1 + \lambda_r h)^2 + (\lambda_i h)^2 \leq 1 \qquad (6.65)$$

Another way to look at Eq. (6.64) is

$$\sigma = \Lambda e^{i\phi}$$

where

$$\Lambda = \sqrt{(1 + \lambda_r h)^2 + (\lambda_i h)^2} \text{ and } \phi = \tan^{-1}\left(\frac{\lambda_i h}{1 + \lambda_r h}\right) \qquad (6.66)$$

Since $|e^{i\phi}| = 1$, the condition $|\sigma| \leq 1$ essentially means $\Lambda \leq 1$ which leads to Eq. (6.65). This equation gives the condition imposed on the time step h for the stability of the Euler forward method and provides the stability region. When plotted on the $\lambda_r h$–$\lambda_i h$ plane, this is the interior of a unit circle centered at $(-1, 0)$. The stability region is shown in Fig. 6.8.

When λ is real and negative (necessary for bounded analytical solution), i.e., $\lambda_i = 0$ and $\lambda_r = -\lambda$, we obtain the range of time step as

$$0 < h < \frac{2}{|\lambda|}$$

for which the method will be stable. At this point, one can relate to Example 6.4. Since $\lambda = 0.6$, the maximum limit of h is 3.33. Instability was observed for $h = 4.0$ but not for 3.0 [Fig. 6.6(a, b)].

For purely imaginary λ, i.e., $\lambda_r = 0$, there is no value of the time step h other than zero that will satisfy Eq. (6.65). This can be seen visually from Fig. 6.8 as the stability region is tangent to the imaginary axis. Therefore, the method is unconditionally unstable for purely imaginary λ.

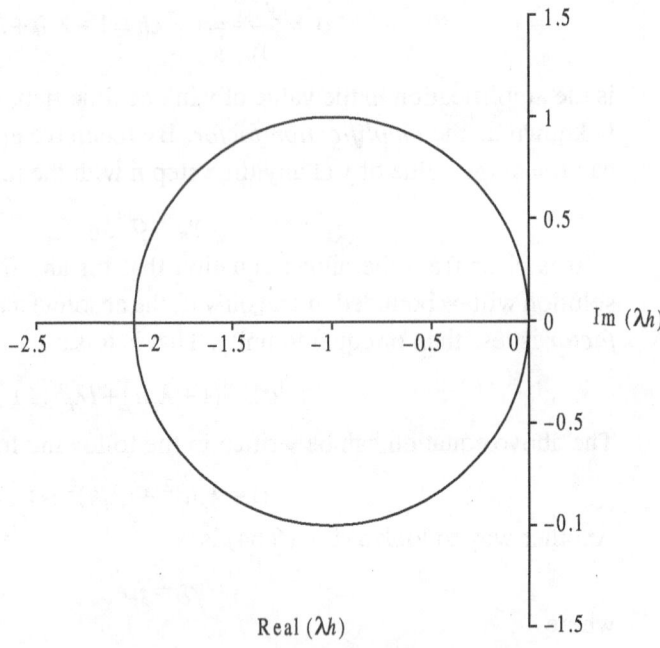

Fig. 6.8   Stability region of Euler forward method. The method is stable inside the circle.

6.3.2.2 *Euler Backward Method*

Application of the Euler backward method on the model problem leads to the following discrete equation:

$$y_{n+1} = y_n + hf_{n+1} = y_n + h\lambda y_{n+1} \tag{6.67}$$

This leads to the following amplification factor:

$$\frac{y_{n+1}}{y_n} = \sigma = \frac{1}{1-\lambda h} = \frac{1}{1-\lambda_r h - i\lambda_i h} = \frac{1}{\Lambda e^{i\phi}} \tag{6.68}$$

where

$$\Lambda = \sqrt{(1-\lambda_r h)^2 + (-\lambda_i h)^2} \quad \text{and} \quad \phi = \tan^{-1}\left(\frac{-\lambda_i h}{1-\lambda_r h}\right)$$

In this case, the stability condition $|\sigma| < 1$ essentially means $\dfrac{1}{\Lambda} < 1$ since, $\left|e^{-i\phi}\right| = 1$.

The stability region is therefore given by the condition:

$$\sqrt{(1-\lambda_r h)^2 + (-\lambda_i h)^2} > 1 \quad \text{or} \quad (\lambda_r h - 1)^2 + (\lambda_i h)^2 > 1 \tag{6.69}$$

The above relation is always true for $\lambda_r < 0$ which is required for the bounded analytical solution of the model problem. Therefore, the Euler backward method is unconditionally stable for the model problem. The stability region also includes some parts on the right side when $\lambda_r > 0$. Equation (6.69) suggests that the method is stable everywhere outside the unit circle whose center is at (1, 0). This is shown in Fig. 6.9. More importantly, the method is stable for all purely imaginary λ.

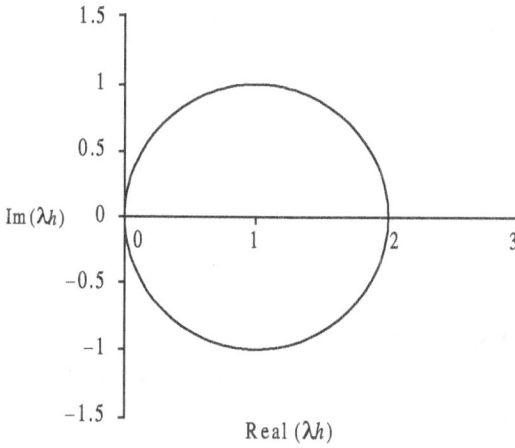

Fig. 6.9 Stability region of Euler backward method. The method is stable outside the circle.

6.3.2.3 *Trapezoidal Method*

Application of trapezoidal method to the model problem leads to the following expression for the amplification factor:

$$\frac{y_{n+1}}{y_n} = \sigma = \frac{1 + \dfrac{1}{2}\lambda h}{1 - \dfrac{1}{2}\lambda h} = \frac{1 + \dfrac{1}{2}\lambda_r h + i\dfrac{1}{2}\lambda_i h}{1 - \dfrac{1}{2}\lambda_r h - i\dfrac{1}{2}\lambda_i h} = \frac{\Lambda_1 e^{i\phi}}{\Lambda_2 e^{i\varphi}} \tag{6.70}$$

where $\Lambda_1 = \sqrt{\left(1 + \dfrac{1}{2}\lambda_r h\right)^2 + \left(\dfrac{1}{2}\lambda_i h\right)^2}$

$$\Lambda_2 = \sqrt{\left(1-\frac{1}{2}\lambda_r h\right)^2 + \left(-\frac{1}{2}\lambda_i h\right)^2}$$

$$\phi = \tan^{-1}\left(\frac{\frac{1}{2}\lambda_i h}{1+\frac{1}{2}\lambda_r h}\right)$$

and

$$\varphi = \tan^{-1}\left(\frac{-\frac{1}{2}\lambda_i h}{1-\frac{1}{2}\lambda_r h}\right)$$

Since, $\left|e^{i\phi}\right| = 1$ and $\left|e^{i\varphi}\right| = 1$, the stability condition $|\sigma| < 1$ leads to the following inequality:

$$\frac{\sqrt{\left(1+\frac{1}{2}\lambda_r h\right)^2 + \left(\frac{1}{2}\lambda_i h\right)^2}}{\sqrt{\left(1-\frac{1}{2}\lambda_r h\right)^2 + \left(-\frac{1}{2}\lambda_i h\right)^2}} < 1 \text{ or } \lambda_r h < 0 \qquad (6.71)$$

For negative λ_r $(\lambda_r < 0)$, the numerator is always less than the denominator except for the case of $\lambda_r = 0$ where it is unity.

For the higher-order methods, it is not easy to visualize the stability conditions so easily or derive explicit expressions as we did for Euler methods and trapezoidal method. We will describe a generalized framework that can be used for all multi-step methods and BDFs to generate the stability region.

6.3.2.4 Adams-Bashforth Methods

Let us derive the equation for stability region of the third order Adams-Bashforth method. Application of the third order Adams-Bashforth method to the model problem leads to the following approximation

$$y_{n+1} = y_n + h\left(\frac{23}{12}\lambda y_n - \frac{4}{3}\lambda y_{n-1} + \frac{5}{12}\lambda y_{n-2}\right) \qquad (6.72)$$

We have defined the amplification factor as $\sigma = y_{n+1}/y_n$ for all values of n. Recursive application of this relation leads to the following identities:

$$\frac{y_{n+1}}{y_{n-2}} = \sigma^3, \quad \frac{y_n}{y_{n-2}} = \sigma^2, \text{ and } \frac{y_{n-1}}{y_{n-2}} = \sigma \tag{6.73}$$

Using these relations, Eq. (6.72) becomes

$$\lambda h = \lambda_r h + i\lambda_i h = \frac{\sigma^3 - \sigma^2}{\left(\dfrac{23}{12}\sigma^2 - \dfrac{4}{3}\sigma + \dfrac{5}{12}\right)} \tag{6.74}$$

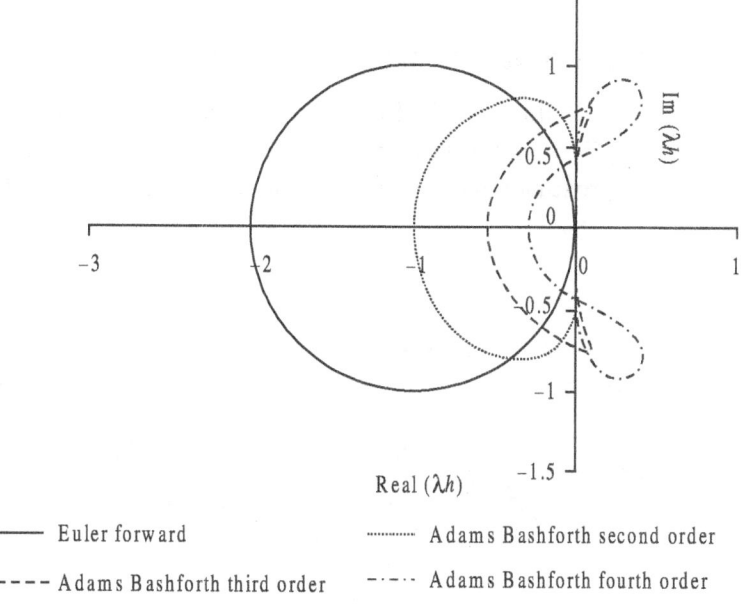

Fig. 6.10 Stability regions of the Adams-Bashforth methods in comparison with Euler forward. The methods are stable inside the enclosed regions.

The stability of the method requires σ to be less than unity. In a complex plane, the boundary of the region defined by $\sigma = 1$ is given by the identity $\sigma = e^{i\theta}$, $\theta \in (0,\ 2\pi)$. Equation (6.74) thus leads to the following stability region:

$$\lambda_r h + i\lambda_i h = \frac{(\cos 3\theta - \cos 2\theta) + i(\sin 3\theta - \sin 2\theta)}{\left(\dfrac{23}{12}\cos 2\theta - \dfrac{4}{3}\cos\theta + \dfrac{5}{12}\right) + i\left(\dfrac{23}{12}\sin 2\theta - \dfrac{4}{3}\sin\theta\right)} \tag{6.75}$$

Equating the real and imaginary parts, one can compute the values of $\lambda_r h$ and $\lambda_i h$. If we chose θ in $(0, 2\pi)$ and plot them, we get the stability region. We show the stability regions of Adams-Bashforth methods up to fourth order (Fig. 6.10). We have also included the stability region of the Euler forward method in the figure for comparison with the multi-step explicit methods.

We leave it to the readers to derive the equations and plot the stability region of the second and fourth order Adams-Bashforth methods following the procedure illustrated above and verify with the figure. It is important to note that the stability regions of the third and fourth order Adams-Bashforth methods include some part of the y-axis which means that they are conditionally stable for certain values of purely imaginary λh. These methods thus becomes usable for purely imaginary λ while Euler explicit method was unusable.

6.3.2.5 Adams-Moulton Methods

We will illustrate the formulation of the equation for stability region of multi-step implicit methods using third order Adams-Moulton method as example. Approximation of the model problem using this method is as follows:

$$y_{n+1} = y_n + h\left(\frac{5}{12}\lambda y_{n+1} + \frac{2}{3}\lambda y_n - \frac{1}{12}\lambda y_{n-1}\right) \tag{6.76}$$

Following a procedure similar to the previous section, we obtain the following identity:

$$\lambda_r h + i\lambda_i h = \frac{\sigma^2 - \sigma}{\left(\dfrac{5}{12}\sigma^2 + \dfrac{2}{3}\sigma - \dfrac{1}{12}\right)} = \frac{(\cos 2\theta - \cos\theta) + i(\sin 2\theta - \sin\theta)}{\left(\dfrac{5}{12}\cos 2\theta + \dfrac{2}{3}\cos\theta - \dfrac{1}{12}\right) + i\left(\dfrac{5}{12}\sin 2\theta + \dfrac{2}{3}\sin\theta\right)} \tag{6.77}$$

The stability regions for the third and fourth order Adams-Moulton methods are shown in Fig. 6.11. The reader should compare the stability regions of explicit and implicit methods of same orders. The stability region of an implicit method is much larger than the explicit method of the same order. This allows larger time step h for the same value of λ. For the multi-step methods (explicit and implicit), size of the stability region decreases with increasing order of the method. Notice that, for Euler forward, Adams Bashforth, and Adams Moulton methods, the stability region is inside the closed regions shown in Figs 6.8, 6.10, and 6.11. In contrast, the stability region of Euler backward method is outside the closed region shown in Fig. 6.9. The BDFs also show similar properties where the stability region is outside a closed region. These are described in the next subsection.

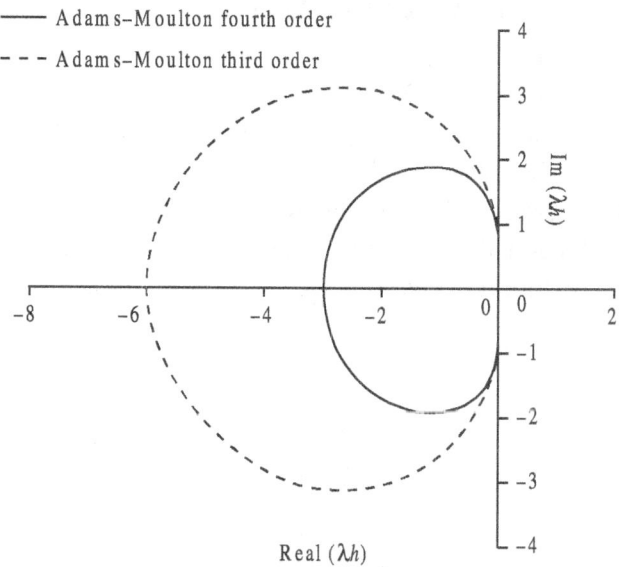

——— Adams-Moulton fourth order

– – – Adams-Moulton third order

Fig. 6.11 Stability regions of third and fourth order Adams-Moulton methods. The methods are stable within the enclosed region.

6.3.2.6 *Backward Difference Formulae*

Recall that the first order BDF is the Euler backward method and we have already seen the stability region for it. Let us illustrate the formulation of equations for stability regions for BDFs with the third order method (Table 6.5). The discrete approximation is given by

$$\frac{11}{6}y_{n+1} - 3y_n + \frac{3}{2}y_{n-1} - \frac{1}{3}y_{n-2} = hf_{n+1} \tag{6.78}$$

Application of this approximation to the model equation followed by recursive application of the relation $\sigma = y_{n+1}/y_n$ leads to,

$$
\begin{aligned}
\lambda_r h + i\lambda_i h &= \left(\frac{11}{6} - \frac{3}{\sigma} + \frac{3}{2\sigma^2} - \frac{1}{3\sigma^3} \right) \\
&= \left(\frac{11}{6} - 3\cos\theta + \frac{3}{2}\cos 2\theta - \frac{1}{3}\cos 3\theta \right) \\
&\quad - i\left(3\sin\theta - \frac{3}{2}\sin 2\theta + \frac{1}{3}\sin 3\theta \right)
\end{aligned}
\tag{6.79}
$$

Following on a similar line, the equations for the stability regions can be formulated for all the BDFs. The stability regions are shown in Fig. 6.12 for BDF up to order 6. In order to get a sense of the relative sizes, all six regions are shown in one plot but regions of orders 1-3 appear too small. In order to get a clearer picture, orders 1-3 are shown once again on a separate plot (Fig. 6.13). Just like the Euler backward method, the stability regions for all the BDFs are outside of the closed curve. Therefore, the stability regions of BDFs shown in Fig. 6.12 can be visualized as holes in the entire two-dimensional space. The BDF of orders 1 and 2 have no part of the curves on the left-hand side of the y-axis. This is not so

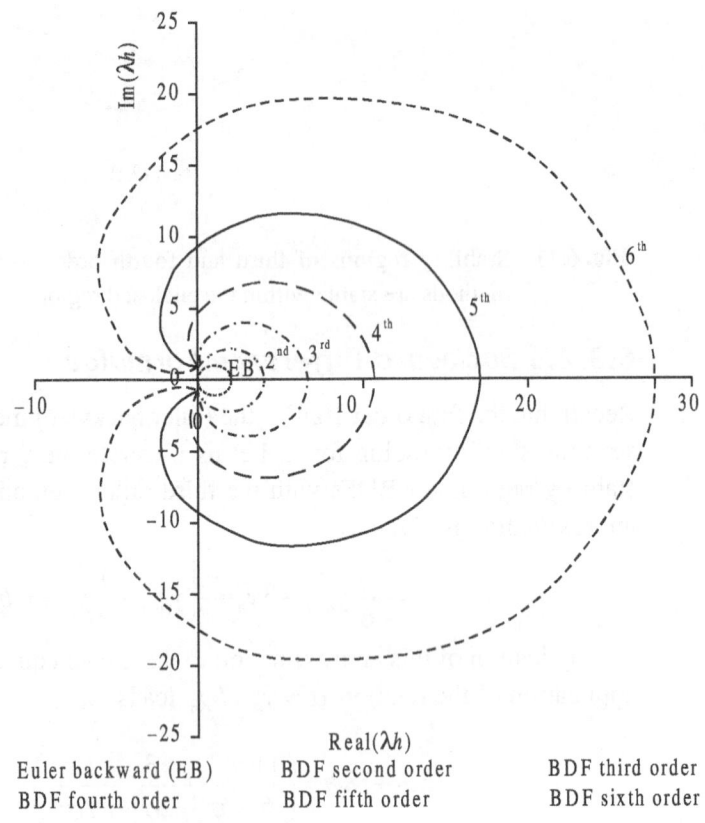

| Euler backward (EB) | BDF second order | BDF third order |
| BDF fourth order | BDF fifth order | BDF sixth order |

Fig. 6.12 Stability regions of backward difference formulae up to order six. All the regions are shown in one plot for comparison. The methods are stable outside the closed curves. These may also be seen as the regions of instability (or holes in the stability region). Size of the hole increases with the order.

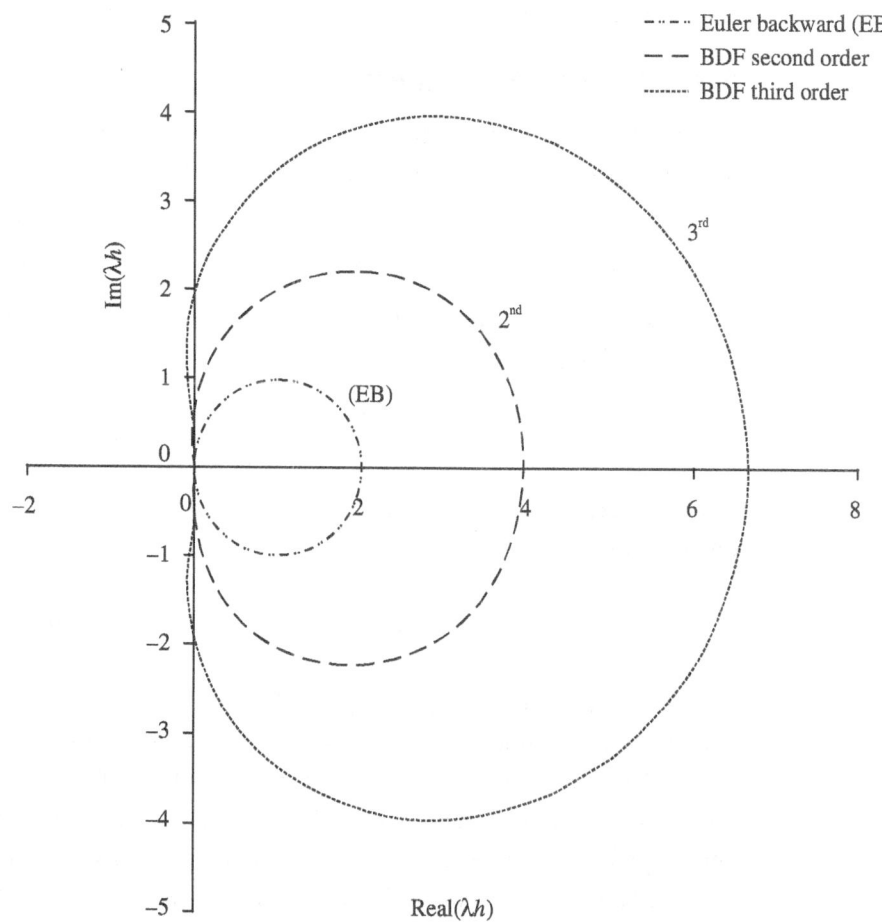

Fig. 6.13 Stability regions of BDF of orders 1–3. No part of the curve is on the left of the
y-axis for the first and second order methods. The third order method includes
some part of the region left of imaginary axis which expands (Fig. 6.12) as we
move up the order of BDF.

for BDF of orders 3-6 where some part of the curve crosses over but note that
purely real values along the axis on the left are excluded.

It is easy to see that out of all the methods described so far, the BDFs have the
best stability properties. Moreover, BDF of up to order 6 are unconditionally stable
for all purely real $\lambda < 0$. Therefore, any time step size becomes usable without

having to worry about the stability. This property makes them especially suitable for a special group of problems known as stiff equations. These will be discussed later in Section 6.5.

6.3.2.7 Runge-Kutta Methods

We will illustrate the formulation of the equation for stability region with a second order Runge-Kutta Method (Table 6.6). The method we choose is given below

$$y_{n+1} = y_n + \frac{1}{2}\phi_0 + \frac{1}{2}\phi_1, \quad \phi_0 = hf(y_n, t_n), \quad \phi_1 = hf(y_n + \phi_0, t_n + h) \qquad (6.80)$$

Application of this method to the model problem leads to the following approximation for y_{n+1}

$$y_{n+1} = y_n + \frac{1}{2}h\lambda y_n + \frac{1}{2}h\lambda(y_n + h\lambda y_n) \qquad (6.81)$$

Using the amplification factor and the stability limit for the amplification factor, one obtains,

$$1 + \lambda h + \frac{1}{2}(\lambda h)^2 = \frac{y_{n+1}}{y_n} = \sigma = e^{i\theta} = \cos\theta + i\sin\theta, \quad \theta \in (0, \ 2\pi) \qquad (6.82)$$

At different values of θ, it is required to solve the polynomial equation for the complex roots. While it is easy for a quadratic equation, the third and fourth order Runge-Kutta method involves solution of third and fourth order polynomials, respectively. Any method described in Chapter 3 for solving non-linear equations and polynomials can be used. Recall that complex roots occur in conjugate pairs. So, determining only half of the distinct roots are enough to plot the full region. Plotting of these roots on the complex plane provides the stability region.

The stability regions for the second and fourth order Runge-Kutta methods are shown in Fig. 6.14. A few points to note are:

- Stability region of the fourth order method includes part of the imaginary axis and therefore, it is stable for purely imaginary λ.
- For all the multi-step methods (explicit and implicit) and BDF discussed so far, the stability region becomes smaller as the order of the method increases. It is opposite for the case of the Runge-Kutta. The fourth order method has a larger stability region compared to the second order (Fig. 6.14).
- The stability regions of Runge-Kutta methods, as expected are smaller compared to same order implicit methods and BDFs. However, they are larger compared to the multi-step explicit methods of same order.

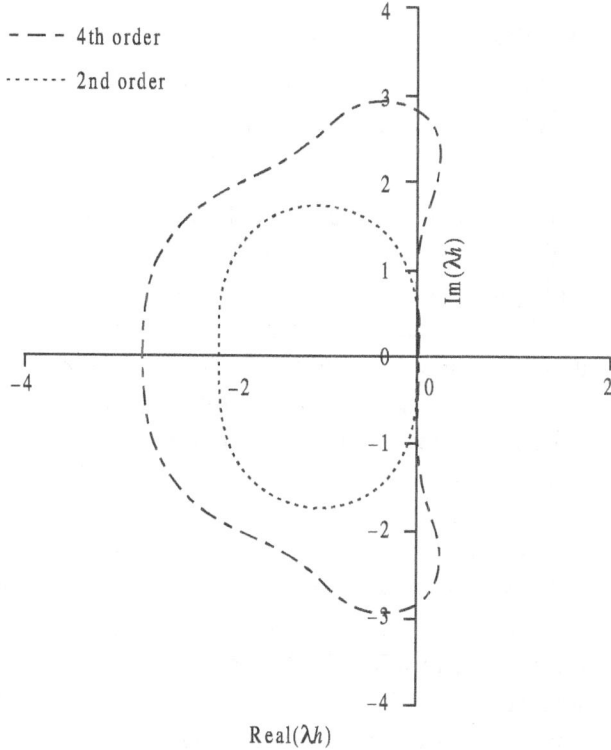

Fig. 6.14 Stability regions of second and fourth order Runge-Kutta methods. The methods are stable inside the closed regions.

We have now learned how to analyze a numerical method for initial-value problems for truncation error and stability. We have shown these analysis for all the groups of methods that were developed in Section 6.2. It is now time to formalize the definition of stability. With the help of consistency and stability, we will also formalize the definition of convergence.

6.3.2.8 *Formal Definition of Stability and Convergence*

Before we venture into formalizing the definition of stability of a numerical method, let us look at the stability of the original IVP given by Eq. (6.3). In the first subsection, we have shown that the solution of the model problem becomes unbounded for $\lambda_r > 0$. Let us first define the stability for a general solution of the original Eq. (6.3). For this, let us consider two well-defined solutions of the original equation in the interval $[x_0, X]$ with two different initial conditions, $y = u(x)$ for $u(x_0)$ $= u_0$ and $y = v(x)$ for $v(x_0) = v_0$. The stability can be defined as follows:

Definition 6.3 The solution $y = u(x)$ is *stable* on the interval $[x_0, X]$ if for every $\varepsilon > 0$, there exists $\delta > 0$ such that for all v_0 satisfying $|u_0 - v_0| < \delta$, the inequality $|u(x) - v(x)| < \varepsilon$ is true for all $x \in [x_0, X]$. The solution is *asymptotically stable* if it is stable in $[x_0, \infty)$ and $\lim_{x \to \infty} |u(x) - v(x)| = 0$.

In simple terms, the above definition is saying that a finite perturbation in the initial condition will produce a finite change in the solution as well and the change will approach zero when the independent variable approaches infinity. In other words, the effect of perturbation in the initial condition diminishes after sufficient number of time steps. It is now easy to see that the solution to the model equation is unstable for $\lambda_r > 0$ and asymptotically stable for $\lambda_r < 0$. Do not confuse about the existence of the solution. The IVP (original as well as the model problem) still possess a unique solution for $\lambda_r > 0$. Now, we shall consider stability of the numerical method in the sense of its capability to simulate the analytical solution as closely as possible.

We shall start with the general linear multi-step method described by Eq. (6.26). We have already seen that it is a general representation of Euler methods, trapezoidal method, Adams-Bashforth, Adams-Moulton, and BDF. For the convenience of writing, we rewrite it here after generalizing the limits of the indices for a general linear multi-step method consisting of $(n + 1)$ steps:

$$\sum_{i=0}^{n+1} \gamma_i y_i = h \sum_{i=0}^{n+1} \beta_i f_i \qquad (6.83)$$

Notice that, i ranges from a future time step $(n + 1)$ to as far back as possible (up to 0). So, to startup this method, we will require the set of values $\{y_0, y_1, ..., y_n\}$ as initial conditions. These have to be computed by a different method or assumed. We shall first consider the sensitivity of this numerical method to small changes in the initial condition, which may occur due to use of different methods for generating these values or input error.

A method is stable if it can attenuate the changes in the initial conditions in the final solution. This is known as *zero stability*. For this purpose, we shall consider solutions in the interval $[x_0, X]$ where a unique solution exists for the original IVP and assume two initial conditions $\{y_0, y_1, ..., y_n\}$ and $\{z_0, z_1, ..., z_n\}$ generated by two different methods. The *zero stability* is defined as follows:

Definition 6.4 The linear multi-step method Eq. (6.83) is *zero stable* if there exists a constant K such that the following inequality holds,

$$|y_{n+1} - z_{n+1}| \le K \max\{|y_0 - z_0|, |y_1 - z_1|, \ldots, |y_n - z_n|\} \quad \text{as } h \to 0$$

Let us now see how we can assess the zero stability for a given method. We have earlier defined the amplification factor $\left|\dfrac{y_{n+1}}{y_n}\right| = \sigma$ for the model problem. For the general problem $y' = f$, let us define another amplification factor for the functional value as $\left|\dfrac{f_{n+1}}{f_n}\right| = \xi$. Note that both σ and ξ can be complex. In terms of the amplification factors, the method can be written as follows:

$$y_0 \sum_{i=0}^{n+1} \gamma_i \sigma^i = h f_0 \sum_{i=0}^{n+1} \beta_i \xi^i \tag{6.84}$$

The y_0 and f_0 are the initial conditions and the slope function value at the initial condition, respectively. Therefore, both are known and constant. For a fixed time step $h > 0$, the above equation defines two polynomials in the complex space involving σ and ξ. These polynomials can be written as follows:

$$P(\sigma) = y_0 \sum_{i=0}^{n+1} \gamma_i \sigma^i = \sum_{i=0}^{n+1} a_i \sigma^i, \text{ where } a_i = y_0 \gamma_i$$

$$Q(\xi) = h f_0 \sum_{i=0}^{n+1} \beta_i \xi^i = \sum_{i=0}^{n+1} b_i \xi^i, \text{ where } b_i = h f_0 \beta_i \tag{6.85}$$

The $P(\sigma)$ and $Q(\xi)$ are known as first and second characteristic polynomials, respectively. It can be shown mathematically (the proof is beyond the scope of this book) that the zero stability implies that all the roots of the first characteristic polynomial lies within a unit disc $(|\sigma| < 1)$ and if any root lies on the unit circle $(|\sigma| = 1)$, it is a simple root. This can now be used easily to evaluate the *zero stability* of any of the linear multi-step methods described in this chapter.

A numerical method is said to be convergent if the computed value at any time step approaches the true solution as $h \to 0$. The *consistency* and *zero stability* are necessary conditions for the convergence of a linear multi-step method. For any consistent linear multi-step method, zero stability is also a sufficient condition for the method to be convergent.

Although *zero stability* helps in assessing the convergence of a linear multi-step method, it is not very useful in the context of practical application because it defines the stability in the asymptotic limit $h \to 0$. So, we need to concentrate on cases with fixed $h > 0$ and assess how the approximate solution behaves with respect to the true solution. For this purpose, we shall apply the general linear multi-step method Eq. (6.26) to our model problem Eq. (6.57),

$$\sum_{i=0}^{n+1} \gamma_i y_i = h \sum_{i=0}^{n+1} \beta_i (\lambda y_i) \tag{6.86}$$

Using the definition of the amplification factor, we obtain the *stability polynomial* as,

$$R(\sigma) = \sum_{i=0}^{n+1} y_0 (\gamma_i - \lambda h \beta_i) \sigma^i = \sum_{i=0}^{n+1} c_i \sigma^i = 0 \text{ where } c_i = y_0 (\gamma_i - \lambda h \beta_i) \tag{6.87}$$

The reader can now compare this polynomial with the ones derived earlier in the section for various methods. For any fixed h, the coefficient c_i's are constants. In that case, the method will be stable if all the roots lie within the unit disc $\left(|\sigma| < 1 \right)$. If any root is outside the disc $\left(|\sigma| > 1 \right)$, the method is unstable. The case $|\sigma| = 1$ indicates the boundary. Any root lying on this boundary must be a simple root for the method to be stable. In the earlier subsections, our interest was to find this boundary and therefore, we had used the values of $|\sigma| = 1$ to draw it. In this subsection, our interest is to classify various kinds of stability that may be observed depending on the nature of the roots of this characteristic polynomial. We will define some of them here. Let us assume $\{\sigma_1, \sigma_2, \ldots, \sigma_n\}$ are the roots of the stability polynomial.

Definition 6.5 A linear multi-step method is *absolutely stable* in an open set C_A of the complex plane if the roots of the stability polynomial satisfy $|\sigma_i| < 1$ for all i and for all $\lambda h \in C_A$. The open set C_A is the *region of absolute stability* for the method.

According to this definition, all the methods described in this chapter are absolutely stable. In Sections 6.3.2.1 to 6.3.2.7, we had drawn the regions of their absolute stability.

At this point, recall our working definitions of stability (unconditional and conditional). We would now like to relate that to our formal definitions.

Definition 6.6 A linear multi-step method is called *A-stable* if its region of *absolute stability* includes the whole left half of the complex plane ($h\lambda_r < 0$) shown as the shaded region in Fig. 6.7.

Now, it is easy to see that the methods we defined as *unconditionally stable* are in fact *A-stable*. Whereas, both *unconditionally stable* and *conditionally stable* are included in the definition of *absolute stability*. Therefore, the *A-stable* methods are subset of the *absolutely stable* methods. For example, none of the explicit methods are *A-stable*. The implicit methods and BDF of order larger than 2 are also not *A-stable*. However, all the methods are *absolutely stable*. In fact, it is possible to show that highest order of an *A-stable* implicit linear multi-step method cannot exceed 2 (Dahlquist 1963). Trapezoidal method is the highest order implicit method with the minimum truncation error constant.

Recognizing that the definition of *A-stability* is too restrictive, there have been many attempts to define a few more relaxed options. One notable amongst them is $A(\alpha)$-stability with $\alpha \in \left(0, \dfrac{\pi}{2} \right)$. We shall refrain from describing all such definitions, as they are beyond the scope of this book. However, we will define one last type called *stiff stability* that we shall use at a later section.

Definition 6.7 A linear multi-step method is *stiffly stable* if its region of *absolute stability* includes the following two sets C_1 and C_2:

$$C_1 = \{\lambda h \in C : \lambda_r h < -r\}$$
$$C_2 = \{\lambda h \in C : -r < \lambda_r h < 0, \ -s < \lambda_i h < s\}$$

If you want to visualize it, think of a wedge on the left half of the complex plane with the tip placed at the origin. If you moved away towards left, from the axis beyond $-r$, the complete imaginary range is included in the region of absolute stability. If you are approaching along the real axis from the left and towards the origin, the region of absolute stability is getting constricted to a maximum limit of $\pm s$ along the imaginary axis as you approach closer than $-r$. In this case, the complete left half of the real axis is included in the region. Now compare and see that all the BDFs of order 2 to 6 are *stiffly stable*.

In addition to the error analysis described above, another kind of analysis known as *phase error analysis* becomes important if one is dealing with equations that have periodic solutions. When the truncation error accumulates in a periodic solution (think of a wave!), it manifests itself into two forms. It affects the amplitude as well as the phase of the wave. The stability is a measure of increase (or decrease) of the amplitude. The phase error is a measure of the phase change of the wave compared to its analytical (true) solution. It is important to note that both are caused by accumulation of truncation error. We only observe their effects in different forms through different analysis. Periodic solutions occur in many science and engineering scenario such as waves, earthquake, prey-predator interactions, etc. We address the analysis of phase error in the next section.

6.3.3 Phase Error

We will introduce the concept of phase error by applying Euler forward method to an example problem similar to the model equation but with purely imaginary λ. Let us consider the following equation:

$$\frac{dy}{dt} = i\lambda y, \ y(0) = y_0 \tag{6.88}$$

The analytical solution of the above equation is of course $y = y_0 e^{i\lambda t}$. Application of the Euler forward method to the problem leads to the following approximate solution:

$$y_{n+1} = y_n (1 + i\lambda h) \tag{6.89}$$

Note that y_n is a complex number. At any time step n, if $y_n = a + ib$, putting this in Eq. (6.89), we obtain,

$$y_{n+1} = (a - \lambda h b) + i(b + \lambda h a) \tag{6.90}$$

The true or analytical solution at any time step is computed as

$$y_n = y_0 e^{i\lambda h n} = y_0 (\cos \lambda h n + i \sin \lambda h n) \tag{6.91}$$

Therefore, for a given λ and y_0, both true and approximate solution by Euler explicit method can be computed from relations in Eqs (6.90) and (6.91). For illustration purposes, we take $\lambda = 1$, $y_0 = 1$, and $h = \pi/10$.

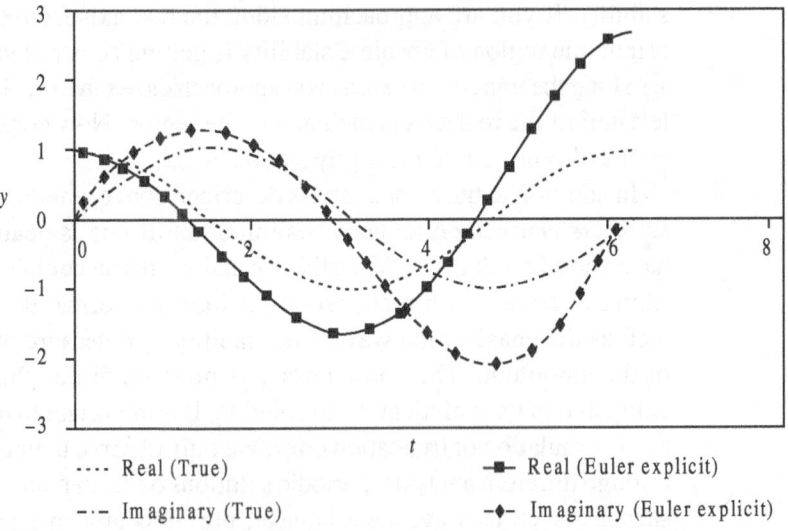

Fig. 6.15 Real and imaginary parts of a periodic solution by Euler forward method compared with true solution.

Figure 6.15 shows the real and imaginary parts of the solution for one full period. The amplitude of the approximate solutions for both real and imaginary parts differs from the true solution and the difference increases with time. This was expected from the stability analysis as Euler explicit method is unconditionally unstable for purely imaginary λ.

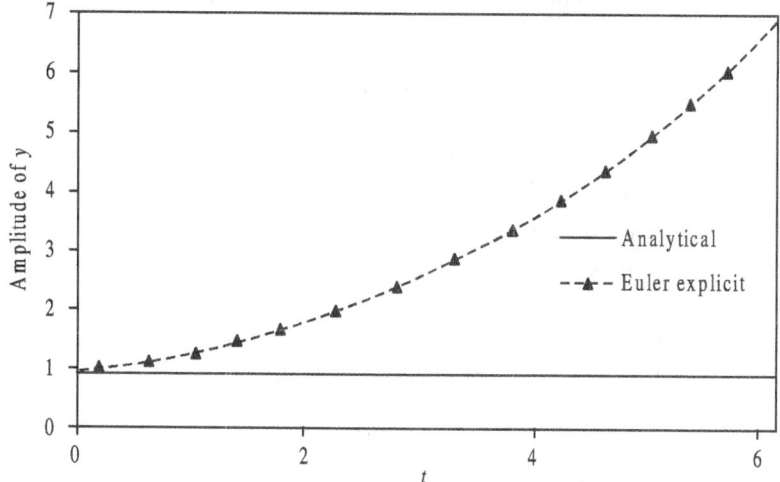

Fig. 6.16   Amplitude error in Euler forward method.

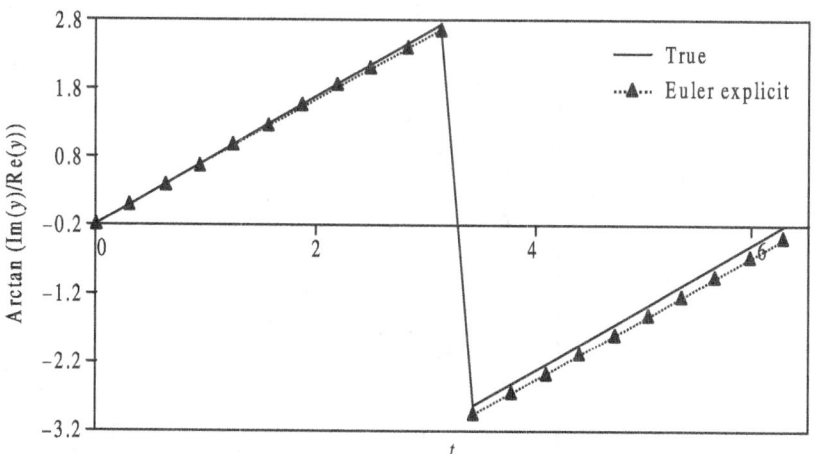

Fig. 6.17   Phase error in Euler forward method.

In fact, the increase in the amplitude can be computed using the amplification factor derived in stability analysis. However, their phases are also different, i.e., crests and troughs are at different locations. This is known as the *phase error* which neither the truncation error analysis nor the stability analysis provided any measure for. Amplitude and phase $\left[\tan^{-1} \dfrac{\text{Im}(y_n)}{\text{Re}(y_n)} \right]$ of the true and approximate solutions are shown in Figs 6.16 and 6.17, respectively.

For the behaviour of a periodic solution, it is important to look at the phase diagram by plotting real vs. imaginary part of the solution (Fig. 6.18). For the analytical solution, it is a unit circle. We observe that the approximate solution becomes a spiral while the true solution is a closed curve (circle in this case). This means that the numerical method has both amplitude and phase errors. If the phase error is absent, the approximate solution will remain as a circle but with larger or smaller radius depending on whether the amplitude error amplifies or attenuates. We will now show how to estimate the phase error for a numerical method.

In order to obtain a measure of the phase error, we shall use Eq. (6.88) as the model equation. Using the analytical solution, we obtain the true amplification factor as:

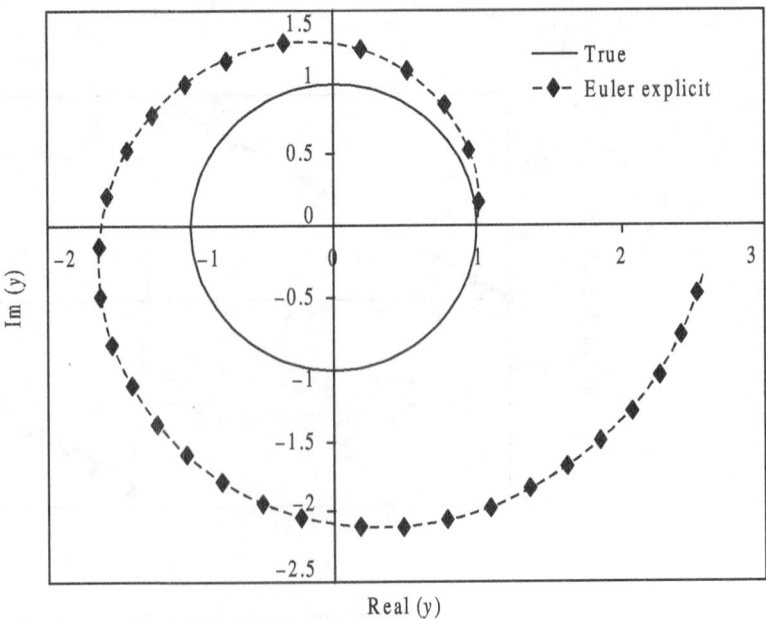

Fig. 6.18 True and approximate solution cycles.

$$\sigma_{\text{True}} = \frac{y_{n+1}}{y_n} = \frac{y_0 e^{i\lambda h(n+1)}}{y_0 e^{i\lambda hn}} = e^{i\lambda h} \tag{6.92}$$

Amplitude of the amplification factor is

$$|\sigma_{\text{True}}| = \sqrt{\cos^2 \lambda h + \sin^2 \lambda h} = 1$$

and the phase is

$$\theta_{\text{True}} = \tan^{-1}\left(\frac{\sin \lambda h}{\cos \lambda h}\right) = \lambda h$$

For the Euler forward method, the amplification factor from Eq. (6.62) can be seen as $\sigma = 1 + i\lambda h$. Therefore, the amplitude of the approximate amplification factor is

$$|\sigma| = \sqrt{1 + (\lambda h)^2} > |\sigma_{\text{True}}| = 1 \tag{6.93}$$

Since, magnitude of the approximate amplification factor is larger than the true amplification factor, the amplitude increases with each time step. This reconfirms the instability of the Euler explicit method for purely imaginary λ. The phase of the approximate amplification factor can be computed as,

$$\theta = \tan^{-1}\left(\frac{\text{Im}(\sigma)}{\text{Re}(\sigma)}\right) = \tan^{-1}(\lambda h) \tag{6.94}$$

We now define phase error (*PE*) as,

$$PE = \theta_{\text{True}} - \theta = \lambda h - \tan^{-1} \lambda h \tag{6.95}$$

Using the Taylor's series, we can write

$$\tan^{-1} \lambda h = \lambda h - \frac{(\lambda h)^3}{3!} + \frac{(\lambda h)^5}{5!} - \frac{(\lambda h)^7}{7!} + \frac{(\lambda h)^9}{9!} \cdots \tag{6.96}$$

Therefore, we obtain the following for the phase error of the Euler explicit method:

$$PE = \frac{(\lambda h)^3}{3!} - \frac{(\lambda h)^5}{5!} + \frac{(\lambda h)^7}{7!} - \frac{(\lambda h)^9}{9!} \cdots \tag{6.97}$$

In the leading order term, λh is raised to the power of 3. Following a convention similar to truncation error, we would refer to this as third order accuracy. Therefore, the Euler explicit method is third order accurate for phase error. Moreover, the first term is positive and summation of the infinite series is also positive. Therefore, θ_{True} is always larger than θ and the waves of the approximate solution will be lagging behind the true solution (Fig. 6.15). This will be referred as *phase lag*.

Let us look at the phase error of a few more methods namely, Euler implicit, trapezoidal, and second order Runge-Kutta. The amplification factors for these methods when applied to the model problem for the phase error can be obtained from the corresponding expressions derived in the stability analysis by replacing λ with $-i\lambda$. The amplification factors, amplitudes and phases using these methods are listed below:

Euler implicit:

$$\sigma = \frac{1}{1-i\lambda h}, \ |\sigma| = \frac{1}{\sqrt{1+(\lambda h)^2}}, \ \theta = \tan^{-1}(\lambda h) \tag{6.98}$$

Trapezoidal method:

$$\sigma = \frac{1+i\frac{1}{2}\lambda h}{1-i\frac{1}{2}\lambda h}, \ |\sigma|=1, \ \theta = 2\tan^{-1}\left(\frac{\lambda h}{2}\right) \tag{6.99}$$

Second order Runge-Kutta:

$$\sigma = 1-\frac{(\lambda h)^2}{2}+i\lambda h, \ |\sigma|=\sqrt{1+\frac{(\lambda h)^4}{4}}, \ \theta = \tan^{-1}\left(\frac{\lambda h}{1-\frac{(\lambda h)^2}{2}}\right) \tag{6.100}$$

From the above relations, the leading order terms for the PE may be calculated as $(\lambda h)^3/3$, $(\lambda h)^3/12$, and $-(\lambda h)^3/6$ for Euler implicit, trapezoidal, and second order Runge-Kutta methods, respectively. Readers should verify these by following a procedure similar to the one shown for Euler forward.

It is easy to see that the magnitude of the amplification factor for Euler implicit method is always less than unity. Therefore, application of this method to a periodic solution will always lead to attenuation of the peaks. Trapezoidal method on the other hand has no amplitude error. The second order Runge-Kutta method was unstable for the purely imaginary λ and leads to increase in amplitude but to a lesser extent compared to the Euler explicit method.

The phase error for the Euler implicit method is same as that of the explicit. Although, the order of the phase error for trapezoidal method is same as the Euler methods but the coefficient is one-fourth. So, the phase error is less but both are positive and leads to phase lag. The second order Runge-Kutta method on the other hand leads to phase lead. Therefore, peaks and troughs of the approximate solution will occur later than the true solution.

Following in similar lines, the phase error of the other methods can also be computed. We leave these to the readers as exercise!

We have looked at ways to analyze the methods for different kinds of error. In the next section, we will look at a few application perspectives of different methods presented in Section 6.2 for different kinds of problems.

EXERCISE 6.3

1. Consider a second order central difference approximation of the derivative in Eq. (6.3) and the slope function evaluated at n. The method is then given by $y_{n+1} = y_{n-1} + 2hf_n$. What is the order of the method? Is it consistent? What is the region of stability for this method?

2. In Exercise 6.2, consider the methods derived in Problem 10 using the Simpson's rules. Evaluate their truncation errors, stability properties, and the phase errors for both the methods.

3. Perform stability analysis and phase error analysis of the third order Runge-Kutta method shown in Table 6.9. Draw the stability boundary for this method and compare with the boundaries of second and fourth order methods given in Fig. 6.14.

4. Consider Problem 2 of Exercise 6.2. By repeated interval halving (i.e., 5, 10, 20, and 40 intervals in one period), numerically determine the order of the amplitude error for both the methods at the maxima and minima of the periodic function. Can you compute the order of the phase error for these methods at the end of one time period from these calculations?

5. For all the multi-step methods (explicit and implicit), the stability region becomes smaller as the order of the methods increase. This is so for the BDFs as well as the hole becomes larger with the order. For the Runge-Kutta method on the other hand, the stability region of fourth order method is larger than that of the second order (Fig. 6.14). Explain why?

6. Find the stability region for the generalized multi-step method derived in Problem 12, Exercise 6.2.

7. Consider Problem 7, Exercise 6.2.
   (a) Compute the maximum value of h that can be used for obtaining stable solution of the equation for Euler forward, Euler backward, fourth order Runge-Kutta method, fourth order AB and fourth order AM methods.
   (b) Using these maximum grid sizes (from stability consideration), compute the true error at 10 km for each method.
   (c) Now approximately compute the maximum grid sizes for each method that will lead to a maximum of 1% error at 10 km.
   (d) Using these maximum grid sizes (from error consideration), compute the

true error at 10 km for each of these methods and observe if 1% error limit at 10 km is satisfied.

(e) Explain your choices for the grid sizes from these experiment.

8. Consider the following methods. Find the consistency for the general IVP (6.3). Compute the stablity using the model problem. Finally comment on their convergence.

(a) $\dfrac{1}{2}\left(\dfrac{y_{n+1}-y_n}{h}+\dfrac{y_n-y_{n-1}}{h}\right)=\left(\dfrac{1}{6}f_{n+1}+\dfrac{2}{3}f_n+\dfrac{1}{6}f_{n-1}\right)$

(b) $\dfrac{y_{n+1}+y_n-2y_{n-1}}{3h}=\dfrac{1}{3}\left(f_{n+1}+f_n+f_{n-1}\right)$

(c) $\dfrac{y_{n+1}+y_{n-1}}{2h}=\dfrac{1}{3}\left(f_{n+1}+2f_n\right)$

9. Compute the phase error of fourth order Runge-Kutta method when applied to the model Problem (6.88).

6.4 APPLICATIONS OF VARIOUS METHODS

In Section 6.2, we have introduced the reader to different numerical methods to solve a first-order ordinary differential equation and in Section 6.3 we have analyzed the errors of these methods due to approximation. We have also shown the ways to compute these errors. In this section, we touch upon some aspects of application of these methods for various types of problems. The following may be used as guidelines for choosing a method for a given problem:

- Decision of the order of the method is guided by how accurate solution one needs. For a preliminary design problem, it often suffices to use a fast lower order scheme with single precision computation. However, for the final solution, one often requires higher-order schemes with more accuracy. For engineering applications, it is generally sufficient to use a scheme with fourth order global truncation error. It is also possible to combine several numerical methods to increase the accuracy. This will be discussed later in this section.
- If the smaller time steps (large amount of computations) are not a concern (typically for small and simple problems), it is easier to use and implement explicit and Runge-Kutta methods. However, often the stability concerns force the time step to be so small that the computation time becomes large even for simple problems.
- Implicit methods and BDFs have much larger stability regions and therefore, allow larger time steps. These can save significant amount of computation times.

- If the solution is periodic and/or imaginary, it is required to choose a method which has at least some parts of its *region of absolute stability* along the imaginary axis and has low phase error.

In the following section, we cover a problem specific to multi-step method that is encountered when one tries to apply these methods.

6.4.1 Startup of Non-Self Starting Methods

We have seen in Example 6.1 that all the multi-step methods of order greater than 2 are *non-self starting*. In a first order IVP, the value of the independent variable is known at the initial point in terms of the initial condition. Typically, we designate it as y_n and try to compute y_{n+1}. However, at the start, n is zero. Thus, only y_0 is known when we try to compute y_1.

Let us illustrate the problem with the following example:

Example 6.5   Solve $\dfrac{dy}{dt} = -0.1y;\ y(0) = 1$ using fourth order Adams-Bashforth method.

Solution   The fourth order Adams-Bashforth method is given by

$$y_{n+1} = y_n + h\left(\frac{55}{24}f_n - \frac{59}{24}f_{n-1} + \frac{37}{24}f_{n-2} - \frac{3}{8}f_{n-3}\right)$$

Using the functional form $f = -0.1y$, the above relation for the first four time steps translates to

$$y_1 = 1 - h\left(\frac{5.5}{24}y_0 - \frac{5.9}{24}y_{-1} + \frac{3.7}{24}y_{-2} - \frac{0.3}{8}y_{-3}\right)$$

$$y_2 = 1 - h\left(\frac{5.5}{24}y_1 - \frac{5.9}{24}y_0 + \frac{3.7}{24}y_{-1} - \frac{0.3}{8}y_{-2}\right)$$

$$y_3 = 1 - h\left(\frac{5.5}{24}y_2 - \frac{5.9}{24}y_1 + \frac{3.7}{24}y_0 - \frac{0.3}{8}y_{-1}\right)$$

$$y_4 = 1 - h\left(\frac{5.5}{24}y_3 - \frac{5.9}{24}y_2 + \frac{3.7}{24}y_1 - \frac{0.3}{8}y_0\right)$$

From the initial condition, $y_0 = y(0) = 1$ is known, but the values at the negative time steps $y_{-1}, y_{-2},$ and y_{-1} are undefined or not known. So, $y_1, y_2,$ and y_3 cannot be computed from the above equations. The expression of y_4 contains y values with positive subscripts but in order to use it, $y_1, y_2,$ and y_3 have to be known. So, the fourth order Adams-Bashforth method becomes un-usable unless we find some other ways to compute the values of y_1, y_2, and y_3. This is the reason all multi-step methods are non-self starting as they require the help of some other method to compute the values of the dependent variable at initial time steps.

Numbers of such values to be computed by other methods vary depending on the type and order of the method. For the Adams-Bashforth methods and BDFs, the number of time steps to be computed by other methods is $(n - 1)$ for $n > 1$ where n is the order of the global truncation error of the method. For the Adams-Moulton methods, the number is $(n - 2)$. There are several ways one can attempt to start these methods. These are discussed below along with their advantages and disadvantages.

6.4.1.1 *Use of Lower-Order Methods*

It is clear from Example 6.5 that if one computes y_1, y_2, and y_3 using lower-order methods, it will be possible to use the fourth order Adams-Bashforth method to compute y_4 and subsequent values of y at future time steps. Such a procedure is shown in Example 6.6.

Example 6.6

Solve the problem of Example 6.1 using fourth order Adams-Bashforth (AB) method with $h = 0.5$. Use Euler forward for the first time step, second order AB for the second, third order AB for the third and fourth order AB for all the successive time steps. Compare the true relative error at $t = 5$ with the error obtained in Example 6.1 using fourth order AB where true solution was used for the first three time steps.

Solution

We already have seen the computation formulae for these methods in Example 6.1. We tabulate the computations in Table 6.14.

Table 6.14

Time (t)	Method used	True solution	Numerical solution	Error (%)
0		1.0000	1.0000	
0.5	EF	0.7408	0.7000	5.51
1	AB2	0.5488	0.5350	2.52
1.5	AB3	0.4066	0.3824	5.95
2	AB4	0.3012	0.3028	0.53
2.5	AB4	0.2231	0.2079	6.81
3	AB4	0.1653	0.1716	3.84
3.5	AB4	0.1225	0.1100	10.20
4	AB4	0.0907	0.0988	8.95
4.5	AB4	0.0672	0.0560	16.68
5	AB4	0.0498	0.0588	18.19

The error at $t = 5.0$ is 18.2% compared to 1.14% obtained using true solution for the first three time steps. It is easy to see the larger error in the three initial time steps.

One obviously sacrifices accuracy since, lower-order methods are less accurate and larger errors incurred in the initial time steps would progress through the solution in the future time steps. However, it is possible to reduce such errors by application of Richardson's extrapolation (Section 5.2.1.1, Chapter 5). From the truncation error analysis, notice that, approximation ($\tilde{y}_{n+1}$) by any method (of local truncation error order p) can be represented as follows in terms of true value (y_{n+1})

$$\tilde{y}_{n+1} = y_{n+1} + c_1 h^p + c_2 h^{p+1} + c_3 h^{p+2} + \cdots \quad (6.101)$$

where c_i's are not functions of h. They only involve constants and higher derivatives evaluated at a fixed point. Some examples of such expression are shown below:

Euler forward: $\tilde{y}_{n+1} = y_{n+1} - \dfrac{h^2}{2} y_n'' - \dfrac{h^3}{6} y_n''' - \dfrac{h^4}{24} y_n'''' + o(h^5)$

Euler backward: $\tilde{y}_{n+1} = y_{n+1} + \dfrac{h^2}{2} y_n'' + \dfrac{h^3}{3} y_n''' + \dfrac{h^4}{8} y_n'''' + o(h^5)$

Third order Adams-Bashforth: $\tilde{y}_{n+1} = y_{n+1} - \dfrac{3}{8} h^4 y_n'''' + o(h^5)$

Let us denote the approximate value $\tilde{y}_{n+1}$ calculated in Eq. (6.101) using a time step h as $\tilde{y}_{n+1}^h$. Similarly we can write,

$$\tilde{y}_{n+1}^{h/2} = y_{n+1} + c_1 \left(\frac{h}{2}\right)^p + c_2 \left(\frac{h}{2}\right)^{p+1} + c_3 \left(\frac{h}{2}\right)^{p+2} + \cdots \quad (6.102)$$

$$\tilde{y}_{n+1}^{h/4} = y_{n+1} + c_1 \left(\frac{h}{4}\right)^p + c_2 \left(\frac{h}{4}\right)^{p+1} + c_3 \left(\frac{h}{4}\right)^{p+2} + \cdots \quad (6.103)$$

Eliminating c_1 from Eqs (6.101) and (6.102), we can get an approximation of y_{n+1} which is $(p + 1)$ order accurate.

$$\tilde{y}_{n+1}^{h,h/2} = \frac{2^p \tilde{y}_{n+1}^{h/2} - \tilde{y}_{n+1}^h}{(2^p - 1)} = y_{n+1} - \left(\frac{1}{2(2^p - 1)}\right) c_2 h^{p+1} - \left(\frac{3}{4(2^p - 1)}\right) c_3 h^{p+2} - \cdots$$

$$(6.104)$$

Since, the new approximation was obtained from two approximations with time steps h and $h/2$, we denote it as $\tilde{y}_{n+1}^{h,h/2}$. Similarly, we can obtain another approximation of y_{n+1} with $(p + 1)$ order accuracy by eliminating c_1 from the pth order accurate approximations with time steps $h/2$ Eq. (6.102) and $h/4$ Eq. (6.103):

$$\tilde{y}_{n+1}^{h/2,h/4} = \frac{2^p \tilde{y}_{n+1}^{h/4} - \tilde{y}_{n+1}^{h/2}}{(2^p - 1)} = y_{n+1} - \left(\frac{1}{2(2^p - 1)}\right) c_2 \left(\frac{h}{2}\right)^{p+1}$$

$$- \left(\frac{3}{4(2^p - 1)}\right) c_3 \left(\frac{h}{2}\right)^{p+2} - \cdots \qquad (6.105)$$

It is now possible to use two $(p + 1)$ order accurate estimates of y_{n+1} to obtain a $(p + 2)$ order accurate estimate by eliminating c_2 from Eqs (6.104) and (6.105)

$$\tilde{y}_{n+1}^{h,h/2,h/4} = \frac{2^{p+1} \tilde{y}_{n+1}^{h/2,h/4} - \tilde{y}_{n+1}^{h,h/2}}{(2^{p+1} - 1)} = y_{n+1} + \frac{3}{8(2^p - 1)(2^{p+1} - 1)} c_3 h^{p+2} + \cdots.$$

$$(6.106)$$

Formulae for the Euler forward method are

$$\tilde{y}_{n+1}^{h,h/2} = \frac{4\tilde{y}_{n+1}^{h/2} - \tilde{y}_{n+1}^{h}}{3}, \quad \tilde{y}_{n+1}^{h/2,h/4} = \frac{4\tilde{y}_{n+1}^{h/4} - \tilde{y}_{n+1}^{h/2}}{3}, \text{ and } \tilde{y}_{n+1}^{h,h/2,h/4} = \frac{8\tilde{y}_{n+1}^{h/2,h/4} - \tilde{y}_{n+1}^{h,h/2}}{7}$$

where $\tilde{y}_{n+1}^{h}$, $\tilde{y}_{n+1}^{h/2}$, and $\tilde{y}_{n+1}^{h/4}$ are approximations of y_{n+1} obtained using Euler forward method with time steps of h, $h/2$, and $h/4$, respectively.

If one also had an approximation $y_{n+1}^{h/8}$ with a time step size of $h/8$, one could obtain another $(p + 2)$ order accurate estimate $\tilde{y}_{n+1}^{h/2,h/4,h/8}$. Using two $(p + 2)$ order accurate estimates $\tilde{y}_{n+1}^{h,h/2,h/4}$ and $\tilde{y}_{n+1}^{h/2,h/4,h/8}$, it would be possible to compute another estimate which is $(p + 3)$ order accurate. This way, it is possible to compute estimates at any time step up to arbitrarily higher-order accuracy.

We had derived the Richardson's extrapolation formulae (6.104–6.106) using the *LTE* of a method. The same expression can be derived using similar steps if one considers p as the *GTE* of the method instead of *LTE*. Recall that for any pth order method, the *LTE* is of order $(p + 1)$. Therefore, for the same method, we can obtain two different Richardson's extrapolation formulae by using *LTE* and *GTE*. The natural question is which error shall one use? By definition, the *LTE* gives the error in one time step if the step is taken from a true point. The order of *LTE* is used when the extrapolated values have to be generated for one time step (or a maximum of 2) from the initial point. This is typically the case for the startup of a second or third order multi-step methods. If the extrapolated values are required at a number of time step, use the order of *GTE*. Overall, if you are in doubt, use the order of *GTE*.

There are two ways to apply the Richardson's extrapolation scheme. In Example

6.6, one is required to calculate y_1, y_2, and y_3 using a lower-order method. It is possible to compute these values at all three times using say Euler forward with time steps of h, $h/2$, and $h/4$. Then, use the computed values for estimation of more accurate values using Richardson's extrapolation. If the extrapolated value at one time step is not used as the starting point for next time step, it is called *passive Richardson's* extrapolation [Fig. 6.19(a)].

In the other application, one computes values of y_1 using the lower-order method with time steps h, $h/2$, $h/4$, etc. and then applies Richardson's extrapolation to

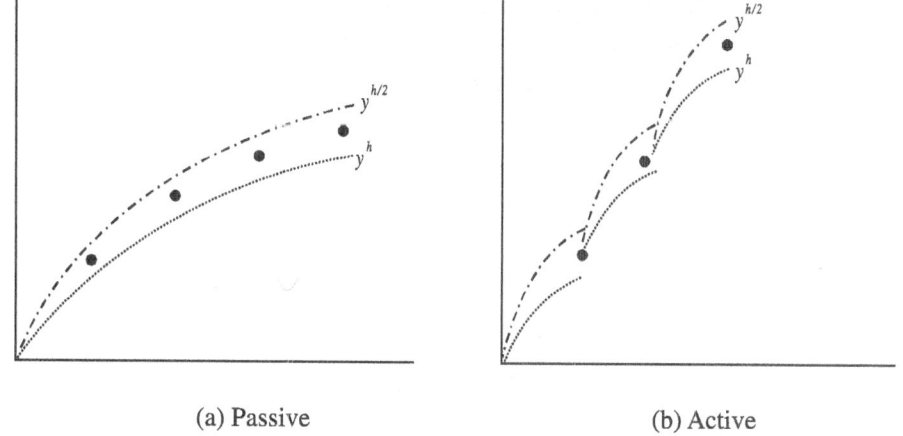

(a) Passive   (b) Active

Fig. 6.19 Schematic showing (a) passive and (b) active Richardson's extrapolation concept. Two different dotted lines showing values obtained using the lower-order method with time step sizes of h and $h/2$. Dots are the higher-order values obtained using Richardson's extrapolation.

compute a more accurate value of y_1. This extrapolated value at y_1 is then used as starting point for computation of y_2 using the lower-order method. This is called the *active Richardson's extrapolation*. A graphical schematic representation of these two schemes are shown in Fig. 6.19. We show the startup using Euler forward method with both active and passive Richardson's extrapolation in Example 6.7.

Example 6.7   Solve Example 6.1 using third order Adams-Bashforth (AB) method with $h = 5.0$. Use Euler forward method along with Richardson's extrapolation (both active and passive) for the first three time steps to obtain same order of local truncation error as the third order Adams-Bashforth method. Compare the errors at $t = 5.0$ with the case of using progressively higher-order method (Example 6.6).

Solution   We observe that the order of *LTE* of Euler forward method is 2 and that of third order AB is 4. Therefore, we will require two time step refinements in order to obtain fourth order approximation using Richardson's extrapolation from Euler forward solution.

If one uses *GTE*, the improvement of orders is from first to third but one would still require two time steps refinements. It is the improvement in orders that decides the number of steps in Richardson's extrapolation, not the absolute magnitude of the orders. The absolute magnitude of the order only determines the extrapolation formulae. Since, the primary time step h is 0.5, we will use 0.25, and 0.125 as the subsequent refined time steps of $h/2$ and $h/4$, respectively.

We shall use the following relationships to obtain third order accurate estimate of y_{n+1}

$$\tilde{y}_{n+1}^{h,h/2} = \frac{4\tilde{y}_{n+1}^{h/2} - \tilde{y}_{n+1}^{h}}{3}$$

$$\tilde{y}_{n+1}^{h/2,h/4} = \frac{4\tilde{y}_{n+1}^{h/4} - \tilde{y}_{n+1}^{h/2}}{3}$$

If we use *GTE*, the formulae will be

$$\tilde{y}_{n+1}^{h,h/2} = 2\tilde{y}_{n+1}^{h/2} - \tilde{y}_{n+1}^{h} \text{ and } \tilde{y}_{n+1}^{h/2,h/4} = 2\tilde{y}_{n+1}^{h/4} - \tilde{y}_{n+1}^{h/2}$$

From the above two estimates, we can formulate the following for the fourth order estimate (recall the *LTE* of third order AB is of order 4)

$$\tilde{y}_{n+1}^{h,h/2,h/4} = \frac{8\tilde{y}_{n+1}^{h/2,h/4} - \tilde{y}_{n+1}^{h,h/2}}{7}$$

If we use the *GTE*, the formula will be $\tilde{y}_{n+1}^{h,h/2,h/4} = \dfrac{4\tilde{y}_{n+1}^{h/2,h/4} - \tilde{y}_{n+1}^{h,h/2}}{3}$.

To start third order AB, we require solutions at $t = 0.5$ and $t = 1.0$ to be evaluated using the Euler forward and Richardson's extrapolation.

Let us first see the application of passive extrapolation. We solve the equation using $h = 0.5$, 0.25, and 0.125 using Euler forward with the given initial condition $y(0) = 1$. Then compute higher-order estimates of y at $t = 0.5$ and $t = 1.0$ using the values at those time steps in the three expressions shown above. The computations are tabulated in Table 6.15.

Table 6.15

t	$y^{(0.5)}$	$y^{(0.25)}$	$y^{(0.125)}$	$y^{(0.5, 0.25)}$	$y^{(0.25, 0.125)}$	$y^{(0.5, 0.25, 0.125)}$
0	1	1	1			
0.125			0.9250			
0.25		0.8500	0.8556			
0.375			0.7915			
0.5	0.7000	0.7225	0.7321	0.7300	0.7353	0.7360
0.625			0.6772			
0.75		0.6141	0.6264			
0.875			0.5794			
1	0.4900	0.5220	0.5360	0.5326	0.5406	0.5417

The computation using active extrapolation has to be done in two steps. First step to $t = 0.5$ is same in active and passive extrapolation and therefore the values from the above table will be used. For the second time step to $t = 1.0$, the Euler forward has to be applied using the extrapolated value at 0.5 as initial condition. This is shown in Table 6.16.

Table 6.16

t	$y^{(0.5)}$	$y^{(0.25)}$	$y^{(0.125)}$	$y^{(0.5,0.25)}$	$y^{(0.25,0.125)}$	$y^{(0.5,0.25,0.125)}$
0.5	0.7360	0.7360	0.7360			
0.625			0.6808			
0.75		0.6256	0.6298			
0.875			0.5825			
1	0.5152	0.5318	0.5389	0.5373	0.5412	0.5418

Using these extrapolated values, calculation of third order Adams-Bashforth method is shown in Table 6.17. The identifications are

$y^{(\text{True})}$: True solution.

$y^{(1)}$: Solution obtained using progressively higher-order explicit methods similar to Example 6.6.

$y^{(2)}$: Solution obtained using passively extrapolated values at $t = 0.5$ and 1.0.

$y^{(3)}$: Solution obtained using actively extrapolated values at $t = 0.5$ and 1.0.

Table 6.17

t	$y^{(\text{True})}$	$y^{(1)}$	Error (%)	$y^{(2)}$	Error (%)	$y^{(3)}$	Error (%)
0	1.0000	1.0000		1.0000		1.0000	
0.5	0.7408	0.7000	5.51	0.7360	0.64	0.7360	0.64
1	0.5488	0.5350	2.52	0.5417	1.29	0.5418	1.28
1.5	0.4066	0.3824	5.95	0.3997	1.70	0.3997	1.70

(contd.)

(contd.)

2	0.3012	0.2890	4.05	0.2945	2.21	0.2946	2.20
2.5	0.2231	0.2089	6.38	0.2173	2.60	0.2173	2.60
3	0.1653	0.1566	5.27	0.1602	3.07	0.1602	3.06
3.5	0.1225	0.1140	6.92	0.1182	3.47	0.1182	3.47
4	0.0907	0.0850	6.34	0.0872	3.92	0.0872	3.91
4.5	0.0672	0.0621	7.55	0.0643	4.32	0.0643	4.32
5	0.0498	0.0461	7.32	0.0474	4.76	0.0474	4.76

When we compare the results at $t = 5.0$, the error has decreased from 7.32% to 4.76% due to the application of Richardson's extrapolation. This brings it much closer to what was obtained (2.99%) using true solution for the first three time steps in Example 6.1. There is not much difference in active and passive extrapolated values in this case because the difference was only for one time step.

6.4.1.2 *Use of Runge-Kutta Method*

In order to avoid the extra error incurred by lowering the time step, one may use the same order Runge-Kutta method for the initial time steps. However, one need to pay attention to the time step size for stability. For example, if one wants to apply the fourth order Adams-Moulton method for solving an ODE and 4th order Runge-Kutta method for start-up, the initial time step chosen should be small enough such that the Runge-Kutta method is stable even though the Adams-Moulton method, being an implicit one may afford a larger time step. One such application is shown in Example 6.8:

Example 6.8 Solve Example 6.1 using fourth order Adams-Moulton method. Use fourth order Runge-Kutta method for start-up.

Solution The iteration formulae for the fourth order Runge-Kutta method is shown in Example 6.3 and that for fourth order Adams-Moulton method in Example 6.1. We tabulate the calculated values in Table 6.18.

Table 6.18

Time (t)	Method used	True solution	Numerical solution	Error (%)
0		1.0000	1.0000	
0.5	RK4	0.7408	0.7408	0.00
1	RK4	0.5488	0.5488	0.01
1.5	AM4	0.4066	0.4066	0.00
2	AM4	0.3012	0.3012	0.01
2.5	AM4	0.2231	0.2231	0.02

(contd.)

(contd.)				
3	AM4	0.1653	0.1653	0.03
3.5	AM4	0.1225	0.1224	0.04
4	AM4	0.0907	0.0907	0.04
4.5	AM4	0.0672	0.0672	0.05
5	AM4	0.0498	0.0498	0.06

In this case, the fourth order Runge-Kutta method gives the best starting option resulting in minimum errors.

6.4.2 Combination of Methods: Predictor Corrector Schemes

One of the problem encountered in the application of any implicit method is the necessity of solving non-linear equations at each time step whenever the right-hand side function f is a non-linear function of the dependent variable. This becomes necessary because one of the function evaluation is at the time step $(n + 1)$. This problem does not occur in the explicit methods. A group of methods combine explicit and implicit methods in such a way so as to utilize the computational simplicity of the explicit methods as well as the stability property of the implicit methods. Basic idea of these methods is

- *Predict* the value of y_{n+1} using an explicit method.
- *Correct* the value of y_{n+1} using the implicit method. For this, the predicted value of y_{n+1} is used to evaluate f_{n+1}. This way, the solution of non-linear equation arising due to evaluation of f at $(n + 1)$ is avoided.

One of the most commonly used single-step predictor corrector method is Heun's method which can be written as follows:

Predictor:
$$y_{n+1}^{p} = y_n + hf(y_n, t_n) \qquad (6.107)$$

Corrector:
$$y_{n+1}^{c} = y_n + \frac{h}{2}[f(y_{n+1}^{p}, t_{n+1}) + f(y_n, t_n)] \qquad (6.108)$$

You may want to compare the above with the second order Runge-Kutta method (Table 6.6). Any combination of linear multistep explicit and implicit methods of same order (Table 6.3) can be used as predictor and corrector, respectively. This essentially means that any combination of Adams-Bashforth and Adams-Moulton methods of same order can serve as a set of predictor and corrector, respectively. They are often referred as *Adams method*.

There are two options for the application of corrector on a predicted value, *single* and *iterative*. In *single* corrector application, the corrector equation is applied only once on the predicted value. However, more commonly, the corrector equation is applied repeatedly (*iteratively*) until the values stop changing or the approximate relative error satisfies the preset error criterion. During each corrector application, the previously corrected value is used as the predicted value. The final truncation error and stability property of a predictor corrector method with repeated application of the corrector is same as the corrector formula provided the time step used for the predictor formula is within its stability limit.

Example 6.9 Solve the IVP

$$\frac{dy}{dt} = t^2 y - 2y; \quad y(0) = 1; \quad t \in (0, 0.5)$$

using (a) Heun's method without single corrector application, and (b) Heun's method with iteration with stopping criterion of 1% approximate relative error. Use $h = 0.25$.

Solution (a) *Heun's method with single corrector application*
Application of Heun's method to the IVP leads to the following equations:

Predictor: $\qquad\qquad y_{n+1}^p = y_n + h(t_n^2 y_n - 2y_n)$

Corrector: $\quad y_{n+1}^c = y_n + \dfrac{h}{2}[(t_{n+1}^2 y_{n+1}^p - 2y_{n+1}^p) + (t_n^2 y_n - 2y_n)]$

Computations are shown in Table 6.19.

Table 6.19

t_n	y_n	f_n	y_{n+1}^p	f_{n+1}^p	y_{n+1}^c
0	1	-2	0.5	-0.9688	0.6289
0.25	0.6289	-1.2185	0.3243	-0.5675	0.4057
0.5	0.4057				

(b) *Heun's method with iterative corrector application*
Application of Heun's method to the IVP leads to the following predictor corrector equations. The index i indicates the iterative sequence of corrector application:

Predictor: $\qquad\qquad y_{n+1}^p = y_n + h(t_n^2 y_n - 2y_n)$

Corrector: $y_{n+1}^{c,i+1} = y_n + \dfrac{h}{2}[(t_{n+1}^2 y_{n+1}^{c,i} - 2y_{n+1}^{c,i}) + (t_n^2 y_n - 2y_n)]; \quad y_{n+1}^{c,0} = y_{n+1}^p$

The computations are shown in Table 6.20. For the first time step,

Table 6.20

t_n	y_n	f_n	i	$y_{n+1}^{c,i}$	$f_{n+1}^{c,i}$	$y_{n+1}^{c,i+1}$	$\varepsilon(\%)$
0.0000	1.0000	-2.0000	0	0.5000	-0.9688	0.6289	
0.0000	1.0000	-2.0000	1	0.6289	-1.2185	0.5977	5.2234

(contd.)

(contd.)

0.0000	1.0000	−2.0000	2	0.5977	−1.1580	0.6052	1.2492
0.0000	1.0000	−2.0000	3	0.6052	−1.1727	0.6034	0.3035
0.2500	0.6034	−1.1691	0	0.3111	−0.5445	0.3892	20.0606
0.2500	0.6034	−1.1691	1	0.3892	−0.6811	0.3721	4.5897
0.2500	0.6034	−1.1691	2	0.3721	−0.6512	0.3759	0.9940
0.5	0.3759						

For the multi-step predictor corrector methods, we will come across the start-up problem once again. Solutions are also similar to the ones discussed earlier. Since, a predictor corrector scheme has the same order of accuracy as that of the corrector equation, one can improve that accuracy by the application of Richardson's extrapolation. Alternatively, one can also use gradually increasing order of predictor corrector method or Runge-Kutta method.

All the methods that we used to solve the single initial-value problems are equally applicable for a system of coupled IVPs that often need to be solved simultaneously in many engineering applications. In the next section, we will illustrate these applications and discuss the special problems that occur in such applications.

EXERCISE 6.4

1. Solve the differential equation $\dfrac{dy}{dt} = -y + e^{-t}$ with the initial condition $y(0) = 1$ using the following methods with $h = 0.1$ for t in $(0,1)$:
   (a) Fourth-order Adams-Bashforth method. For strart-up, use trapezoidal method with Richardson's extrapolation to obtain fourth-order accuracy.
   (b) Predictor corrector with fourth-order *Adams* methods. Use fourth-order Runge-Kutta for start-up.
2. A physical phenomenon is governed by the differential equation:
$$\frac{dv}{dt} = -0.2v - 2v^2 \cos(2t)$$
   subject to the initial condition $v(0) = 1$:
   (a) Using fourth-order Adams-Bashforth and Adams-Moulton predictor-corrector method, compute v at $t = 1$ with a time step of 0.1. Use Heun's method along with Richardson's extrapolation for start-up. Use the corrector iteratively.
   (b) Using Heun's method with iterative application of corrector, solve the system with $h = 0.05, 0.1, 0.2, 0.25, 0.5$. Numerically determine the order of the method with error at $t = 1$.
3. One major advantage of the predictor corrector methods is seen when applied

to IVPs with non-linear slope function where it gives the stability benefits of an implict method without having to iteratively solve a non-linear equation at every time step. Let us explore this numerically. Solve the following problems with non-linear slope functions using a first-order explicit method with limited stability properties (Euler forward), a second-order predictor corrector (*Heun's*) method, first- and second-order implicit *A-stable* methods (*Euler backward* and *trapezoidal*), a fourth-order explicit method (*Runge-Kutta*) and a fourth-order predctor-corrector (*Adams*) method. Compare them with respect to ease of application, accuracy and stability. Numerically explore the time step at which the methods start to produce oscillations. Apply corrector repeatedly for both *Heun's* and *Adams* methods.

(a) $3\dfrac{dy}{dx} - \dfrac{y}{x} - \dfrac{x^2}{y^2} = 0; \quad y(1) = 1; \quad x \in (1,2); \quad h = 0.2$

(b) $\dfrac{dy}{dx} - \dfrac{1}{x+y^2}; \quad y(0) = 0; \quad x \in (0,1); \quad h = 0.1$

4. Consider Problem 3(a). For the accuracy, we would like to use the sixth-order BDF to solve it using $h = 0.05$. Let us explore combinations of a number of lower-order explicit and implicit methods along with active and passive Richardson's extrapolation. Compare the following strategies for ease of application and accuracy. Obtain the true error at each time step from the analytical solution of the equation.

   (a) Euler forward with active Richardson's extrapolation.
   (b) Euler forward with passive Richardson's extrapolation.
   (c) Trapezoidal method with active Richardson's extrapolation.
   (d) Trapezoidal method with passive Richardson's extrapolation.
   (e) Second and fourth-order Runge-Kutta method with active Richardson's extrapolation.
   (f) Second and fourth-order Runge-Kutta method with passive Richardson's extrapolation.

5. Prove that a predictor corrector equation has the truncation error and stability properties of the corrector equation.

6.5 HIGHER-ORDER AND SYSTEM OF IVPs

Many engineering applications involve simultaneous solution of a set of ordinary differential equations of first order. These typically occur when one analyzes some properties of a network, such as flow, pressure, current, concentration, etc. However, these can also occur from an ODE of higher-order as we shall see later in the section.

To illustrate such an application, let us recall Example 2.6 involving inter-connected tanks. In Chapter 2, we solved the steady-state problem. What if we are interested in the unsteady state of the system and analyze for the time system takes to reach steady state after a shock loading. In that case, we will need to solve the following set of ordinary differential equations

$$V_1 \frac{dC_1}{dt} = QC_0 + W_1 + Q_{21}C_2 - Q_{12}C_1 - k_1 V_1 C_1$$

$$V_2 \frac{dC_2}{dt} = Q_{12}C_1 + W_2 + Q_{32}C_3 - Q_{21}C_2 - Q_{23}C_2 - k_2 V_2 C_2$$

$$V_3 \frac{dC_3}{dt} = Q_{23}C_2 + W_3 - Q_{32}C_3 - k_3 V_3 C_3 - QC_3$$

The three ODEs shown above cannot be solved in isolation since, the right-hand side functions are dependent on each other through the dependent variables C_1, C_2, and C_3. This set of equation can be written as follows:

$$\frac{dC}{dt} = AC + b \tag{6.109}$$

where $C = \begin{bmatrix} C_1 \\ C_2 \\ C_3 \end{bmatrix}$, $A = \begin{bmatrix} \left(-\dfrac{Q_{12}}{V_1} - k_1 \right) & \dfrac{Q_{21}}{V_1} & 0 \\[3mm] \dfrac{Q_{12}}{V_2} & \left(-\dfrac{Q_{21}}{V_2} - \dfrac{Q_{23}}{V_2} - k_2 \right) & \dfrac{Q_{32}}{V_2} \\[3mm] 0 & \dfrac{Q_{23}}{V_3} & \left(-\dfrac{Q_{32}}{V_3} - \dfrac{Q}{V_3} - k_3 \right) \end{bmatrix}$

and $$b = \begin{bmatrix} \dfrac{W_2 + QC_0}{V_1} \\[3mm] \dfrac{W_2}{V_2} \\[3mm] \dfrac{W_3}{V_3} \end{bmatrix}$$

In this example, the functions on the right-hand side were linear functions of the dependent variable and therefore, it was easy to express them in matrix form. However, it is easily possible that the functions are non-linear. For example, if the reaction in each tank is second order, the above set of equations becomes

$$V_1 \frac{dC_1}{dt} = QC_0 + W_1 + Q_{21}C_2 - Q_{12}C_1 - k_1 V_1 C_1^2$$

$$V_2 \frac{dC_2}{dt} = Q_{12}C_1 + W_2 + Q_{32}C_3 - Q_{21}C_2 - Q_{23}C_2 - k_2 V_2 C_2^2$$

$$V_3 \frac{dC_3}{dt} = Q_{23}C_2 + W_3 - Q_{32}C_3 - k_3 V_3 C_3^2 - QC_3$$

It is not possible to write these equations in the matrix form anymore. Therefore, a more general way of writing such a set of equations with m dependent variables will be

$$\frac{dy_1}{dt} = f_1(y_1, y_2, \ldots, y_m, t)$$

$$\frac{dy_2}{dt} = f_2(y_1, y_2, \ldots, y_m, t)$$

$$\vdots \qquad\qquad\qquad\qquad (6.110)$$

$$\frac{dy_m}{dt} = f_m(y_1, y_2, \ldots, y_m, t)$$

In the vector form, they can be written as

$$\frac{dy}{dt} = f(y, t) \qquad\qquad (6.111)$$

where y is the vector of the dependent variables and f is the vector containing the right-hand side functions.

The higher-order initial value problems can always be expressed as a system of equations. Let us consider the following example of a second-order initial-value problem:

$$x^2 y'' - xy' + y = \frac{1}{x}, \quad \text{where } x \in (1, \infty), \ y(1) = 0, \text{ and } y'(1) = 0$$

Let us define two variables u and v as follows:

$$u = y \quad \text{and} \quad v = y'$$

Using this definition, the second order initial-value problem can be decomposed into a system of equation as follows:

$$u' = v$$

$$v' = \frac{v}{x} - \frac{u}{x^2} + \frac{1}{x^3}$$

$$u(1) = 0, \ v(1) = 0$$

This is a complete system of IVPs in variables $u(x)$ and $v(x)$ with the initial condition defined appropriately. The values of $u(x)$ will provide the required solution. An added advantage is that the gradient is calculated simultaneously as $v(x)$ which may be required in many engineering problems. The method illustrated in the above example can be generalized for an IVP of arbitrary order. Let us consider an IVP of order m. We shall denote the mth derivative as $y^{[m]}$. A general form of such an IVP is

$$y^{[m]} = f(y^{[m-1]}, y^{[m-2]}, ..., y', y; t) \qquad (6.112)$$

with m initial conditions as $y = a_0$, $y' = a_1$, $y'' = a_2$, ..., $y^{[m-1]} = a_{m-1}$ where $a_0, a_1, ..., a_{m-1}$ are constants or functions of the independent variable t.

Let us first define a set of variable $\{u_1, u_2, u_3, ..., u_m\}$ as follows:

$$u_1 = y, \quad u_2 = y', \quad u_3 = y'', ..., u_m = y^{[m-1]} \qquad (6.113)$$

Using the new variables, the mth order IVP Eq. (6.112) yields the following system of first order IVPs:

$$
\begin{aligned}
u_1' &= u_2 \\
u_2' &= u_3 \\
u_3' &= u_4 \\
&\vdots \\
u_{m-1}' &= u_m \\
u_m' &= f(u_1, u_2, u_3, ..., u_m; t)
\end{aligned}
\qquad (6.114)
$$

Initial conditions: $u_1^0 = a_0$, $u_2^0 = a_1$, $u_3^0 = a_2, ..., u_m^0 = a_{m-1}$

The above set of equation is of the same form as the set of Eq. (6.110). Therefore, using this framework, an arbitrary mth order initial-value problem can be decomposed into a system of m first order initial value problems.

In the vector form, Eq. (6.111) is of the same form as our general single variable equation given by Eq. (6.3). Therefore, all the methods that have been derived for the single variable can be directly applied to Eq. (6.111) as well. The only difference is that the scalar operations are now replaced by the vector operations. In order to illustrate this, let us apply Euler forward method to Eq. (6.111):

$$y^{n+1} = y^n + hf^n \qquad (6.115)$$

In the expanded variable form, the above can be written as

$$\begin{bmatrix} y_1^{n+1} \\ y_2^{n+1} \\ \cdots \\ y_m^{n+1} \end{bmatrix} = \begin{bmatrix} y_1^n \\ y_2^n \\ \cdots \\ y_m^n \end{bmatrix} + h \begin{bmatrix} f_1(y_1^n, y_2^n \cdots y_m^n, t^n) \\ f_2(y_1^n, y_2^n \cdots y_m^n, t^n) \\ \cdots \\ f_m(y_1^n, y_2^n \cdots y_m^n, t^n) \end{bmatrix} \qquad (6.116)$$

The time step identifier in the above expressions is indicated in the superscript, since the subscript is occupied by the variable identifier. This notational structure will be used throughout this section.

The initial condition of these problems is known in the form of a vector as follows:

$$y^0 = \begin{bmatrix} y_1^0 \\ y_2^0 \\ \vdots \\ y_m^0 \end{bmatrix} \qquad (6.117)$$

Using the initial condition, one can easily compute the solution vector at the later time steps through a series of vector operation. One such application is shown in Example 6.10.

Example 6.10   Solve the following IVP using Euler forward and fourth order Runge-Kutta methods with $h = 0.1$:

$$f''' + \alpha ff'' + \beta(1 - f'^2) = 0; \quad f(0) = 0, \ f'(0) = 0, \ f''(0) = 10.0, \ x \in (0,1), \quad \alpha = 1$$

and $\beta = 1$. Graphically compare the solutions obtained by two methods.

Solution   Let us assume $u = f$, $v = f'$, and $w = f''$. Using these, the third order IVP is transformed into the following system of equations:

$$u' = v$$
$$v' = w$$
$$w' = -\alpha uw - \beta(1 - v^2)$$

Initial conditions: $u(0) = 0$, $v(0) = 0$, and $w(0) = 10.0$

Let us denote the vectors of the independent variables and slope functions as

$$y = \begin{bmatrix} u \\ v \\ w \end{bmatrix} \text{ and } f(y, x) = \begin{bmatrix} v \\ w \\ -\alpha uw - \beta(1 - v^2) \end{bmatrix}$$

(a) The Euler forward method can be written as: $y_{n+1} = y_n + h f_n$. This translates to,

$$
\begin{bmatrix} u_{n+1} \\ v_{n+1} \\ w_{n+1} \end{bmatrix} = \begin{bmatrix} u_n \\ v_n \\ w_n \end{bmatrix} + h \begin{bmatrix} v_n \\ w_n \\ -\alpha u_n w_n - \beta(1 - v_n^2) \end{bmatrix} = \begin{bmatrix} u_n + h v_n \\ v_n + h w_n \\ w_n - h\alpha u_n w_n - h\beta(1 - v_n^2) \end{bmatrix}
$$

Given the initial conditions in the form of u_n, v_n, and w_n, the values of u_{n+1}, v_{n+1}, and w_{n+1} can be computed explicitly. We tabulate the computed values in Table 6.21.

Table 6.21

x	u	v	w
0.0000	0.0000	0.0000	10.0000
0.1000	0.0000	1.0000	9.9000
0.2000	0.1000	1.9900	9.9000
0.3000	0.2990	2.9800	10.0970
0.4000	0.5970	3.9897	10.5831
0.5000	0.9960	5.0480	11.4431
0.6000	1.5008	6.1923	12.7517
0.7000	2.1200	7.4675	14.5724
0.8000	2.8668	8.9247	16.9594
0.9000	3.7592	10.6207	19.9626
1.0000	4.8213	12.6169	23.6381

(b) The fourth order Runge-Kutta method in the vector form can be written as:

$$
y_{n+1} = y_n + h\left[\frac{1}{6}\phi_0 + \frac{1}{3}(\phi_1 + \phi_2) + \frac{1}{6}\phi_3 \right]
$$

where ϕ_0, ϕ_1, ϕ_2, and ϕ_3 are vectors. These can be computed as outlined below:

$$
\phi_0 = f(y_n, x_n) = \begin{bmatrix} v_n \\ w_n \\ -\alpha u_n w_n - \beta(1 - v_n^2) \end{bmatrix}
$$

$$
\phi_1 = f\left(y_n + \frac{1}{2} h\phi_0, x_n + \frac{1}{2} h \right)
$$

The new vector of the independent variables (y) at which the vector $\boldsymbol{\phi}_1$ is evaluated can be computed as follows:

$$y_n + \frac{1}{2} h \phi_0 = \begin{bmatrix} u_n \\ v_n \\ w_n \end{bmatrix} + \frac{h}{2} \begin{bmatrix} v_n \\ w_n \\ -\alpha u_n w_n - \beta(1-v_n^2) \end{bmatrix} =$$

$$\begin{bmatrix} u_n + \dfrac{h}{2} v_n \\ v_n + \dfrac{h}{2} w_n \\ w_n - \dfrac{h}{2} \alpha u_n w_n - \dfrac{h}{2} \beta(1-v_n^2) \end{bmatrix} = \begin{bmatrix} u_{n1} \\ v_{n1} \\ w_{n1} \end{bmatrix} \text{ (say)}$$

Given the initial conditions in the form of u_n, v_n, and w_n, the values of u_{n1}, v_{n1}, and w_{n1} can be computed explicitly. The vector $\boldsymbol{\phi}_1$ can then be computed as follows:

$$\phi_1 = \begin{bmatrix} v_{n1} \\ w_{n1} \\ -\alpha u_{n1} w_{n1} - \beta(1-v_{n1}^2) \end{bmatrix}$$

Similarly, $\phi_2 = f\left(y_n + \dfrac{1}{2} h \phi_1, x_n + \dfrac{1}{2} h \right)$.

The new vector of the independent variables (y) at which the vector $\boldsymbol{\phi}_2$ is evaluated can be computed as follows:

$$y_n + \frac{1}{2} h \phi_1 = \begin{bmatrix} u_n \\ v_n \\ w_n \end{bmatrix} + \frac{h}{2} \begin{bmatrix} v_{n1} \\ w_{n1} \\ -\alpha u_{n1} w_{n1} - \beta(1-v_{n1}^2) \end{bmatrix} =$$

$$\begin{bmatrix} u_n + \dfrac{h}{2} v_{n1} \\ v_n + \dfrac{h}{2} w_{n1} \\ w_n - \dfrac{h}{2} \alpha u_{n1} w_{n1} - \dfrac{h}{2} \beta\left(1-v_{n1}^2\right) \end{bmatrix} = \begin{bmatrix} u_{n2} \\ v_{n2} \\ w_{n2} \end{bmatrix} \text{ (say)}$$

Once again, the values of u_{n2}, v_{n2}, and w_{n2} can be computed explicitly using the previously computed values. The vector $\boldsymbol{\phi}_2$ can then be computed as follows:

$$\boldsymbol{\phi}_2 = \begin{bmatrix} v_{n2} \\ w_{n2} \\ -\alpha u_{n2} w_{n2} - \beta\left(1 - v_{n2}^2\right) \end{bmatrix}$$

Lastly, $\boldsymbol{\phi}_3 = f\left(\mathbf{y}_n + h\boldsymbol{\phi}_2, x_n + h\right)$.

The new vector of the independent variables ($\mathbf{y}$) at which the vector $\boldsymbol{\phi}_3$ is evaluated can be computed as follows:

$$\mathbf{y}_n + h\boldsymbol{\phi}_2 = \begin{bmatrix} u_n \\ v_n \\ w_n \end{bmatrix} + h\begin{bmatrix} v_{n2} \\ w_{n2} \\ -\alpha u_{n2} w_{n2} - \beta\left(1 - v_{n2}^2\right) \end{bmatrix} =$$

$$\begin{bmatrix} u_n + hv_{n2} \\ v_n + hw_{n2} \\ w_n - h\alpha u_{n2} w_{n2} - h\beta\left(1 - v_{n2}^2\right) \end{bmatrix} = \begin{bmatrix} u_{n3} \\ v_{n3} \\ w_{n3} \end{bmatrix} \text{ (say)}$$

The values of u_{n3}, v_{n3}, and w_{n3} can be computed explicitly using the previously computed values. The vector $\boldsymbol{\phi}_3$ can then be computed as follows:

$$\boldsymbol{\phi}_3 = \begin{bmatrix} v_{n3} \\ w_{n3} \\ -\alpha u_{n3} w_{n3} - \beta(1 - v_{n3}^2) \end{bmatrix}$$

Using the values of the vectors $\boldsymbol{\phi}_0$, $\boldsymbol{\phi}_1$, $\boldsymbol{\phi}_2$, and $\boldsymbol{\phi}_3$, and the vector of the initial condition ($\mathbf{y}_n$), the $\mathbf{y}_{n+1}$ can be computed. The values are tabulated in Table 6.22.

Table 6.22

x	u	v	w
0.0000	0.0000	0.0000	10.0000
0.1000	0.0499	0.9954	9.9166
0.2000	0.1990	1.9866	9.9320
0.3000	0.4475	2.9883	10.1425

(*contd.*)

(contd.)

0.4000	0.7978	4.0247	10.6405
0.5000	1.2547	5.1288	11.5126
0.6000	1.8272	6.3422	12.8382
0.7000	2.5285	7.7138	14.6895
0.8000	3.3771	9.2997	17.1337
0.9000	4.3976	11.1624	20.2389
1.0000	5.6212	13.3719	24.0831

The graphical comparison of the solutions obtained by two methods is shown in Fig. 6.20.

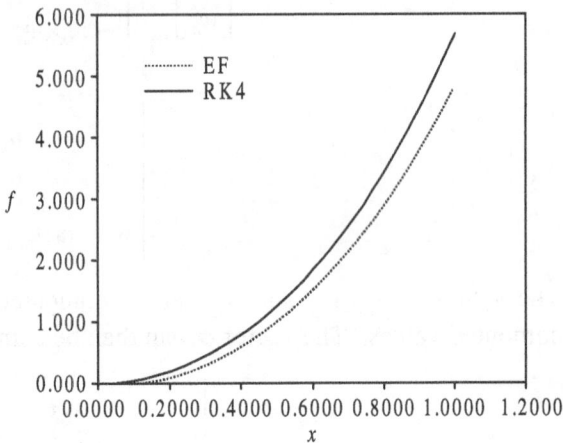

Fig. 6.20

The multi-step explicit method will experience start-up problems similar to the single variable case but the same strategies (Section 6.4.1) can be applied with vector operations.

Let us now attempt to apply an implicit method to a set of equation. We will use the Euler backward method for simplicity, but it will illustrate all the problems typically associated with such applications. Application of Euler backward method to Eq. (6.111) gives

$$y^{n+1} = y^n + hf^{n+1} \tag{6.118}$$

In the expanded vector form, this looks like

$$
\begin{bmatrix} y_1^{n+1} \\ y_2^{n+1} \\ \vdots \\ y_m^{n+1} \end{bmatrix} = \begin{bmatrix} y_1^n \\ y_2^n \\ \vdots \\ y_m^n \end{bmatrix} + h \begin{bmatrix} f_1(y_1^{n+1}, y_2^{n+1} \cdots y_m^{n+1}, t^n) \\ f_2(y_1^{n+1}, y_2^{n+1} \cdots y_m^{n+1}, t^n) \\ \vdots \\ f_m(y_1^{n+1}, y_2^{n+1} \cdots y_m^{n+1}, t^n) \end{bmatrix} \tag{6.119}
$$

The above is similar to Eq. (3.47) where one can solve for the vector y^{n+1} using any of the methods described in Section 3.7. However, before we proceed with the solution, let us look at the cases when the functions in f are linear and when they are non-linear, separately. For the linear functions, we have seen Eq. (6.109) that the set of equations can be expressed as

$$
\frac{dy}{dt} = Ay + b \tag{6.120}
$$

Application of Euler backward method results in,

$$
y^{n+1} = y^n + hAy^{n+1} + hb^{n+1} \tag{6.121}
$$

Notice that the vector b not only has constants but may also contain functions of the independent variable t. In that case, the value of t has to be set at $(n + 1)$ time step. Some algebraic manipulation leads to the following matrix equation

$$
[I - hA] y^{n+1} = y^n + hb^{n+1} \tag{6.122}
$$

This set of linear equation needs to be solved at every step where the coefficient matrix $(I - hA)$ remain constant for uniform time steps. Therefore, if a LU decomposition (Section 2.2.1.3) of the coefficient matrix is carried out once, solutions at all the time steps (as long as h remains constant) can be easily computed by solving two triangular matrix equations.

For non-linear functions, there is no option other than solving the non-linear set of Eq. (6.119). One can apply either fixed point iteration or Newton-Raphson method. For faster convergence, it is a common practice to use the latter. Application of Newton-Raphson method to Eq. (6.119) gives,

$$
[I - hJ^{n+1,k}] (y^{n+1,k+1} - y^{n+1,k}) = -y^{n+1,k} + hf^{n+1,k} + y^n \tag{6.123}
$$

where k is the Newton-Raphson iteration index and J is the Jacobian given by

$$
J = \begin{bmatrix} \dfrac{\partial f_1}{\partial y_1} & \dfrac{\partial f_1}{\partial y_2} & \cdots & \dfrac{\partial f_1}{\partial y_m} \\[2mm] \dfrac{\partial f_2}{\partial y_1} & \dfrac{\partial f_2}{\partial y_2} & \cdots & \dfrac{\partial f_2}{\partial y_m} \\[2mm] \vdots & \vdots & \vdots & \vdots \\[2mm] \dfrac{\partial f_m}{\partial y_1} & \dfrac{\partial f_m}{\partial y_2} & \cdots & \dfrac{\partial f_m}{\partial y_m} \end{bmatrix} .
$$

The Jacobian has to be evaluated at $y^{n+1,k} = [y_1^{n+1,k} \, y_2^{n+1,k} \cdots y_m^{n+1,k}]^{\mathrm{T}}$. At each time step, the Newton-Raphson iterations are started by setting $y^{n+1, \, 0} = y^n$. The Jacobian is evaluated at this point. Thereafter, most common implementations do not alter the Jacobian at every Newton-Raphson iterations. It is recalculated only intermittently during every time step. If the Newton-Raphson iteration slows down, it is an indication to recalculate the Jacobian. Near the convergence, the Newton-Raphson is slow even after recalculating the Jacobian. Since the Jacobian is always calculated near the convergence at every time step, the same Jacobian can also be used for starting the iteration at the next time step if the start-up condition $y^{n+1, \, 0} = y^n$ is used. If the Jacobian is held constant for a few iterations, it is always more economical (with respect to computation time) to perform a *LU* decomposition. The following example illustrates one such application.

Example 6.11 Solve the problem of Example 6.10 with $f''(0) = 5.0$ using Euler backward method with Newton-Raphson iterations for the system of equation. Use a tolerance of 0.1% as the maximum norm of approximate relative error vector. Show one complete iteration.

Solution The problem in the vector form can be written as

$$y' = f$$

$$y = \begin{bmatrix} u \\ v \\ w \end{bmatrix}, \quad f = \begin{bmatrix} v \\ w \\ -\alpha u w - \beta(1 - v^2) \end{bmatrix}$$

The Jacobian for the functions is

$$J = \begin{bmatrix} 0 & 1 & 0 \\ 0 & 0 & 1 \\ -\alpha w & 2\beta v & -\alpha u \end{bmatrix}$$

Applying the Euler backward time stepping (superscript index n) along with Newton-Raphson iterations (subscript index k) similar to Eq. (6.123), we obtain

$$\begin{bmatrix} 1 & -h & 0 \\ 0 & 1 & -h \\ \alpha h w^{n+1,k} & -2\beta h v^{n+1,k} & 1+\alpha h u^{n+1,k} \end{bmatrix} \begin{bmatrix} u^{n+1,k+1} - u^{n+1,k} \\ v^{n+1,k+1} - v^{n+1,k} \\ w^{n+1,k+1} - w^{n+1,k} \end{bmatrix} =$$

$$\begin{bmatrix} -u^{n+1,k} + h v^{n+1,k} + u^n \\ -v^{n+1,k} + h w^{n+1,k} + v^n \\ -w^{n+1,k} - \alpha h u^{n+1,k} w^{n+1,k} - \beta h \left[1 - \left(v^{n+1,k} \right)^2 \right] + w^n \end{bmatrix}$$

The above is in the form of $Ax = b$ where the notations are as follows:

$$A = \begin{bmatrix} 1 & -h & 0 \\ 0 & 1 & -h \\ \alpha h w^{n+1,k} & -2\beta h v^{n+1,k} & 1+\alpha h u^{n+1,k} \end{bmatrix}$$

$$x = \begin{bmatrix} u^{n+1,k+1} - u^{n+1,k} \\ v^{n+1,k+1} - v^{n+1,k} \\ w^{n+1,k+1} - w^{n+1,k} \end{bmatrix}$$

$$b = \begin{bmatrix} -u^{n+1,k} + h v^{n+1,k} + u^n \\ -v^{n+1,k} + h w^{n+1,k} + v^n \\ -w^{n+1,k} - \alpha h u^{n+1,k} w^{n+1,k} - \beta h \left[1 - \left(v^{n+1,k} \right)^2 \right] + w^n \end{bmatrix}$$

Notice that the vector x is the incremental improvement of the solution at every Newton-Raphson iterations. To start Newton-Raphson iterations, we set $u^{n+1,0} = u^n$, $v^{n+1,0} = v^n$, and $w^{n+1,0} = w^n$. Two such iterations are shown below. We leave the other iterations for the readers to fill up.

Initial vector is given as $y^0 = \begin{bmatrix} 0 \\ 0 \\ 5 \end{bmatrix}$. With this, the initial Jacobian matrix is calculated as

$$J = \begin{bmatrix} 0 & 1 & 0 \\ 0 & 0 & 1 \\ -5 & 0 & 0 \end{bmatrix}$$

Since, the matrix $A = I - hJ$, we obtain

$$A = \begin{bmatrix} 1 & -0.05 & 0 \\ 0 & 1 & -0.05 \\ 0.25 & 0 & 1 \end{bmatrix}$$

We are not going to recalculate Jacobian until the solution vector becomes stagnant, i.e., difference between successive iterations is less than error tolerance. Therefore, we compute an LU decomposition of A using Doolittle's algorithm:

$$L = \begin{bmatrix} 1 & 0 & 0 \\ 0 & 1 & 0 \\ 0.25 & 0.0125 & 1 \end{bmatrix}, \quad U = \begin{bmatrix} 1 & -0.05 & 0 \\ 0 & 1 & -0.05 \\ 0 & 0 & 1.000625 \end{bmatrix}$$

Using the initial conditions as starting values for the Newton-Raphson iterations, we compute the vector,

$$b = \begin{bmatrix} 0 \\ 0.25 \\ -0.05 \end{bmatrix}$$

Now, we are ready to compute the improvement vector x using the LU decomposition with the right-side vector b. We also compute the new y vector by adding x to the existing y the error vector containing the approximate relative errors. These are

$$x = \begin{bmatrix} 0.0124 \\ 0.2473 \\ -0.0531 \end{bmatrix}, \quad y^{1,1} = \begin{bmatrix} 0.0124 \\ 0.2473 \\ 4.9469 \end{bmatrix}, \quad \text{and } e = \begin{bmatrix} NA \\ NA \\ 1.06 \end{bmatrix}$$

Notice that the first two errors denoted by NA cannot be computed at this step.

Since the error criterion is not satisfied, we recompute the vector b using the new values of y as

$$b = \begin{bmatrix} 0 \\ 2.78 \times 10^{-17} \\ 0.0031 \end{bmatrix}$$

Once again, using the LU decomposition we compute vectors x, y, and e:

$$x = \begin{bmatrix} 7.72 \times 10^{-6} \\ 0.0002 \\ 0.0031 \end{bmatrix}, \quad y^{1,2} = \begin{bmatrix} 0.0124 \\ 0.2475 \\ 4.95 \end{bmatrix}, \quad \text{and} \quad e = \begin{bmatrix} 0.0625 \\ 0.0625 \\ 0.0625 \end{bmatrix}$$

Although the error criterion is satisfied, we cannot exit the Newton-Raphson iterations until we recompute the Jacobian and verify whether the convergence has really taken place. Or else, we have to reiterate with the new Jacobian. So, the new Jacobian was recomputed as:

$$J = \begin{bmatrix} 0 & 1 & 0 \\ 0 & 0 & 1 \\ -4.95 & 0.495 & -0.0124 \end{bmatrix}$$

The new A, L, and U were recomputed using the new Jacobian:

$$A = \begin{bmatrix} 1 & -0.05 & 0 \\ 0 & 1 & -0.05 \\ 0.2475 & -0.02475 & 1.0006 \end{bmatrix}$$

$$L = \begin{bmatrix} 1 & 0 & 0 \\ 0 & 1 & 0 \\ 0.2475 & -0.0124 & 1 \end{bmatrix}, \quad U = \begin{bmatrix} 1 & -0.05 & 0 \\ 0 & 1 & -0.05 \\ 0 & 0 & 1 \end{bmatrix}$$

Using the new y and the initial conditions, we recomputed the vector

$$b = \begin{bmatrix} 1.73 \times 10^{-18} \\ -2.78 \times 10^{-17} \\ 1.93 \times 10^{-6} \end{bmatrix}$$

Next we compute vectors x, y, and e as:

$$x = \begin{bmatrix} 4.83 \times 10^{-9} \\ 9.66 \times 10^{-8} \\ 1.93 \times 10^{-6} \end{bmatrix}, \quad y^{1,3} = \begin{bmatrix} 0.0124 \\ 0.2475 \\ 4.95 \end{bmatrix}, \quad \text{and} \quad e = \begin{bmatrix} 3.9 \times 10^{-5} \\ 3.9 \times 10^{-5} \\ 3.9 \times 10^{-5} \end{bmatrix}$$

Since the error satisfies the specified criterion, we terminate the Newton-Raphson iterations and set

$$y^1 = \begin{bmatrix} 0.0124 \\ 0.2475 \\ 4.95 \end{bmatrix}$$

We proceed similarly to compute y^2 and so on. Notice that the last computed Jacobian could be used as the starting Jacobian for the next time step.

The problem of start-up of the multi-step method also occurs in this case and is addressed similar to single equations (Section 6.3.1). The predictor corrector method can save significant amount of computation time for the system of IVPs. In a typical application, the explicit method is used for the prediction of y^{n+1} and the predicted vector $\left(y_p^{n+1}\right)$ is used as a starting point for the Newton-Raphson iteration for implicit methods, i.e., $y^{n+1,0} = y_p^{n+1}$.

We have seen the application of the methods to systems of IVPs. However, we also need to understand how the error analysis, shown for single variable IVPs translate to multiple equations. Intuitively, it would be easy to visualize that the order of truncation error of any method would remain the same. In the next section, we will show the stability analysis for a system of IVPs and in the process, introduce the concept of *stiff equations*.

6.5.1 Stability of a System of IVPs and Stiff Systems

First we need a model problem for the analysis of stability for a system of IVPs. Following the same logic (Section 6.3.2) as of single variable IVP, we formulate a system of IVPs where the left-hand side consists of the derivatives of dependent variables and the right-hand side functions are *linear* combinations of all the dependent variables. We neglect all terms consisting of functions of independent variable and constants. We also do not consider the non-linear terms of the dependent variables. This can be expressed as follows:

$$\frac{dy}{dt} = Ay \tag{6.124}$$

Application of Euler forward method to the model problem yields,

$$y^{n+1} = [I + hA]y^n \quad \text{or} \quad y^{n+1} = [I + hA]^{n+1}y^0 \tag{6.125}$$

For the above iteration to be stable and eventually converge, the limit $\lim_{n \to \infty}[I + hA]^{n+1}$ must approach zero. Following the logic of Section 2.3.2, this would mean that all the eigenvalues of the matrix are less than unity. This is automatically satisfied if the spectral radius or the largest eigenvalue of the matrix $[I + hA]$ is less than unity.

Alternatively, taking the norms of the vectors and the matrix, one can define an amplification factor similar to the single IVP:

$$\left| y^{n+1} \right| \leq \left\| I + hA \right\| \left| y^n \right| \quad \text{or} \quad \sigma = \frac{\left| y^{n+1} \right|}{\left| y^n \right|} \leq \left\| I + hA \right\| \tag{6.126}$$

The amplification factor according to the definition of stability must be less than unity. Since $\sigma \leq \left\| I + hA \right\|$, using the relation between the spectral radius and matrix norms [Gelfund's formula, Eq. (2.38)], it is possible to write a necessary condition for convergence as follows:

$$\rho(I + hA) \leq 1 \tag{6.127}$$

If the maximum eigenvalue of the matrix A is λ_{max}, the above relation translates to

$$\left| 1 + h\lambda_{max} \right| \leq 1 \tag{6.128}$$

Since the eigenvalue can be real or imaginary, the above relation is similar to Eq. (6.64) and defines a similar stability region. To illustrate the difficulty with stiff system, let us assume that the λ_{max} is real. In that case, the stability condition becomes

$$h \leq \frac{2}{\lambda_{max}} \tag{6.129}$$

It is easy to see that the time step for the whole system of equation is restricted by λ_{max}. In many engineering systems, this may happen due to one or two equations being too restrictive while rest of the equations allow much larger time steps. Such a system is typically characterized by a large $(\lambda_{max}/\lambda_{min})$ ratio and is called a *stiff system*. Although there is no universally accepted limit that can be placed on this ratio, in general, a system may be stiff if $\lambda_{max}/\lambda_{min} > 100$.

For a general problem where the right-hand side functions are non-linear, a first order approximation (linearization) leads to Jacobian as the equivalent of A matrix of the model problem. Therefore, one can estimate the largest and smallest eigenvalues of the Jacobian at the initial point to evaluate the stiffness before one sets out to solve the system. Before we proceed to discuss the solution methods for stiff systems, let us look at an example to get a visual understanding of a stiff system and the problems associated with them. We consider the following system of equations:

$$u' = -50u$$
$$v' = -50u - 0.1v + t$$
$$u(0) = 1, \quad v(0) = 0, \quad t \in (0, \infty)$$

The above set of equations can be expressed in the form of Eq. (6.109). The eigenvalues of the matrix A are 50 and 0.1. Therefore, $\lambda_{max}/\lambda_{min} = 500$.

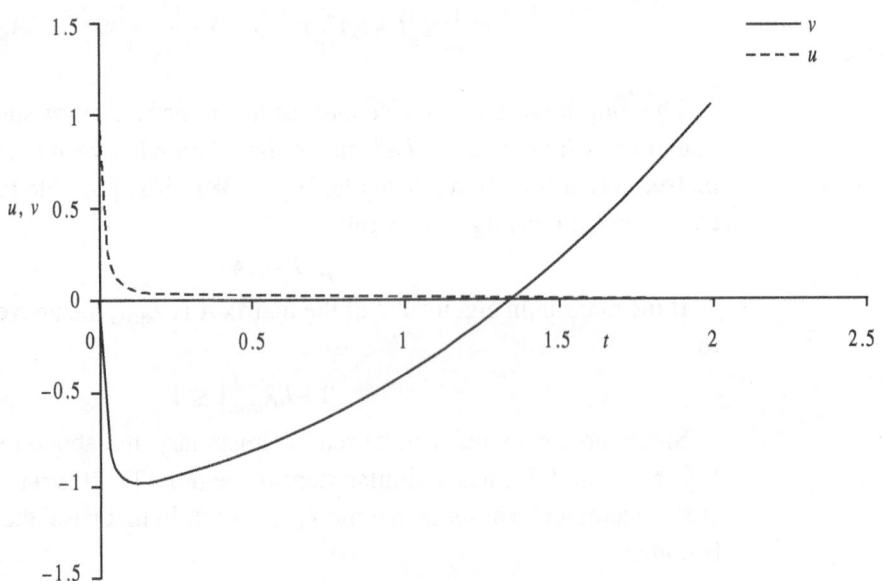

Fig. 6.21   Analytical solution of a stiff system consisting of one (u, v) rapidly decaying solution and another slowly growing solution in v.

To have a visual sense of a *stiff system*, let us plot the analytical solutions (Fig. 6.21). Analytical solution of the system is

$$u = e^{-50t}$$

and
$$v = 1.002e^{-50t} + 98.998e^{-0.1t} + 10t - 100$$

Both u and v initially decrease rapidly. Thereafter, v increases and u decreases slowly. Essentially, there are two different time scales, one very fast and the other, slow. To get an understanding of how fast the initial decrease is, let us look at a few values (Table 6.23).

Table 6.23   Values of u and v of Fig. 6.21 at the initial stages

t	u	v
0.0000	1.0000	0.0000
0.0100	0.6065	−0.3932

(contd.)

(contd.)

0.0200	0.3679	−0.6312
0.0300	0.2231	−0.7750
0.0400	0.1353	−0.8616
0.0500	0.0821	−0.9135
0.0600	0.0498	−0.9443
0.0700	0.0302	−0.9623
0.0800	0.0183	−0.9725
0.0900	0.0111	−0.9779
0.1000	0.0067	−0.9803

Within 0.05, the values of both u and v have decreased by about 92%. For rest of the time, u decreased only by 8% and v increased gradually. In order to simulate the rapid initial decrease, we will require a few closely spaced grid points (or small time steps) within the first 0.05 time units. Thereafter, the grid points may be sparsed or the time steps may be large.

If we want to apply Euler forward method to solve it, maximum allowable time step throughout the time span will be given by $2/50 = 0.04$. To compute a solution that is free from any oscillation, we need to use an even smaller time step. This we have observed in Example 6.4 that oscillations start to occur at time steps much below the stability limit although they are damped out eventually. We will not be able to increase the time step beyond 0.04 even after the initial rapid decrease stage because of the stability limitation of the numerical method used although the actual problem will allow a larger time step after 0.05 time units.

It is clear from the above example that in a stiff system, the step size requirement is guided by one critical time scale and this severely affects the computational efficiency of the whole system. Most of the stiff systems require smaller time steps in the initial phase (when the solution has sharp gradients) but can easily use a larger time step if the numerical method allows it from the stability condition. The explicit method such as Euler forward do not allow larger time steps even when the problem might.

The solution, of course, would be to use methods that are capable of handling changing time steps and stable for a wide range of time step values. Cleary, all *A-stable* methods would permit this. The Euler backward and trapezoidal methods are *A-stable* and can be used for stiff problems. Recall that the *stiffly stable* methods also allow larger time steps as their stability region is wedge shaped on the left half of the plane. The region of stability for *stiffly stable* methods expands as we move away from the origin towards $-\infty$ along the real axis. The whole group of BDFs up to order six are *stiffly stable* and are used to solve the stiff system with highest efficiency. These are also known as *Gear methods*. All of them allow changing time steps to a large extent.

It is possible to start with the first-order BDF with very small time step. Gradually, time steps can be increased as well as higher-order BDF can be used. When used judiciously, these allow increasing time steps as well as solve the start-up problem of the multi-step BDFs. For example, if we start with a first-order BDF with a time step of h, we need to use it for *at least* two time steps before we can use a second order BDF with a time step of $2h$. Subsequently, the second-order BDF have to be used for *at least* 3 time steps before one can use a third order BDF with a time step of $4h$ and the third order BDF with a time step of $4h$ have to be used for *at least* 4 time steps before one can use the fourth order BDF with a time step of $8h$. In general, if one wants to increase the time step by a factor of k when going to a higher-order method, the first order method have to be used for *at least* k times. Thereafter, each method has to be used for *at least* $(m + 1)$ times where m is the order of the method.

However, one often needs to continue with the smaller time steps for longer than the minimum required number. Essentially, the following may be used as guidelines for changing time steps:

- Have sufficient number of closely spaced grid points where the solution is changing rapidly.
- When changing to a higher-order method with larger time step, make sure the values at the required past time nodes are available from the lower-order computation with smaller time steps.
- The BDF are essentially weighted average approximation of values at different time nodes. When changing to higher-order multi-step method, avoid using values from the nodes where the solution is changing rapidly. This may result in oscillation (not instability!). It is wise to continue with smaller time steps for some more steps so that enough values from the slowly changing region are generated for the start-up of the multi-step method.

An application of BDF with change of time step and method is shown in Example 6.12.

It is, of course, possible to change time steps whenever the problem allows it but in that case, the BDF shown in Table 6.4 will not be applicable. Because in such a case, different time step sizes will be included in the same formulae. For example, $(t_{n-3} - t_{n-2})$ and $(t_{n-2} - t_{n-1})$ will be different if the time step was changed at t_{n-2}.

Recall that the BDF in Table 6.4 was derived based on the assumption of uniform grid sizes. They were formulated by approximating dy/dt on a uniform mesh involving a number of past grid points starting from $(n + 1)$.

We need to recompute new coefficients for the BDF for non-uniform grid sizes by obtaining a backward difference approximation of the derivative on a non-uniform grid. Computation of new coefficients has to be done at every time step. Most of the modern implementations of the *Gear's method* are based on recomputation of coefficients on a non-uniform grid at every time step. We will discuss the difference approximation on a non-uniform grid in Chapter 8.

Before we can start the solution, we will need to decide on an initial time step h_0 for a stiff problem to be able to clearly depict the rapidly changing part of the solution. There are several suggestions in the literature. For the stiff problems, a reasonable choice is, [Enright et al. (1975)]:

$$h_0 = \frac{1}{|\lambda_{max}|} \tag{6.130}$$

where λ_{max} is the eigenvalue of maximum absolute magnitude of matrix A if the functions are linear [Eq. (6.120)] or of the Jacobian for a general non-linear problem.

Example 6.12   Solve the following system of IVPs using Gears method or BDF up to order 6 (Table 6.5). Solve up to $t = 4.0$ and graphically compare the solution with the analytical solution.

$$u' = -50u$$
$$v' = -50u - 0.1v + t$$
$$u(0) = 1, \quad v(0) = 0, \quad t \in (0, \infty)$$

Solution   We will first write the computation schemes using BDF of up to sixth order for the above problem. We denote the kth order BDF by BDFk:

BDF1:

$$u^{n+1} - u^n = h(-50u^{n+1})$$

$$u^{n+1} = \frac{u^n}{(1+50h)}$$

$$v^{n+1} - v^n = h[-50u^{n+1} - 0.1v^{n+1} + t^{n+1}]$$

$$v^{n+1} = \frac{v^n + h[-50u^{n+1} + t^{n+1}]}{(1+0.1h)}$$

BDF2:

$$\frac{3}{2}u^{n+1} - 2u^n + \frac{1}{2}u^{n-1} = h(-50u^{n+1})$$

$$u^{n+1} = \frac{2u^n - \frac{1}{2}u^{n-1}}{\left(\frac{3}{2} + 50h\right)}$$

$$\frac{3}{2}v^{n+1} - 2v^n + \frac{1}{2}v^{n-1} = h[-50u^{n+1} - 0.1v^{n+1} + t^{n+1}]$$

$$v^{n+1} = \frac{2v^n - \frac{1}{2}v^{n-1} + h[-50u^{n+1} + t^{n+1}]}{\left(\frac{3}{2} + 0.1h\right)}$$

BDF3:

$$\frac{11}{6}u^{n+1} - 3u^n + \frac{3}{2}u^{n-1} - \frac{1}{3}u^{n-2} = h(-50u^{n+1})$$

$$u^{n+1} = \frac{3u^n - \frac{3}{2}u^{n-1} + \frac{1}{3}u^{n-2}}{\left(\frac{11}{6} + 50h\right)}$$

$$\frac{11}{6}v^{n+1} - 3v^n + \frac{3}{2}v^{n-1} - \frac{1}{3}v^{n-2} = h[-50u^{n+1} - 0.1v^{n+1} + t^{n+1}]$$

$$v^{n+1} = \frac{3v^n - \frac{3}{2}v^{n-1} + \frac{1}{3}v^{n-2} + h[-50u^{n+1} + t^{n+1}]}{\left(\frac{11}{6} + 0.1h\right)}$$

BDF4:

$$\frac{25}{12}u^{n+1} - 4u^n + 3u^{n-1} - \frac{4}{3}u^{n-2} + \frac{1}{4}u^{n-3} = h(-50u^{n+1})$$

$$u^{n+1} = \frac{4u^n - 3u^{n-1} + \frac{4}{3}u^{n-2} - \frac{1}{4}u^{n-3}}{\left(\frac{25}{12} + 50h\right)}$$

$$\frac{25}{12}v^{n+1} - 4v^n + 3v^{n-1} - \frac{4}{3}v^{n-2} + \frac{1}{4}v^{n-3} = h[-50u^{n+1} - 0.1v^{n+1} + t^{n+1}]$$

$$v^{n+1} = \frac{4v^n - 3v^{n-1} + \frac{4}{3}v^{n-2} - \frac{1}{4}v^{n-3} + h\left[-50u^{n+1} + t^{n+1}\right]}{\left(\frac{25}{12} + 0.1h\right)}$$

BDF5:

$$\frac{137}{60}u^{n+1} - 5u^n + 5u^{n-1} - \frac{10}{3}u^{n-2} + \frac{5}{4}u^{n-3} - \frac{1}{5}u^{n-4} = h(-50u^{n+1})$$

$$u^{n+1} = \frac{5u^n - 5u^{n-1} + \frac{10}{3}u^{n-2} - \frac{5}{4}u^{n-3} + \frac{1}{5}u^{n-4}}{\left(\frac{137}{60} + 50h\right)}$$

$$\frac{137}{60}v^{n+1} - 5v^n + 5v^{n-1} - \frac{10}{3}v^{n-2} + \frac{5}{4}v^{n-3} - \frac{1}{5}v^{n-4} = h\left[-50u^{n+1} - 0.1v^{n+1} + t^{n+1}\right]$$

$$v^{n+1} = \frac{5v^n - 5v^{n-1} + \frac{10}{3}v^{n-2} - \frac{5}{4}v^{n-3} + \frac{1}{5}v^{n-4} + h\left[-50u^{n+1} + t^{n+1}\right]}{\left(\frac{137}{60} + 0.1h\right)}$$

BDF6:

$$\frac{49}{20}u^{n+1} - 6u^n + \frac{15}{2}u^{n-1} - \frac{20}{3}u^{n-2} + \frac{15}{4}u^{n-3} - \frac{6}{5}u^{n-4} + \frac{1}{6}u^{n-5} = h(-50u^{n+1})$$

$$u^{n+1} = \frac{6u^n - \frac{15}{2}u^{n-1} + \frac{20}{3}u^{n-2} - \frac{15}{4}u^{n-3} + \frac{6}{5}u^{n-4} - \frac{1}{6}u^{n-5}}{\left(\frac{49}{20} + 50h\right)}$$

$$\frac{49}{20}v^{n+1} - 6v^n + \frac{15}{2}v^{n-1} - \frac{20}{3}v^{n-2} + \frac{15}{4}v^{n-3} - \frac{6}{5}v^{n-4} + \frac{1}{6}v^{n-5}$$
$$= h\left[-50u^{n+1} - 0.1v^{n+1} + t^{n+1}\right]$$

$$v^{n+1} = \frac{6v^n - \frac{15}{2}v^{n-1} + \frac{20}{3}v^{n-2} - \frac{15}{4}v^{n-3} + \frac{6}{5}v^{n-4} - \frac{1}{6}v^{n-5} + h\left[-50u^{n+1} + t^{n+1}\right]}{\left(\frac{49}{20} + 0.1h\right)}$$

Let us now choose an initial time step size. The set of equations can be written similar to Eq. (6.109) as,

$$\begin{bmatrix} u' \\ v' \end{bmatrix} = - \begin{bmatrix} 50 & 0 \\ 50 & 0.1 \end{bmatrix} \begin{bmatrix} u \\ v \end{bmatrix} + \begin{bmatrix} 0 \\ t \end{bmatrix}$$

Characteristics equation for the coefficient matrix on the right is given by,

$$(50 - \lambda)(0.1 - \lambda) = 0$$

The eigenvalues are therefore, 50 and 0.1. According to Eq. (6.130), a suitable choice for the initial time step would be $1/50 = 0.02$. Computations are tabulated in Table 6.24.

Table 6.24

Method	h	t	u	v
		0	1.00E + 00	0.0000
BDF1	0.02	0.02	5.00E – 01	–0.4986
	0.02	0.04	2.50E – 01	–0.7463
BDF2	0.04	0.08	0.00E – 00	–0.9903
	0.04	0.12	–3.57E – 02	–1.0181
	0.04	0.16	–2.04E – 02	–0.9932
BDF3	0.08	0.24	4.66E – 02	–0.9024
	0.08	0.32	2.92E – 02	–0.8900
	0.08	0.4	1.88E – 03	–0.8815
	0.08	0.48	–3.89E – 03	–0.8451
BDF4	0.16	0.64	–3.77E – 02	–0.7767
	0.16	0.8	–9.44E – 03	–0.6222
	0.16	0.96	6.24E – 03	–0.4570
	0.16	1.12	3.92E – 04	–0.2904
	0.16	1.28	–2.01E – 03	–0.0977
BDF5	0.32	1.6	–1.98E – 04	0.3607
	0.32	1.92	4.53E – 03	0.9086
	0.32	2.24	8.73E – 05	1.5310
	0.32	2.56	–1.05E – 03	2.2378
	0.32	2.88	5.08E – 04	3.0257
	0.32	3.2	1.29E – 04	3.8877
BDF6	0.64	3.84	–4.81E – 03	5.8257
	0.64	4.48	–1.45E – 03	8.0484

The comparison plot with the true solution is shown in Figs 6.22 and 6.23.

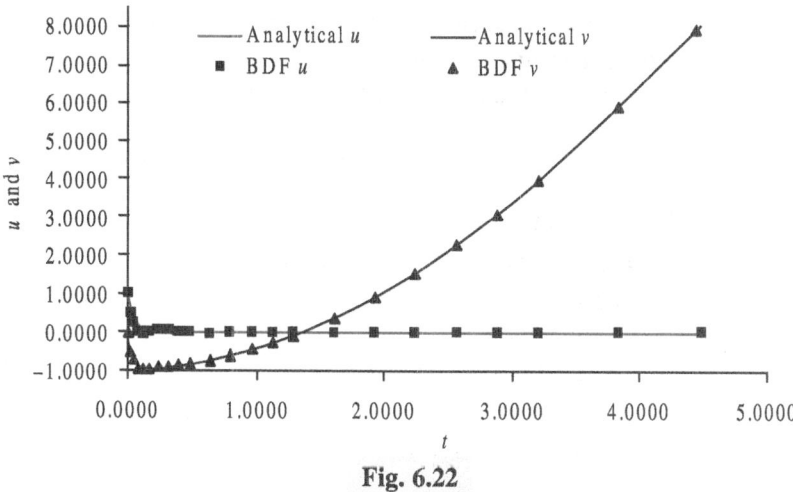

Fig. 6.22

In order to see how it simulated the steeply varying solutions of u and v, an enlarged view of the initial portion (0,0.6) of the solution is shown below,

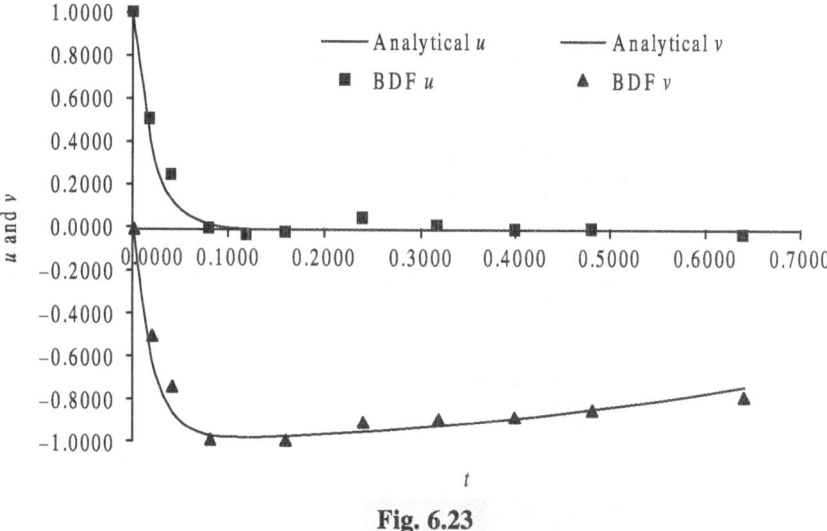

Fig. 6.23

In this example, solution of up to $t = 4.48$ was obtained in 22 time steps with doubling of time steps and increasing the order of BDF. It is easy to see, uniform time step of 0.02 would have required computations in 224 time steps.

So far, we have discussed ordinary differential equations where the independent variable was either time or behaved like time in the sense that the variable had an initial starting point and progressed in one direction. All the initial conditions for the higher-order differential equations were specified at the starting point. This is not always practical. More often than not, conditions are known at two or more different spatial location by actual measurements or by physical location of a boundary. In the next section, we look at these higher-order ODEs.

EXERCISE 6.5

1. Solve the following systems of IVP using Heun's method with repeated application of corrector and fourth order Runge-Kutta method:

   (i) $u' = u + v$
   $v' = 9u + v$
   $u(0) = 1; v(0) = 1$

   (ii) $u' = u - 2v$
   $v' = 2u + v$
   $u(0) = 1; v(0) = 0$

   (iii) $u' = w$
   $v' = 3u + 7v - 9w$
   $w' = 2v - w$
   $u(0) = 1; v(0) = 0; w(0) = 0$

   (iv) $u' = 2u + v - w$
   $v' = -4u - 3v - w$
   $w' = 4u + 4v + 2w$
   $u(0) = 1; v(0) = 0; w(0) = -1$

2. Solve the following initial value problems using the trapezoidal method for 10 time steps using $h = 0.1$:

   (i) $\dfrac{d^2 y}{dx^2} = \left(\dfrac{dy}{dx}\right)^3 + \dfrac{dy}{dx};\quad y(0) = 0$ and $\left.\dfrac{dy}{dx}\right|_{x=1} = 0$

   (ii) $x^2 \dfrac{d^2 y}{dx^2} - x\dfrac{dy}{dx} - 3y = x^2;\quad y(1) = 0$ and $\left.\dfrac{dy}{dx}\right|_{x=1} = 0$

   (iii) $x^2 \dfrac{d^2 y}{dx^2} + 2x\dfrac{dy}{dx} - 2y = 6x;\quad y(1) = 0$ and $\left.\dfrac{dy}{dx}\right|_{x=1} = 0$

   (iv) $\dfrac{d^2 y}{dx^2} - 2\dfrac{dy}{dx} = e^x \sin x;\quad y(0) = 0$ and $\left.\dfrac{dy}{dx}\right|_{x=0} = 1$

   (v) $\dfrac{d^2 y}{dx^2} + 2\dfrac{dy}{dx} + y = \dfrac{e^{-x}}{x};\quad y(1) = 1$ and $\left.\dfrac{dy}{dx}\right|_{x=1} = 1$

   (vi) $\dfrac{d^2 y}{dx^2} + 2\dfrac{dy}{dx} + y = e^{-x} \ln;\quad y(1) = 1$ and $\left.\dfrac{dy}{dx}\right|_{x=1} = 0$

3. The organic pollutant concentration (p) and dissolved oxygen deficit (y) in a river (of Problem 7, Exercise 6.2) can also be expressed as a coupled set of equations as follows:

$$u\frac{dp}{dx} = -k_d p$$

$$u\frac{dy}{dx} + k_a y = k_d p$$

where the stream velocity $u = 10$ km/day, $k_d = 0.5$ / day and $k_a = 0.2$ / day. The initial conditions are $p(0) = 12$ mg/l and $y(0) = 3$ mg/l. The x is measured along the length of the river. Use *trapezoidal* method to compute y for a stretch of 10 km by taking $h = 0.2$ km. Graphically compare the approximate solution with the analytical solution.

4. Consider the pendulum shown in Fig. 2.24. Equating the forces, one obtains

$$ml\frac{d^2\theta}{dt^2} = -mg\sin\theta$$

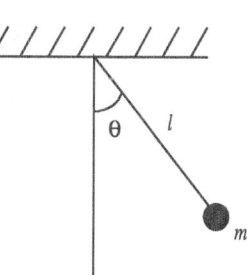

For small angles θ, $\sin\theta \approx \theta$ and the linearized equation of motion is the Newton's equation:

$$\frac{d^2\theta}{dt^2} + \omega^2\theta = 0$$

Fig. 6.24

where $\omega = \sqrt{\dfrac{g}{l}}$ is the frequency of the pendulum.

The accleration due to gravity is $g = 9.81$ m/sec^2 and $l = 0.6$ m. Assume that the pendulum starts from rest $[\theta'(0) = 0]$ with $\theta(0) = 10°$. Solve the equation using the following methods for five complete periods of oscillation. Use a uniform grid size with six equal divisions in each time period:

(a) True solution by analytical method.
(b) First order explicit and implicit methods (*Euler*).
(c) Second order explicit and implicit methods (second order Runge-Kutta and *trapezoidal*)
(d) Third order explicit and implicit methods (third order Adams-Bashforth and Adams-Moulton). Use of third order Runge-Kutta method for startup.
(e) Fourth order explicit and predictor corrector methods (fourth order Runge-Kutta and fourth order *Adams*).
(f) Compare the numerical solutions with the true solution and analyze each method for amplitude and phase errors. Which method produces the best results? Which method requires minimum computation? Rank these methods according to usability (accuracy and ease of application). Would

your conclusion be different if a finer time step (12 equal divisions in a period) was used?

5. Newton's equation for simple pendulums can be modified to incorporate the effect of damping. If we assume that the frictional force is proportional to the velocity the equation for the damped motion becomes:

$$\frac{d^2\theta}{dt^2} + \gamma\frac{d\theta}{dt} + w^2\theta = 0 \text{ where } \gamma \text{ is frictional coefficient}$$

Assuming the same initial conditions and, values of g and l as the previous problem, solve the damped equation for five complete periods with $\gamma = 0.5w$, $\gamma = w$, $\gamma = 2w$, and $\gamma = 3w$, using fourth order *Runge-Kutta* method. Take a uniform grid with 10 equal intervals in each period. Compare these solutions with the true solutions at these γ values including $\gamma = 0$ from the previous problem. The motion is called critically damped when $\gamma = 2w$. Can you physically interpret from your results what happens for $\gamma = 2w$.

6. Let us consider a LCR circuit where the current is governed by the following equation:

$$L\frac{d^2i}{dt^2} + R\frac{di}{dt} + \frac{1}{C}i = -220\sin t$$

Given are the values of inductance $L = 12$ H, resistance $R = 18\ \Omega$ and conductance $C = 0.75$F. The initial condition is $i = 0$ and $\frac{di}{dt} = 0$ at $t = 0$. Use fourth order Runge-Kutta method with a suitable time step for stability.

7. Displacement in a cable suspended under self-weight between two horizontal points is given by

$$\frac{d^2y}{dx^2} = \left[1 + \left(\frac{dy}{dx}\right)^2\right]^{1/2} ;\ y(0) = 1;\ y'(0) = 0$$

Solve the equation using fourth order *Adams* predictor corrector method (repeated application of corrector) in $x \in (-1,1)$ with $h = 0.1$. Use fourth order *Runge-Kutta* method for start-up. Can you recognize the *catenary*? Compare graphically with the analytical solution.

8. The enzyme (E) – substrate (S) reaction can generally be represented as follows:

$$E + S \underset{k_2}{\overset{k_1}{\rightleftharpoons}} ES \xrightarrow{k_3} E + P$$

The set of equation is given by

$$\frac{d[E]}{dt} = -k_1[E][S] + (k_2 + k_3)[ES]$$

$$\frac{d[S]}{dt} = -k_1[E][S] + k_2[ES]$$

$$\frac{d[ES]}{dt} = k_1[E][S] - (k_2 + k_3)[ES]$$

$$\frac{d[P]}{dt} = k_3[ES]$$

For the constants $k_1 = 1.0 \times 10^4$, $k_2 = 1.0 \times 10^{-3}$, and $k_3 = 1.0 \times 10^1$, the system is stiff. The initial conditions are $[E] = 1.0 \times 10^{-5}\,M$; $[S] = 1.0\,M$; $[ES] = 0$ and $[P] = 1.0\,M$ at $t = 0$. Solve the system of equation using *Gears* method with BDF of up to order four and gradually increasing the time step as well as the order of the method. Choose suitable initial time step. Proceed until the concentration of substrate (S) becomes 1% of the initial concentration.

9. Using a steady state assumption $\left(\dfrac{d[ES]}{dt} = 0 \right)$, the above problem can be reduced to Michaelis-Menten equation which is given by

$$-\frac{d[S]}{dt} = \frac{v_{max}[S]}{K_M + [S]} \quad \text{where } v_{max} = k_3[E]_T \text{ and } K_M = \frac{k_2 + k_3}{k_1}. \quad [E]_T = \text{Total}$$

enzyme which is equal to the concentration of enzyme (E) at $t = 0$. Solve this equation using the fourth order Runge-Kutta method and compare your solution graphically with the solution of Problem 7. Comment on the validity of the steady state assumption.

10. A simple Lotka-Volterra equation for the prey–predator relationship in the ecological model can be written as follows:

$$\frac{dn_1}{dt} = \alpha n_1 - \beta n_1 n_2$$

$$\frac{dn_2}{dt} = -\gamma n_2 + \delta n_1 n_2$$

where n_1 is the prey population and n_2 is the the predator population. The parameter values are given as $\alpha = 1.2$, $\beta = 0.1$, $\gamma = 0.6$ and $\delta = 0.025$. At time $t = 0$, the population of prey $n_1 = 50$ and the population of predator $n_2 = 15$. The above gives periodic solution. Therefore, the phase error becomes very important.

(a) Solve the set of equation using fourth order Runge-Kutta and fourth order Adams predictor-corrector methods. Proceed for sufficient time such that both prey and predator population completes at least one complete cycle.

(b) Plot n_1 vs. n_2 for both methods. Do you see a cycle?

(c) Eliminate the time variable by dividing the first equation with the second and solve the equation analytically. Now plot the analytical n_1 vs. n_2 and compare with the numerical solution. Which method is the best for this problem? Is there a better method over the two used above?

11. Two masses (m_1 and m_2) are fixed between two rigid support with three springs between them as shown in Fig. 6.25.

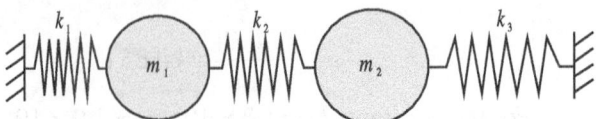

Fig. 6.25

The equation for displacement of the masses are given by

$$m_1 \frac{d^2x_1}{dt^2} = -(k_1 + k_2)x_1 + k_2 x_2$$

$$m_2 \frac{d^2x_2}{dt^2} = k_2 x_1 - (k_2 + k_3)x_2$$

Both the masses start from rest. The mass m_1 has been given a unit displacement at time $t = 0$. For $k_1 = 2.0 \times 103$, $k_2 = 1.0 \times 10^{-3}$, and $k_3 = 2.0 \times 10^{-2}$, is the system stiff? Solve the system using *Gears* method with BDF of up to sixth order until the motion stabilizes to smoothness. Choose appropriate time steps.

6.6 BOUNDARY VALUE PROBLEMS

In this section, we will look at the ordinary differential equations of order 2 or more where the conditions are specified at more than one point. The points at which these conditions are specified are typically the solution boundaries or the ranges of the independent variable. Therefore, the independent variable is a space variable on which a grid can be laid and the conditions are specified at the two grid points at the boundaries. These are commonly known as boundary-value problems (BVPs). For example, a second-order general boundary value problem may look like,

$$p(x, y)\frac{d^2y}{dx^2} + q(x, y)\frac{dy}{dx} + r(x, y) = 0 \qquad (6.131)$$

Boundary conditions: $y(0) = y_0$, $y(L) = y_L$.

If p and q are only functions of x and r is a linear function of y, it will be a linear boundary value problem. For a non-linear boundary value problem, the functions p and q can also contain derivative of y. Any one of the boundary conditions may also be specified in terms of derivative of y. Boundary value problem can also be of higher-order, e.g., in Example 6.10 we may have the third condition as the single derivative f' specified at infinity.

There are two fundamentally different approaches to solving these equations.

One involves decomposing the equation into a set of first order IVP and known as *shooting method*. The second approach involves laying a grid on the solution space and approximating the equation and the boundary conditions by finite difference approximations derived in Chapter 5. The latter is known as *direct method*. Both of these approaches rely on the methods already derived and tested in various chapters of this book.

6.6.1 Shooting Method

We have seen in Section 6.5 that any higher-order ODE can be decomposed into a system of first order ODE. Let us illustrate how we can use that to solve a boundary value problem. Consider the following second order linear boundary value problem:

$$p(x)y'' + q(x)y' + r(x)y = s(x) \tag{6.132}$$

where $x \in (0, l)$ and the boundary conditions are $y(0) = a$ and $y(l) = b$.

Using $u = y$ and $v = y'$, we can decompose the original equation as follows:

$$u' = v$$

$$v' = -\frac{q(x)v}{p(x)} - \frac{r(x)u}{p(x)} + \frac{s(x)}{p(x)} \tag{6.133}$$

$$u(0) = a \quad v(0) = ?$$

It is clear that we require the second initial condition $v(0)$ that we do not have for the solution of the set of the ODEs. Instead, we have a condition $u(l) = b$ that we cannot use directly to start the solution. The principle of shooting method is to assume a $v(0)$ and estimate $u(l)$. For an arbitrary choice of $v(0)$, the estimate of $u(l)$ is unlikely to match b (*If it does, you may head for Las Vegas with some money and come back a millionaire!*). However, one thing is easy to see that different choice of $v(0)$ will lead to different estimates of $u(l)$. So, one can say that $u(l)$ is a function of the independent variable $v(0)$. Therefore, the problem boils down to computing the value of the independent variable $v(0)$ for which the function $u(l) - b$ is zero. This is equivalent to computing the root of the equation $u(l) - b = 0$. Theoretically, any method of Chapter 3 can be used for this purpose. We shall use secant method in the shooting method.

Let us start with two initial guesses of the independent variables as $v_1(0)$ and $v_2(0)$ and assume the corresponding estimates as $u_1(l)$ and $u_2(l)$, respectively. One can use any of the methods described in this chapter for the solution of the system of IVPs as long as the time steps are within the limits of stability. One can compute the next guess, $v_3(0)$ as follows:

$$v_3(0) = v_2(0) - \frac{v_2(0) - v_1(0)}{u_2(l) - u_1(l)} [u_2(l) - b]$$

The general iteration scheme is as follows:

$$v_{k+1}(0) = v_k(0) - \frac{v_k(0) - v_{k-1}(0)}{u_k(l) - u_{k-1}(l)} [u_k(l) - b] \qquad (6.134)$$

The iteration can be terminated when the true relative error $\left| \frac{b - u_k(l)}{b} \right| \times 100$ satisfies the preset error criterion. If the condition specified in the original problem at l is a gradient condition, i.e., $y'(l) = b$, the method can be easily modified. In that case, the original condition $v(l) = b$ is to be matched for various values of $v(0)$. So, the dependent variable $u(l)$ is replaced by $v(l)$ in the secant method iteration scheme. An application of shooting method is shown in Example 6.13.

The problem considered to illustrate the shooting method is linear and as a result, the decomposed system of IVPs is also linear. If the BVP is non-linear, the system of IVPs is also non-linear. Once an initial condition is assumed, the shooting method essentially follows the same methods described in the previous section for solving the set of equation. Once solved, the secant method have to be applied to compute the next initial condition irrespective of whether the system is linear or non-linear.

Example 6.13　Solve the following boundary-value problem using shooting method with second order Runge-Kutta method (Ralston's) and $h = 0.25$. Stop the secant method when the true absolute error in the boundary condition is less than 10^{-6}.

$$y'' + y + x = 0; \qquad x \in [0,1]; \qquad y(0) = y(1) = 0$$

Solution　The equation can be decomposed into the following system of two equations if we define $u = y$ and $v = y'$. The equation can then be written as,

$$u' = v$$
$$v' = -u - x$$
$$u(0) = u(1) = 0$$

In the vector form, the above equation can be written as $y' = f$ where

$$y = \begin{bmatrix} u \\ v \end{bmatrix} \text{ and } f = \begin{bmatrix} v \\ -u - x \end{bmatrix}$$

Ralston's method for the above system translates to

$$y_{n+1} = y_n + h\left[\frac{1}{3}\phi_0 + \frac{2}{3}\phi_1\right]$$

where

$$\phi_0 = f(y_n, x_n) = \begin{bmatrix} v_n \\ -u_n - x_n \end{bmatrix}$$

$$\phi_1 = f\left(y_n + \frac{3}{4}h\phi_0, x_n + \frac{3}{4}h\right) = \begin{bmatrix} v_n - \frac{3}{4}hu_n - \frac{3}{4}hx_n \\ -u_n - \frac{3}{4}hv_n - x_n - \frac{3}{4}h \end{bmatrix}$$

because, $y_n + \frac{3}{4}h\phi_0 = \begin{bmatrix} u_n \\ v_n \end{bmatrix} + \frac{3}{4}h\begin{bmatrix} v_n \\ -u_n - x_n \end{bmatrix} = \begin{bmatrix} u_n + \frac{3}{4}hv_n \\ v_n - \frac{3}{4}hu_n - \frac{3}{4}hx_n \end{bmatrix}$

Therefore, given the initial conditions in the form of u_n and v_n, the vectors ϕ_0, ϕ_1, and y_{n+1} can be computed explicitly. Computations are tabulated.

The initial condition for the system is $u(0) = 0$ but the $v(0)$ is not known. We need to assume two values for this initial condition for applying secant method. Let us assume 0 and 1 as the starting values. We show the computations for these initial conditions in Tables 6.25 and 6.26.

Initial conditions $u(0) = 0$ and $v_1(0) = 0$

Table 6.25

x	u	v
0	0.0000	0.0000
0.25	0.0000	–0.0313
0.5	–0.0156	–0.1240
0.75	–0.0618	–0.2725
1	–0.1514	–0.4673

Initial conditions $u(0) = 0$ and $v_2(0) = 1$

Table 6.26

x	u	v
0	0.0000	1.0000
0.25	0.2500	0.9375
0.5	0.4688	0.7520
0.75	0.6265	0.4550
1	0.6972	0.0654

Using the secant method of Eq. (6.134), we can compute the new initial condition. Notice that the right hand boundary condition (b in Eq. (6.134)) is zero

$$v_3(0) = 1 - \frac{1-0}{0.6972-(-0.1514)}[0.6972-0] = 0.1784$$

The solution with the new initial conditions $u(0) = 0$ and $v_3(0) = 0.1784$ are given in Table 6.27.

Table 6.27

x	u	v
0	0.0000	0.1784
0.25	0.0446	0.1416
0.5	0.0708	0.0323
0.75	0.0610	−0.1427
1	0.0000	−0.3722

Since, the error is less than the tolerance, we can stop. The u column contains the required solution. We will compare this solution with the direct method and true solution in Example 6.14.

6.6.2 Direct Method

Philosophy of this method is approximation of the original equation using difference approximations of the derivatives derived in Chapter 5. Before starting this method, two decisions have to be made based on the practical considerations. These are as follows:

- *Grid Size* The choice is governed by how closely spaced are the points where the solution is required in the entire solution space. This can be evenly spaced (uniform) or unevenly spaced (non-uniform). Near the boundary, it is often required to use closely spaced grid points because of rapidly changing solution (high gradient). The entire solution space is divided into integer number of segments.

- *Order of Approximation* This is governed by the accuracy of the solution desired. It is not entirely independent of the grid size. A lower-order approximation with closely spaced grid points may yield the same accuracy as that of a higher-order method with sparser grid points.

   Once the approximation is made, the boundary value problem transforms into a system of linear algebraic equation that can be solved using any of the methods described in Chapter 2. Let us outline the method using second order approximation for the derivatives in Eq. (6.132) and a grid size of h such that $l = nh$, where n is an integer.

The grid is shown in Fig. 6.26.

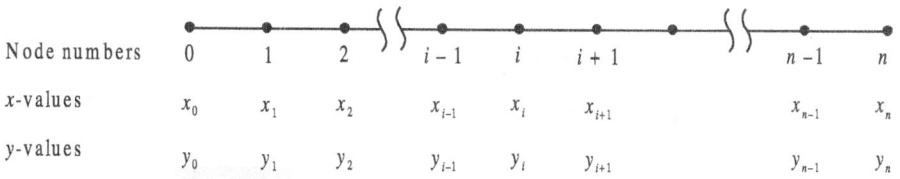

Node numbers	0	1	2	$i-1$	i	$i+1$	$n-1$	n
x-values	x_0	x_1	x_2	x_{i-1}	x_i	x_{i+1}	x_{n-1}	x_n
y-values	y_0	y_1	y_2	y_{i-1}	y_i	y_{i+1}	y_{n-1}	y_n

Fig. 6.26 A typical grid along with nomenclatures.

The grid points are numbered as 0 to n. The independent variable values at the grid points are $\{x_0, x_1, x_2, ..., x_n\}$ and the dependent variable values are $\{y_0, y_1, y_2, ..., y_n\}$. Therefore, $x_i = x_0 + ih$, where $i = 0$, 1, 2, ..., n. Notice that $y_0 = a$ and $y_n = b$ are known from the boundary conditions. Therefore, the problem is to determine the values of $\{y_1, y_2, ..., y_{n-1}\}$. For the grid point i, the following second order central difference approximation can be written using the approximations derived in Chapter 5 for double and single derivatives:

$$p(x_i)\frac{y_{i+1}-2y_i+y_{i-1}}{h^2}+q(x_i)\frac{y_{i+1}-y_{i-1}}{2h}+r(x_i)y_i=s(x_i) \qquad (6.135)$$

This equation can be rearranged to form:

$$\left(\frac{p(x_i)}{h^2}-\frac{q(x_i)}{2h}\right)y_{i-1}+\left(-\frac{2p(x_i)}{h^2}+r(x_i)\right)y_i+\left(\frac{p(x_i)}{h^2}+\frac{q(x_i)}{2h}\right)y_{i+1}=s(x_i)$$

$$(6.136)$$

For any point i, the above equation involves values at one preceding nodal point at $(i-1)$ and at one successive nodal point at $(i+1)$. Therefore, the above equation can be written only for the interior nodes, i.e., $i = 2$ to $(n-1)$. This is convenient because the equation is valid at the interior nodes while at the boundary nodes, one should apply the boundary conditions. For convenience, we denote the coefficients of Eq. (6.136) as

$$\alpha_i = \frac{p(x_i)}{h^2}-\frac{q(x_i)}{2h}, \ \beta_i = -\frac{2p(x_i)}{h^2}+r(x_i), \text{ and } \gamma_i = \frac{p(x_i)}{h^2}+\frac{q(x_i)}{2h} \qquad (6.137)$$

The set of equations for the interior nodes becomes

$$i=1: \ \alpha_1 y_0 + \beta_1 y_1 + \gamma_1 y_2 = s(x_1)$$
$$i=2: \ \alpha_2 y_1 + \beta_2 y_2 + \gamma_2 y_3 = s(x_2)$$
$$\vdots \qquad\qquad\qquad\qquad\qquad (6.138)$$
$$i=n-1: \ \alpha_{n-1} y_{n-2} + \beta_{n-1} y_{n-1} + \gamma_{n-1} y_n = s(x_{n-1})$$

One can now add the two boundary conditions, $y_0 = a$ and $y_n = b$ to the above set of equations to yield the following matrix equation:

$$
\begin{bmatrix}
\beta_1 & \gamma_1 & 0 & \cdots & 0 & 0 \\
\alpha_2 & \beta_2 & \gamma_2 & \cdots & 0 & 0 \\
0 & \alpha_3 & \beta_3 & \cdots & 0 & 0 \\
\vdots & \vdots & \vdots & \vdots & \vdots & \vdots \\
0 & 0 & 0 & \alpha_{n-2} & \beta_{n-2} & \gamma_{n-2} \\
0 & 0 & 0 & 0 & \alpha_{n-1} & \beta_{n-1}
\end{bmatrix}
\begin{bmatrix}
y_1 \\ y_2 \\ y_3 \\ \vdots \\ y_{n-2} \\ y_{n-1}
\end{bmatrix}
=
\begin{bmatrix}
s(x_1) - \alpha_1 a \\
s(x_2) \\
s(x_3) \\
\vdots \\
s(x_{n-2}) \\
s(x_{n-1}) - \gamma_{n-1} b
\end{bmatrix}
\tag{6.139}
$$

The coefficient matrix is tridiagonal and therefore, the Thomas algorithm can be used for solution. Since the equation considered is in the general form, the second order central difference approximation of a second order boundary value problem generally yields a tridiagonal matrix. The tridiagonal structure is disrupted if one uses forward or backward difference second order approximations instead of the central difference.

If one expects a steep downward gradient in the solution domain, a backward difference approximation typically produces better result, while a steep upward gradient calls for a forward difference approximation. This is because the central difference puts equal weights on both sides of the central node, which may lead to erroneous result when a steep gradient levels out sharply to become parallel to the axis. Let us now look at an application of direct method to solve a boundary value problem.

Example 6.14   Solve the boundary value problem of Example 6.13 using direct methods with second order central difference approximation and a grid size of 0.25. Compute the true relative errors at the interior node for the solutions obtained by direct method and by the shooting method in Example 6.13.

Solution   With a grid size of 0.25, we will have four intervals and five nodal points in the entire domain (0,1). Let us number the nodes as $0-4$. The variable values are y_0, y_1, y_2, y_3, and y_4. Out of these, two values are known in terms of boundary conditions. These are: $y_0 = y_4 = 0$. We have to determine the other three values.

We will now write a second order central difference approximation of the differential equation for an arbitrary interior node i.

$$
\frac{y_{i+1} - 2y_i + y_{i-1}}{(0.25)^2} + y_i + x_i = 0
$$

or
$$16y_{i-1} - 31y_i + 16y_{i+1} = -x_i$$

The equations for node numbers 1–3 using the boundary condition values $y_0 = y_4 = 0$ are as follows:

$$-31y_1 + 16y_2 \qquad = -0.25$$
$$16y_1 - 31y_2 + 16y_3 = -0.5$$
$$16y_2 - 31y_3 = -0.75$$

The tridiagonal system of equation can be solved using Thomas algorithm. The calculation is given in Table 6.28.

Table 6.28

I	d	u	b	α	β	y
	−31	16	− 0.25	−31.0000	− 0.2500	0.0443
16	−31	16	− 0.5	−22.7419	− 0.6290	0.0702
16	−31		− 0.75	−19.7433	− 1.1926	0.0604

The true solution of the equation is given by $y = \dfrac{\sin x}{\sin 1} - x$. The true relative errors in the shooting method and direct method are shown in Table 6.29.

Table 6.29

x	y	True	Shooting	Error (%)	Direct	Error (%)
0	y_0	0	0	0	0	0
0.25	y_1	0.044013654	0.044602079	1.336913042	0.044274014	0.591542815
0.5	y_2	0.069746964	0.070791527	1.497647491	0.070155902	0.586317053
0.75	y_3	0.060056166	0.061018808	1.60290267	0.060403046	0.577592438
1	y_4	0	2.77556E–17	0	0	0

In this particular problem, direct method provides more accurate solution compared to shooting method, although both central difference and Ralston's method are second order accurate and both the methods were used with same h. However, this is not a universal rule that the direct method would always produce better result. It depends on the type of the problem.

One last thing we will discuss in this section is the case of derivative boundary conditions. Let us assume the right-hand side boundary condition involves derivative of the dependent variable, i.e., $y'_n = b$.

The set of equations for the interior nodes Eq. (6.138) remain unaltered. To this, we add the left-hand boundary condition $y_0 = a$ and a difference approximation of the right-hand boundary condition $y'_n = b$. There are many ways to make the difference approximation. We list a few here.

First order backward difference

$$\frac{y_n - y_{n-1}}{h} = b \tag{6.140}$$

This preserves the tridiagonal structure of the coefficient matrix but changes (reduces) the order of approximation at the boundary node.

Second order backward difference

$$\frac{y_{n-2} - 4y_{n-1} + 3y_n}{2h} = b \tag{6.141}$$

This preserves the order of the approximation but tridiagonal structure is disturbed and the Thomas algorithm cannot be applied any more.

Ghost Node

Let us put an imaginary node $(n + 1)$ at a distance h from the node n (Fig. 6.22). The nth node has now become an interior node and thus the approximation of the ODE can be written for this node:

$$\alpha_n y_{n-1} + \beta_n y_n + \gamma_n y_{n+1} = s(x_n) \tag{6.142}$$

We also know that the boundary condition $y'_n = b$ is applicable at the nth node. Therefore, we can write a second order central difference approximation of the boundary condition,

$$\frac{y_{n+1} - y_{n-1}}{2h} = b \tag{6.143}$$

Combining the above two equations, we obtain

$$(\alpha_n + \gamma_n)y_{n-1} + \beta_n y_n = s(x_n) - 2hb\gamma_n \tag{6.144}$$

The above equation can be added to the set of Eq. (6.138) to yield,

$$
\begin{bmatrix}
\beta_1 & \gamma_1 & 0 & \cdots & 0 & 0 \\
\alpha_2 & \beta_2 & \gamma_2 & \cdots & 0 & 0 \\
0 & \alpha_3 & \beta_3 & \cdots & 0 & 0 \\
\vdots & \vdots & \vdots & \vdots & \vdots & \vdots \\
0 & 0 & 0 & \alpha_{n-1} & \beta_{n-1} & \gamma_{n-1} \\
0 & 0 & 0 & 0 & \alpha_n + \gamma_n & \beta_n
\end{bmatrix}
\begin{bmatrix}
y_1 \\
y_2 \\
y_3 \\
\vdots \\
y_{n-1} \\
y_n
\end{bmatrix}
=
\begin{bmatrix}
s(x_1) - \alpha_1 a \\
s(x_2) \\
s(x_3) \\
\vdots \\
s(x_{n-1}) \\
s(x_n) - 2hb\gamma_n
\end{bmatrix}
$$

$$(6.145)$$

This preserves the tridiagonal structure as well as the order of approximation but increases the dimension of the matrix equation to be solved by one.

If the derivative boundary condition is at the left side boundary (at x_0), the above procedures can be modified by replacing backward difference approximations by the forward differences or assuming a ghost node to the left at x_{-1}. We illustrate the formulation of the set of equation with gradient boundary condition in Example 6.15.

Example 6.15   Formulate the system of equations with second order backward difference as well as with ghost node for the ODE in Example 6.13 with the following boundary conditions,

$$y(0) = 0, \quad y'(1) = -0.3722$$

Solution   The unknown variable values are y_1, y_2, y_3, and y_4. We outline both backward difference and ghost node approach for formulating the equations:

Second Order Backward Difference
The equations for nodes 1–3 are similar to Example 6.14 as

$$
\begin{aligned}
-31y_1 + 16y_2 &= -0.25 \\
16y_1 - 31y_2 + 16y_3 &= -0.5 \\
16y_2 - 31y_3 + 16y_4 &= -0.75
\end{aligned}
$$

The fourth equation is obtained by writing the second order backward difference approximation of the right-hand boundary condition $y'(1) = -0.3722$ as

$$y_2 - 4y_3 + 3y_4 = 2(0.25)(-0.3722) = -0.1861$$

Therefore, the set of equations to solve is

$$
\begin{aligned}
-31y_1 + 16y_2 &= -0.25 \\
16y_1 - 31y_2 + 16y_3 &= -0.5 \\
16y_2 - 31y_3 + 16y_4 &= -0.75 \\
y_2 - 4y_3 + 3y_4 &= -0.1861
\end{aligned}
$$

Notice that the matrix is not tridiagonal anymore, and the Thomas algorithm cannot be used. We leave it to the readers to solve it using a suitable algorithm from Chapter 2.

Ghost Node

We assume a ghost node (5) at distance 0.25 from the right-hand boundary node (4). The assumed value of the variable at this node is y_5. Therefore, the set of equation for the interior nodes 1–4 can be written as

$$
\begin{aligned}
-31y_1 + 16y_2 & = -0.25 \\
16y_1 - 31y_2 + 16y_3 & = -0.5 \\
16y_2 - 31y_3 + 16y_4 & = -0.75 \\
16y_3 - 31y_4 + 16y_5 & = -1
\end{aligned}
$$

We now write a second order central difference approximation of the boundary condition at node 4:

$$
y_5 - y_3 = 2(0.25)(-0.3722) = -0.1861
$$

or

$$
y_5 = y_3 - 0.1861
$$

Using the value of y_5, in the last equation, we get the following set of equations:

$$
\begin{aligned}
-31y_1 + 16y_2 & = -0.25 \\
16y_1 - 31y_2 + 16y_3 & = -0.5 \\
16y_2 - 31y_3 + 16y_4 & = -0.75 \\
32y_3 - 31y_4 & = 1.9776
\end{aligned}
$$

We get back the tridiagonal structure of the equations. We leave it to the readers to solve both the systems of equation and verify which one provides better solution.

EXERCISE 6.6

1. A commonly used equation in many fields of engineering and science (colloid, intermolecular interactions, biophysics, etc.) is known as Poisson Boltzman equation. Consider the following Poisson Boltzman equation in one-dimension

    $$
    \frac{d^2\psi}{dx^2} = \sinh\psi; \quad \psi(0) = 1; \quad \psi(x) \to 0 \text{ and } \frac{d\psi}{dx} \to 0 \text{ as } x \to 0
    $$

    (a) Solve the equation using the shooting method with fourth order Runge Kutta method.

    (b) Solve using the direct method with the second order central difference scheme and second order backward difference scheme. Use the result of (a) to locate the right-hand boundary.

    (c) Compare three solutions with the analytical solution given by

$$\tanh\frac{\psi}{4} = \left(\tanh\frac{1}{4}\right)e^{-x}.$$

2. Determine approximations of the smallest characteristic value of λ for the following *Sturm-Liouville* problem using second order central difference scheme,

$$y'' + \lambda y = 0, \quad y(0) = y(1) = 0$$

(a) $h = 1/2$ and solve analytically.
(b) $h = 1/3$ and solve analytically.
(c) $h = 1/5$ and use inverse power method with an initial vector of $(1, 1, 1)$ and relative error in eigenvalue, $\varepsilon < 0.01\%$.
(d) Compare each of the three results with the true value of π^2 and comment on your results.

3. Let us consider the following differential equation:

$$\frac{d^2T}{dx^2} + a(x)\frac{dT}{dx} + b(x)T = f(x)$$

where $a(x)$, $b(x)$, and $f(x)$ are functions given by,

$$a(x) = x^2,\ b(x) = \frac{1}{x},\ \text{and}\ f(x) = e^{-x}$$

x is between $(0,1)$. $T(0) = 10$ and $dT/dx = 0$ at $x = 1$. Discretize the above equation using second order finite difference approximation and formulate the set of linear simultaneous equations. Incorporate the boundary conditions such that the accuracy of the scheme is preserved. Use $\Delta x = 0.1$. Plot the solution.

4. Solve the differential equation $d^2y/dx^2 - dy/dx - 2y + 2x = 3$ with the boundary conditions $y(0) = 0$ and $y(1) = 1$ using the shooting method with (a) fourth order Runge-Kutta, and (b) Heun's predictor corrector method. For both cases, use $\Delta x = 0.2$.

5. Consider the problem of Example 6.10. Change the third boundary condition to $f''(\infty) = 0$. Use

$$\alpha = \frac{1}{2}(m+1) \text{ and } \beta = m$$

This becomes the well known Falkner-Skan equation. Solve it for $m = 1$ (Hiemenz flow towards a plane stagnation point) using shooting method with fourth order Runge-Kutta method.

SUMMARY

The goal of this chapter was to initiate the readers with the concept of solution of ODEs. This chapter introduced the commonly used finite difference methods for solving ODEs and was not intended to be a comprehensive repository of the methods. Our intention was to provide practical information for the use of the methods in science and engineering applications. We have carefully avoided (or stayed away from) many methods found in the literature. However, majority of them will fall into one of the categories presented here. The basic understanding developed here should help the reader to comprehend many more methods that exist for the solution of IVP but not included in this chapter. Importance was given to understand the basics of stability and different kinds of errors that creeps in various applications. However, the error (of all kinds!) analysis presented here is by no means exhaustive.

We have established that a numerical method developed for a basic first-order IVP is good enough to solve ordinary differential equation of arbitrarily higher order by decomposing the higher order ODEs into system of IVPs. Lastly, the concept of stiff system was introduced and some insight was given for the ways of solving them. Understanding of the numerical methods for IVPs developed here will once again form the basis for the solution of partial differential equations in the next chapter.

7

Partial Differential Equations

7.1 INTRODUCTION

Batman goes paragliding. While flying around with his *Batwings*, he found out that some of the para gliders were maneuvering better than him by using the thermals in the mountain terrain. So, he decided to learn paragliding to improve his flying and turning techniques. He realized with *Batmanesque* instinct that it is important to precisely locate the thermals. While studying the temperature distribution in atmosphere, he came across a number of partial differential equations. For example, the energy balance equation takes the following form:

$$\rho C_v \left(\frac{\partial T}{\partial t} + u \frac{\partial T}{\partial x} + v \frac{\partial T}{\partial y} + w \frac{\partial T}{\partial z} \right) - \alpha \left(\frac{\partial^2 T}{\partial x^2} + \frac{\partial^2 T}{\partial y^2} + \frac{\partial^2 T}{\partial z^2} \right)$$

$$= -p \left(\frac{\partial u}{\partial x} + \frac{\partial v}{\partial y} + \frac{\partial w}{\partial z} \right) + d \pm s \tag{7.1}$$

where six state variables are u, v, w, p, ρ, and T. The variables, u, v, and w are wind velocities in the directions x, y, and z, respectively, p is the pressure, ρ is the density, and T is the temperature. The parameters and constants are heat capacity at constant volume C_v, thermal conductivity α, heat generated due to dissipation d and any source or sink s.

In order to solve for six state variables, he needed five more equations. These are provided by continuity equation (one) and momentum equations (three) written for compressible Newtonian fluid in a gravitational field, and a constitutive relationship for ideal gas relating the pressure and density as

$$p\frac{m}{\rho} = RT$$

Batman quickly learned all the atmospheric fluid mechanics. If you want to learn that too, you may consult other books, for example, Seinfeld (1986). However, in this book, we shall concentrate on numerical solution of these equations.

In the beginning, the full set of equations was too complicated even for the *Batman*. So, he needed some *spherical cow* like approximations. With *Boussinesq approximations* and adiabatic equilibrium assumption, one obtains the temperature distribution equation as

$$\rho C_p\left(\frac{\partial T}{\partial t} + u\frac{\partial T}{\partial x} + v\frac{\partial T}{\partial y} + w\frac{\partial T}{\partial z}\right) - \alpha\left(\frac{\partial^2 T}{\partial x^2} + \frac{\partial^2 T}{\partial y^2} + \frac{\partial^2 T}{\partial z^2}\right) = \pm s \qquad (7.2)$$

This is the equation for heat transfer in an incompressible fluid. Near the mountain ridge, he was interested in getting the lift along vertical planes parallel to the ridge. In these cases, he was only interested in thermal gradients along the vertical planes running parallel to the ridge. So, he could get rid of one of the horizontal dimension perpendicular to the ridge whenever he wanted to simulate the temperature contours near the ridge and obtain:

$$\rho C_p\left(\frac{\partial T}{\partial t} + u\frac{\partial T}{\partial x} + w\frac{\partial T}{\partial z}\right) - \alpha\left(\frac{\partial^2 T}{\partial x^2} + \frac{\partial^2 T}{\partial z^2}\right) = \pm s \qquad (7.3)$$

Lastly, when he wanted to know only the vertical distribution of temperature as he rose from the ground, he only needed to keep the vertical dimension leading to

$$\rho C_p\left(\frac{\partial T}{\partial t} + w\frac{\partial T}{\partial z}\right) - \alpha\frac{\partial^2 T}{\partial z^2} = \pm s \qquad (7.4)$$

If the flying condition was quiescent (no wind) for a long time such that a steady state temperature distribution was established, the equations in three- and two-dimensions for such cases were

$$\alpha\left(\frac{\partial^2 T}{\partial x^2} + \frac{\partial^2 T}{\partial y^2} + \frac{\partial^2 T}{\partial z^2}\right) = \mp s$$

$$\alpha\left(\frac{\partial^2 T}{\partial x^2} + \frac{\partial^2 T}{\partial z^2}\right) = \mp s \qquad (7.5)$$

Furthermore, if there were no source or sink in the region for the heat energy, he needed only to solve the classic Laplace equations:

$$\frac{\partial^2 T}{\partial x^2} + \frac{\partial^2 T}{\partial y^2} + \frac{\partial^2 T}{\partial z^2} = 0$$

$$\frac{\partial^2 T}{\partial x^2} + \frac{\partial^2 T}{\partial z^2} = 0 \tag{7.6}$$

These latter equations must be looking familiar, as you come across similar equations in many engineering disciplines. Advection-dispersion equations [Eqs (7.2)–(7.4)] and Laplace equations [Eqs (7.5)–(7.6)] are some of the most common PDEs in engineering applications.

Having learned the equations, *Batman* now wants to program his Batwing computer for their solutions so that the thermals or the temperature contours in the region are displayed on his screen as he is flying. He requires fast and accurate numerical algorithms to do that. And as you can guess, here he is with us, studiously learning again.

Before we venture into any numerical methods, let us recall some of the commonly used terms in PDE:

1. *Order* of a PDE is the highest-order of the derivative present in the equation.
2. *Degree* of a PDE is the highest power to which the highest-order term is raised in the equation.
3. A PDE of *degree one* with no term(s) containing the product(s) of dependent variable and its derivative(s) is *linear*. Otherwise, it is *non-linear*.
4. A PDE is *homogeneous* if it does not contain any term which is a constant or a function of only independent variables. *Non-homogeneous* PDE, on the other hand, contains a term involving only the independent variable(s) or a constant.

Some examples of the categorization of PDEs are given below:

Second order, first degree, linear, homogeneous PDE:

$$x^2 \frac{\partial^2 \phi}{\partial x \partial t} + e^{-t} \frac{\partial \phi}{\partial x} + \phi = 0 \tag{7.7}$$

First order, second degree, non-linear, homogeneous PDE:

$$\left(\frac{\partial \phi}{\partial x}\right)^2 + \left(\frac{\partial \phi}{\partial y}\right)^2 = 0 \tag{7.8}$$

Second order, first degree, non-linear, non-homogeneous PDE:

$$\frac{\partial \phi}{\partial t} + \phi \frac{\partial \phi}{\partial x} - \alpha \frac{\partial^2 \phi}{\partial x^2} = f(x,t) \tag{7.9}$$

Second order, first degree, linear, non-homogeneous PDE:

$$\frac{\partial^2 \phi}{\partial x^2} + \frac{\partial^2 \phi}{\partial y^2} = f(x, y) \tag{7.10}$$

First hurdle for the *Batman* was to recognize the natural boundary conditions for the specific problems and the number of boundary conditions required for obtaining a meaningful solution for a PDE. In order to familiarize the readers, we shall first present a brief review of the characteristic analysis of PDEs. Then, we shall venture into methods for solving some of the commonly encountered form of PDEs in the engineering problems. In this, we will only consider the methods for linear PDEs. We will show how to analyse the methods for stability and convergence. Lastly, we will address how one can use the methods for linear PDEs in order to solve non-linear problems.

7.2 CHARACTERISTICS OF PDE

Partial differential equations typically describe a physical process in the space-time continuum. Every physical process often has well-defined boundaries in space and suitably chosen landmarks in time where the conditions are known. These serve as natural boundary and initial conditions. However, one often wonders whether it is possible to solve a PDE, given a set of boundary or initial conditions. Conversely, where and what kind of boundary and initial conditions are required to solve a particular PDE in a certain space-time domain? The analysis of characteristics helps us to understand this question.

In this section, we will introduce the readers to the concept of characteristics and how it can help to determine the type and location of initial and boundary conditions required to solve a PDE. We shall limit ourselves to introduction because a detailed analysis is beyond the scope of an introductory numerical analysis book and there are many good texts available for detailed concept of characteristics for various types of PDEs, e.g., Evans (2010), John (1981).

The characteristics of a PDE are non-intersecting lines or planes or hyper-planes (depending on the number of independent variables) along which the solution remain constant. Therefore, the information propagates along the characteristics and no information propagates across them. It is then easy to visualise that one must specify the initial and boundary conditions in such a way that they cut across the characteristics, and information can flow into the domain of interest from every point of the boundary. Alternatively, if the boundary happens to run along the characteristics, no information will flow from this boundary into the domain which lies across.

Let us introduce the concept of characteristics with a first-order PDE as follows:

$$\frac{\partial \phi}{\partial t} + u \frac{\partial \phi}{\partial x} = 0 \qquad (7.11)$$

Recall the definition of characteristics along which the information propagates from the boundary. Let us denote these curves as $\psi(x,t) = K$, where K is a constant. Therefore, the change *along any of these curves is zero*. Mathematically, this can be expressed as

$$d\psi = 0 = \psi_t dt + \psi_x dx \quad \text{or} \quad \frac{dt}{dx} = -\frac{\psi_x}{\psi_t} \qquad (7.12)$$

A small change in the state variable ϕ can be expressed in terms of its partial derivative as

$$d\phi = \phi_t dt + \phi_x dx \qquad (7.13)$$

If the change in ϕ is considered *along the curve* $\psi(x,t) = K$, then using Eqs (7.11) and (7.12) in Eq. (7.13), one obtains

$$\left(1 + u \frac{\psi_x}{\psi_t}\right)\phi_x = \frac{d\phi}{dx}\bigg|_{\psi(x,t)=K} \qquad (7.14)$$

If the above equation has a solution, that would mean specifying a condition along $\psi(x,t) = K$ would enable computation of ϕ_x at any point. By definition of characteristics, this cannot be true if $\psi(x,t) = K$ is indeed a characteristic. This would mean that the coefficient on the left must be zero for the equation not to have a solution. This gives the equation of characteristics as

$$\left(1 + u \frac{\psi_x}{\psi_t}\right) = 0 \quad \text{or} \quad \psi_t + u\psi_x = 0 \qquad (7.15)$$

Equation (7.15) has solutions of $x - ut = K$, where K *is some constant*. In a time space domain, these are parallel straight lines with a slope of u. Interesting is the case of $K = 0$. Along this line, $x = ut$ or $t = x/u$. An initial condition which is specified for all x at $t = 0$ will influence the solution in the region $x > ut$. The boundary condition which is specified for all t at $x = 0$ will influence the solution in the region $t > x/u$. These regions are schematically shown in Fig. 7.1. Therefore, both initial and boundary conditions are required to obtain solution in the complete space–time continuum.

If any condition is specified along $x - ut = K$, we will not be able to obtain solution for this PDE at all times and at all locations, as no information will propagate across these lines. As seen in Fig. 7.1, the characteristics must intersect the curves along which the conditions are specified so as to pick the information from these curves and propagate them into the domain while satisfying the PDE.

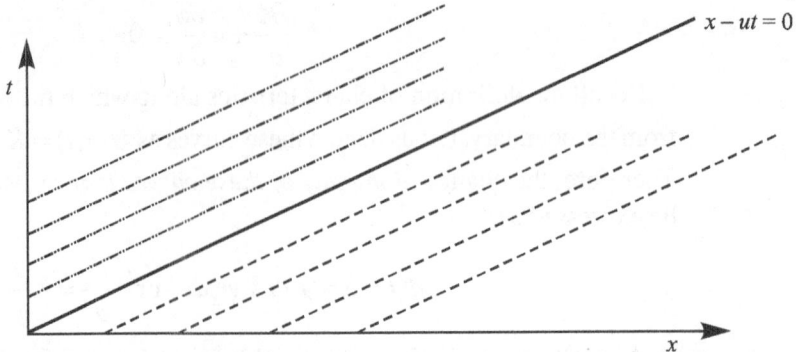

Fig. 7.1   Influence of boundary and initial conditions in the solution domain. Initial condition is specified along x-axis at $t = 0$. Dashed line shows the region of influence of the initial condition. Boundary condition is specified along t-axis at $x = 0$. Region of influence of the boundary condition is shown by the chain dotted line.

Let us now consider a general second-order PDE with two independent variables:

$$\alpha\phi_{xx} + 2\beta\phi_{xy} + \gamma\phi_{yy} + \theta\phi_x + \omega\phi_y + \rho(\phi, x, y) = 0 \qquad (7.16)$$

Recall in Chapter 6, we converted the higher-order IVPs to first order by defining the lower-order derivatives as new variables.

Let us define similar variables to reduce the above PDE to first order. We define as follows:

$$\xi = \phi_x \quad \text{and} \quad \eta = \phi_y \qquad (7.17)$$

This automatically leads to,

$$\xi_y = \eta_x \qquad (7.18)$$

Using these, the PDE can be written as

$$\alpha\xi_x + \beta\xi_y + \beta\eta_x + \gamma\eta_y + \theta\phi_x + \omega\phi_y + \rho(\phi, x, y) = 0 \qquad (7.19)$$

Similar to the characteristics analysis shown for the first-order problem, let us once again consider $\psi(x, y) = K$ as the characteristics such that Eq. (7.12) is true. The changes in the three variables of this PDE can be expressed similar to Eq. (7.13):

$$d\phi = \phi_x dx + \phi_y dy \qquad (7.20)$$

$$d\xi = \xi_x dx + \xi_y dy \qquad (7.21)$$

$$d\eta = \eta_x dx + \eta_y dy \qquad (7.22)$$

By considering the changes along $\psi(x, y) = K$, one can derive a relation

Eq. (7.23) similar to Eq. (7.12). Using this relation in Eqs (7.20)–(7.22), we obtain the relations in Eqs (7.24)–(7.26)

$$\frac{dy}{dx} = -\frac{\psi_x}{\psi_y} \tag{7.23}$$

$$\phi_x = \phi_y \frac{\psi_x}{\psi_y} + \frac{d\phi}{dx}\bigg|_{\psi(x,y)=K} \tag{7.24}$$

$$\xi_x = \xi_y \frac{\psi_x}{\psi_y} + \frac{d\xi}{dx}\bigg|_{\psi(x,y)=K} \tag{7.25}$$

$$\eta_x = \eta_y \frac{\psi_x}{\psi_y} + \frac{d\eta}{dx}\bigg|_{\psi(x,y)=K} \tag{7.26}$$

Now it is possible to eliminate the partial derivatives with respect to x from Eq. (7.19) and only retain the partial derivatives with respect to y as unknowns. Substituting the relations (7.24)–(7.26) in Eq. (7.19), we obtain

$$\left(\beta + \alpha\frac{\psi_x}{\psi_y}\right)\xi_y + \left(\gamma + \beta\frac{\psi_x}{\psi_y}\right)\eta_y + \left(\omega + \theta\frac{\psi_x}{\psi_y}\right)\phi_y + \rho(\phi, x, y)$$

$$= -\rho(\phi, x, y) - \alpha\frac{d\xi}{dx}\bigg|_{\psi(x,y)=K} - \beta\frac{d\eta}{dx}\bigg|_{\psi(x,y)=K} - \theta\frac{d\phi}{dx}\bigg|_{\psi(x,y)=K} \tag{7.27}$$

From Eqs (7.17) and (7.26), we can also write

$$\xi_y - \eta_y \frac{\psi_x}{\psi_y} = \frac{d\eta}{dx}\bigg|_{\psi(x,y)=K} \tag{7.28}$$

Lastly, the definition

$$\phi_y = \eta \tag{7.29}$$

From Eqs (7.27)–(7.29), the partial derivatives of ϕ, ξ, and η with respect to y cannot be solved if the determinant of the coefficient matrix is zero. This will define the characteristics as

$$\begin{vmatrix} \beta + \alpha\dfrac{\psi_x}{\psi_y} & \gamma + \beta\dfrac{\psi_x}{\psi_y} & \omega + \theta\dfrac{\psi_x}{\psi_y} \\[2mm] 1 & -\dfrac{\psi_x}{\psi_y} & 0 \\[2mm] 0 & 0 & 1 \end{vmatrix} = 0 \tag{7.30}$$

This leads to the equation

$$\alpha \left(\frac{\psi_x}{\psi_y} \right)^2 + 2\beta \frac{\psi_x}{\psi_y} + \gamma = 0 \tag{7.31}$$

The solution or the roots are

$$\frac{\psi_x}{\psi_y} = \frac{-\beta \pm \sqrt{\beta^2 - \alpha\gamma}}{\alpha} \tag{7.32}$$

In Eq. (7.19), if we eliminated the partial derivatives with respect to y and solved for the partial derivatives with respect to x, we will obtain similar solution for the characteristics as Eq. (7.32) for ψ_y/ψ_x with γ in the denominator. The characteristics depend on the values of α, β, and γ, the coefficients of second-order partial derivatives in Eq. (7.16).

$\beta^2 - \alpha\gamma > 0$: Two unique real characteristics and the PDE is called *hyperbolic*. Examples are second order wave equations, where $\alpha = -c^2$, $\gamma = 1$, and $\beta = 0$.

$\beta^2 - \alpha\gamma = 0$: One unique real characteristics and the PDE is called *parabolic*. Examples are diffusion and advection-diffusion equations, where $\alpha = K$, $\gamma = 0$, and $\beta = 0$.

$\beta^2 - \alpha\gamma < 0$: Two complex conjugate characteristics and the PDE is *elliptic*. Example is Laplace equation, where $\alpha = 1$, $\gamma = 1$, and $\beta = 0$.

From the exact values of α, β, and γ, one can determine the characteristics and ensure that the boundaries intersect them. Detailed discussions of characteristics for specific equations are beyond the scope of this book. Interested readers, may refer to Celia and Gray (1992) for elaborate presentations of boundary condition requirements based on characteristics analysis. In the following sections, we will develop methods for the solution of PDEs and demonstrate some analysis techniques for assuring convergence.

EXERCISE 7.2

1. Using Eq. (7.32), determine the characteristics of the Laplace equation. Draw the region indicating what type of boundary conditions is necessary to solve the equation.
2. Determine the characteristics of the advection-diffusion Eq. (7.34). Now set $\alpha = 0$ and determine the characteristics again. This new sets of characteristics are termed as secondary characteristics of the equation. State under what circumstances the primary or secondary characteristics would influence the solution.
3. In Problem 2, set $u = 0$ and obtain the characteristics for pure diffusion problem.

7.3 NUMERICAL METHODS FOR PDEs

In this section, we shall consider PDEs commonly encountered in the engineering problems. In order to develop the numerical methods, we will draw heavily on the concepts of the earlier chapters. Readers are advised to familiarise themselves on these concepts before venturing onto solving PDEs.

Specifically, we shall use concepts of ODEs developed in Chapter 6, finite difference approximation of derivatives developed in Chapter 5, approximation theory of Chapter 4, and solution of linear systems of equation developed in Chapter 2. The goal is to obtain a meaningful solution at discrete locations that satisfies the PDE as well as the boundary and initial conditions. PDEs will be first decomposed into ODEs and further onto linear algebraic equations. For the analysis, we will use the concept of Fourier representation of a function developed in Chapter 4. We divide the section into subsections covering different types of PDEs.

7.3.1 Diffusion and Advective-Diffusion Equation

In this section, we will develop numerical methods for the solution of two linear parabolic equations namely, diffusion equation and advective-diffusion (also called advection-dispersion) equation. These are shown below:

$$\frac{\partial \phi}{\partial t} = \alpha \frac{\partial^2 \phi}{\partial x^2} \tag{7.33}$$

$$\frac{\partial \phi}{\partial t} + u \frac{\partial \phi}{\partial x} = \alpha \frac{\partial^2 \phi}{\partial x^2} \tag{7.34}$$

where u and α may be functions of the independent variables. For the solution, each equation also requires two boundary conditions and one initial condition. These equations contain two independent variables in the form of time (t) and space (x). These two independent variables have very different physical significance. For example, range of the space variable is the entire real line ($-\infty, \infty$) while for most of the practical problems, the time progresses forward from any arbitrary initial starting point, that is in ($0, \infty$). Therefore, while one can use all the forward, backward and central difference approximations developed in Chapter 5 for derivatives involving the space variable, the methods for initial value problems developed in Chapter 6 need to be used for the derivative involving time variable. Since, space and time both are independent variables, it is possible to decouple the approximations in these two variables.

In order to understand this, let us consider a finite domain ($0, L$) for the space variable x. For all approximations of the dependent variable, we shall use the subscript index for the space variable and the superscript index for the time variable.

Therefore, the notation ϕ_i^n would indicate the value of ϕ at ith location at nth time step. In order to write difference approximations, we need to lay a grid on the domain (in other words, choose the points in the domain at which the approximate solutions will be obtained). Let us choose a finite number $(m + 1)$ of equally spaced points (Fig. 7.2). Therefore, the length of each interval between two successive points is $\Delta x = L/m$.

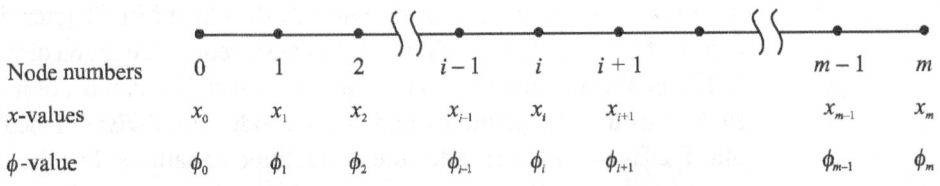

Node numbers	0	1	2	$i-1$	i	$i+1$	$m-1$	m
x-values	x_0	x_1	x_2	x_{i-1}	x_i	x_{i+1}	x_{m-1}	x_m
ϕ-value	ϕ_0	ϕ_1	ϕ_2	ϕ_{i-1}	ϕ_i	ϕ_{i+1}	ϕ_{m-1}	ϕ_m

Fig. 7.2 A grid for the space variable.

One can now write finite difference approximation for the space derivative in Eqs (7.33) and (7.34). If we choose second-order central difference scheme for approximations, we can write the following for each interior points (i) in the grid:

Diffusion equation:

$$\frac{d\phi_i}{dt} = \alpha_i \frac{\phi_{i+1} - 2\phi_i + \phi_{i-1}}{\Delta x^2} \tag{7.35}$$

Advection-diffusion equation:

$$\frac{d\phi_i}{dt} = -u_i \frac{\phi_{i+1} - \phi_{i-1}}{2\Delta x} + \alpha_i \frac{\phi_{i+1} - 2\phi_i + \phi_{i-1}}{\Delta x^2} \tag{7.36}$$

Notice that we converted the time derivatives into total derivative. This is because, we have already made discrete approximation of the space derivatives and each ϕ_i is only a function of time. The partial differential equations are now converted to a system of coupled initial-value problems with ϕ_i's as dependent variables.

In order to observe this, let us divide the solution domain into four equal divisions $(m = 4)$ such that we have 5 nodes (numbered 0-4), 2 at the boundaries (0 and 4) and 3 in the interiors (1, 2, 3). The time dependent variables are then ϕ_0, ϕ_1, ϕ_2, ϕ_3, and ϕ_4. For illustration purpose, we shall also assume two boundary conditions as $\phi(0,t) = c_0$ and $\phi(L,t) = c_L$. These boundary conditions essentially mean, $\phi_0 = c_0$ and $\phi_4 = c_L$ at all times, i.e., the values at two boundaries remain constant at all times.

Later on we will discuss in detail various types of boundary conditions and how to incorporate them in the discrete form. At present, we shall write the discrete forms of Eqs (7.33) and (7.34) by incorporating the above boundary conditions:

Diffusion equation:

$$
\begin{bmatrix} \dfrac{d\phi_1}{dt} \\[2ex] \dfrac{d\phi_2}{dt} \\[2ex] \dfrac{d\phi_3}{dt} \end{bmatrix} = \begin{bmatrix} -\dfrac{2\alpha_1}{\Delta x^2} & \dfrac{\alpha_1}{\Delta x^2} & 0 \\[2ex] \dfrac{\alpha_2}{\Delta x^2} & -\dfrac{2\alpha_2}{\Delta x^2} & \dfrac{\alpha_2}{\Delta x^2} \\[2ex] 0 & \dfrac{\alpha_3}{\Delta x^2} & -\dfrac{2\alpha_3}{\Delta x^2} \end{bmatrix} \begin{bmatrix} \phi_1 \\[2ex] \phi_2 \\[2ex] \phi_3 \end{bmatrix} + \begin{bmatrix} \dfrac{c_0\alpha_1}{\Delta x^2} \\[2ex] 0 \\[2ex] \dfrac{c_L\alpha_3}{\Delta x^2} \end{bmatrix} \tag{7.37}
$$

Advection-diffusion equation:

$$
\begin{bmatrix} \dfrac{d\phi_1}{dt} \\[2ex] \dfrac{d\phi_2}{dt} \\[2ex] \dfrac{d\phi_3}{dt} \end{bmatrix} = \begin{bmatrix} -\dfrac{2\alpha_1}{\Delta x^2} & -\dfrac{u_1}{2\Delta x}+\dfrac{\alpha_1}{\Delta x^2} & 0 \\[2ex] \dfrac{u_2}{2\Delta x}+\dfrac{\alpha_2}{\Delta x^2} & -\dfrac{2\alpha_2}{\Delta x^2} & -\dfrac{u_2}{2\Delta x}+\dfrac{\alpha_2}{\Delta x^2} \\[2ex] 0 & \dfrac{u_3}{2\Delta x}+\dfrac{\alpha_3}{\Delta x^2} & -\dfrac{2\alpha_3}{\Delta x^2} \end{bmatrix} \begin{bmatrix} \phi_1 \\[2ex] \phi_2 \\[2ex] \phi_3 \end{bmatrix} + \begin{bmatrix} \dfrac{u_1 c_0}{2\Delta x}+\dfrac{c_0\alpha_1}{\Delta x^2} \\[2ex] 0 \\[2ex] -\dfrac{u_3 c_L}{2\Delta x}+\dfrac{c_L\alpha_3}{\Delta x^2} \end{bmatrix}
$$

$$\tag{7.38}$$

We can write both the equations in the form:

$$
\frac{d\bar{\phi}}{dt} = A\bar{\phi} + b \tag{7.39}
$$

where

$$
\bar{\phi} = \begin{bmatrix} \phi_1 \\ \phi_2 \\ \phi_3 \end{bmatrix}
$$

This constitutes a system of IVP similar to Eq. (6.120) where the initial vector is given by the initial condition of PDE. Therefore, through discretization of space variable, we have reduced a PDE into a system of IVP. This is commonly termed as *semi-discretization*, since we have discretized only one of the two independent variables in the original PDE.

For time discretization, one can use any of the methods described in the previous chapter. For example, application of Euler forward leads to Eq. (6.115) and Euler backward yields Eq. (6.121). Instead of repeating the same equations here, let us write a weighted parameterized form that becomes Euler's methods or trapezoidal method depending on the parameter value. Using the superscript for the time step index, we write the parameterized scheme for Eq. (7.39) as follows:

$$\bar{\phi}^{n+1} = \bar{\phi}^n + \Delta t \left[\mu\{A^n\bar{\phi}^n + b^n\} + (1-\mu)\{A^{n+1}\bar{\phi}^{n+1} + b^{n+1}\} \right] \quad (7.40a)$$

This is generalisation of Euler and trapezoidal methods developed in Chapter 6 for the solution of ODEs. The slope function (Fig. 6.3) is evaluated as the weighted average of two slopes, one at the present time and the other at the future time. The value of μ is the weight for the slope at the present time (n) and $(1 - \mu)$ is the weight for slope at the future time ($n + 1$). Therefore, giving full weight to the slope at the present time ($\mu = 1$) leads to an *explicit method*. On the other hand, giving zero weight ($\mu = 0$) at the present time leads to *fully implicit method*. For all the intermediate values of μ, the method is multi-step. For $\mu = 1$, 1/2, and 0, the above yields *Euler forward*, *trapezoidal*, and *Euler backward* method, respectively. Very often, the reader may come across the above method written in the following form:

$$\bar{\phi}^{n+1} = \bar{\phi}^n + \Delta t[(1-\theta)\{A^n\bar{\phi}^n + b^n\} + \theta\{A^{n+1}\bar{\phi}^{n+1} + b^{n+1}\}] \quad (7.40b)$$

Both the methods are essentially same, and θ and μ are related by $\theta = 1 - \mu$. Some may find Eq. (7.40b) easier to remember intuitively, since $\theta = 1$ essentially puts full weightage at the end of the time step ($n + 1$) and $\theta = 0$ puts full weightage at the beginning of the time step (n). In this chapter, we will present all analysis (truncation error, stability, etc.) using Eq. (7.40a). However, the readers only need to substitute $\mu = (1 - \theta)$ in order to obtain equivalent expressions for Eq. (7.40b).

The coefficient matrix A and the vector b contain u and α, which may be functions of both the independent variables. Therefore, the time step index is also shown on A and b but these can be computed exactly for all time steps since they are functions of independent variables only. The system of linear algebraic equation to be solved at each time step is then given by

$$\left[I - \Delta t(1-\mu)A^{n+1} \right]\bar{\phi}^{n+1} = \left[I + \mu\Delta t A^n \right]\bar{\phi}^n + \Delta t \left[(1-\mu)b^{n+1} + \mu b^n \right] \quad (7.41)$$

For $\mu = 1$, the coefficient matrix on the left-hand side is identity and therefore does not require solution of a system of equations. This is expected for Euler forward method. For all other values of μ, the system of equations needs to be solved. If u and α are not functions of time, the coefficient matrix becomes constant and a decomposition such as LU or Thomas algorithm (for tridiagonal matrices) saves substantial computation time.

Instead of discretization in two steps (first space, and then time), we could have discretized the variable ϕ in both time and space together. This is commonly termed as *full-discretization*. In this case, we will have to choose both space and time discretization schemes. For example, let us choose central difference (CD) scheme for spatial discretization and the weighted parameter (μ) scheme for time variable. This we shall refer to as μ-CD scheme for PDE. Application of this scheme to the diffusion and advective-diffusion equations leads to following discrete equations for an interior node (i):

Diffusion equation:

$$\frac{\phi_i^{n+1} - \phi_i^n}{\Delta t} = \mu \alpha_i^n \frac{\phi_{i+1}^n - 2\phi_i^n + \phi_{i-1}^n}{\Delta x^2} + (1-\mu)\alpha_i^{n+1} \frac{\phi_{i+1}^{n+1} - 2\phi_i^{n+1} + \phi_{i-1}^{n+1}}{\Delta x^2} \quad (7.42)$$

Advection-diffusion equation:

$$\frac{\phi_i^{n+1} - \phi_i^n}{\Delta t} = \mu \left[-u_i^n \frac{\phi_{i+1}^n - \phi_{i-1}^n}{2\Delta x} + \alpha_i^n \frac{\phi_{i+1}^n - 2\phi_i^n + \phi_{i-1}^n}{\Delta x^2} \right]$$
$$+ (1-\mu) \left[-u_i^{n+1} \frac{\phi_{i+1}^{n+1} - \phi_{i-1}^{n+1}}{2\Delta x} + \alpha_i^{n+1} \frac{\phi_{i+1}^{n+1} - 2\phi_i^{n+1} + \phi_{i-1}^{n+1}}{\Delta x^2} \right] \quad (7.43)$$

We leave it to the reader to show that for a given problem, if the same set of schemes is applied for time and space discretization, both *semi* and *full-discretization* lead to the exact same set of linear algebraic equations to solve. For $\mu = \frac{1}{2}$, the numerical scheme is known *as Crank-Nicolson* method. We will now discuss, how to incorporate various types of boundary conditions into discrete equations. Three types of boundary conditions are commonly encountered for PDEs. These are as follows:

7.3.1.1 *First Type or Dirichlet Condition*

The value of dependent variable is specified at the boundaries. The example boundary conditions considered in this section were of this type. Since exact values are specified at the boundary, this can be directly associated with the boundary nodes and replaced in the discrete equations as in Eqs (7.37) and (7.38).

7.3.1.2 *Second Type or Neumann Condition*

The flux or gradient (derivative) of the dependent variable may be specified at one or more boundaries. For Eqs (7.33) and (7.34), this takes the form

$$\left. \frac{d\phi}{dx} \right|_{(0,t) \text{ and/or } (L,t)} = c \quad (7.44)$$

7.3.1.3 *Third Type or Robin Condition*

A linear combination of dependent variable and its gradient (derivative) may be specified at one or more boundary. For the PDEs in Eqs (7.33) and (7.34), this is of the form

$$a\frac{d\phi}{dx} + b\phi = c \quad \text{at } (0, t) \text{ or } (L, t) \quad (7.45)$$

The coefficients a, b, and c in the second and third type boundary conditions can be functions of independent variables. The readers may have noticed by now that the space derivatives of the PDE are discretized similar to the direct methods for BVP (Section 6.5.2). Therefore, the second and third type boundary conditions can be incorporated in the discrete equations using either a ghost node or backward/forward difference similar to the boundary value problems. We demonstrate this in Example 7.2.

Example 7.1   Consider the one-dimensional convection-diffusion equation:

$$\frac{\partial T}{\partial t} + u \frac{\partial T}{\partial x} = \alpha \frac{\partial^2 T}{\partial x^2}, \qquad 0 \le x \le 1$$

where

$$u = 0.1; \quad \alpha = 0.01; \quad T(0,t) = 0; \quad T(1,t) = 0; \quad T(x,0) = 50\sin(\pi x)$$

Show solution for one time step using the following methods:
(a) Euler forward in time and central difference in space (EF-CD),
(b) Euler backward in time and central difference in space (EB-CD),
(c) Crank-Nicolson method,
(d) A second order Runge-Kutta method in time and central difference approximation in space. Use $\Delta x = 0.25$ and $\Delta t = 0.5$.

Solution   All the methods use central difference approximation in space. Let us first perform a semi-discretization in space using central difference and convert the PDE into a system of ODE. Using $\Delta x = 0.25$, five equidistant nodes in space are shown in the figure below:

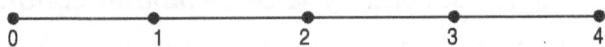

For any node j, the discrete approximation of the equation is

$$\frac{dT_j}{dt} + u \frac{T_{j+1} - T_{j-1}}{2\Delta x} = \alpha \frac{T_{j+1} - 2T_j + T_{j-1}}{\Delta x^2}$$

or

$$\frac{dT_j}{dt} = \left(\frac{u}{2\Delta x} + \frac{\alpha}{\Delta x^2} \right) T_{j-1} + \left(-\frac{2\alpha}{\Delta x^2} \right) T_j + \left(-\frac{u}{2\Delta x} + \frac{\alpha}{\Delta x^2} \right) T_{j+1}$$

Using the values of the constants, one obtains

$$\frac{dT_j}{dt} = 0.36 T_{j-1} - 0.32 T_j - 0.04 T_{j+1}$$

Writing the above equations for T_1, T_2, and T_3, and using two boundary conditions, we obtain the following system of IVPs:

$$\frac{dT}{dt} = AT$$

where $\qquad T = \begin{bmatrix} T_1 \\ T_2 \\ T_3 \end{bmatrix}$ and $A = \begin{bmatrix} -0.32 & -0.04 & 0 \\ 0.36 & -0.32 & -0.04 \\ 0 & 0.36 & -0.32 \end{bmatrix}$

From the initial condition:

$$T^0 = \begin{bmatrix} T_1^0 \\ T_2^0 \\ T_2^0 \end{bmatrix} = \begin{bmatrix} 35.3553 \\ 50.0 \\ 35.3553 \end{bmatrix}$$

(a) Euler forward

$$T^{n+1} = [I + \Delta t A] T^n$$

or $\qquad T^{0.5} = \begin{bmatrix} T_1^{0.5} \\ T_2^{0.5} \\ T_2^{0.5} \end{bmatrix} = \begin{bmatrix} 0.84 & -0.02 & 0 \\ 0.18 & 0.84 & -0.02 \\ 0 & 0.18 & 0.84 \end{bmatrix} \begin{bmatrix} 35.3553 \\ 50.0 \\ 35.3553 \end{bmatrix} = \begin{bmatrix} 28.6985 \\ 47.6568 \\ 38.6985 \end{bmatrix}$

(b) Euler backward

$$[I - \Delta t A] T^{n+1} = T^n$$

or $\qquad \begin{bmatrix} 1.16 & 0.02 & 0 \\ -0.18 & 1.16 & 0.02 \\ 0 & -0.18 & 1.16 \end{bmatrix} \begin{bmatrix} T_1^{0.5} \\ T_2^{0.5} \\ T_2^{0.5} \end{bmatrix} = \begin{bmatrix} 35.3553 \\ 50.0 \\ 35.3553 \end{bmatrix}$

Solve using Thomas algorithm to obtain

$$T^{0.5} = \begin{bmatrix} T_1^{0.5} \\ T_2^{0.5} \\ T_2^{0.5} \end{bmatrix} = \begin{bmatrix} 29.6674 \\ 47.0556 \\ 37.7804 \end{bmatrix}$$

(c) Crank-Nicolson method

$$[I - 0.5\Delta t A] T^{n+1} = [I + 0.5\Delta t A] T^n$$

$$\begin{bmatrix} 1.08 & 0.01 & 0 \\ -0.09 & 1.08 & 0.02 \\ 0 & -0.09 & 1.08 \end{bmatrix} \begin{bmatrix} T_1^{0.5} \\ T_2^{0.5} \\ T_2^{0.5} \end{bmatrix} = \begin{bmatrix} 0.92 & -0.01 & 0 \\ 0.09 & 0.92 & -0.01 \\ 0 & 0.09 & 0.92 \end{bmatrix} \begin{bmatrix} 35.3553 \\ 50.0 \\ 35.3553 \end{bmatrix} = \begin{bmatrix} 32.0269 \\ 48.8284 \\ 37.0269 \end{bmatrix}$$

Solve using Thomas algorithm to obtain

$$T^{0.5} = \begin{bmatrix} T_1^{0.5} \\ T_2^{0.5} \\ T_2^{0.5} \end{bmatrix} = \begin{bmatrix} 29.2166 \\ 47.2923 \\ 38.2252 \end{bmatrix}$$

(d) Second order Runge-Kutta method

Three different forms of second order Runge-Kutta method are shown in Table 6.6. Let us use *Heun's predictor corrector* form (second in Table 6.6). We leave the other two forms of second order Runge-Kutta method for the readers to apply on this problem. Let us use subscripts p for the predictor and c for the corrector. Applying the predictor formula (Eq. 6.107) on the problem leads to

$$T_p^{n+1} = [I + \Delta t A] T^n$$

Since this is same as the Euler forward method, the solution of (a) is the predictor. The corrector is then given by

$$T_c^{n+1} = T^n + \frac{\Delta t}{2} \left[A T_p^{n+1} + A T^n \right] = T^n + \frac{\Delta t}{2} A \left[T_p^{n+1} + T^n \right]$$

$$T_c^{0.5} = \begin{bmatrix} T_{1c}^{0.5} \\ T_{2c}^{0.5} \\ T_{2c}^{0.5} \end{bmatrix} = \begin{bmatrix} 35.3553 \\ 50.0 \\ 35.3553 \end{bmatrix} + \begin{bmatrix} -0.08 & -0.01 & 0 \\ 0.09 & -0.08 & -0.01 \\ 0 & 0.09 & -0.08 \end{bmatrix} \left\{ \begin{bmatrix} 28.6985 \\ 47.6568 \\ 38.6985 \end{bmatrix} + \begin{Bmatrix} 35.3553 \\ 50.0 \\ 35.3553 \end{Bmatrix} \right\}$$

$$= \begin{bmatrix} 29.2544 \\ 47.2118 \\ 38.2201 \end{bmatrix}$$

We have seen how to discretize and obtain solution for a diffusion and advective-diffusion equation. Let us now try to look at the discretization along with the physics of the system. For any interior node i, the discrete equations for the PDEs [Eqs (7.42) and 7.43)] can be rearranged into following forms:

Diffusion equation:

$$\left[-(1-\mu)\alpha_i^{n+1} \frac{\Delta t}{\Delta x^2} \right]\phi_{i+1}^{n+1} + \left[1+2(1-\mu)\alpha_i^{n+1} \frac{\Delta t}{\Delta x^2} \right]\phi_i^{n+1} + \left[-(1-\mu)\alpha_i^{n+1} \frac{\Delta t}{\Delta x^2} \right]\phi_{i-1}^{n+1}$$

$$= \left[\mu\alpha_i^n \frac{\Delta t}{\Delta x^2} \right]\phi_{i+1}^n + \left[1-2\mu\alpha_i^n \frac{\Delta t}{\Delta x^2} \right]\phi_i^n + \left[\mu\alpha_i^n \frac{\Delta t}{\Delta x^2} \right]\phi_{i-1}^n$$

(7.46)

Equation (7.46) essentially shows that the coefficient matrix will always be symmetric tridiagonal unless disturbed by incorporation of boundary conditions. Moreover, if α is constant which is true more often than not, the coefficient matrix only depends on the weighting parameter μ and the group $\alpha\Delta t/\Delta x^2$.

Advection-diffusion equation:

$$\left[(1-\mu)\left(u_i^{n+1} \frac{\Delta t}{2\Delta x} - \alpha_i^{n+1} \frac{\Delta t}{\Delta x^2} \right) \right]\phi_{i+1}^{n+1} + \left[1+2(1-\mu)\alpha_i^{n+1} \frac{\Delta t}{\Delta x^2} \right]\phi_i^{n+1}$$

$$+ \left[(1-\mu)\left(-u_i^{n+1} \frac{\Delta t}{2\Delta x} - \alpha_i^{n+1} \frac{\Delta t}{\Delta x^2} \right) \right]\phi_{i-1}^{n+1}$$

$$= \left[\mu\left(-u_i^n \frac{\Delta t}{2\Delta x} + \alpha_i^n \frac{\Delta t}{\Delta x^2} \right) \right]\phi_{i+1}^n + \left[1-2\mu\alpha_i^n \frac{\Delta t}{\Delta x^2} \right]\phi_i^n$$

$$+ \left[\mu\left(u_i^n \frac{\Delta t}{2\Delta x} + \alpha_i^n \frac{\Delta t}{\Delta x^2} \right) \right]\phi_{i-1}^n$$

(7.47)

Once again, we observe that the resulting matrix is tridiagonal but the symmetry is missing. The asymmetry in the coefficient matrix is the result of advection. For constant u and α, the coefficient matrix can be generated using values of μ, $u\Delta t/\Delta x$, and $\alpha\Delta t/\Delta x^2$.

The *Peclet number* in an advection-diffusion system is defined as the dimensionless number relating advection and diffusion in the system:

$P_e = uL/\alpha$, where u is the advection velocity, L is the characteristic length scale, and α is the diffusion coefficient which is the thermal diffusion in case of heat transfer and mass diffusion or dispersion in case of mass transfer.

In case of numerical solution, using the finite difference schemes described so far, we will be unable to get any information about the state variable at a spatial resolution less than the grid length Δx. The solutions are obtained only at the grid points. For values at any intermediate points, we will have to resort to interpolation.

Therefore, using the spatial grid length as the characteristic length, we can define a dimensionless number similar to the Peclet number and we shall call it the *grid peclet number* (P_g):

$$P_g = \frac{u\Delta x}{\alpha} \tag{7.48}$$

For pure advection systems, another important dimensionless quantity is the CFL number named after Courant, Friedrich, and Lewy. It is defined as

$$C = u\frac{\Delta t}{\Delta x} \tag{7.49}$$

For constant u and α, both the discrete equations can now be expressed in terms of these two dimensionless numbers. For example, the discrete advection-dispersion equation takes the following form:

$$\left[(1-\mu)\left(\frac{C}{2} - \frac{C}{P_g}\right)\right]\phi_{i+1}^{n+1} + \left[1 + 2(1-\mu)\frac{C}{P_g}\right]\phi_i^{n+1} + \left[(1-\mu)\left(-\frac{C}{2} - \frac{C}{P_g}\right)\right]\phi_{i-1}^{n+1}$$

$$= \left[\mu\left(-\frac{C}{2} + \frac{C}{P_g}\right)\right]\phi_{i+1}^{n} + \left[1 - 2\mu\frac{C}{P_g}\right]\phi_i^{n} + \left[\mu\left(\frac{C}{2} + \frac{C}{P_g}\right)\right]\phi_{i-1}^{n}$$

$$\tag{7.50}$$

Let us now analyse these schemes to understand their consistency and stability properties. We shall demonstrate this using μ-CD scheme applied to the diffusion Eq. (7.46) with constant α

$$\left[-(1-\mu)\alpha\frac{\Delta t}{\Delta x^2}\right]\phi_{i+1}^{n+1} + \left[1 + 2(1-\mu)\alpha\frac{\Delta t}{\Delta x^2}\right]\phi_i^{n+1} + \left[-(1-\mu)\alpha\frac{\Delta t}{\Delta x^2}\right]\phi_{i-1}^{n+1}$$

$$= \left[\mu\alpha\frac{\Delta t}{\Delta x^2}\right]\phi_{i+1}^{n} + \left[1 - 2\mu\alpha\frac{\Delta t}{\Delta x^2}\right]\phi_i^{n} + \left[\mu\alpha\frac{\Delta t}{\Delta x^2}\right]\phi_{i-1}^{n} \tag{7.51}$$

We first remind the readers the following Taylor series expansions for a function of two independent variables:

$$\phi_{i\pm1}^{n+1} = \phi_i^n + \left.\left(\Delta t\frac{\partial}{\partial t} \pm \Delta x\frac{\partial}{\partial x}\right)\phi\right|_i^n + \left.\frac{1}{2!}\left(\Delta t\frac{\partial}{\partial t} \pm \Delta x\frac{\partial}{\partial x}\right)^2\phi\right|_i^n$$

$$+ \left.\frac{1}{3!}\left(\Delta t\frac{\partial}{\partial t} \pm \Delta x\frac{\partial}{\partial x}\right)^3\phi\right|_i^n + \text{HOT} \tag{7.52}$$

$$\phi_{i\pm1}^n = \phi_i^n \pm \Delta x \left.\frac{\partial \phi}{\partial x}\right|_i^n + \frac{\Delta x^2}{2!}\left.\frac{\partial^2 \phi}{\partial x^2}\right|_i^n \pm \frac{\Delta x^3}{3!}\left.\frac{\partial^3 \phi}{\partial x^3}\right|_i^n + \frac{\Delta x^4}{4!}\left.\frac{\partial^4 \phi}{\partial x^4}\right|_i^n + \text{HOT} \qquad (7.53)$$

$$\phi_i^{n+1} = \phi_i^n + \Delta t \left.\frac{\partial \phi}{\partial t}\right|_i^n + \frac{\Delta t^2}{2!}\left.\frac{\partial^2 \phi}{\partial t^2}\right|_i^n + \frac{\Delta t^3}{3!}\left.\frac{\partial^3 \phi}{\partial t^3}\right|_i^n + \text{HOT} \qquad (7.54)$$

We can now substitute these expansions into the finite difference approximation Eq. (7.51) of the diffusion equation:

$$\left[-(1-\mu)\alpha\frac{\Delta t}{\Delta x^2}\right]\left\{ \phi_i^n + \left.\left(\Delta t\frac{\partial}{\partial t} + \Delta x\frac{\partial}{\partial x}\right)\phi\right|_i^n + \frac{1}{2!}\left.\left(\Delta t\frac{\partial}{\partial t} + \Delta x\frac{\partial}{\partial x}\right)^2\phi\right|_i^n + \frac{1}{3!}\left.\left(\Delta t\frac{\partial}{\partial t} + \Delta x\frac{\partial}{\partial x}\right)^3\phi\right|_i^n + \frac{1}{4!}\left.\left(\Delta t\frac{\partial}{\partial t} + \Delta x\frac{\partial}{\partial x}\right)^4\phi\right|_i^n + \text{HOT} \right\}$$

$$+ \left[1 + 2(1-\mu)\alpha\frac{\Delta t}{\Delta x^2}\right]\left\{ \phi_i^n + \Delta t\left.\frac{\partial \phi}{\partial t}\right|_i^n + \frac{\Delta t^2}{2!}\left.\frac{\partial^2 \phi}{\partial t^2}\right|_i^n + \frac{\Delta t^3}{3!}\left.\frac{\partial^3 \phi}{\partial t^3}\right|_i^n + \text{HOT} \right\} +$$

$$\left[-(1-\mu)\alpha\frac{\Delta t}{\Delta x^2}\right]\left\{ \phi_i^n + \left.\left(\Delta t\frac{\partial}{\partial t} - \Delta x\frac{\partial}{\partial x}\right)\phi\right|_i^n + \frac{1}{2!}\left.\left(\Delta t\frac{\partial}{\partial t} - \Delta x\frac{\partial}{\partial x}\right)^2\phi\right|_i^n + \frac{1}{3!}\left.\left(\Delta t\frac{\partial}{\partial t} - \Delta x\frac{\partial}{\partial x}\right)^3\phi\right|_i^n + \frac{1}{4!}\left.\left(\Delta t\frac{\partial}{\partial t} - \Delta x\frac{\partial}{\partial x}\right)^4\phi\right|_i^n + \text{HOT} \right\}$$

$$= \left[\mu\alpha\frac{\Delta t}{\Delta x^2}\right]\left\{ \phi_i^n + \Delta x\left.\frac{\partial \phi}{\partial x}\right|_i^n + \frac{\Delta x^2}{2!}\left.\frac{\partial^2 \phi}{\partial x^2}\right|_i^n + \frac{\Delta x^3}{3!}\left.\frac{\partial^3 \phi}{\partial x^3}\right|_i^n + \frac{\Delta x^4}{4!}\left.\frac{\partial^4 \phi}{\partial x^4}\right|_i^n + \text{HOT} \right\}$$

$$+\left[1-2\mu\alpha\frac{\Delta t}{\Delta x^2}\right]\phi_i^n +\left[\mu\alpha\frac{\Delta t}{\Delta x^2}\right]\left\{\begin{array}{l}\phi_i^n -\Delta x\left.\frac{\partial\phi}{\partial x}\right|_i^n +\frac{\Delta x^2}{2!}\left.\frac{\partial^2\phi}{\partial x^2}\right|_i^n -\frac{\Delta x^3}{3!}\left.\frac{\partial^3\phi}{\partial x^3}\right|_i^n \\ +\frac{\Delta x^4}{4!}\left.\frac{\partial^4\phi}{\partial x^4}\right|_i^n +\text{HOT}\end{array}\right\} \quad (7.55)$$

Grouping similar terms and making algebraic manipulations, one obtains

$$\left.\frac{\partial\phi}{\partial t}\right|_i^n -\alpha\left.\frac{\partial^2\phi}{\partial x^2}\right|_i^n =-\frac{\Delta t}{2}\left.\frac{\partial^2\phi}{\partial t^2}\right|_i^n +(1-\mu)\Delta t\alpha\left.\frac{\partial^3\phi}{\partial t\partial x^2}\right|_i^n -\frac{\Delta t^2}{6}\left.\frac{\partial^3\phi}{\partial t^3}\right|_i^n$$

$$+(1-\mu)\frac{\Delta t^2}{2}\alpha\left.\frac{\partial^4\phi}{\partial t^2\partial x^2}\right|_i^n +\alpha\frac{\Delta x^2}{12}\left.\frac{\partial^4\phi}{\partial x^4}\right|_i^n +\text{HOT} \qquad (7.56)$$

The left side of this equation contains the original PDE [Eq. (7.33)]. If the approximation was accurate, the right-hand side would be zero. The leftover terms on the right are therefore the truncation error due to approximation. Using our standard definition of error, i.e., Error = True – Approximation (value), we obtain the truncation error as

$$\text{TE} =\frac{\Delta t}{2}\left.\frac{\partial^2\phi}{\partial t^2}\right|_i^n -(1-\mu)\Delta t\alpha\left.\frac{\partial^3\phi}{\partial t\partial x^2}\right|_i^n +\frac{\Delta t^2}{6}\left.\frac{\partial^3\phi}{\partial t^3}\right|_i^n$$

$$-(1-\mu)\frac{\Delta t^2}{2}\alpha\left.\frac{\partial^4\phi}{\partial t^2\partial x^2}\right|_i^n -\alpha\frac{\Delta x^2}{12}\left.\frac{\partial^4\phi}{\partial x^4}\right|_i^n +\text{HOT} \qquad (7.57)$$

Now, let us recognize an identity using the original PDE,

$$\alpha\frac{\partial^3\phi}{\partial t\partial x^2} =\frac{\partial}{\partial t}\left(\alpha\frac{\partial^2\phi}{\partial x^2}\right) =\frac{\partial}{\partial t}\left(\frac{\partial\phi}{\partial t}\right) =\frac{\partial^2\phi}{\partial t^2} \quad\text{and}\quad \alpha\frac{\partial^4\phi}{\partial t^2\partial x^2} =\frac{\partial^3\phi}{\partial t^3} \qquad (7.58)$$

Using this identity in Eq. (7.57) and retaining only similar terms (in derivative), we obtain

$$\text{TE} =\left(\mu -\frac{1}{2}\right)\Delta t\left.\frac{\partial^2\phi}{\partial t^2}\right|_i^n -\alpha\frac{\Delta x^2}{12}\left.\frac{\partial^4\phi}{\partial x^4}\right|_i^n +\left(\mu -\frac{2}{3}\right)\frac{\Delta t^2}{2}\left.\frac{\partial^3\phi}{\partial t^3}\right|_i^n +\text{HOT} \qquad (7.59)$$

The first term of the truncation error is first order in time (Δt) but it vanishes for $\mu = 1/2$. Rest of the terms are second order or higher in both time and space.

Therefore, the scheme is second order in time and space for $\mu = 1/2$ and we denote it as $o(\Delta t^2, \Delta x^2)$. For all other values of μ, the scheme is first order in time and second order in space, i.e., $o(\Delta t, \Delta x^2)$.

Let us try to interpret this from our knowledge of ODE methods. We have used the central difference scheme for the spatial derivatives, which is second order. Therefore, it comes as no surprise that the accuracy in space is always second order. For the time discretization, we have used a weighting factor μ. For $\mu = 0$ and 1, we get the Euler methods (backward and forward, respectively). Since, both of these are first order methods, we obtain first order accuracy in time for PDE as well. For $\mu = 1/2$, the method becomes trapezoidal which is known to be second order. Therefore, the results are in the expected lines. However, in some cases, we can increase the order of truncation error for this scheme by judicious choice of time step size and the spatial grid size, which was not possible for the ODE. In order to see this, let us consider the following identity:

$$\frac{\partial^2 \phi}{\partial t^2} = \frac{\partial}{\partial t}\left(\frac{\partial \phi}{\partial t}\right) = \frac{\partial}{\partial t}\left(\alpha\frac{\partial^2 \phi}{\partial x^2}\right) = \alpha\frac{\partial^2}{\partial x^2}\left(\frac{\partial \phi}{\partial t}\right) = \alpha\frac{\partial^2}{\partial x^2}\left(\alpha\frac{\partial^2 \phi}{\partial x^2}\right) = \alpha^2\frac{\partial^4 \phi}{\partial x^4}$$

(7.60)

Using this identity, the truncation error becomes

$$\text{TE} = \left\{\left(\mu - \frac{1}{2}\right)\Delta t\alpha^2 - \alpha\frac{\Delta x^2}{12}\right\}\left.\frac{\partial^4 \phi}{\partial x^4}\right|_i^n + \frac{\Delta t^2}{6}\left.\frac{\partial^3 \phi}{\partial t^3}\right|_i^n + \text{HOT}$$

(7.61)

The first term on the right will vanish if the coefficient in the curly bracket is set to zero. This essentially increases the order of the method even when μ is not equal to 1/2. The required identity is

$$\alpha\frac{\Delta t}{\Delta x^2} = \frac{1}{12\left(\mu - \frac{1}{2}\right)}$$

(7.62)

Note that on the left, we have a familiar term group namely the ratio C/P_g. This ratio can be suitably chosen (by adjusting Δt and Δx) for all values of $\mu > 1/2$. For $\mu = 1/2$, the right-hand side is infinite. This can be achieved only if Δx is zero which is not possible. However for $\mu = 1/2$, the method is second order in time anyway [Eq. (7.59)] and therefore, there is no need to adjust these parameters. For $\mu < 1/2$, the ratio of C/P_g has to be negative for the time and space derivative terms to cancel each other. Since, α, Δt, and Δx all are greater than zero, negative value is not possible. Therefore, for fully implicit method like Euler backward, it is not possible to cancel the time and space terms in order to achieve a higher order of accuracy. For Euler forward method, $\mu = 1$ and the relation is

$$\alpha \frac{\Delta t}{\Delta x^2} = \frac{1}{6} \qquad (7.63)$$

We shall see in Section 7.3 that the above condition is within the stability limit of the Euler forward method for the diffusion equation and therefore, it is possible to achieve the condition. However, this puts more severe restriction than the stability on the choice of time step for any chosen Δx.

Let us now perform similar truncation error analysis on the discrete advective-diffusion Eq. (7.47). For constant u and α, the equation is as follows:

$$\left[(1-\mu)\left(u\frac{\Delta t}{2\Delta x} - \alpha\frac{\Delta t}{\Delta x^2} \right) \right]\phi_{i+1}^{n+1} + \left[1 + 2(1-\mu)\alpha\frac{\Delta t}{\Delta x^2} \right]\phi_i^{n+1}$$

$$+ \left[(1-\mu)\left(-u\frac{\Delta t}{2\Delta x} - \alpha\frac{\Delta t}{\Delta x^2} \right) \right]\phi_{i-1}^{n+1}$$

$$= \left[\mu\left(-u\frac{\Delta t}{2\Delta x} + \alpha\frac{\Delta t}{\Delta x^2} \right) \right]\phi_{i+1}^n + \left[1 - 2\mu\alpha\frac{\Delta t}{\Delta x^2} \right]\phi_i^n + \left[\mu\left(u\frac{\Delta t}{2\Delta x} + \alpha\frac{\Delta t}{\Delta x^2} \right) \right]\phi_{i-1}^n$$

$$(7.64)$$

Using the expansions of Eqs (7.53) and (7.54) and performing algebraic operations similar to diffuion equation, we obtain,

$$\left.\frac{\partial \phi}{\partial t}\right|_i^n + u\left.\frac{\partial \phi}{\partial x}\right|_i^n - \alpha\left.\frac{\partial^2 \phi}{\partial x^2}\right|_i^n = -u^2\Delta t\left(\mu - \frac{1}{2} \right)\left.\frac{\partial^2 \phi}{\partial x^2}\right|_i^n + 2u\alpha\Delta t\left(\mu - \frac{1}{2} \right)\left.\frac{\partial^3 \phi}{\partial x^3}\right|_i^n$$

$$- \alpha^2\Delta t\left(\mu - \frac{1}{2} \right)\left.\frac{\partial^4 \phi}{\partial x^4}\right|_i^n + \frac{u^3\Delta t^2}{6}\left(\frac{\mu}{2} - \frac{1}{3} \right)\left.\frac{\partial^3 \phi}{\partial x^3}\right|_i^n$$

$$- \frac{u\Delta x^2}{6}\left.\frac{\partial^3 \phi}{\partial x^3}\right|_i^n + \text{HOT} \qquad (7.65)$$

Therefore, the μ-CD scheme is first order in time and second order in space except for $\mu = 1/2$ where all the first order time term disappear from the truncation error leaving the second order as the highest-order terms in the truncation error. For all values of μ except 1/2, the first term on the right-hand side is similar to the diffusion term on the left-hand side containing second-order space derivative. In fact, one can write the above equation in the following form:

$$\left.\frac{\partial \phi}{\partial t}\right|_i^n + u\left.\frac{\partial \phi}{\partial x}\right|_i^n - \left[\alpha - u^2 \Delta t\left(\mu - \frac{1}{2}\right)\right]\left.\frac{\partial^2 \phi}{\partial x^2}\right|_i^n = 2u\alpha\Delta t\left(\mu - \frac{1}{2}\right)\left.\frac{\partial^3 \phi}{\partial x^3}\right|_i^n$$

$$- \alpha^2 \Delta t\left(\mu - \frac{1}{2}\right)\left.\frac{\partial^4 \phi}{\partial x^4}\right|_i^n$$

$$+ \frac{u^3 \Delta t^2}{6}\left(\frac{\mu}{2} - \frac{1}{3}\right)\left.\frac{\partial^3 \phi}{\partial x^3}\right|_i^n \quad (7.66)$$

$$- \frac{u\Delta x^2}{6}\left.\frac{\partial^3 \phi}{\partial x^3}\right|_i^n + \text{HOT}$$

This essentially shows that the effect of the highest-order truncation error term is to increase or decrease the net diffusion (or dispersion) coefficient of the physical problem depending on the value of μ. This extra diffusion (or dispersion) term is purely a numerical artifact and will be called *numerical diffusion*. Simplest way to minimize the effect of numerical diffusion is to use a modified artificial diffusion coefficient (α') in place of the original one. For the μ-CD scheme, this modification rule is

$$\alpha' = \alpha + u^2 \Delta t\left(\mu - \frac{1}{2}\right) \quad (7.67)$$

Use of this modified diffusion coefficient does not increase the order of accuracy of the method, but it reproduces the effect of diffusion in the numerical solution equivalent to the original diffusion coefficient in the analytical equation. In order to avoid using this artificial diffusion coefficient, one can change the numerical scheme such that the numerical diffusion does not appear in the truncation error. One such commonly used method is *upwind scheme*. In this, the first-order spatial derivative with the velocity term u is discretized using a backward difference instead of central difference. This backward difference can be first, second, or third order. Higher the order, less is the numerical diffusion. We leave it to the readers to show this.

We shall use the upwind scheme in Section 7.3.3 for first order hyperbolic wave equation where the advection term has the opposite sign and backward difference will be used instead of the forward difference to achieve the same objective of reducing numerical diffusion. Let us illustrate the numerical diffusion with the help of an example.

Example 7.2   Solve the following one-dimensional advection-diffusion equation using EF-CD scheme with original and modified diffusion coefficient. Also, solve it using Crank-Nicolson scheme with original diffusion coefficient. Compare three results graphically at $t = 3$.

$$\frac{\partial C}{\partial t} + u\frac{\partial C}{\partial x} = \alpha\frac{\partial^2 C}{\partial x^2}, \qquad 0 \le x \le 1$$

where $u = 0.2$, $\alpha = 0.01$, $C(0,t) = 1$, $\frac{\partial C}{\partial x}(1,t) = 0$, $C(x,0) = 0$
Use $\Delta x = 0.2$ and $\Delta t = 0.5$.

Solution   With $\Delta x = 0.2$, we have five equal intervals in the domain which leads to 6 nodes including the boundary nodes. Let us number them 0–5 with node zero at $x = 0$ and node 5 at $x = 1$.

The equations for an interior node (i) can be formulated using Eq. (7.64) for both EF-CD and Crank-Nicolson schemes with appropriate values of μ. Let us first formulate the equations for the EF-CD scheme, $\mu = 1$.

With original diffusion coefficient:

$$\frac{u\Delta t}{2\Delta x} = 0.25, \quad \frac{\alpha\Delta t}{\Delta x^2} = 0.125$$

$$C_i^{n+1} = 0.375C_{i-1}^n + 0.75C_i^n - 0.125C_{i+1}^n$$

We can now write the following set of equations for the values at the interior nodes by using the boundary condition $C_0 = 1$:

$$C_1^{n+1} = 0.75C_1^n - 0.125C_2^n + 0.375$$
$$C_2^{n+1} = 0.375C_1^n + 0.75C_2^n - 0.125C_3^n$$
$$C_3^{n+1} = 0.375C_2^n + 0.75C_3^n - 0.125C_4^n$$
$$C_4^{n+1} = 0.375C_3^n + 0.75C_4^n - 0.125C_5^n$$

In the last equation, C_5 may be eliminated by making a second-order backward difference approximation of the boundary condition at node 5,

$$\frac{C_3 - 4C_4 + 3C_5}{2\Delta x} = 0 \text{ or } C_5 = \frac{4}{3}C_4 - \frac{1}{3}C_5$$

Using this relation, the last equation becomes

$$C_4^{n+1} = 0.4167C_3^n + 0.5833C_4^n$$

Now, we can compute $C_1 - C_4$ at all the time steps using the above equation with the initial conditions,

$$C_1^0 = C_2^0 = C_3^0 = C_4^0 = 0$$

The modified diffusion coefficient computed using Eq. (7.67) is $\alpha' = 0.02$. This gives

$$\frac{\alpha' \Delta t}{\Delta x^2} = 0.25$$

The new equation for the node (i) is

$$C_i^{n+1} = 0.5C_i^n + 0.5C_{i-1}^n$$

Notice that the value C_{i+1} is absent from the above equation. This essentially reduced the formulation to upwind scheme. So, use of modified diffusion coefficient and the upwind scheme are equivalent in this case. Because of this, we do not even need the boundary condition to the right. We can write the set of equations as follows:

$$C_1^{n+1} = 0.5C_1^n + 0.5$$

$$C_2^{n+1} = 0.5C_2^n + 0.5C_1^n$$

$$C_3^{n+1} = 0.5C_3^n + 0.5C_2^n$$

$$C_4^{n+1} = 0.5C_4^n + 0.5C_3^n$$

Lastly, the equation for node (i) for the Crank-Nicolson scheme is

$$-0.1875C_{i-1}^{n+1} + 1.125C_i^{n+1} + 0.0625C_{i+1}^{n+1} = 0.1875C_{i-1}^n + 0.875C_i^n - 0.0625C_{i+1}^n$$

Using the left hand boundary condition and second order backward difference approximation of the right hand boundary condition, we obtain the following system of equation for values at the interior nodes:

$$\begin{bmatrix} 1.125 & 0.0625 & 0 & 0 \\ -0.1875 & 1.125 & 0.0625 & 0 \\ 0 & -0.1875 & 1.125 & 0.0625 \\ 0 & 0 & -0.2083 & 1.2083 \end{bmatrix} \begin{bmatrix} C_1^{n+1} \\ C_2^{n+1} \\ C_3^{n+1} \\ C_4^{n+1} \end{bmatrix} =$$

$$\begin{bmatrix} 0.875 & -0.0625 & 0 & 0 \\ 0.1875 & 0.875 & -0.0625 & 0 \\ 0 & 0.1875 & 0.875 & -0.0625 \\ 0 & 0 & 0.2083 & 0.7917 \end{bmatrix} \begin{bmatrix} C_1^n \\ C_2^n \\ C_3^n \\ C_4^n \end{bmatrix} + \begin{bmatrix} 0.375 \\ 0 \\ 0 \\ 0 \end{bmatrix}$$

All three methods can now be solved using any spreadsheet or MATLAB. We show the result in Fig. 7.3.

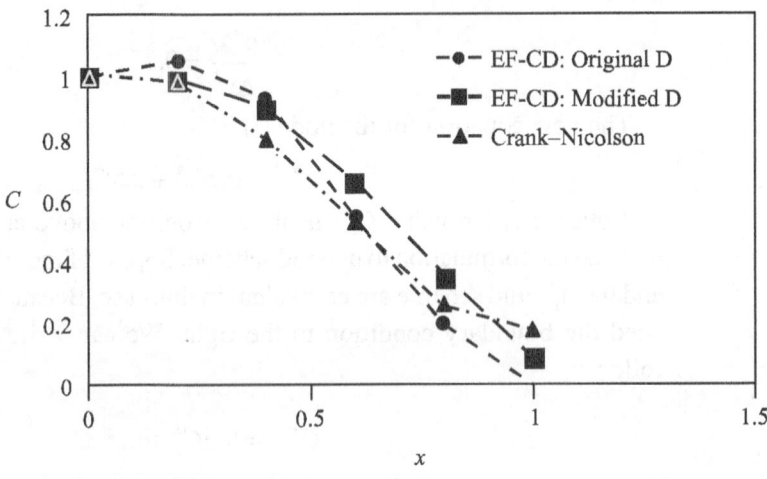

Fig. 7.3

In the solution using EF-CD scheme with original diffusion coefficient, the concentration goes above 1 at node 1 and negative at node 5. The solution is better behaved with the modified diffusion coefficient. The solution with Crank-Nicolson scheme is good as expected. The value at node 5 is a function of the difference scheme used for discretization of the boundary condition. The second-order backward difference scheme has put weights on the previous two nodes. This would be good for an advection-dominated problem but for diffusion problems, it is affecting the value at node 4 as well because of the replacement. A first-order backward difference approximation in this case may work better.

Lastly, we limited the number of unknown nodes to 4 in order to illustrate the solution method manually. In reality, the problem requires finer grids in the spatial dimension to adequately resolve the effects of advection. We will see the pure advection part of this problem with finer grid in Section 7.3.3.

We outlined the solution method for parabolic equations using the μ-CD scheme in time and central difference in space. Both the schemes use the same concept as was described in the previous chapter for ODEs. We have also shown that the PDE is reduced to a system of IVP once the spatial discretization is done. Then, one can use any other method for the solution of system of IVP, such as Runge-Kutta, Gears method, etc.

Especially, the reader may try to apply fourth order Runge-Kutta method for the

Example Problem 7.2 and compare the solution. The fourth order Runge-Kutta method combines high order of accuracy with decent stability property. Its stability region also encompasses some part of purely imaginary λ (Fig. 6.14) which is important for advection-dominated problems (Section 7.3.3). As a result, the method has found widespread application in the solution of ODEs as well as PDEs.

7.3.2 Laplace Equation

The most commonly encountered PDE in engineering problems is the Laplace equation. It is encountered in stream function, flow potential, temperature distribution, steady state concentration, stress distribution, etc. to name a few. We will illustrate the solution methods with a two-dimensional problem which is given by Eq. (7.6). We rewrite it in terms of our state variable ϕ:

$$\frac{\partial^2 \phi}{\partial x^2} + \frac{\partial^2 \phi}{\partial y^2} = 0 \qquad (7.68)$$

Both the independent variables are in spatial dimension. Therefore, we need to have finite boundaries for both the variables to obtain a numerical solution although for analytical solution, infinite boundaries may be allowed. The boundaries are governed by the physics of the problem. In most of the practical problems, these are physical boundaries and therefore finite. However, in a very large domain, if we are concerned about variation or perturbation at a particular point, we need to put the finite boundary far enough from this point so that the effect of perturbation is negligible and can be treated as infinity of the analytical problem.

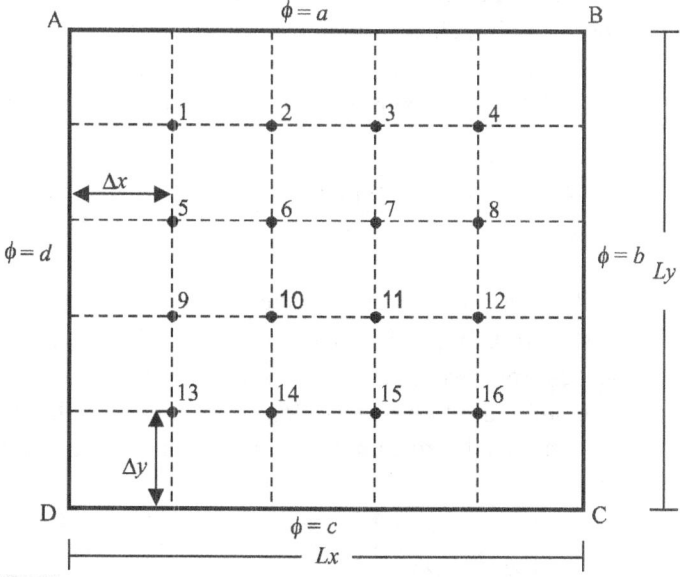

Fig. 7.4 A finite boundary in Cartesian geometry with uniform grid sizes.

For illustration, we shall start by using first type boundary conditions at all the boundaries. This however will not affect our description of the solution of the PDE and we will show how to incorporate other types of BCs later on. Once the boundary is defined, the first job is to lay a grid on x and y directions. In order to illustrate the solution method, let us consider a finite boundary in Cartesian geometry with first type BC (Fig. 7.4). The problem shown in Fig. 7.4 may be mathematically stated as follows:

Solve Laplace equation in the region $x \in (0, L_x)$ and $y \in (0, L_y)$ with the boundary conditions $\phi(x,0) = c$, $\phi(x,L_y) = a$, $\phi(0,y) = d$, and $\phi(L_x,y) = b$.

We have laid uniform grids in both x and y directions with grid sizes Δx and Δy, respectively. However, Δx may not be equal to Δy. In Fig. 7.4, we have laid out five equally spaced intervals in each direction. This essentially resulted in a total of 36 intersection points. However, 20 of these 36 points are on the boundaries. The values of ϕ are known at these 20 points from the boundary conditions. So, we only need to determine the unknown values of ϕ at 16 interior points. These points are numbered as shown in Fig. 7.4. Let us denote the unknown values of the state variable as $\phi_1 - \phi_{16}$. We will now write discrete approximation of the Laplace equation at each of these 16 points and obtain 16 linear algebraic equations. It will then be easy to compute the required values using the methods of Chapter 2.

Let us choose 2nd order central difference scheme for approximating the partial derivatives in the Laplace equation. We will use i and j as the space variable indices for x and y, respectively. Following our convention, we will put both the indices in the subscript, since both are space variables.

Therefore, we can write the following general expressions for the 2nd order central difference approximation of the double derivatives in a regular grid:

$$\left.\frac{\partial^2 \phi}{\partial x^2}\right|_{i,j} = \frac{\phi_{i+1,j} - 2\phi_{i,j} + \phi_{i-1,j}}{\Delta x^2} \quad \text{and} \quad \left.\frac{\partial^2 \phi}{\partial y^2}\right|_{i,j} = \frac{\phi_{i,j+1} - 2\phi_{i,j} + \phi_{i,j-1}}{\Delta y^2} \quad (7.69)$$

A visual representation of the values of different nodes used in these expressions is shown in Fig. 7.5. Approximation of the partial derivative at a point requires values from two adjacent nodes in each direction.

Combining two independent expressions of the partial derivatives Eq. (7.69), we obtain a discrete approximation of the Laplace equation for the node (i, j):

$$\left(\frac{\partial^2 \phi}{\partial x^2} + \frac{\partial^2 \phi}{\partial y^2}\right)_{i,j} = \frac{\phi_{i+1,j} - 2\phi_{i,j} + \phi_{i-1,j}}{\Delta x^2} + \frac{\phi_{i,j+1} - 2\phi_{i,j} + \phi_{i,j-1}}{\Delta y^2} = 0 \quad (7.70)$$

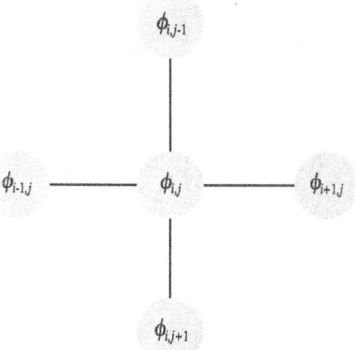

Fig. 7.5 Approximation molecule for the CD scheme applied to the Laplace equation.

$$\left(\frac{1}{\Delta y^2}\right)\phi_{i,\,j-1} +\left(\frac{1}{\Delta x^2}\right)\phi_{i-1,\,j} +\left(-\frac{2}{\Delta x^2}-\frac{2}{\Delta y^2}\right)\phi_{i,\,j}$$

$$+\left(\frac{1}{\Delta x^2}\right)\phi_{i+1,\,j} +\left(\frac{1}{\Delta y^2}\right)\phi_{i,\,j+1} =0 \tag{7.71}$$

Therefore, an approximation at node (i, j) requires values from four neighbouring nodes. This, we will term as the approximation molecule (Fig. 7.5) for the node (i, j) using 2nd order central difference scheme. The molecule will be different for the same equation if we change the discretization scheme in any direction, e.g., forward or backward difference in place of central difference. This will be required for the boundary nodes as we will be missing nodes on one of the sides depending on which side the boundary is located.

Now, let us apply this scheme for the problem shown in Fig. 7.4. For Node 1, two of the neighbouring nodes are on the boundary and therefore the values at these nodes are known from the boundary condition. Nodes 4, 13, and 16 are similar to Node 1. On some boundary points such as 2, 3, 5, 8, 9, 12, 14, and 15, only one of the neighbouring node is on the boundary. For the interior points 6, 7, 10, and 11, all the nodes in the molecule are interior nodes with unknown values. Two examples of approximation molecules for Nodes 1 (with neighbouring boundary nodes) and 6 (neighbouring interior nodes) are shown in Fig. 7.6. The approximate equations for these nodes are

Node 1: $\left(\frac{1}{\Delta y^2}\right)a +\left(\frac{1}{\Delta x^2}\right)d +\left(-\frac{2}{\Delta x^2}-\frac{2}{\Delta y^2}\right)\phi_1 +\left(\frac{1}{\Delta x^2}\right)\phi_2 +\left(\frac{1}{\Delta y^2}\right)\phi_5 =0$

$$\tag{7.72}$$

or $\qquad \left(-\dfrac{2}{\Delta x^2}-\dfrac{2}{\Delta y^2}\right)\phi_1+\left(\dfrac{1}{\Delta x^2}\right)\phi_2+\left(\dfrac{1}{\Delta y^2}\right)\phi_5=-\left(\dfrac{1}{\Delta y^2}\right)a-\left(\dfrac{1}{\Delta x^2}\right)d$

$$(7.73)$$

Node 6: $\qquad \left(\dfrac{1}{\Delta y^2}\right)\phi_2+\left(\dfrac{1}{\Delta x^2}\right)\phi_5+\left(-\dfrac{2}{\Delta x^2}-\dfrac{2}{\Delta y^2}\right)\phi_6+\left(\dfrac{1}{\Delta x^2}\right)\phi_7+\left(\dfrac{1}{\Delta y^2}\right)\phi_{10}=0$

$$(7.74)$$

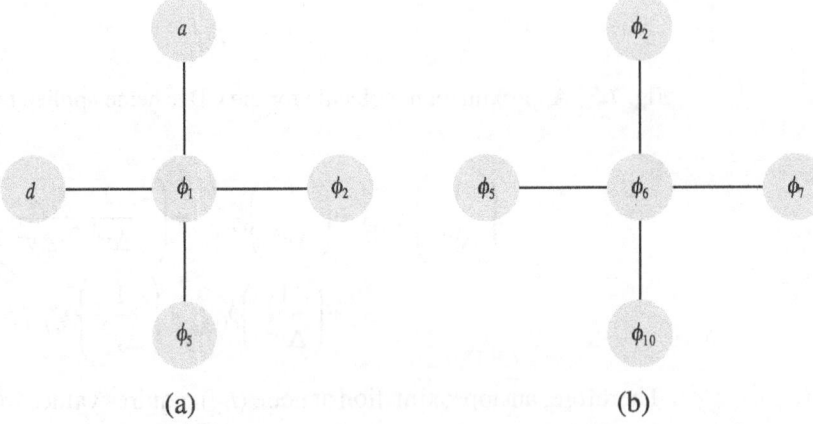

(a) $\qquad\qquad\qquad\qquad$ (b)

Fig. 7.6 Approximation molecules for (a) Node 1 and (b) Node 6.

Writing similar equations for all 16 nodes, one obtains 16 linear algebraic equations in the form of $A\Phi = b$. The matrix A contains the coefficients of the equations similar to Eqs (7.73) and (7.74), Φ is the vector containing unknowns $\phi_1 - \phi_{16}$, and b is a vector containing the known constants originating from the boundary conditions contained in the equations for the nodes adjacent to the boundaries [Eqs (7.72) and (7.73)].

The easiest way to generate matrix A and vector b would be to initiate a square matrix and a vector of size 16 with zero entries. Then for each node, change the entries corresponding to the nodes contained in that equation with appropriate values. Let us denote the values of matrix A as a_{ij} and those of vector b as b_i. Then for Node 1 [Eq. (7.72)], we will set the following:

$$a_{11}=\left(-\dfrac{2}{\Delta x^2}-\dfrac{2}{\Delta y^2}\right),\quad a_{12}=\left(\dfrac{1}{\Delta x^2}\right),\quad a_{15}=\left(\dfrac{1}{\Delta y^2}\right),\quad b_1=-\left(\dfrac{1}{\Delta y^2}\right)a-\left(\dfrac{1}{\Delta x^2}\right)d$$

$$(7.75)$$

Once Δx and Δy are chosen, the entries in the matrix A (coefficients of $\phi_{i,j}$) are constant [Eqs (7.73) and (7.74)]. One can generate these values from Fig. 7.4 without having to write Eqs (7.73) and (7.74) explicitly. For any approximation molecule, the coefficient for the node at the centre is $-\left(\dfrac{2}{\Delta x^2}+\dfrac{2}{\Delta y^2}\right)$

coefficient of two neighbouring nodes in x direction is $(1/\Delta x^2)$ irrespective of whether they are on the left or right of the center and coefficient of the neighbouring nodes in y direction is $(1/\Delta y^2)$. If the approximation molecule contains one of the nodes at the boundary, the term corresponding to that node contributes an entry in the vector b. For example, Node 9 contributes the following entries:

$$a_{9\,5}=\left(\frac{1}{\Delta y^2}\right), a_{9\,9}=\left(-\frac{2}{\Delta x^2}-\frac{2}{\Delta y^2}\right), a_{9\,10}=\left(\frac{1}{\Delta x^2}\right), a_{9\,13}=\left(\frac{1}{\Delta y^2}\right), b_9=-\left(\frac{1}{\Delta x^2}\right)d$$

(7.76)

For Node 7, the entries are

$$a_{7\,3}=\left(\frac{1}{\Delta y^2}\right), a_{7\,6}=\left(\frac{1}{\Delta x^2}\right), a_{7\,7}=\left(-\frac{2}{\Delta x^2}-\frac{2}{\Delta y^2}\right), a_{7\,8}=\left(\frac{1}{\Delta x^2}\right), a_{7\,11}=\left(\frac{1}{\Delta y^2}\right)$$

(7.77)

The entries are made on the kth row for the equation (or approximation molecule) corresponding to node k. The node numbers in the kth approximation molecules are the column index where the entries are made in that row. The coefficient of the node at the centre is therefore, always a diagonal element of the matrix. Since, the approximation molecule for central difference contains five nodes, a maximum of five entries are non-zero in any row. The rest are all zeros. This essentially leads to a sparse matrix A. Therefore, the iterative methods described in Section (2.2.2) are most appropriate for solution of the system of equation. Since a_{kk} entries are the largest in absolute magnitude [Eq. (7.71)], the coefficient matrix is diagonal dominant and the convergence is assured.

An oft-used method to accelerate the convergence of the solution is known as *multi-grid* method. This will be briefly described in the next chapter. We shall now describe how to incorporate a derivative boundary condition. If in the problem of Fig. 7.4, the gradient of ϕ is specified along boundaries BC and CD, the problem will be defined as:

Solve Laplace equation in the region $x \in (0, L_x)$ and $y \in (0, L_y)$ with the boundary conditions:

$$\left.\frac{\partial \phi}{\partial y}\right|_{(x,0)} = c, \quad \phi(x, L_y) = a, \quad \phi(0, y) = d \quad \text{and} \quad \left.\frac{\partial \phi}{\partial x}\right|_{(L_x, y)} = b$$

The values of ϕ at the nodes along boundaries BC and CD are also unknown. So, the number of unknowns now increases to 25 (Fig. 7.7). The obvious solution is to write the discrete approximation of Laplace equation for all the interior nodes and discrete approximations of boundary condtions for the boundary nodes. Since we will be able to write one equation for each node, this will yield the same number of equations as the number of unknowns.

At nodes 5, 10, 15, 20, 21, 22, 23, and 24, one can write a second order backward difference or ghost node approximation of the boundary condition similar to the boundary value problems (Section 6.5). For example, 2nd order backward difference equations for Nodes 5 and 21 are shown below:

Node 5:
$$\frac{\phi_3 - 4\phi_4 + 3\phi_5}{2\Delta x} = b$$

or
$$\left(\frac{1}{2\Delta x}\right)\phi_3 + \left(-\frac{2}{\Delta x}\right)\phi_4 + \left(\frac{3}{2\Delta x}\right)\phi_5 = b \qquad (7.78)$$

Node 21:
$$\frac{\phi_{11} - 4\phi_{16} + 3\phi_{21}}{2\Delta y} = c$$

or
$$\left(\frac{1}{2\Delta y}\right)\phi_{11} + \left(-\frac{2}{\Delta y}\right)\phi_{16} + \left(\frac{3}{2\Delta y}\right)\phi_{21} = c \qquad (7.79)$$

Entries corresponding to these nodes are

Node 5: $\quad a_{53} = \left(\frac{1}{2\Delta x}\right)$, $\quad a_{54} = \left(-\frac{2}{\Delta x}\right)$, $\quad a_{55} = \left(\frac{3}{2\Delta x}\right)$, and $b_5 = b$ $\qquad (7.80)$

Node 21: $\quad a_{21\ 11} = \left(\frac{1}{2\Delta y}\right)$, $\quad a_{21\ 16} = \left(-\frac{2}{\Delta y}\right)$, $\quad a_{21\ 21} = \left(\frac{3}{2\Delta x}\right)$, and $b_{21} = c$ $\quad (7.81)$

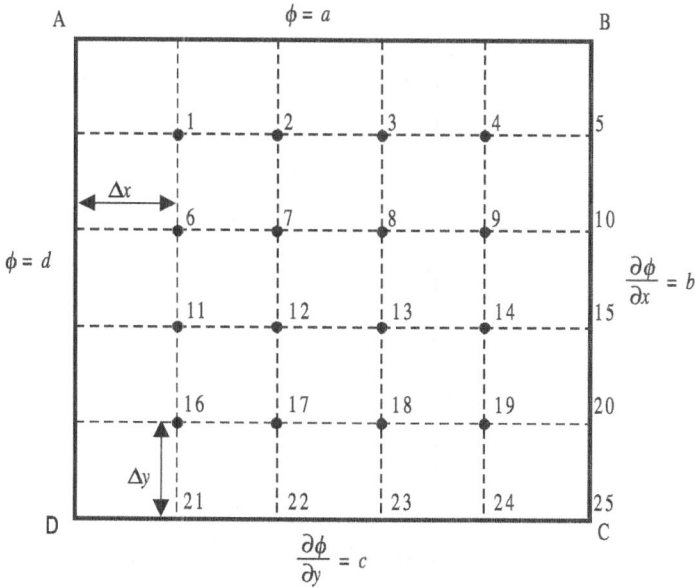

Fig. 7.7 New node numbers with gradient boundary conditions for the problems in Fig. 7.4.

For Node 25, both the boundary conditions are valid. So, one may wonder which equation to write. However, it is easy to see that Node 25 does not belong to the approximation molecule of any other node. Therefore, the entry in the coefficient matrix corresponding to this node is zero in every other row except Row 25. Therefore, this node can be eliminated from the unknown, which essentially reduces the unknown vector size to 24. The flux in both x and y directions at Node 25 can be computed from the values at nodes 15, 20, 23, and 24 using second-order backward difference.

You must have realized by now that changing the boundary conditions has increased the size of the matrix equation to be solved to 24 from 16. In addition, node numbers of the interior nodes have changed. Therefore, location of the entries corresponding to these nodes in the coefficient matrix also changed. This essentially means, the complete matrix have to be regenerated. It would be nice, if we could keep the existing matrix of the original boundary conditions and only change the entries in the rows corresponding to the nodes adjacent to the boundary. This will also keep the size of the system of equations to 16. This can be accomplished by keeping the original node numbers and give some temporary numbers to the boundary nodes where the ϕ values are unknown (Fig. 7.8).

The discrete approximation of the Laplace equation at the interior nodes adjacent to the boundary would contain these temporary node variables. These can be

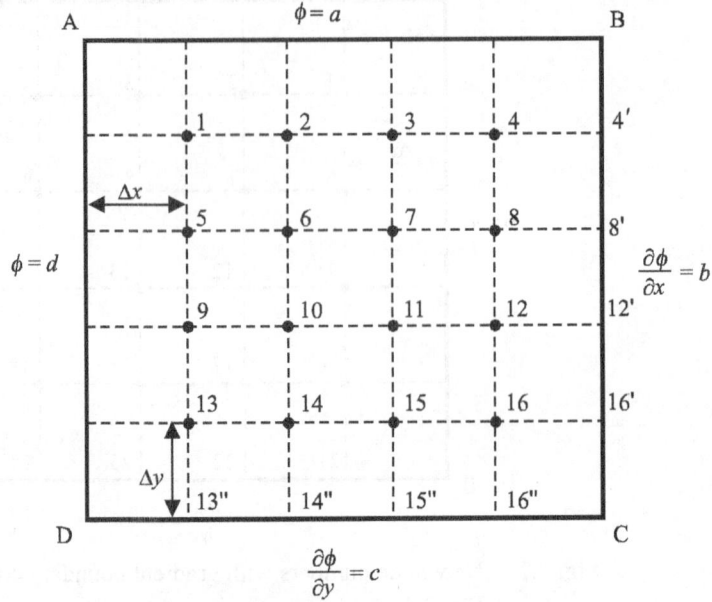

Fig. 7.8 Temporary node numbers along the boundary for derivative boundary conditions for the problem of Fig. 7.4.

eliminated by using the boundary condition approximations of Eqs (7.78) and (7.79) at the boundary nodes. We will illustrate this using the example of Node 16 (Figs 7.7 and 7.8). With the first type boundary conditions (Fig. 7.4), the equation for Node 16 was

$$\left(\frac{1}{\Delta y^2}\right)\phi_{12} + \left(\frac{1}{\Delta x^2}\right)\phi_{15} + \left(-\frac{2}{\Delta x^2} - \frac{2}{\Delta y^2}\right)\phi_{16} = -\frac{b}{\Delta x^2} - \frac{c}{\Delta y^2} \qquad (7.82)$$

With the second type boundary conditions, the approximation molecule of Node 16 has two temporary unknown nodes at the boundary. One can write the following discrete approximations for Laplace equation as Node 16 and boundary conditions at nodes 16′ and 16″:

Node 16:
$$\left(\frac{1}{\Delta y^2}\right)\phi_{12} + \left(\frac{1}{\Delta x^2}\right)\phi_{15} + \left(-\frac{2}{\Delta x^2} - \frac{2}{\Delta y^2}\right)\phi_{16}$$
$$+ \left(\frac{1}{\Delta x^2}\right)\phi_{16'} + \left(\frac{1}{\Delta y^2}\right)\phi_{16''} = 0 \qquad (7.83)$$

Node 16' :
$$\frac{\phi_{15} - 4\phi_{16} + 3\phi_{16'}}{2\Delta x} = b$$

or
$$\left(\frac{1}{2\Delta x}\right)\phi_{15} + \left(-\frac{2}{\Delta x}\right)\phi_{16} + \left(\frac{3}{2\Delta x}\right)\phi_{16'} = b \qquad (7.84)$$

Node 16" :
$$\frac{\phi_{12} - 4\phi_{16} + 3\phi_{16''}}{2\Delta y} = c$$

or
$$\left(\frac{1}{2\Delta y}\right)\phi_{12} + \left(-\frac{2}{\Delta y}\right)\phi_{16} + \left(\frac{3}{2\Delta y}\right)\phi_{16''} = c \qquad (7.85)$$

Using Eqs (7.84) and (7.85) in Eq. (7.83), we obtain the following equation for Node 16:

$$\left(\frac{2}{3\Delta y^2}\right)\phi_{12} + \left(\frac{2}{3\Delta x^2}\right)\phi_{15} + \left(-\frac{2}{3\Delta x^2} - \frac{2}{3\Delta y^2}\right)\phi_{16} = -\frac{2b}{3\Delta x} - \frac{2c}{3\Delta y} \qquad (7.86)$$

By changing boundary condition, the original equation for Node 16 [Eq. (7.82) will be replaced by Eq. (7.86)]. Similarly, equations corresponding to all the nodes adjacent to the boundary(ies) where the condition(s) has (have) changed will have to be replaced with new ones. Entries corresponding to all other nodes will remain unaltered. For the original example problem (Fig. 7.4), the entries will be changed in the rows 4, 8, and 12 – 16 as a result of changing boundary conditions along two of the boundaries (Fig. 7.7).

Only 7 out of 16 (43.75%) rows will change while the majority of the matrix will remain unaltered. The fraction of rows that actually changes gets smaller as the total number of nodes increase. For example, for the same problem, if we divide each direction into 10 intervals instead of five, we will have a total of 81 interior nodes. Similar change of boundary conditions will lead to change of 18 out of 81 rows (22.22%). For 100 intervals in each direction, the change is in 198 rows out of total 9801 rows (nodes), which is only 2.02%.

Once $\phi_1 - \phi_{16}$ are computed from the matrix equation, the values at the temporary nodes can be computed from the boundary conditions approximations used to replace these variables. For example, values at 16' and 16" can be computed from 7.84 and 7.85, respectively.

Solution of Laplace equation with different types of boundary conditions is shown in Example 7.3.

Example 7.3 Compute the steady state temperature (T) distribution on plate of size $L_x = 2$ m and $L_y = 8$ m, with the boundary conditions

$$\left.\frac{\partial T}{\partial x}\right|_{(0,y)} = 0,\ T(2,y) = 50\sin\left(\frac{\pi y}{8}\right),\ T(x,0) = 0 \text{ and } T(x,8) = 0$$

Use second-order finite difference approximations with $\Delta x = 1$ m and $\Delta y = 2$ m.

Solution Steady state temperature distribution in a plate is governed by the Laplace equation. The plate with grid is shown in Fig. 7.9.

We have not numbered any node on the boundary where the values of temperature (T) are known or in other words, the boundary condition is of the first type. Nodes 1, 2, and 3 are interior nodes. Therefore, *Laplace equation* needs to be solved for these nodes. Using Eq. (7.70), we can write the following approximation for Node 1 with $\Delta x = 1$ and $\Delta y = 2$:

$$\frac{T_4 - 2T_1 + 50\sin\left(\dfrac{6\pi}{8}\right)}{1^2} + \frac{0 - 2T_1 + T_2}{2^2} = 0$$

or

$$-\frac{5}{2}T_1 + \frac{1}{4}T_2 + T_4 = -35.3553$$

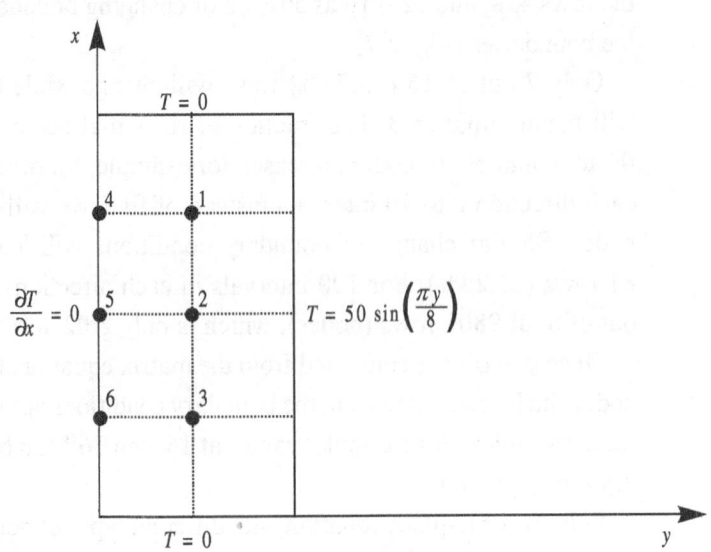

Fig. 7.9

Similarly, we can write the following two equations for Nodes 2 and 3, respectively:

$$\frac{1}{4}T_1 - \frac{5}{2}T_2 + \frac{1}{4}T_3 + T_5 = -50$$

$$\frac{1}{4}T_2 - \frac{5}{2}T_3 + T_6 = -35.3553$$

Now, for the boundary Nodes 4, 5, and 6, we can write the second-order forward difference approximation for the derivative boundary condition similar to Eq. (7.84). In Eq. (7.84), we wrote a backward difference because the boundary was on the right and the unknown interior nodes were on the left of the boundary. In this case, since the boundary is on the left and the interior nodes are to the right of the boundary, we have to write forward difference approximation. The approximations for Node 4 is

$$\frac{-3T_4 + 4T_1 - 50\sin\left(\dfrac{6\pi}{8}\right)}{2(1)} = 0$$

or
$$T_4 = \frac{4}{3}T_1 - 11.7851$$

Similarly, for Nodes 5 and 6, we obtain

$$T_5 = \frac{4}{3}T_2 - 16.6667$$

$$T_6 = \frac{4}{3}T_3 - 11.7851$$

Replacing the values of T_4, T_5, and T_6 in the equations for Nodes 1, 2, and 3, we obtain the following set of equations:

$$-\frac{7}{6}T_1 + \frac{1}{4}T_2 = -23.5702$$

$$\frac{1}{4}T_1 - \frac{7}{6}T_2 + \frac{1}{4}T_3 = -33.3333$$

$$\frac{1}{4}T_2 - \frac{7}{6}T_3 = -23.5702$$

This is a tridiagonal system of equation. This was solved by Thomas algorithm and the solution is

$$T_1 = T_3 = 28.9876 \quad \text{and} \quad T_2 = 40.9947$$

Using these, the temperatures at Nodes 4, 5, and 6 were computed from the equations for these nodes as

$$T_4 = T_6 = 26.865 \quad \text{and} \quad T_5 = 37.9929$$

This problem could actually be solved by much less number of nodes by utilizing the symmetry of the problem and the boundary condition. For example, if we draw a horizontal line through Nodes 2 and 5, this constitutes an axis of symmetry. All the boundary conditions are also symmetric with respect to this axis. This is important. It is not just the geometry of the plate that matters. The line joining Nodes 1-2-3 is not an axis of symmetry.

At the true axis of symmetry, one can always assume a no flux condition. Therefore, we can put a pseudo boundary through Nodes 2-5 and assume a no flux condition, i.e., $\partial T / \partial y = 0$. Then, only interior node is the Node 1. So, we can solve the whole problem by solving only one equation since in that case, T_2 can also be replaced in the equation for Node 1 by making a backward difference approximation of this new boundary condition.

7.3.3 First Order Wave Equation

The simplest form of first order wave equation is obtained if we put $\alpha = 0$ in the advective-diffusion Eq. (7.34). It is written as

$$\frac{\partial \phi}{\partial t} + u \frac{\partial \phi}{\partial x} = 0 \tag{7.87}$$

Application of our μ-CD scheme to this equation leads to the following discrete equation:

$$\frac{\phi_i^{n+1} - \phi_i^n}{\Delta t} = -\mu u_i^n \frac{\phi_{i+1}^n - \phi_{i-1}^n}{2\Delta x} - (1-\mu)u_i^{n+1} \frac{\phi_{i+1}^{n+1} - \phi_{i-1}^{n+1}}{2\Delta x} \tag{7.88}$$

or

$$(1-\mu)u_i^{n+1} \frac{\Delta t}{2\Delta x} \phi_{i+1}^{n+1} + \phi_i^{n+1} - (1-\mu)u_i^{n+1} \frac{\Delta t}{2\Delta x} \phi_{i-1}^{n+1}$$

$$= -\mu u_i^n \frac{\Delta t}{2\Delta x} \phi_{i+1}^n + \phi_i^n + \mu u_i^n \frac{\Delta t}{2\Delta x} \phi_{i-1}^n \tag{7.89}$$

For constant u, the above translates to

$$(1-\mu)\frac{C}{2}\phi_{i+1}^{n+1} + \phi_i^{n+1} - (1-\mu)\frac{C}{2}\phi_{i-1}^{n+1} = -\mu\frac{C}{2}\phi_{i+1}^n + \phi_i^n + \mu\frac{C}{2}\phi_{i-1}^n \tag{7.90}$$

Let us now analyse the scheme for accuracy and consistency. Using the expansions [Eqs (7.52) to (7.54)] and identities derived from the original PDE [Eq. (7.87)] relating the time derivative with space derivative [similar to Eqs (7.58) and (7.60)], we obtain the following:

$$\frac{\partial \phi}{\partial t}\bigg|_i^n + u \frac{\partial \phi}{\partial x}\bigg|_i^n = -\left(\mu - \frac{1}{2}\right) u^2 \Delta t \frac{\partial^2 \phi}{\partial x^2}\bigg|_i^n + \left(\frac{\mu}{2} - \frac{1}{3}\right) u^3 \Delta t^2 \frac{\partial^3 \phi}{\partial x^3}\bigg|_i^n - \frac{\Delta x^2}{6} \frac{\partial^3 \phi}{\partial x^3}\bigg|_i^n + \text{HOT}$$

(7.91)

The accuracy of the scheme is first order in time and second order in space as expected except for $\mu = 1/2$, when it is second order in both time and space. Also, for all values of μ other than 1/2, there is numerical diffusion. We will see in Section 7.6 that for $\mu = 1$ (Euler forward), the method is unconditionally unstable in addition to having numerical diffusion.

A possible way out of numerical diffusion while preserving the explicit structure is to numerically simulate the first term on the right in the discrete equation. If we add a central difference approximation of the first residual term on the right with the existing method and set $\mu = 1$ (explicit), the numerical diffusion term is cancelled in the simulation. This reduces the numerical diffusion as well as provides the advantage of explicit computation. The resulting scheme is

$$\frac{\phi_i^{n+1} - \phi_i^n}{\Delta t} = -u \frac{\phi_{i+1}^n - \phi_{i-1}^n}{2\Delta x} + \frac{u^2 \Delta t}{2} \frac{\phi_{i+1}^n - 2\phi_i^n + \phi_{i-1}^n}{\Delta x^2}$$

(7.92)

or

$$\phi_i^{n+1} = \phi_i^n - \frac{C}{2}(\phi_{i+1}^n - \phi_{i-1}^n) + \frac{C^2}{2}(\phi_{i+1}^n - \phi_i^n + \phi_{i-1}^n)$$

(7.93)

This is known as *Lax-Wendorf* scheme. It is conditionally stable (Exercise 7.4). Another commonly used explicit method to counter numerical diffusion is the upwind scheme. In this, the advection term is simulated by a backward difference approximation in place of central difference. The scheme is

$$\frac{\phi_i^{n+1} - \phi_i^n}{\Delta t} + u_i^n \frac{\phi_i^n - \phi_{i-1}^n}{\Delta x} = 0$$

(7.94)

For a constant u, the above yields

$$\phi_i^{n+1} = \phi_i^n - C(\phi_i^n - \phi_{i-1}^n)$$

(7.95)

If we perform a truncation error analysis, the upwind scheme results in,

$$\frac{\partial \phi}{\partial t}\bigg|_i^n + u \frac{\partial \phi}{\partial x}\bigg|_i^n = \left(\frac{u\Delta x}{2} - \frac{u^2 \Delta t}{2}\right) \frac{\partial^2 \phi}{\partial x^2}\bigg|_i^n + \left(\frac{u\Delta x^2}{6} + \frac{u^3 \Delta t^2}{6}\right) \frac{\partial^3 \phi}{\partial x^3}\bigg|_i^n + \text{HOT}$$

(7.96)

The numerical diffusion is eliminated if we choose $C = 1$. At this value of C, one may note from Eq. (7.95) that the values in one node is translated to the next node with progression of time steps. However, the exact choice depends on the stability condition of the scheme, which we shall learn in Section 7.3. If the velocity in the equation is less than zero, i.e., $u < 0$, the equation becomes

$$\frac{\partial \phi}{\partial t} - u \frac{\partial \phi}{\partial x} = 0 \tag{7.97}$$

The upwind scheme for this case is

$$\frac{\phi_i^{n+1} - \phi_i^n}{\Delta t} - u_i^n \frac{\phi_{i+1}^n - \phi_i^n}{\Delta x} = 0 \tag{7.98}$$

Notice that the backward difference is replaced by the forward difference. In fact, for any moving front, approximation involves a point behind the front. For a forward moving wave ($u > 0$), the point ($i - 1$) is taken and for the receding wave ($u < 0$), the point ($i + 1$) is taken. We leave it to the readers to show the latter also yields a similar truncation error as in the case of $u > 0$. We have shown the upwind scheme with first order forward/backward differences. For increasing the accuracy or order of the method in space, one may incorporate higher-order forward / backward difference approximations for the space derivatives involving more points.

The analytical solution of first-order wave equation [Eq. (7.87)] requires one initial condition and one boundary condition. However, for numerical solution using central difference scheme in space (μ-CD and *Lax-Wendorf*), it requires a second boundary condition. This is because, adjacent to both the boundaries, the difference approximation in spatial dimension involves a point on the boundary. This can be avoided by incorporating a backward or forward difference approximation adjacent to the open boundary. The latter is automatically achieved in the upwind schemes. Example application of various schemes are shown (Example 7.4) demonstrating numerical diffusion.

Example 7.4 Consider the one-dimensional advection-diffusion equation of Example 7.2. The equation is typically used for contaminant transport in wide rivers and estuaries. For narrow rivers, the dispersion can be neglected reducing the advection-diffusion equation to one-dimensional wave equation. Solve it using the upwind scheme with $\Delta x = 0.1$, $\Delta t = 0.1$, and 0.5. Compare the results graphically. The equation is given by

$$\frac{\partial C}{\partial t} + u \frac{\partial C}{\partial x} = 0, \qquad 0 \le x < \infty$$

where $u = 0.2$, $C(0,t) = 1$, $C(x,0) = 0$.

Solution The upwind scheme is given by Eq. (7.94), which for the present case of constant velocity, we write as follows:

$$C_i^{n+1} = C_i^n - \frac{u\Delta t}{\Delta x}(C_i^n - C_{i-1}^n)$$

for

$$\Delta t = 0.1, \quad \frac{u\Delta t}{\Delta x} = 0.2$$

This can be explicitly solved for all i given the initial conditions of $C_i^n = 0$ for $i = 1, 2, 3, \ldots, \infty$ and the boundary condition $C_0^0 = 0$. For example, at the first time step, $t = 0.1$, the concentrations are

$$C_1^{0.1} = 0 - 0.2(0 - 1) = 0.2$$

$$C_2^{0.1} = C_3^{0.1} = \cdots = C_\infty^{0.1} = 0$$

At the second time step, $t = 0.2$, the concentrations are

$$C_1^{0.2} = 0.2 - 0.2(0.2 - 1) = 0.36$$

$$C_2^{0.2} = 0 - 0.2(0 - 0.2) = 0.04$$

$$C_3^{0.1} = \cdots = C_\infty^{0.1} = 0$$

We can compute the solutions for $\Delta t = 0.5$ similarly. The progress of solution with time is shown in Fig. 7.10 for $\Delta t = 0.1$.

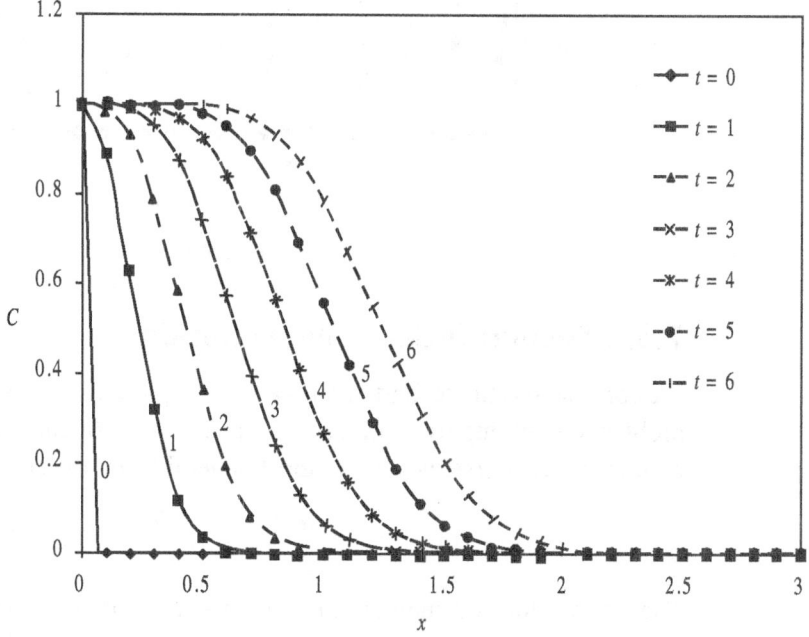

Fig. 7.10

We see that a significant numerical diffusion is added and as a result, the expected progression of sharp front is diffused. This is expected with upwind scheme for CFL number less than unity [Eq. (7.96)].

For $\Delta t = 0.5$, the CFL number is unity, and we expect the numerical diffusion to disappear. This indeed happens as we can see in Fig. 7.11 where the sharp front progresses with time. The line at $t = 0$ is supposed to drop vertically from 1 to 0 but for the plotting program, which joins the point at $x = 0$ and Δx with a straight line. This is a common artefact of nearly all computer plotting programs, and one needs to be careful while interpreting the plotted result.

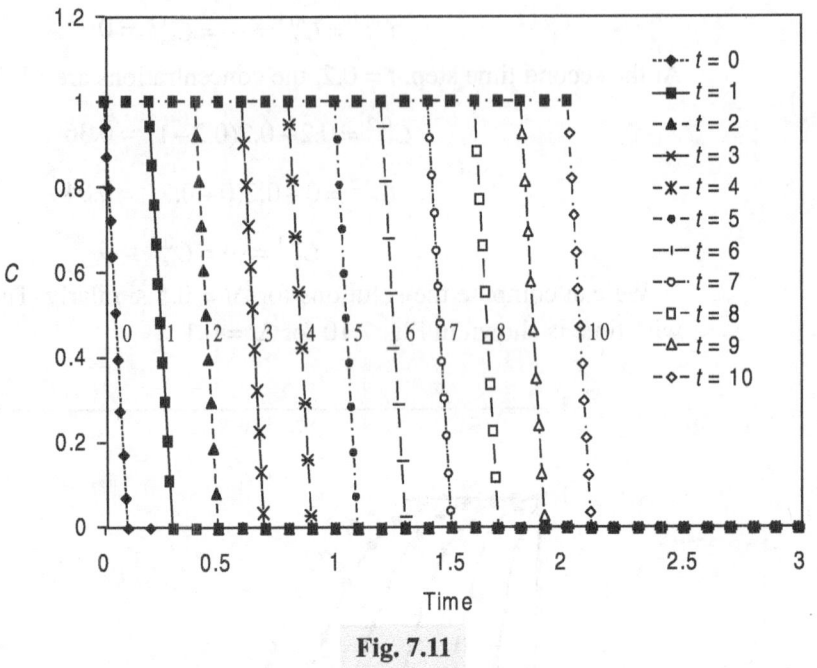

Fig. 7.11

7.3.4 Second Order Wave Equation

We come across the second order wave propagation equation in many engineering problems involving oscillations and waves. Areas of earthquake engineering and hydraulics use it extensively. In most general form, the equation appears as

$$\frac{\partial^2 \phi}{\partial t^2} = u(x)^2 \frac{\partial^2 \phi}{\partial x^2} \tag{7.99}$$

The wave velocity u may or may not be a function of x. For solution, we require two initial conditions and two boundary conditions. The initial conditions generally look like,

$$\phi(x,0) = a \quad \text{and} \quad \left.\frac{\partial \phi}{\partial t}\right|_{(x,0)} = b \tag{7.100}$$

The boundary conditions in physical problems are often specified asymptotically at infinity, i.e., $-\infty < x < \infty$. However, for the solution using a numerical method, we have to restrict the boundaries to finite space and specify meaningful conditions at the boundaries. This is generally governed by the physics of the problem. Let us denote these finite boundaries as $(-L, L)$.

In order to understand the behaviour of Eq. (7.99), let us factorize it as follows:

$$\left(\frac{\partial}{\partial t} + u\frac{\partial}{\partial x}\right)\left(\frac{\partial}{\partial t} - u\frac{\partial}{\partial x}\right)\phi = 0 \tag{7.101}$$

It is easy to see that Eq. (7.101) describes two first-order wave equations, one right moving wave and one left moving wave. Therefore, for $(0, L)$, the solution will follow the right moving wave,

$$\frac{\partial \phi}{\partial t} + u\frac{\partial \phi}{\partial x} = 0 \tag{7.102}$$

and for $(-L, 0)$, the solution will involve the left moving wave,

$$\frac{\partial \phi}{\partial t} - u\frac{\partial \phi}{\partial x} = 0 \tag{7.103}$$

This provides a rational way to specify boundary conditions at the finite boundaries as follows:

$$\left.\frac{\partial \phi}{\partial t}\right|_{(L,t)} = -u(L)\left.\frac{\partial \phi}{\partial x}\right|_{(L,t)} \quad \text{and} \quad \left.\frac{\partial \phi}{\partial t}\right|_{(-L,t)} = u(-L)\left.\frac{\partial \phi}{\partial x}\right|_{(-L,t)} \tag{7.104}$$

For the ease of numerical solution, let us break down PDE into two first order PDEs in time. This will enable us to apply the methods already learned for IVPs in Chapter 6 and in the earlier sections for diffusion equation. This can be accomplished as follows:

$$\frac{\partial \phi}{\partial t} = \psi$$

$$\frac{\partial \psi}{\partial t} = u(x)^2 \frac{\partial^2 \phi}{\partial x^2} \tag{7.105}$$

The second PDE is a diffusion equation which we already solved in Section 7.3.1. Using central difference for the spatial discretization, this PDE can be converted to a system of IVPs. For spatial discretization, the boundary conditions are either given by the physics of the problem or obtained from Eq. (7.104). The first PDE is already in the form of a system of IVP. Both the system of IVPs can then be solved using any of the schemes described in Chapter 6. Notice that the initial conditions for both the ODEs are known from Eq. (7.100). Once again, the popular scheme for this is fourth order Runge-Kutta method which combines the computational advantage of an explicit method with high order of accuracy and decent stability property.

An alternative to decomposition of the original PDE is to use direct discretization in time and space. There are a number of such methods in the numerical modelling literature. One commonly used scheme for Eq. (7.99) with second order accuracy in both space and time is given below

$$\frac{\phi_j^{n+1} - 2\phi_j^n + \phi_j^{n-1}}{\Delta t^2} = u_j^2 \left[\begin{array}{l} \dfrac{1}{4} \dfrac{\phi_{j+1}^{n+1} - 2\phi_j^{n+1} + \phi_{j-1}^{n+1}}{\Delta x^2} + \dfrac{1}{2} \dfrac{\phi_{j+1}^n - 2\phi_j^n + \phi_{j-1}^n}{\Delta x^2} \\[2ex] + \dfrac{1}{4} \dfrac{\phi_{j+1}^{n-1} - 2\phi_j^{n-1} + \phi_{j-1}^{n-1}}{\Delta x^2} \end{array} \right] \tag{7.106}$$

Equation (7.106) is the approximation of the PDE at spatial node j. Therefore, this approximation can be written at all the spatial nodes and the resulting set of equation can be solved to obtain the values of ϕ at various time steps. Incorporation of the boundary and initial conditions are similar to those described in the earlier sections. It is easy to see that the left-hand side is a central difference approximation of the double derivative in time and the right-hand side is the weighted average of the central difference approximation of the double derivative in space at three difference time steps. Applications of the methods described in Eqs (7.105) and (7.106) are shown in Example 7.5.

So far, we have used one spatial dimension in all the PDEs which contained time derivative (diffusion, wave etc.). We will describe one equation in the next section for multiple spatial dimensions along with the time derivative.

Example 7.5 Transverse displacement (y) of a vibrating string is governed by the following equation:

$$\frac{\partial^2 y}{\partial t^2} = 4 \frac{\partial^2 y}{\partial x^2}, \quad 0 < x < 6, \ t \geq 0$$

The string was fixed at two ends, i.e., $y(0,t) = y(6,t) = 0$ and initially at rest, i.e., $y(x,0) = 0$. In order to start the vibration, the following velocity was applied to the string by striking appropriately:

$$\left. \frac{\partial y}{\partial t} \right|_{(x,0)} = \sin\left(\frac{\pi x}{2} \right)$$

Solve the PDE using the following methods:

(a) Decompose into two first order PDEs similar to Eq. (7.105). Using central difference scheme for semi-discretization with $\Delta x = 1$, formulate the system of ODEs to be solved for the solution. Describe the steps of solution by incorporating the initial and boundary conditions appropriately.

(b) Discretize the PDE using Eq. (7.106) with $\Delta x = 1$ and time step h. Incorporate the initial and boundary conditions in such a way that the method remains second order accurate in both time and space at all time steps. Describe the steps of solution.

Solution

(a) *Decomposing into two first order PDE [Eq. (7.105)]*

Let us define a new variable $z(x, t)$ as,

$$\frac{\partial y}{\partial t} = z$$

The PDE then transforms to

$$\frac{\partial z}{\partial t} = 4\frac{\partial^2 y}{\partial x^2}$$

Using $\Delta x = 1$, we have 7 nodes in the spatial domain. Let us number them as 0–6 with 0 and 6 on the boundary. So, from the boundary conditions, we have $y_0 = y_6 = 0$. Therefore, the goal is to obtain $y_1 - y_5$. Semi-discretization of the above set of PDEs lead to the following equation for any arbitrary interior node (i):

$$\frac{dy_i}{dt} = z_i$$

$$\frac{dz_i}{dt} = 4\frac{y_{i+1} - 2y_i + y_{i-1}}{\Delta x^2} = 4y_{i-1} - 8y_i + 4y_{i+1}$$

Writing the above set of equations for all the interior nodes 1–5, we obtain the following set of ODEs:

$$\frac{d}{dt}\begin{bmatrix} y_1 \\ y_2 \\ y_3 \\ y_4 \\ y_5 \end{bmatrix} = \begin{bmatrix} z_1 \\ z_2 \\ z_3 \\ z_4 \\ z_5 \end{bmatrix} \quad \text{or} \quad \frac{dy}{dt} = z$$

$$\frac{d}{dt}\begin{bmatrix} z_1 \\ z_2 \\ z_3 \\ z_4 \\ z_5 \end{bmatrix} = \begin{bmatrix} -8 & 4 & 0 & 0 & 0 \\ 4 & -8 & 4 & 0 & 0 \\ 0 & 4 & -8 & 4 & 0 \\ 0 & 0 & 4 & -8 & 4 \\ 0 & 0 & 0 & 4 & -8 \end{bmatrix}\begin{bmatrix} y_1 \\ y_2 \\ y_3 \\ y_4 \\ y_5 \end{bmatrix} \quad \text{or} \quad \frac{dz}{dt} = Ay$$

The initial conditions for the two systems of ODEs are provided by two initial conditions given in the problem, as follows:

First initial condition $y(x,0) = 0$ leads to

$$y^0 = \begin{bmatrix} y_1^0 \\ y_2^0 \\ y_3^0 \\ y_4^0 \\ y_5^0 \end{bmatrix} = \begin{bmatrix} 0 \\ 0 \\ 0 \\ 0 \\ 0 \end{bmatrix}$$

Second initial condition $\left.\dfrac{\partial y}{\partial t}\right|_{(x,0)} = \sin\left(\dfrac{\pi x}{2}\right)$ leads to $z^0 = \begin{bmatrix} z_1^0 \\ z_2^0 \\ z_3^0 \\ z_4^0 \\ z_5^0 \end{bmatrix} = \begin{bmatrix} 1 \\ 0 \\ -1 \\ 0 \\ 1 \end{bmatrix}$

These now consist of two linear system of IVPs that are coupled through the right-hand side functions. Since the initial conditions are known for both the systems, application of explicit methods is easy. Linearity of the right-hand side function makes application of the Runge Kutta methods easier. We leave the reader to practice the solution using explicit method since, we have already shown such solutions at many other places (Examples 6.10, 6.11, 6.12, 7.1 etc.).

However, from the point of view of stability, it may be desirable to apply implicit methods such as, Euler backward and trapezoidal. For such applications, one will need to rely on iterative improvement at every time step. For example, application of trapezoidal method to the system will lead to following iterative systems:

$$y^{n+1,k+1} = y^n + \frac{\Delta t}{2}(z^{n+1,k} + z^n)$$

$$z^{n+1,k+1} = z^n + \frac{\Delta t}{2} A(y^{n+1,k} + y^n)$$

where n is the time step index and k is the iteration index at each time step. The iteration at each time step (n to $n+1$) is started by assuming $y^{n+1,0} = y^n$ and $z^{n+1,0} = z^n$, and stopped when both $\left\| y^{n+1,k+1} - y^{n+1,k} \right\| \le \varepsilon$ and $\left\| z^{n+1,k+1} - z^{n+1,k} \right\| \le \varepsilon$ are satisfied with user specified tolerance $\varepsilon > 0$.

(b) Direct discretization in time and space [Eq. (7.106)]

Using $\Delta x = 1$, we have 7 nodes in the spatial domain. Let us number them as 0–6 with 0 and 6 on the boundary. From the boundary conditions, we have $y_0 = y_6 = 0$. Therefore, we only need to write approximation using Eq. (7.106) for the interior nodes 1 to 5. For any interior node j, the approximation of the original equation may be written as:

$$\frac{y_j^{n+1} - 2y_j^n + y_j^{n-1}}{h^2} = \left(y_{j+1}^{n+1} - 2y_j^{n+1} + y_{j-1}^{n+1}\right) + 2\left(y_{j+1}^n - 2y_j^n + y_{j-1}^n\right) + \left(y_{j+1}^{n-1} - 2y_j^{n-1} + y_{j-1}^{n-1}\right)$$

Writing the above approximation for nodes $j = 1, 2, \dots 5$, using the boundary conditions $y_0 = y_6 = 0$ in the equations for nodes 1 and 5, and rearranging lead to the following matrix iteration equation:

$$
\begin{bmatrix}
1+2h^2 & -h^2 & 0 & 0 & 0 \\
-h^2 & 1+2h^2 & -h^2 & 0 & 0 \\
0 & -h^2 & 1+2h^2 & -h^2 & 0 \\
0 & 0 & -h^2 & 1+2h^2 & -h^2 \\
0 & 0 & 0 & -h^2 & 1+2h^2
\end{bmatrix}
\begin{bmatrix}
y_1^{n+1} \\
y_2^{n+1} \\
y_3^{n+1} \\
y_4^{n+1} \\
y_5^{n+1}
\end{bmatrix}
=
$$

$$
\begin{bmatrix}
2(1-2h^2) & 2h^2 & 0 & 0 & 0 \\
2h^2 & 2(1-2h^2) & 2h^2 & 0 & 0 \\
0 & 2h^2 & 2(1-2h^2) & 2h^2 & 0 \\
0 & 0 & 2h^2 & 2(1-2h^2) & 2h^2 \\
0 & 0 & 0 & 2h^2 & 2(1-2h^2)
\end{bmatrix}
\begin{bmatrix}
y_1^n \\
y_2^n \\
y_3^n \\
y_4^n \\
y_5^n
\end{bmatrix}
+
$$

$$
\begin{bmatrix}
-(1+2h^2) & h^2 & 0 & 0 & 0 \\
h^2 & -(1+2h^2) & h^2 & 0 & 0 \\
0 & h^2 & -(1+2h^2) & h^2 & 0 \\
0 & 0 & h^2 & -(1+2h^2) & h^2 \\
0 & 0 & 0 & h^2 & -(1+2h^2)
\end{bmatrix}
\begin{bmatrix}
y_1^{n-1} \\
y_2^{n-1} \\
y_3^{n-1} \\
y_4^{n-1} \\
y_5^{n-1}
\end{bmatrix}
$$

First initial condition gives to $\begin{bmatrix} y_1^0 \\ y_2^0 \\ y_3^0 \\ y_4^0 \\ y_5^0 \end{bmatrix} = \begin{bmatrix} 0 \\ 0 \\ 0 \\ 0 \\ 0 \end{bmatrix}$

There is a startup problem because of y_j^{n-1} terms which are not defined at the start. Once we are able to take one time step and values of y_j^0 and y_j^1 are available, all subsequent time steps can be taken using the above set of equation. In order to solve the startup problem of the first time step, i.e., to estimate y_j^1 from y_j^0 we make use of the second initial condition. Since, it is an *initial condition*, it can only be used at the first time step. In order to preserve the second order accuracy in time, we make a central difference approximation of the second initial condition for any spatial location j as follows:

$$\frac{y_j^{n+1} - y_j^{n-1}}{2h} = \sin\left(\frac{\pi j}{2}\right)$$

Writing the above approximation for $j = 1, 2, \ldots 5$, we obtain the vector form as:

$$\begin{bmatrix} y_1^{n-1} \\ y_2^{n-1} \\ y_3^{n-1} \\ y_4^{n-1} \\ y_5^{n-1} \end{bmatrix} = \begin{bmatrix} y_1^{n+1} \\ y_2^{n+1} \\ y_3^{n+1} \\ y_4^{n+1} \\ y_5^{n+1} \end{bmatrix} - \begin{bmatrix} 2h \\ 0 \\ -2h \\ 0 \\ 2h \end{bmatrix}$$

Substituting this relation in the matrix iteration equation yields the equation for first time step as:

$$\begin{bmatrix} 2(1+2h^2) & -2h^2 & 0 & 0 & 0 \\ -2h^2 & 2(1+2h^2) & -2h^2 & 0 & 0 \\ 0 & -2h^2 & 2(1+2h^2) & -2h^2 & 0 \\ 0 & 0 & -h^2 & 2(1+2h^2) & -2h^2 \\ 0 & 0 & 0 & -2h^2 & 2(1+2h^2) \end{bmatrix} \begin{bmatrix} y_1^{n+1} \\ y_2^{n+1} \\ y_3^{n+1} \\ y_4^{n+1} \\ y_5^{n+1} \end{bmatrix} =$$

$$
\begin{bmatrix}
2(1-2h^2) & 2h^2 & 0 & 0 & 0 \\
2h^2 & 2(1-2h^2) & 2h^2 & 0 & 0 \\
0 & 2h^2 & 2(1-2h^2) & 2h^2 & 0 \\
0 & 0 & 2h^2 & 2(1-2h^2) & 2h^2 \\
0 & 0 & 0 & 2h^2 & 22(1-2h^2)
\end{bmatrix}
\begin{bmatrix} y_1^n \\ y_2^n \\ y_3^n \\ y_4^n \\ y_5^n \end{bmatrix}
-
\begin{bmatrix} 4h(1-2h^2) \\ 0 \\ -4h(1-2h^2) \\ 0 \\ 4h(1-2h^2) \end{bmatrix}
$$

Using the above equation y_j^1 can be computed from the known values of y_j^0 for any time step h. For all subsequent time steps, the matrix iteration equation has to be used. Choice of time step is governed by the stability criterion.

7.3.5 Diffusion Equation in 2-D

Almost all the physical problems of nature contain three spatial dimensions [Eq. (7.1)]. However, in many cases, some reasonable, practical assumption leads to elimination of one of the spatial dimension leading to a two-dimensional representation of the physical world [Eq. (7.3)].

In Laplace equation, we have seen such a process and how to solve the equation using numerical method. However, Laplace equation is time invariant process or *steady state* of a physical process. In the transition state, the same process will be represented by the following equation:

$$
\frac{\partial \phi}{\partial t} = \alpha \left(\frac{\partial^2 \phi}{\partial x^2} + \frac{\partial^2 \phi}{\partial y^2} \right)
\tag{7.107}
$$

Solution of this equation requires an initial condition in addition to the boundary conditions similar to Laplace equation. Comparing with Eq. (7.33), it is easy to see that it is a two-dimensional diffusion equation. Therefore, the methods described in Section 7.3.1 can be applied for the solution of this equation as well. If we perform spatial discretization, we can write the following IVP for node (i, j):

$$
\frac{d\phi_{i,j}}{dt} = \alpha \left(\frac{\phi_{i+1,j} - 2\phi_{i,j} + \phi_{i-1,j}}{\Delta x^2} + \frac{\phi_{i,j+1} - 2\phi_{i,j} + \phi_{i,j-1}}{\Delta y^2} \right)
\tag{7.108}
$$

After incorporating the boundary conditions, we have seen for the Laplace equation that the right-hand side can be expressed as a matrix equation in terms of values of ϕ at the nodes [Eq. (7.71)]. Following similar notations, the above can be expressed as,

$$
\frac{d\bar{\phi}}{dt} = A\bar{\phi} + b
\tag{7.109}
$$

In fact, for a transition state temperature distribution problem for the plate shown in Fig. 7.4 with same boundary conditions, the matrix A and vector b remains the same as in the Laplace equation except for the coefficient α and sign of b. We only need to specify the initial condition. This is a system of linear IVP similar to Eq. (7.39) that can be solved using any of the method described in Chapter 6 given the initial condition. The simplest explicit method is the Euler forward which in matrix representation, translates to

$$\bar{\phi}^{n+1} = \bar{\phi}^n + \Delta t \left[A \bar{\phi}^n + b \right] \tag{7.110}$$

Each nodal equation in the above set of equation is a discrete form of the original PDE [Eq. (7.107)]:

$$\frac{\phi_{i,j}^{n+1} - \phi_{i,j}^n}{\Delta t} = \alpha \left(\frac{\phi_{i+1,j}^n - 2\phi_{i,j}^n + \phi_{i-1,j}^n}{\Delta x^2} + \frac{\phi_{i,j+1}^n - 2\phi_{i,j}^n + \phi_{i,j-1}^n}{\Delta y^2} \right) \tag{7.111}$$

$$\phi_{i,j}^{n+1} = \phi_{i,j}^n + \alpha \frac{\Delta t}{\Delta y^2} \phi_{i,j-1}^n + \alpha \frac{\Delta t}{\Delta x^2} \phi_{i-1,j}^n - 2\left(\alpha \frac{\Delta t}{\Delta x^2} + \alpha \frac{\Delta t}{\Delta y^2} \right) \phi_{i,j}^n$$

or $\qquad\qquad\qquad\qquad\qquad\qquad\qquad\qquad\qquad\qquad\qquad\qquad\qquad$ (7.112)

$$+ \alpha \frac{\Delta t}{\Delta x^2} \phi_{i+1,j}^n + \alpha \frac{\Delta t}{\Delta y^2} \phi_{i,j+1}^n$$

Computational effort required in solving this is not much more than its one-dimensional form. However, as we know, use of the explicit method such as Euler forward is severely restricted by its stability property. One can also use fourth order Runge-Kutta for increasing accuracy. However, one is often interested in the implicit methods for better stability properties. Let us see application of Euler backward method to Eq. (7.107):

$$\frac{\phi_{i,j}^{n+1} - \phi_{i,j}^n}{\Delta t} = \alpha \left(\frac{\phi_{i+1,j}^{n+1} - 2\phi_{i,j}^{n+1} + \phi_{i-1,j}^{n+1}}{\Delta x^2} + \frac{\phi_{i,j+1}^{n+1} - 2\phi_{i,j}^{n+1} + \phi_{i,j-1}^{n+1}}{\Delta y^2} \right) \tag{7.113}$$

$$-\alpha \frac{\Delta t}{\Delta y^2} \phi_{i,j-1}^{n+1} - \alpha \frac{\Delta t}{\Delta x^2} \phi_{i-1,j}^{n+1} + \left[1 + 2\left(\alpha \frac{\Delta t}{\Delta x^2} + \alpha \frac{\Delta t}{\Delta y^2} \right) \right] \phi_{i,j}^{n+1}$$

or

$$-\alpha \frac{\Delta t}{\Delta x^2} \phi_{i+1,j}^{n+1} - \alpha \frac{\Delta t}{\Delta y^2} \phi_{i,j+1}^{n+1} = \phi_{i,j}^n \tag{7.114}$$

In the matrix notation, this is essentially a discrete form of the system of Eq. (7.109):

$$[I - \Delta t A] \bar{\phi}^{n+1} = \bar{\phi}^n + \Delta t \, b \tag{7.115}$$

Recall from Section 7.3.2 that the vector b consists of the terms representing the known values at the boundary node. In this case also, the contributions will come from the known nodal values in Eq. (7.114). It is now easy to see that we need to solve this system of equation using an iterative method as described for Laplace equation. The difference is that, in Laplace equation, we have to solve it once but here, we have to solve it at each time step. This will always be the case for any implicit method irrespective of the order of method. One can combine Eqs (7.111) and (7.113) to form a μ-CD scheme.

The coefficient matrix in the application of an implicit method for this problem is sparse. As we have seen for the Laplace equation, each row can contain up to a maximum of five elements. There is a way to formulate it in the form of a *block tridiagonal* matrix where each element is a small matrix. One can then apply a modified form of *Thomas algorithm* to solve it. This reduces the computational effort at each time step and the solution can be found using a finite number of computations as opposed to the iterative methods.

A second option is to modify the time stepping scheme such that the coefficient matrix is a tridiagonal. Then, one can solve it at each time step using vector operations of *Thomas algorithm* described in Chapter 2. Such a method is called *alternating direction implicit* (ADI) scheme. In this method, one takes one time step in two half steps. In one-half step, the x-direction space derivative is implicit and in the next half step, the y-direction space derivative is implicit. The scheme is shown as follows:

First half step:

$$\phi_{i,j}^{n+\frac{1}{2}} = \phi_{i,j}^{n} + \alpha \frac{\Delta t}{2\Delta x^2}\left(\phi_{i+1,j}^{n+\frac{1}{2}} - 2\phi_{i,j}^{n+\frac{1}{2}} + \phi_{i-1,j}^{n+\frac{1}{2}}\right)$$
$$+ \alpha \frac{\Delta t}{2\Delta y^2}\left(\phi_{i,j+1}^{n} - 2\phi_{i,j}^{n} + \phi_{i,j-1}^{n}\right) \tag{7.116}$$

or

$$-\alpha \frac{\Delta t}{2\Delta x^2}\phi_{i+1,j}^{n+\frac{1}{2}} + \left(1 + \alpha \frac{\Delta t}{2\Delta x^2}\right)\phi_{i,j}^{n+\frac{1}{2}} - \alpha \frac{\Delta t}{2\Delta x^2}\phi_{i-1,j}^{n+\frac{1}{2}}$$
$$= \alpha \frac{\Delta t}{2\Delta y^2}\phi_{i,j+1}^{n} + \left(1 - \alpha \frac{\Delta t}{2\Delta y^2}\right)\phi_{i,j}^{n} + \alpha \frac{\Delta t}{2\Delta y^2}\phi_{i,j-1}^{n} \tag{7.117}$$

Second half step:

$$\phi_{i,j}^{n+1} = \phi_{i,j}^{n+\frac{1}{2}} + \alpha \frac{\Delta t}{2\Delta x^2}\left(\phi_{i+1,j}^{n+\frac{1}{2}} - 2\phi_{i,j}^{n+\frac{1}{2}} + \phi_{i-1,j}^{n+\frac{1}{2}}\right)$$
$$+ \alpha \frac{\Delta t}{2\Delta y^2}\left(\phi_{i,j+1}^{n+1} - 2\phi_{i,j}^{n+1} + \phi_{i,j-1}^{n+1}\right) \tag{7.118}$$

or

$$-\alpha \frac{\Delta t}{2\Delta y^2} \phi_{i,j+1}^{n+1} + \left(1 + \alpha \frac{\Delta t}{\Delta y^2}\right)\phi_{i,j}^{n+\frac{1}{2}} - \alpha \frac{\Delta t}{2\Delta y^2}\phi_{i,j-1}^{n+\frac{1}{2}}$$

$$= \alpha \frac{\Delta t}{2\Delta x^2}\phi_{i+1,j}^{n+\frac{1}{2}} + \left(1 - \alpha \frac{\Delta t}{\Delta x^2}\right)\phi_{i,j}^{n+\frac{1}{2}} + \alpha \frac{\Delta t}{2\Delta x^2}\phi_{i-1,j}^{n+\frac{1}{2}} \qquad (7.119)$$

In order to avoid accumulation of errors in one direction, it is recommended to change the order of the half steps. For example, in Eqs (7.116)–(7.119), x-direction is implicit for the first half step and y-direction is implicit for the next half step. This will be reversed in the next time step. It is easy to see that in both the half steps, the coefficient matrix is tridiagonal. This makes the solution of the equation simpler.

So far, we have presented a lot of numerical methods for the solution of a number of important engineering PDEs. During presentation of the methods, we have made sure that all the methods are consistent, i.e., in the asymptotic limit of $\Delta t \rightarrow 0$ and $\Delta x \rightarrow 0$, the discrete approximations converge to the original PDEs. This was ensured by the fact that the minimum power of Δt and Δx in the truncation error was 1. This is not true for all the methods one can design. We shall see in Problem 3 of Exercise 7.4 a method called *Dufort-Frankel scheme*, which is not consistent because a term $(\Delta t^2/\Delta x^2)$ appears in the truncation error. This term will go to zero conditionally depending on which of the Δt or Δx goes to zero faster. If they go to zero at the same rate, the ratio becomes a constant. The term is zero only if Δt approaches zero faster than Δx.

Recall that for convergence (Section 6.3.2.8), the method has to be *consistent* as well as *stable*. In the next section, we will learn to analyse the methods for stability.

EXERCISE 7.3

1. The one-dimensional heat equation with a source term is

$$\frac{\partial T}{\partial t} = \alpha \frac{\partial^2 T}{\partial x^2} + S(x)$$

where

$T(x, 0) = 0; \; T(0, t) = 0; \; T(1, t) = T_{\text{Steady}} (1);$

$S(x) = -(x^2 - 4x + 2)e^{-x}; \; T_{\text{steady}}(x) = x^2 e^{-x}.$

Discretize the above equation using Crank-Nicolson scheme (μ-method in time with $\mu = 1/2$ and central difference in space). Formulate the matrix equation using $\alpha = 1$, $\Delta x = 0.25$, and $\Delta t = 0.1$. Solve it for one time step.

2. Consider the one-dimensional convection diffusion equation:

$$\frac{\partial T}{\partial t} + u\frac{\partial T}{\partial x^2} = a\frac{\partial^2 T}{\partial x^2}, \quad 0 \le x \le 1$$

where

$$u = 0.08; \ a = 0.001; \ T(0, t) = 0; \ T(1, t) = 0;$$

$$T(x, 0) = \begin{cases} 1 - (10x - 1)^2, & \text{for } 0 \le x \le 0.25 \\ 0, & \text{for } 0.25 \le x \le 1 \end{cases}$$

Using second order Runge-Kutta method in time and central difference approximation in space with $\Delta x = 0.25$ and $\Delta t = 0.1$, solve for one time step.

3. Consider the following non-homogeneous heat equation:

$$\frac{\partial T}{\partial t} = \alpha\frac{\partial^2 T}{\partial x^2} + (\pi^2 - 1)e^{-1}\sin\pi x, \ 0 \le x \le 1; \ t \ge 0$$

with initial and boundary condition $T(0, t) = T(1, t) = 0$ and $T(x, 0) = \sin \pi x$.

(a) Write a computer program to solve the equation using Euler explicit-central difference approximations, for $\alpha = 1$, $\Delta x = 0.05$, and $\Delta t = 0.001$. Plot $T(x)$ vs. x at $t = 0.0, 0.5, 1.0, 1.5$, and 2.0 in one plot.

(b) Take new $\Delta t = 0.0015$ and solve the equation for the same α and Δx. Plot $T(x)$ vs. x in the second plot at $t = (0.0, 0.075, 0.15, 0.153, 0.1545, 0.156)$.

(c) Explain the result obtained in (a) and (b).

4. Toxic pollutant transport in a river is governed by the following equation:

$$\frac{\partial c}{\partial t} + v\frac{\partial c}{\partial x} - D\frac{\partial^2 c}{\partial x^2} + kc = 0; \ 0 \le x \le 1; \ c_0; \frac{\partial c}{\partial x}\Big|_{(1,t)} = 0; c(x,0) = x^2 e^{-x}$$

Discretize the above equation using μ-CD scheme and express in terms of CFL number and Grid Peclet number. The velocity $v = 0.5$ m/sec and the dispersion coefficient $D = 0.1$ m²/sec. Normalized concentration at the inlet $c_0 = 1$ mg/L. Solve the equation with $\mu = 0, 1/2$, and 1 using $\Delta x = 0.1$ and $\Delta t = 0.1, 0.2$, and 0.3. Write a program to solve the equation for all these cases. For each μ and Δt combination (9 plots), plot the concentration vs. x at times $t = 0, 0.6$ and 1.2. Comment on the effect of μ and Δt on the numerically observed transport behaviour of the pollutant. Can you explain some of the behaviour in the light of analysis done in this chapter (numerical diffusion, stability, etc.)?

5. Steady state temperature (T) distribution in a plate is governed by the Laplace equation. The boundary conditions are as shown in Fig. 7.12. Calculate the temperatures at points a, b, and c. Given $\Delta x = \Delta y = 0.25$ m.

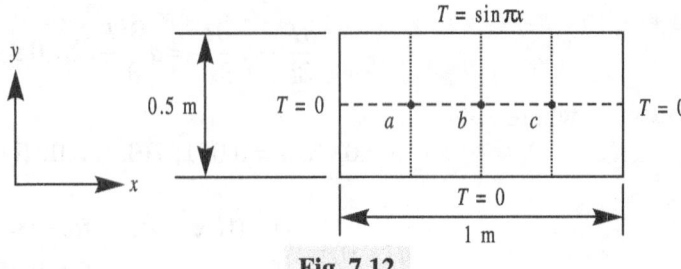

Fig. 7.12

6. Consider the fire clay slab of the dimension 1 m × 1 m shown in Fig. 7.13.

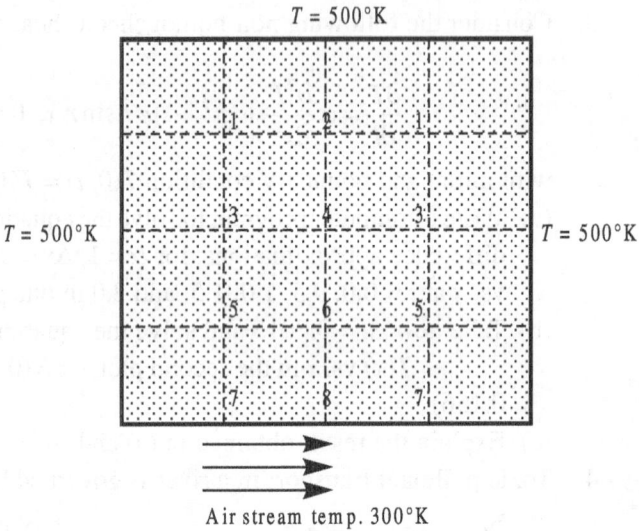

Fig. 7.13

Three surfaces of the slab are maintained at 500°K while the remaining surface is exposed to an air stream of temperature 300°K and heat transfer coefficient (*h*) of 10 W/m² °K. Thermal conductivity (*k*) of the slab is 1 W/m °K. The steady state temperature (*T*) distribution in the slab is governed by two-dimensional Laplace equation. Use a grid size of $\Delta x = \Delta y = 0.25$ m, formulate the matrix equation, and obtain temperature at eight nodal points shown in the above figure.

Formulation of the BC's at the bottom boundary: The algebric sum of the conduction heat flux and convective heat flux along the boundary (say Node 7) is zero. In order to calcualate the conduction flux at Node 7, one has to calculate the energy flux from Node 5 to Node 7 by multiplying the thermal conductivity (*k*) of the solid with the temperature gradient. This quantity will be positive at Node 7. For the convective flux, one has to consider, Newton's law of cooling. The film heat transfer coefficient, *h* has to be multiplied by the temperature difference between Node 7 and the ambient temperature (fluid temperature). This energy will move out from the node.

7. Temperature distribution in the quarter circular disc of unit radius shown in Fig. 7.14 is governed by,

$$\frac{1}{r}\frac{\partial}{\partial r}\left(r\frac{\partial T}{\partial r}\right)+\frac{1}{r^2}\frac{\partial^2 T}{\partial \theta^2}=0$$

$T(r, 0) = 10$, $T(r, \pi/2) = 10$, $T(0, \theta) = 10$, and $-\dfrac{\partial T(1,\theta)}{\partial r}=10$. Use $\Delta r = 1/2$, $\Delta\theta = \pi/6$.

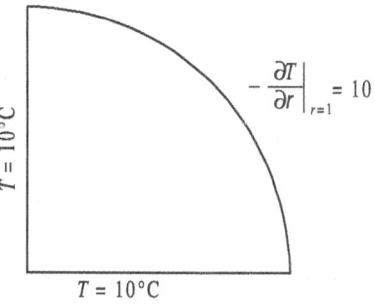

$$-\left.\frac{\partial T}{\partial r}\right|_{r=1} = 10$$

$T = 10°C$ (left side label)

$T = 10°C$

Fig. 7.14

8. A square plate 3 cm × 3 cm is initially at a temperature of 0°C. Suddenly, the temperature along the boundary $x = 3$ cm and $y = 3$ cm is raised to 100°C. Solve the heat-conduction equation:

$$\frac{\partial T}{\partial t}=k\left(\frac{\partial^2 T}{\partial x^2}+\frac{\partial^2 T}{\partial y^2}\right)$$

at half-time step ($\Delta t/2$) and full step (Δt) using the ADI scheme for $\Delta x = \Delta y = 1$ cm, $k = 1$ cm^2/sec and $\Delta t = 1$ sec.

9. Consider the following wave equation:

$$\frac{\partial^2 u}{\partial t^2}=\frac{\partial^2 u}{\partial x^2}\quad \text{in } x\in[0,1] \text{ for } t\geq 0$$

subject to $u(0,t) = \sin 2\pi t$, $u_x(1,t) = 0$, $u(x,0) = 0$, $u_t(x,0) = 2\pi\cos 2\pi x$.

(a) Discretize the equation and express in the matrix form using the $O(\Delta x^2, \Delta t^2)$ accurate implicit scheme shown in Eq. (7.106). Incorporate the time derivative boundary condition in such a manner, which preserves the accuracy of the scheme. Find the solution using $\Delta x = 0.25$ and $\Delta t = 0.1$ at $t = 0.5$.

(b) Decompose the PDE into two first order PDEs in time, similar to Eq. (7.105) and Example 7.5. Solve it using fourth order Runge-Kutta method in time and central difference in space with the same Δx and Δt as (a).

(c) Compare the solutions by two methods at $t = 0.5$.

7.4 STABILITY ANALYSIS

The PDEs such as diffusion, advection-diffusion, and first order wave equation were reduced to a system of IVPs by *semi-discretization*. Therefore, we should be able to analyse for stability using the matrix method described in Section 6.5.1. Let us illustrate this using the diffusion equation with Euler forward for time and central difference approximation for space derivatives (EF-CD scheme). After application of the central difference scheme for the space derivative, one can write an ODE for each node (j) as follows:

$$\frac{d\phi_j}{dt} = \alpha \frac{\phi_{j+1} - 2\phi_j + \phi_{j-1}}{\Delta x^2} \tag{7.120}$$

Let us use two, first type zero boundary conditions for the ease of illustration (Eq. 7.37). Then, the above equation can be written in the matrix form as:

$$\frac{d\bar{\phi}}{dt} = A\bar{\phi} \tag{7.121}$$

where A is a tridiagonal matrix whose diagonal elements are $-2\alpha / \Delta x^2$ and the off-diagonal elements are $\alpha / \Delta x^2$ and $\bar{\phi}$ is the vector containing the values of the state variable ϕ at the interior nodes. This is in the same form as the model problem for the stability analysis of a system of IVP (Section 6.5.1).

If we used non-zero boundary conditions, a vector b containing the constants from the boundary conditions will appear on the right-hand side. We have already seen that the constant additive vector b has no effect on the stability of the system (Section 6.3.2). Therefore, the eigenvalues of the coefficient matrix is of interest. The ratio of maximum to minimum eigenvalue will determine the stiffness of the system and the magnitude of the maximum eigenvalue will determine the stability for the application of Euler forward method (Section 6.5.1). In order to see the importance of the eigenvalues, let us use Theorem 2.8 to diagonalize matrix A using a matrix X whose columns are the eigenvectors:

$$A = X \Lambda X^{-1} \quad \text{or} \quad \Lambda = X^{-1}AX \tag{7.122}$$

where Λ is a diagonal matrix containing the eigenvalues (λ_k's). Using this relation, Eq. (7.121) can be written as

$$\frac{d(X^{-1}\bar{\phi})}{dt} = \Lambda(X^{-1}\bar{\phi}) \tag{7.123}$$

Let us denote a new vector as $\bar{\psi} = X^{-1}\bar{\phi}$. Then, the above equation is

$$\frac{d\bar{\psi}}{dt} = \Lambda \bar{\psi} \tag{7.124}$$

Each equation in the above system of equation looks as follows:

$$\frac{d\psi_k}{dt} = \lambda_k \psi_k \tag{7.125}$$

where ψ_k is related to ϕ_k as $\phi_k = X^{(k)}\psi_k$ and $X^{(k)}$ is the kth column of the matrix X. Equation (7.125) is the model equation for single IVP [Eq. (6.57)]. The solution of course is $\psi_k = \psi_k(0)e^{\lambda_k t}$ and the initial condition on ψ_k can be easily obtained from the original initial conditions specifying ϕ_k's using their relation through matrix X. It is now clear that the stability of any time stepping scheme for the solution of the PDE depends on whether it is stable for the IVP (Eq. 7.125) for all k.

We already know from Chapter 6 that for real and negative λ_k's, most of the methods will have a finite stability limit while some are unconditionally stable. For purely imaginary λ_k's, the method will be stable if its stability region contains some part of the imaginary axis. When λ_k's have both negative real and imaginary components, the stability will depend on whether the chosen time step value is within the stability region. So, the problem now boils down to computing the λ_k's of matrix A of Eq. (7.121) and compute the stability limit for various time stepping schemes in terms of these eigenvalues.

Remember that the λ_k's obtained from Eq. (7.121) is applicable only for the central difference scheme applied to the space derivative. If the discretization scheme for the space derivative is changed, the matrix A will also change and so are the λ_k's.

Let us represent the tridiagonal matrix A in Eq. (7.121) as

$$A = \frac{\alpha}{\Delta x^2} \begin{bmatrix} -2 & 1 & 0 & 0 & \cdots & \cdots & 0 & 0 \\ 1 & -2 & 1 & 0 & \cdots & \cdots & 0 & 0 \\ 0 & 1 & -2 & 1 & \cdots & \cdots & 0 & 0 \\ 0 & 0 & 1 & -2 & \cdots & \cdots & \cdots & 0 \\ \cdots & \cdots & \cdots & \cdots & \cdots & \cdots & \cdots & \cdots \\ 0 & 0 & \cdots & \cdots & \cdots & -2 & 1 & 0 \\ 0 & 0 & \cdots & \cdots & \cdots & 1 & -2 & 1 \\ 0 & 0 & \cdots & \cdots & \cdots & 0 & 1 & -2 \end{bmatrix} \tag{7.126}$$

The tridiagonal matrix is of the form where all the diagonal elements are equal (d), all the upper diagonal elements are equal (u) and all the lower diagonal elements are also equal (l). Let us denote this matrix as $B[l, d, u]$. The eigenvalues of such a square matrix of size n can be computed in the closed form as (Gregory and Karney 1969):

$$\lambda_k = d + 2\sqrt{lu} \cos \frac{k\pi}{n+1} \; ; \; k = 1, 2, \ldots, n \tag{7.127}$$

For our problem, if we divide the spatial domain $(0, L)$ in m equal segments such that $m\Delta x = L$, there are $(m + 1)$ nodes including two boundary nodes where the values of ϕ are known from the first type boundary conditions. Therefore, the matrix A is of size $(m - 1)$. Using the values of $l, d,$ and u, we obtain the eigenvalues (λ_k's) of matrix A as

$$\lambda_k = \frac{\alpha}{\Delta x^2} \left(-2 + 2\cos \frac{k\pi}{m} \right); \; k = 1, 2, \ldots, m-1 \tag{7.128}$$

First thing to notice is that all the eigenvalues are negative and real since $\cos(k\pi/m) \leq 1$ and $\alpha > 0$. Therefore, any time stepping scheme that includes at least some part of the negative real axis in its stability region can be applied to solve this system as long as values chosen are within the stability limits. The largest and smallest absolute values of λ_k would define the two limits of stability as well as the stiffness of the system.

Theoretically, it is easy to see from Eq. (7.128) that the maximum and minimum absolute values of λ_k would be achieved when $\cos(k\pi/m)$ approaches -1, i.e., $(k\pi/m) = \pi$ and 1, i.e., $(k\pi/m) = 0$, respectively. For large m or small Δx, $(m-1)/m \approx 1$ and the eigenvalue with largest absolute magnitude is given by

$$\lambda_{m-1} \approx -\frac{4\alpha}{\Delta x^2} \tag{7.129}$$

The other limit of zero can never be achieved but the closest to zero is given by $k = 1$. Therefore, the eigenvalue of smallest absolute magnitude is

$$\lambda_1 = \frac{2\alpha}{\Delta x^2} \left(\cos \frac{\pi}{m} - 1 \right) \tag{7.130}$$

Therefore, the ratio of maximum to minimum eigenvalue is given by

$$\left| \frac{\lambda_{m-1}}{\lambda_1} \right| \approx \left| \frac{2}{\cos(\pi/m) - 1} \right| \tag{7.131}$$

If we approximate the quantity in the denominator using first two terms of the series representation of the cosine, the ratio is given by

$$\left| \frac{\lambda_{m-1}}{\lambda_1} \right| \approx \frac{4m^2}{\pi^2} \tag{7.132}$$

It is now clear that larger the m is (smaller Δx), stiffer the system becomes. The stability criterion of the Euler forward method to this system is given by Eq. (6.129) as

$$\Delta t \leq \frac{2}{|\lambda_{m-1}|} \quad \text{or} \quad \alpha \frac{\Delta t}{\Delta x^2} \leq \frac{1}{2} \qquad (7.133)$$

We leave it to the readers to compute this limit for other schemes, such as *AB*, *AM*, *RK*, etc. For example, the limit for second order Runge-Kutta and central-difference scheme applied to the diffusion equation would be the same as [Eq. (7.133)] and that of fourth order Runge-Kutta and central difference scheme:

$$\Delta t \leq \frac{2.785}{|\lambda_{m-1}|} \qquad (7.134)$$

Let us now apply similar analysis to the first order wave equation. Application of central difference scheme for the spatial derivative yields

$$\frac{d\phi_j}{dt} = -u \left(\frac{\phi_{j+1} - \phi_{j-1}}{2\Delta x} \right) \qquad (7.135)$$

Once again, use of first type zero boundary conditions lead to a system of equation similar to Eq. (7.121) with the coefficient matrix now defined as follows:

$$A = -\frac{u}{2\Delta x} B[-1, \ 0, \ 1] \qquad (7.136)$$

Using Eq. (7.127), the eigenvalues of matrix A can be computed as

$$\lambda_k = -i \frac{u}{\Delta x} \cos \frac{k\pi}{m}; \quad k = 1, \ 2, \ ..., \ m-1 \qquad (7.137)$$

First thing to notice that all the eigenvalues are purely imaginary which essentially means that the coefficient of the model Eq. (7.125) are purely imaginary. Therefore, any time stepping scheme which does not include a part of the imaginary axis in its stability region will not be stable. Euler backward and trapezoidal scheme can be applied. The fourth order Runge-Kutta method is applicable because it includes some part of the imaginary axis. The eigenvalue of maximum absolute magnitude for large m is given by,

$$\lambda_{m-1} \approx i \frac{u}{\Delta x} \qquad (7.138)$$

The stability region of fourth order Runge-Kutta method (Fig. 6.14) crosses the ordinate approximately at ± 2.83. Therefore, the stability limit is

$$|\lambda_{m-1}\Delta t| \leq 2.83 \quad \text{or} \quad C \leq 2.83 \qquad (7.139)$$

We leave it to the readers to explore the stability of the upwind schemes using matrix analysis (Exercise 7.3).

In this section, we have seen that it is possible to analyse the methods applied to PDEs for stability using the procedures described in Chapter 6 for the system of IVPs. However, we had made some assumptions regarding the boundary conditions during the semi-discretization in spatial variable. If we relax this assumption and include the second type boundary condition, the tridiagonal matrix may not have the properties such as equality of all diagonal and off diagonal terms (Section 7.2). If only first and last diagonal terms are different, a closed form solution for the eigenvalues can still be obtained easily (Yueh 2005). However, if some of the off-diagonal terms are also unequal (which may occur due to forward or backward difference approximation of the derivative boundary conditions), obtaining a closed form solution for the eigenvalue may be very difficult and it may be easier to compute the maximum and minimum eigenvalues using *power and inverse power* methods, respectively. In the next section, we present a technique to analyse the numerical methods for stability in full discretized form.

7.4.1 Von Neumann Stability Analysis

The Von Neumann stability analysis can be applied on the fully discrete form of any numerical method applied to a linear PDE. The analysis technique relies on two basic assumptions:

- Separation of variable is possible, i.e., the solution can be expressed as products of functions of single independent variables.
- The functions of space variables in the separated variables form can be represented in terms of orthogonal basis functions e^{ikx}, i.e., Fourier series representation.

None of these assumptions are overly restrictive. A large number of PDEs in science and engineering problems yield to solution by separation of variables. Since, the solutions of the practical problems are expected to be continuous or piecewise continuous and finite (absolutely integrable), the Fourier series representation would be possible.

On the basis of these assumptions, one may write the solution of the diffusion, advection-diffusion, or wave equations $\phi(x, t)$ as

$$\phi(x,t) = \sum_{k=-\infty}^{\infty} \Gamma_k(t) e^{ikx} \tag{7.140}$$

where k is the wave number.

In order to see the Fourier series, let us take the example of one-dimensional wave Eq. (7.87). Substituting the expression of Eq. (7.140) in Eq. (7.87), we obtain

$$\sum_{k=-\infty}^{\infty} \left(\frac{d\Gamma_k(t)}{dt} + iku\Gamma_k(t) \right) e^{ikx} = 0 \tag{7.141}$$

We know from Chapter 4 that the functions e^{ikx} are orthogonal. As a result, for the above to be true, coefficient for each k must go to zero. This essentially means

$$\frac{d\Gamma_k(t)}{dt} + iku\Gamma_k(t) = 0 \quad \text{or} \quad \Gamma_k(t) = a_k e^{-ikut} \tag{7.142}$$

Using this in Eq. (7.140), we obtain the Fourier series solution of the one-dimensional advection equation as

$$\phi(x,t) = \sum_{k=-\infty}^{\infty} a_k e^{ik(x-ut)} \tag{7.143}$$

where a_k are the coefficient of the Fourier series which can be computed using the initial condition. Thus, the expression in Eq. (7.140) is indeed a solution of the PDE. Similar analysis can also be done with the other equations presented in this chapter. In fact, this is valid for all linear PDEs. This means that the $\Gamma_k(t)$ in Eq. (7.140) is the amplitude of the unit waves in the Fourier series. Therefore, one can define an amplification factor for the analytical problem as

$$\sigma_k^A = \frac{\Gamma_k(t_{n+1})}{\Gamma_k(t_n)} \tag{7.144}$$

The solution will remain bounded if and only if the amplitude is attenuated for each wave number k. In that case, their linear combination will also remain bounded. This means that the analytical solution is bounded if $\sigma_k^A \leq 1$. In this section, we will only consider the problems which have bounded analytical solution.

For a numerical method, the value of ϕ at the jth node at nth time step can be expressed using Eq. (7.140) as

$$\phi_j^n = \sum_{k=-\infty}^{\infty} \Gamma_k(t_n) e^{ikx_j} \tag{7.145}$$

We will represent each term in a numerical approximation using the convention of Eq. (7.145). Let us illustrate Von Neumann analysis through application of μ-CD scheme to the advection-diffusion equation. The fully discrete form is shown in Eq. (7.47).

Applying the Fourier representation of Eq. (7.145) to the fully discrete form for node j with constant u and α (7.43):

$$\sum_{k=-\infty}^{\infty} \left\{ \begin{array}{l} \dfrac{\Gamma_k(t_{n+1})e^{ikx_j} - \Gamma_k(t_n)e^{ikx_j}}{\Delta t} + \mu u \dfrac{\Gamma_k(t_n)e^{ikx_{j+1}} - \Gamma_k(t_n)e^{ikx_{j-1}}}{2\Delta x} \\[3mm] -\mu\alpha \dfrac{\Gamma_k(t_n)e^{ikx_{j+1}} - 2\Gamma_k(t_n)e^{ikx_j} + \Gamma_k(t_n)e^{ikx_{j-1}}}{\Delta x^2} \\[3mm] +(1-\mu)\alpha \dfrac{\Gamma_k(t_{n+1})e^{ikx_{j+1}} - \Gamma_k(t_{n+1})e^{ikx_{j-1}}}{2\Delta x} \\[3mm] -(1-\mu)\alpha \dfrac{\Gamma_k(t_{n+1})e^{ikx_{j+1}} - 2\Gamma_k(t_{n+1})e^{ikx_j} + \Gamma_k(t_{n+1})e^{ikx_{j-1}}}{\Delta x^2} \end{array} \right\} = 0 \qquad (7.146)$$

By definition of uniform spatial grid length Δx, we can write

$$x_{j+1} - x_j = x_j - x_{j-1} = \Delta x \qquad (7.147)$$

Using this relation in Eq. (7.146), we obtain

$$\sum_{k=-\infty}^{\infty} \left\{ \begin{array}{l} \dfrac{\Gamma_k(t_{n+1}) - \Gamma_k(t_n)}{\Delta t} + \mu u \dfrac{\Gamma_k(t_n)e^{ik\Delta x} - \Gamma_k(t_n)e^{-ik\Delta x}}{2\Delta x} \\[3mm] -\mu\alpha \dfrac{\Gamma_k(t_n)e^{ik\Delta x} - 2\Gamma_k(t_n) + \Gamma_k(t_n)e^{-ik\Delta x}}{\Delta x^2} \\[3mm] +(1-\mu)\alpha \dfrac{\Gamma_k(t_{n+1})e^{ik\Delta x} - \Gamma_k(t_{n+1})e^{-ik\Delta x}}{2\Delta x} \\[3mm] -(1-\mu)\alpha \dfrac{\Gamma_k(t_{n+1})e^{ik\Delta x} - 2\Gamma_k(t_{n+1}) + \Gamma_k(t_{n+1})e^{-ik\Delta x}}{\Delta x^2} \end{array} \right\} e^{ikx_j} = 0 \qquad (7.148)$$

Since e^{ikx_j} is orthogonal, the coefficient must go to zero for each k. Therefore, we can write

$$\begin{aligned} &\frac{\Gamma_k(t_{n+1}) - \Gamma_k(t_n)}{\Delta t} + \mu u \frac{\Gamma_k(t_n)e^{ik\Delta x} - \Gamma_k(t_n)e^{-ik\Delta x}}{2\Delta x} \\ &\qquad -\mu\alpha \frac{\Gamma_k(t_n)e^{ik\Delta x} - 2\Gamma_k(t_n) + \Gamma_k(t_n)e^{-ik\Delta x}}{\Delta x^2} \\ &\qquad +(1-\mu)\alpha \frac{\Gamma_k(t_{n+1})e^{ik\Delta x} - \Gamma_k(t_{n+1})e^{-ik\Delta x}}{2\Delta x} \\ &\qquad -(1-\mu)\alpha \frac{\Gamma_k(t_{n+1})e^{ik\Delta x} - 2\Gamma_k(t_{n+1}) + \Gamma_k(t_{n+1})e^{-ik\Delta x}}{\Delta x^2} = 0 \end{aligned} \qquad (7.149)$$

For the numerical solution to be stable, the amplitude of the unit waves for each wave number must attenuate with the progression of time. This will ensure that their linear combination also remain bounded. Therefore, we can define an amplification factor similar to the analytical problem as

$$\sigma_k = \frac{\Gamma_k(t_{n+1})}{\Gamma_k(t_n)} \tag{7.150}$$

Then for stability, one must satisfy $|\sigma_k| \le 1$ for all k. The overall stability condition for the method will then be governed by wave number for which the absolute value of the amplification factor is maximum.

Dividing Eq. (7.149) by $\Gamma_k(t_n)$ and using the definition of Eq. (7.150), we obtain

$$\frac{\sigma_k - 1}{\Delta t} = -\mu u \frac{e^{ik\Delta x} - e^{-ik\Delta x}}{2\Delta x} + \mu \alpha \frac{e^{ik\Delta x} - 2 + e^{-ik\Delta x}}{\Delta x^2}$$
$$-(1-\mu)u \frac{\sigma_k e^{ik\Delta x} - \sigma_k e^{-ik\Delta x}}{2\Delta x} + (1-\mu)\alpha \frac{\sigma_k e^{ik\Delta x} - 2\sigma_k + \sigma_k e^{-ik\Delta x}}{\Delta x^2} \tag{7.151}$$

Using Euler's formula, one can write the following identities

$$\frac{e^{ik\Delta x} - e^{-ik\Delta x}}{2} = i\sin(k\Delta x) \quad \text{and} \quad \frac{e^{ik\Delta x} + e^{-ik\Delta x}}{2} = \cos(k\Delta x) \tag{7.152}$$

Using these identities [Eq. (7.152)] and definition of C and P_g [Eqs (7.48) and (7.49)] in Eq. (7.151), we obtain the following expression for the amplification factor

$$\sigma_k = \frac{1 + 2\mu \dfrac{C}{P_g}[\cos(k\Delta x) - 1] - i\mu C \sin(k\Delta x)}{1 - 2(1-\mu)\dfrac{C}{P_g}[\cos(k\Delta x) - 1] + i(1-\mu)C \sin(k\Delta x)} \tag{7.153}$$

Let us define,

$$\xi_k = 2\frac{C}{P_g}[\cos(k\Delta x) - 1] - iC\sin(k\Delta x) \tag{7.154}$$

Therefore, the amplification factor σ is

$$\sigma_k = \frac{1 + \mu\xi_k}{1 - (1-\mu)\xi_k} \tag{7.155}$$

Let us verify if this indeed is the stability condition with the known case of Eq. (7.133) for application of *Euler forward central difference* scheme to purely diffusion problem. Set $u = 0$ for purely diffusion problem and $\mu = 1$ for Euler forward scheme. For this case,

$$\sigma_k = 1 + \xi_k \quad \text{and} \quad \xi_k = 2\alpha \frac{\Delta t}{\Delta x^2} [\cos(k\Delta x) - 1] \tag{7.156}$$

Notice the similarity of ξ_k in Eq. (7.156) with λ_k in Eq. (7.128) and that of σ_k with the amplification factor of Euler forward scheme for a system of ODE in Eq. (6.128). The $\xi_k = \Delta t \lambda_k$ with $\Delta x = \pi/m$. Using the stability condition $|\sigma_k| \leq 1$, we obtain,

$$\alpha \frac{\Delta t}{\Delta x^2} \leq \frac{1}{1 - \cos(k\Delta x)} \tag{7.157}$$

The most critical condition arises for the wave number k when $\cos(k\Delta x) = -1$. Therefore, the condition is

$$\alpha \frac{\Delta t}{\Delta x^2} \leq \frac{1}{2} \tag{7.158}$$

For Euler backward method applied to the diffusion equation with central difference scheme, the ξ_k remains the same but $\mu = 0$ gives the following for the amplification factor:

$$\sigma_k = \frac{1}{1 - \xi_k} \tag{7.159}$$

Since $\xi_k \leq 0$ for all k, the denominator is always less than or equal to unity. So, the scheme is unconditionally stable. We leave it the readers to derive the conditions for the advection-diffusion equation (Exercise 7.3, Problem 2).

In order to obtain a general condition for μ-CD scheme, let us denote the real and imaginary parts of ξ as

$$\xi_k = \xi_{Rk} + i\xi_{Ik} \tag{7.160}$$

Using this in Eq. (7.155) and applying the stability condition, one obtains the following (after some algebraic operations)

$$\left(\xi_{Rk} + \frac{1}{2\mu - 1} \right)^2 + (\xi_{Ik})^2 \leq \left(\frac{1}{2\mu - 1} \right)^2 \tag{7.161}$$

where ξ_{Rk} and ξ_{Ik} are the real and imaginary components of ξ_k, respectively

Let us now see an example application of the Von Neumann analysis.

Example 7.6   Analyse the upwind scheme applied to the first order wave equation [Eq. (7.94)] for stability using Von Neumann analysis.

Solution   Using the representation of Eq. (7.145) in Eq. (7.94), we obtain

$$\Gamma_k(t_{n+1})e^{ikx_j} = \Gamma_k(t_n)e^{ikx_j} - C[\Gamma_k(t_n)e^{ikx_j} - \Gamma_k(t_n)e^{ikx_{j-1}}]$$

The amplification factor is given by

$$\sigma_k = 1 + C\{\cos(k\Delta x) - 1\} - iC\sin(k\Delta x)$$

Using the stability condition leads to

$$[1 + C\{\cos(k\Delta x) - 1\}]^2 + [C\sin(k\Delta x)]^2 \le 1$$

After some algebraic operations, we arrive at the condition of stability as $C \le 1$.

Readers, may have noticed by now that the application of Von Neumann analysis can be made on fully discrete form of any numerical schemes described in this chapter following the principle and example described in this section. We leave it for the readers to explore and analyse other schemes as exercise.

EXERCISE 7.4

1. The following numerical method has been proposed to solve $\dfrac{\partial u}{\partial t} = c\dfrac{\partial u}{\partial x}$:

$$\frac{1}{\Delta t}\left[u_j^{n+1} - \frac{1}{2}(u_{j+1}^n + u_{j-1}^n)\right] = \frac{c}{2\Delta x}\left[u_{j+1}^n - u_{j-1}^n\right]$$

(a) Find the range of CFL number for which the method is stable using Von Neumann analysis.
(b) Is the method consistent (i.e., does it reduce to the original PDE as Δx, $\Delta t \to 0$)?

2. Consider the Lax Wendorf scheme [Eq. (7.93)] for the solution of one-dimensional convection-diffusion equation. What is the order of accuracy of this scheme? Is the scheme consistent? Obtain the stability criteria for this scheme.

3. *Du Fort-Frankel* scheme for the diffusion equation

$$\frac{\partial T}{\partial t} = \alpha\frac{\partial^2 T}{\partial x^2}; \ \alpha > 0$$

is given by

$$\frac{T_j^{n+1} - T_j^{n-1}}{2\Delta t} = \alpha\left[\frac{T_{j+1}^n - T_j^{n+1} - T_j^{n-1} + T_{j-1}^n}{\Delta x^2}\right]$$

(a) Obtain three leading order terms of the truncation error.

(b) In the limit $(\Delta t, \Delta x \to 0)$, what is the modified form of the original equation one obtains from the discretized equation? Based on this, comment on the consistency of the scheme.

(c) Obtain the stability criteria for this scheme.

SUMMARY

In this chapter, we presented numerical methods for solution of PDEs. The emphasis was on utilization of the concepts developed in the earlier chapters on finite difference approximation of derivatives (Chapter 5) and solution of initial and boundary-value problems (Chapter 6). We have observed how an initial-value problem and a boundary-value problem combine to form a PDE. At the same time, we have observed how the time and space derivatives can interact and their residuals can cancel each other to increase the order of the method.

The Von Neumann stability analysis introduced in this chapter is a powerful tool and finds many applications in the literature. While we have introduced the basic concepts for the finite difference approach of solution of PDE, the literature is too rich to do full justice within the scope of an introductory numerical analysis book. We hope that the readers have gained enough encouragement from this book to explore further for the more advanced methods for the solution of PDEs.

8

Advanced Topics

8.1 INTRODUCTION

Batman went through our book and was confident that he could handle any matter concerning numerical methods. However, one fine morning, he came to us virtually in tears. It so happened that he had gone to meet Santa and boasted about his abilities to deal with any *function*. Santa was quick to introduce him to a function which *Batman* could not approximate accurately. It may have been an isolated case but earlier today *Batman* had gone to watch an athletics event and encountered a function when he saw an athlete trying to jump over a horizontal bar with the help of a long stick. Again, he was not able to deal with this function and came running to us. We, being as smart as we are, immediately saw the problem. We have equipped him to analyse various types of functions but the methods discussed in Chapter 4 for approximating a function would not work very well (for functions which have poles[1]) near *poles* (whether North pole or in pole vault). We then decided to provide a very brief introduction of some topics which, we thought, may be useful in some situations. The aim is to provide a working knowledge of these topics rather than a detailed overview.

In this chapter, we discuss a diverse collection of topics which are thought to be a little advanced for an undergraduate course. The sequence of topics follows roughly the order of the previous chapters, but it is rather disjointed for a chapter.

---

[1]Recall that a function $f(x)$ has a pole of order n at a if n is the smallest positive integer for which $(x - a)^n f(x)$ is analytic at a.

We provide a brief introduction to these, since we believe that some of these are important enough and may find their way in the curriculum in the near future. The emphasis is on the methods rather than analysis, and references are provided to direct the interested reader to more rigorous treatment of the topics.

8.2 APPROXIMATION USING RATIONAL FUNCTIONS

In Chapter 4, we concentrated on interpolation and regression on discrete data and approximation of functions using polynomial (and, sometimes, trigonometric) functions. In some cases, a polynomial may not be able to represent the true nature of the function to be approximated. For example, if a function has a pole, polynomial interpolation may result in large errors near the pole. However, if we use the ratio of two polynomials, the poles of the function may be easily taken care of, since these would correspond to the zeros of the denominator. Using rational function in such cases results in a higher accuracy as shown in the following example.

Example 8.1 The function $f(x) = \tan x$ has been sampled at four points near its pole ($x = \pi/2$) as follows:

x	1.4	1.45	1.5	1.55
$f(x)$	5.79788	8.23809	14.1014	48.0785

Estimate the value of the function at $x = 1.51$ (true value is 16.4281) by polynomial (cubic) and rational (using a ratio of linear and quadratic polynomials) approximation.

Solution We may use any of the methods described in Chapter 4 and obtain the interpolating cubic polynomial. Here, we use the divided difference form and obtain,

$$f_3(x) = 5.79788 + 48.8042(x-1.4) + 684.62(x-1.4)(x-1.45)$$
$$+ 32920.92(x-1.4)(x-1.45)(x-1.5)$$

The estimated value of the function at $x = 1.51$ is, therefore, 17.8576 (an error of −8.7%).

For the rational interpolation, we use the ratio of a linear and a quadratic polynomial. Note that this rational function would have five coefficients, but since the ratio is unchanged if we multiply both the numerator and denominator with the same constant, there are effectively only four coefficients which have to be determined using the four data points. Assuming the interpolating rational function to be

$$R_{12}(x) = \left(\frac{a_0 + a_1 x}{1 + b_1 x + b_2 x^2} \right)$$

we get the following four equations:

$$a_0 + 1.4a_1 = 5.79788\,(1 + 1.4b_1 + 1.4^2 b_2)$$

$$a_0 + 1.45a_1 = 8.23809\,(1 + 1.45b_1 + 1.45^2 b_2)$$

$$a_0 + 1.5a_1 = 14.1014\,(1 + 1.5b_1 + 1.5^2 b_2)$$

$$a_0 + 1.55a_1 = 48.0785\,(1 + 1.55b_1 + 1.55^2 b_2)$$

which are linear in the coefficients and may be solved using any of the methods described in Chapter 2 to obtain $a_0 = 0.649825$, $a_1 = -0.544071$, $b_1 = -1.47764$, $b_2 = 0.535405$. The estimated value of the function at $x = 1.51$ is obtained as 16.4179 (an error of only 0.06%!). Figure 8.1 shows a plot of the function, and its polynomial and rational approximations. We could have assumed $a_0 = 1$ and use b_0 in the denominator. The resulting rational function would be identical to the one obtained here on dividing by the computed value of a_0, i.e., 0.649825. Note that it does not account for the case when $b_0 = 0$. Another commonly used option is to use a monic polynomial as the numerator (i.e., the coefficient of the largest power in the numerator is taken as unity). The general technique described later in this section uses this option.

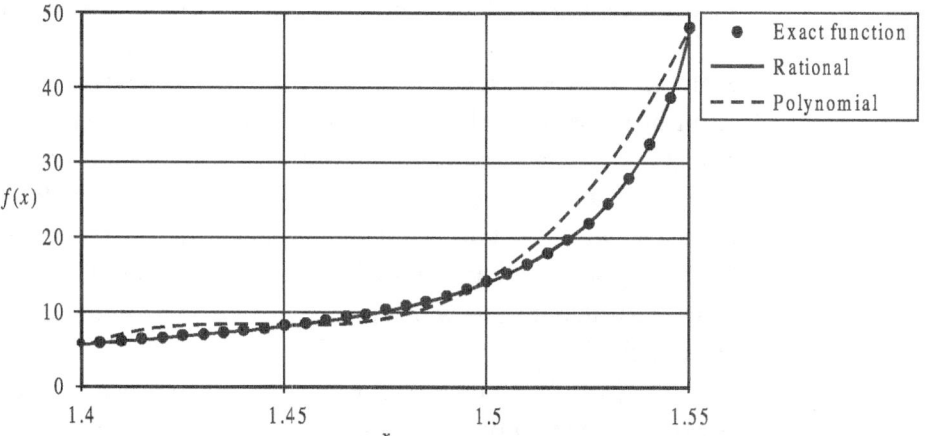

Fig. 8.1 Comparison of polynomial and rational approximations.

The suitability of rational approximation is clearly seen in Fig. 8.1.

As discussed in Chapter 4, if the function to be approximated is sampled at discrete points, we may opt for interpolation or regression and if the function is known, we may approximate it by another (simpler) function. In this section, we look at the use of rational polynomial functions to perform these operations. We start with interpolation, move on to regression, and finally describe approximation of functions.

8.2.1 Interpolation

In Example 8.1, the rational approximation was easily obtained by solving a set of four linear simultaneous equations. However, for higher-order interpolation, it would be cumbersome to use this technique. Moreover, the system of equations becomes more ill-conditioned as the order of the rational function increases. Hence, a technique similar to the divided difference method is generally used.

We denote the data points by the set of distinct values $\{(x_k, f(x_k)), k = 0,1,\ldots,n\}$ and assume the interpolating rational function to be

$$R_{n1,n2} = \frac{N_{n1}(x)}{D_{n2}(x)}$$

where N and D are the numerator and the denominator, respectively, which are polynomials and the subscript represents its *maximum* degree.

As discussed earlier, any *one* of the coefficients of the polynomials N_{n1} and D_{n2} may be arbitrary. Therefore, we must have $n = n1 + n2$ since the number of equality conditions are $(n + 1)$ and the number of free coefficients are $(n1 + n2 + 1)$. We also assume that this problem of interpolation does have a solution. As opposed to the polynomial interpolation which is always solvable for distinct grid points, the rational interpolation may not exist even when $n = n1 + n2$. For example, with $n = 1$, $n1 = 0$, and $n2 = 1$, the grid points $(0, 0)$ and $(1, 1)$ result in $R_{0,1} = 0$ which does not pass through the second point! In such cases, providing a small perturbation in one of the grid points may lead to a solution (e.g., by taking the function value as ε at $x = 0$, we will get $R_{0,1} = \varepsilon/[1 - (1 - \varepsilon)x]$. However, we will not discuss these degenerate cases. The methodology followed for arriving at the rational approximation is similar to the divided difference method, in the sense that it starts with a point and adds an additional point to get a higher-order interpolation.

Starting with the point x_0 at which the function value is given as $f(x_0)$, it is obvious that the rational function $R_{0,0} = f(x_0)$ would be the desired interpolating function. Now, adding another point $[x_1, f(x_1)]$, it is readily seen that the interpolating function would be,

$$R_{1,0} = f(x_0) + \cfrac{x - x_0}{\cfrac{x_1 - x_0}{f(x_1) - f(x_0)}} \tag{8.1}$$

At the next step, we add the point $[x_2, f(x_2)]$ and look for the interpolating rational $R_{1,1}$. Similar to the divided difference method, but in a slightly different form, we may write,

$$R_{1,1} = f(x_0) + \cfrac{x - x_0}{\cfrac{x_1 - x_0}{f(x_1) - f(x_0)} + C_{1,1}(x - x_1)} \qquad (8.2)$$

where $C_{1,1}$ is a constant which would be obtained by matching the function value at x_2 (It is obvious that $R_{1,1}$ interpolates the function value at the points x_0 and x_1.) Thus,

$$C_{1,1} = \frac{1}{x_2 - x_1}\left[\frac{x_2 - x_0}{f(x_2) - f(x_0)} - \frac{x_1 - x_0}{f(x_1) - f(x_0)}\right] \qquad (8.3)$$

We now define *inverse divided differences* as,
First inverse divided difference:

$$\chi\left[x_j, x_i\right] = \frac{x_j - x_i}{f(x_j) - f(x_i)}\left(= \chi\left[x_i, x_j\right]\right)$$

Second inverse divided difference:

$$\chi\left[x_k, x_j, x_i\right] = \frac{x_k - x_j}{\chi\left[x_k, x_i\right] - \chi\left[x_j, x_i\right]} \left(= \chi\left[x_j, x_k, x_i\right] \neq \cdots \neq \chi\left[x_i, x_j, x_k\right]\right)$$

Third inverse divided difference:

$$\chi\left[x_l, x_k, x_j, x_i\right] = \frac{x_l - x_k}{\chi\left[x_l, x_j, x_i\right] - \chi\left[x_k, x_j, x_i\right]}$$

nth inverse divided difference:

$$\chi\left[x_n, x_{n-1}, \ldots, x_1, x_0\right] = \frac{x_n - x_{n-1}}{\chi\left[x_n, x_{n-2}, \ldots, x_0\right] - \chi\left[x_{n-1}, x_{n-2}, \ldots, x_0\right]} \qquad (8.4)$$

and note that

$$R_{1,0} = f(x_0) + \frac{x - x_0}{\chi[x_1, x_0]}$$

$$R_{1,1} = f(x_0) + \cfrac{x - x_0}{\chi[x_1, x_0] + \cfrac{x - x_1}{\chi[x_2, x_1, x_0]}}$$

Similarly, writing

$$R_{2,1} = f(x_0) + \cfrac{x - x_0}{\chi[x_1, x_0] + \cfrac{x - x_1}{\chi[x_2, x_1, x_0] + C_{2,1}(x - x_2)}}$$

it can be shown that $C_{2,1} = \dfrac{1}{\chi[x_3, x_2, x_1, x_0]}$.

Continuing to $R_{2,2}, R_{3,2}, \ldots$ till all the points are exhausted, we get the continued fraction representation of the rational interpolation as,

$$R_{n1,n2} = f(x_0) + \cfrac{x - x_0}{\chi[x_1, x_0] + \cfrac{x - x_1}{\chi[x_2, x_1, x_0] + \cfrac{x - x_2}{\chi[x_3, x_2, x_1, x_0] + \cfrac{x - x_3}{\chi[x_4, x_3, x_2, x_1, x_0] + \cdots}}}}$$

(8.5)

From the way it is derived, it is obvious that $n1$ will be equal to $n2$ (for even n), or would be one more than $n2$ (for odd n). Also note that some of the inverse divided difference may become infinite, since the denominator may become zero. Another difference compared to the divided difference is that the inverse divided differences are not, in general, symmetric with respect to their arguments.

Example 8.2   Obtain the rational interpolation for the data given in Example 8.1.

Solution   Table 8.1 shows the inverse divided differences:

Table 8.1

i	x_i	$f(x_i)$	$\chi[x_i, x_0]$	$\chi[x_i, x_1, x_0]$	$\chi[x_i, x_2, x_1, x_0]$
0	1.40	5.79788			
1	1.45	8.23809	0.0204900		
2	1.50	14.1014	0.0120431	−5.91929	
3	1.55	48.0785	0.00354772	−5.90238	2.95646

The interpolating rational is, therefore,

$$R_{2,1} = 5.79788 + \cfrac{x - 1.4}{0.0204900 + \cfrac{x - 1.45}{-5.91929 + \cfrac{x - 1.5}{2.95646}}} = \frac{-0.511721 - 3.14015x + x^2}{-4.67618 + 2.97695x}$$

and the value at $x = 1.51$ is predicted as 16.4281, the error being negligible. The plot of this function is not shown, but it is indistinguishable from that of the rational function shown in Fig. 8.1. Also note that both these rational interpolating functions have a pole at $x = \pi/2$ (the given function, tan x, does have a pole at this point, although this information is not explicitly incorporated in the interpolation process). This rational approximation could as well be written as

$$\frac{0.109431 + 0.671520x - 0.213850x^2}{1 - 0.636620x} \quad \text{or} \quad \frac{0.335915x^2 - 1.05482x - 0.171894}{x - 1.57080}$$

the latter form clearly showing the pole near $x = \pi/2$.

If we are only interested in estimating the value of the function at a point and are not concerned with the coefficients of the interpolating rational, a formula similar to Neville's method (Section 4.4.1.3) may be obtained (see Stoer and Bulirsch 2002).

8.2.2 Regression

As discussed in Chapter 4, sometimes we may not want the approximate function to pass through every observed data point. To perform regression using a rational polynomial function, we aim at minimizing some norm of the residual which is defined as,

$$r_i = f(x_i) - R_{n1,n2}(x_i), \qquad i = 0, 1, 2, \ldots, n$$

with $n1 + n2 < n$. Although other norms could be used, the L_2 norm is generally preferred and gives rise to the least squares rational polynomial approximation. The example below illustrates a method of arriving at the best rational approximation in the least square sense.

Example 8.3   Estimate the value of the function sampled in Example 8.1 at $x = 1.51$ by rational (using a ratio of two linear polynomials) regression.

Solution   Let the approximating polynomial be written as,

$$R_{1,1}(x) = \frac{a_0 + a_1 x}{1 + b_1 x}$$

The sum of the squares of the residuals is written as,

$$\sum_{i=0}^{3} r_i^2 = \sum_{i=0}^{3} \left[f(x_i) - \frac{a_0 + a_1 x_i}{1 + b_1 x_i} \right]^2 \tag{8.6}$$

Equating the partial derivatives with respect to a_0, a_1, and b_1, to 0, we get the following equations:

$$\sum_{i=0}^{3} \left[f(x_i) - \frac{a_0 + a_1 x_i}{1 + b_1 x_i} \right] \frac{1}{1 + b_1 x_i} = 0$$

$$\sum_{i=0}^{3} \left[f(x_i) - \frac{a_0 + a_1 x_i}{1 + b_1 x_i} \right] \frac{x_i}{1 + b_1 x_i} = 0$$

$$\sum_{i=0}^{3} \left[f(x_i) - \frac{a_0 + a_1 x_i}{1 + b_1 x_i} \right] \frac{x_i}{(1 + b_1 x_i)^2} = 0$$

This is a set of three *nonlinear* equations in the three unknowns and may be solved using any of the techniques discussed in Section 3.7. However, it must be kept in mind that the solution is not only difficult but may also be non-unique. Using the Newton-Raphson method with a starting guess of $a_0 = a_1 = 0$ and $b_1 = -2/\pi$ (the solution process is very sensitive to the initial guess of b_1, hence we start with the theoretical value), we obtain the desired best rational approximation as

$$R_{1,1} = \frac{0.555924 + 0.053591x}{1 - 0.636587x}$$

The estimated value at $x = 1.51$ is 16.4330.

This example depicts the difficulties associated with the regression, and it is obvious that for higher-order rational polynomials, the level of difficulty increases significantly. Some analysts argue that the process could be simplified by defining the residual as,

$$r_i = D_{n2}(x_i) f(x_i) - N_{n1}(x_i), \qquad i = 0, 1, 2, \ldots, n \qquad (8.7)$$

and then minimizing the sum of squares. The solution will, of course, be different from the one obtained using the residual defined in Eq. (8.6). For this example problem, the normal equations with $N_1 = a_0 + a_1 x$ and $D_1 = 1 + b_1 x$ are obtained as,

$$\sum_{i=0}^{3} [a_0 + x_i a_1 - x_i f(x_i) b_1 - f(x_i)] = 0$$

$$\sum_{i=0}^{3} [x_i a_0 + x_i^2 a_1 - x_i^2 f(x_i) b_1 - x_i f(x_i)] = 0$$

$$\sum_{i=0}^{3} \{ x_i f(x_i) a_0 + x_i^2 f(x_i) a_1 - x_i^2 [f(x_i)]^2 b_1 - x_i [f(x_i)]^2 \} = 0$$

which are *linear* equations in the three unknowns. The *unique* solution is obtained as

$$R_{1,1} = \frac{0.551490 + 0.0566931x}{1 - 0.636581x}$$

which is not very different from the previously obtained approximation (the estimated value at $x = 1.51$ is obtained as 16.4361).

Since the coefficients are obtained directly, we may use a strategy wherein the nonlinear set of equations could be solved with the linear model values of the parameters used as an initial guess. The iterations are then likely to converge rapidly. As discussed in Chapter 4, the normal equations for a linear least squares problem are typically ill-conditioned. An alternative technique based on orthogonal polynomials has been suggested for faster and more robust computations (see Van Barel and Bultheel 1992).

8.2.3 Approximation of a Given Function

To approximate a given function by another (simpler) function over a specified range, we minimize some norm of the residual (see Sections 4.2 and 4.3). For approximation using the rational polynomial function, the L_∞ norm is generally used since it leads to a simple iterative method (Remes algorithm) for obtaining the approximation. However, if the function may be expressed as a power series, an alternative form (Padé approximation) is preferred. Here, we describe (very briefly) these two techniques.

8.2.3.1 *Exchange Algorithm (or Modified Remes Second Algorithm)*

This algorithm is based on the fact that a minimax rational polynomial approximation, $R_{n1,\,n2}$ $(=N_{n1}/D_{n2})$ of a given function over the interval $[a, b]$ follows the Tchebycheff alternation theorem (Theorem 4.3) in that the residual $(=f(x) - R_{n1,n2}(x))$ achieves $n1 + n2 + 2$ extrema in $a \le x \le b$ which are of equal magnitude and alternating signs. (We assume that there are no common factors of the numerator and the denominator of this rational function). The algorithm is as follows [Note that the algorithm will also work with minimax polynomial approximation of a given function]:

1. Choose an initial set of $n1 + n2 + 2$ points in $[a,b]$ where the residual is assumed to be equal and alternating in sign. Label these points as $x_i^{(0)}$, $i = 0,1,2,...,N+1$ such that $a \le x_0^{(0)} < x_1^{(0)} < x_2^{(0)} \cdots < x_N^{(0)} < x_{N+1}^{(0)} \le b$ and $N = n_1 + n_2$.

2. For $k = 0,1,...$ till convergence:

    (a) Solve the set of $N + 2$ equations (k denoting the iteration level)

    $$f(x_i^{(k)}) - R_{n1,n2}^{(k)} = (-1)^i \Delta^{(k)} \tag{8.8}$$

    to obtain the $N + 1$ coefficients of the rational approximation and the residual Δ (equal at all these points).

    (b) For this rational approximation, let the maximum magnitude of the residual occur at a point $x_{\max}^{(k)}$. If this point coincides with any of the $x_i^{(k)}$, we have our desired approximation. If not, we keep *all but one* of the grid points same (i.e. $x_i^{(k+1)} = x_i^{(k)}$) and *exchange* one with $x_{max}^{(k)}$ in such a way that the residual still alternates in sign at the new set of points $x_i^{(k+1)}$. For example (Fig. 8.2), if $x_{max}^{(k)}$ lies between $x_m^{(k)}$ and $x_{m+1}^{(k)}$, and the sign of the residual at $x_m^{(k)}$ and $x_{max}^{(k)}$ is the same, we use $x_m^{(k+1)} = x_{max}^{(k)}$ (if these residuals have opposite signs then $x_{m+1}^{(k+1)} = x_{max}^{(k)}$).

In the special cases where $x_{max}^{(k)}$ lies in the first or the last interval, we need a slight modification in the exchange algorithm. If $x_{max}^{(k)}$ lies between a and $x_0^{(k)}$ and the residual at $x_{max}^{(k)}$ and $x_0^{(k)}$ have the same sign, we use $x_0^{(k+1)} = x_{max}^{(k)}$. If these residuals have opposite signs, we again use $x_0^{(k+1)} = x_{max}^{(k)}$ but also keep $x_0^{(k)}$ as $x_1^{(k+1)}$, increase the subscript of all remaining points by 1, and delete $x_{N+1}^{(k)}$. Similar logic is used in the case when $x_{max}^{(k)}$ lies between $x_{N+1}^{(k)}$ and b.

(c) Go to step (a)

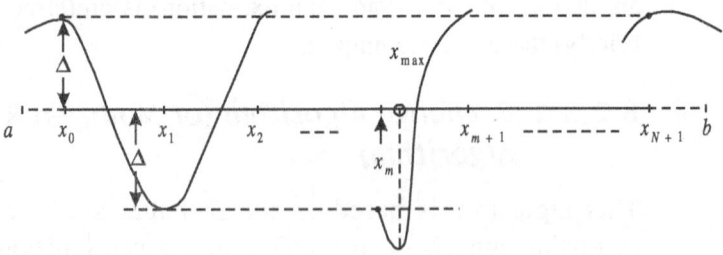

Fig. 8.2 Exchange algorithm for the alternation points.

The following example illustrates the application of the exchange algorithm.

Example 8.4   Obtain the minimax rational approximation, $R_{1,1}$, of the function $\tan(x)$ over the interval [1.40,1.55]. Estimate the value at $x = 1.51$.

Solution   We assume the minimax rational approximation to be,

$$R_{1,1}(x) = \frac{a_0 + a_1 x}{1 + b_1 x}$$

To start the exchange algorithm, we need four alternation points. Although various methods are available for choosing the initial set of points (Ralston and Wilf 1960, Psarakis and Moustakides 2003), we do not go into the details and arbitrarily take these points as 1.40, 1.45, 1.50, and 1.55. The resulting equations are

$$\tan(1.4) - \frac{a_0 + 1.4a_1}{1 + 1.4b_1} = \Delta$$

$$\tan(1.45) - \frac{a_0 + 1.45a_1}{1 + 1.45b_1} = -\Delta$$

$$\tan(1.5) - \frac{a_0 + 1.5a_1}{1 + 1.5b_1} = \Delta$$

$$\tan(1.55) - \frac{a_0 + 1.55a_1}{1 + 1.55b_1} = -\Delta$$

One way of solving these non-linear equations is to write an iterative scheme as

$$a_0 + x_i a_1 - x_i f(x_i) b_1 + (-1)^i (1 + x_i b_1')\Delta = f(x_i) \qquad (8.9)$$

where b_1' represents the value of b_1 at the previous iteration. Starting with $b_1' = 0$, we obtain $a_0 = 0.551188$, $a_1 = 0.0569911$, $b_1 = -0.636577$, and $\Delta = -0.00020902$. In the second iteration (i.e., using $b_1' = -0.636577$), we obtain $a_0 = 0.555097$, $a_1 = 0.0542284$, $b_1 = -0.636584$, and $\Delta = -0.00284353$. One more iteration results in the final values of $a_0 = 0.555097$, $a_1 = 0.0542280$, $b_1 = -0.636584$, and $\Delta = -0.00284395$. The maximum magnitude of residual with these rational polynomial coefficients is 0.02077 and it occurs at $x = 1.538$ *with a negative sign*. The residual at the four guessed alternation points are, obviously, -0.00284395 at $x = 1.40$, 0.00284395 at $x = 1.45$, -0.00284395 at $x = 1.50$, and 0.00284395 at $x = 1.55$. For the next cycle, therefore, we exchange the point $x = 1.50$ and the four alternation points become 1.40, 1.45, 1.538, and 1.55, which have residuals alternating in sign (but not equal in magnitude).

After the next cycle, the values of a_0, a_1, b_1, and Δ are 0.566459, 0.0461718, -0.636599, and -0.00470927, respectively. The maximum magnitude of the residual occurs at $x = 1.483$ where its value is 0.007227. The new set of alternation points become 1.40, 1.483, 1.538, and 1.55. Continuing this process, we get the final solution as $a_0 = 0.567209$, $a_1 = 0.0457308$, $b_1 = -0.636598$, $\Delta = -0.0058444$ (the alternation points are 1.4, 1.482, 1.54, and 1.55).

Another choice of the starting alternating points could be the zeros of the Tchebycheff polynomial $T_4(x)$. Starting from this set (1.4057, 1.4463, 1.5037, 1.5443), we again get the same solution. Thus, the minimax rational approximation is

$$R_{1,1}(x) = \frac{0.567209 + 0.0457308x}{1 - 0.636598x}$$

8.2.3.2 Padé Approximation

We know that the power series of $\tan(x)$ is given by,

$$\tan(x) = x + \frac{1}{3}x^3 + \frac{2}{15}x^5 + \frac{17}{315}x^7 + \cdots \tag{8.10}$$

It may be readily seen that if we use only these four terms of the series (actually nine terms since the coefficients of x^0, x^2, x^4, x^6, x^8 may be thought of as zero) there would be a large error near the pole $\pi/2$. Now, if we want to approximate it by a rational polynomial function, $R_{n1,n2}$, we may stipulate that the power series of the function and the approximant agree to the highest possible order (e.g., if we use nine terms of the series, we could use any combination of $n1$ and $n2$ such that $n1 + n2 = 8$). These are known as the *Padé approximants* and there are various techniques available for evaluation of the coefficients of the rational polynomial. For example, let the given function be expressed as a power series

$$f(x) = \sum_{i=0}^{\infty} c_i x^i \tag{8.11}$$

We could think of this as the Maclaurin series and equate it with that of $R_{n1,n2}$ to obtain the following conditions:

$$\left. \frac{d^i}{dx^i} R_{n1,n2}(x) \right|_{x=0} = \left. \frac{d^i}{dx^i} f(x) \right|_{x=0} \quad ; \quad i = 0,1,2,\ldots,(n1+n2) \tag{8.12}$$

The coefficients of the Padé approximant could be conveniently evaluated by writing,

$$\sum_{i=0}^{\infty} c_i x^i = \frac{\displaystyle\sum_{j=0}^{n1} a_j x^j}{\displaystyle\sum_{k=0}^{n2} b_k x^k} \Rightarrow \sum_{i=0}^{\infty} c_i \left(\sum_{k=0}^{n2} b_k x^{i+k} \right) = \sum_{j=0}^{n1} a_j x^j \tag{8.13}$$

and equating the coefficients of the powers of x^0, x^1,...,x^{n1+n2}. This results in the following equations:

$$\sum_{j=0}^{\min(i,n2)} c_{i-j} b_j = a_i, \qquad i = 0,1,2,\ldots,n1 \tag{8.14a}$$

$$\sum_{j=0}^{\min(i,n2)} c_{i-j} b_j = 0, \qquad i = n1+1, n1+2, \ldots, n1+n2 \tag{8.14b}$$

As before, we must pre-assign one of the coefficients. Conventionally, we use $b_0 = 1$ and Eq. (8.14b) then form a set of $n2$ linear equations in the $n2$ unknown coefficients $b_1, b_2,\ldots,b_{n2}$ in terms of the known coefficients c of the power series.

After solving for the b's, Eq. (8.14a) are used to obtain the $(n1+1)$ coefficients a. The following example illustrates the procedure.

Example 8.5   Obtain the Padé approximation, $R_{4,4}$, of the function $\tan(x)$ using the power series given in Eq. (8.10) and estimate the value at $x = 1.51$.

Solution   From the given power series, we have: $c_0 = c_2 = c_4 = c_6 = c_8 = 0$, $c_1 = 1$, $c_3 = 1/3$, $c_5 = 2/15$, $c_7 = 17/315$. The b's are obtained by taking $b_0 = 1$ and writing Eq. (8.14b) as (note that $n1 = n2 = 4$),

$$c_5 b_0 + c_4 b_1 + c_3 b_2 + c_2 b_3 + c_1 b_4 = 0$$
$$c_6 b_0 + c_5 b_1 + c_4 b_2 + c_3 b_3 + c_2 b_4 = 0$$
$$c_7 b_0 + c_6 b_1 + c_5 b_2 + c_4 b_3 + c_3 b_4 = 0$$
$$c_8 b_0 + c_7 b_1 + c_6 b_2 + c_5 b_3 + c_4 b_4 = 0$$

The solution is obtained as $b_1 = 0$, $b_2 = -3/7$, $b_3 = 0$, and $b_4 = 1/105$. Equation (8.14a) then results in the value of a's as $a_0 = 0$, $a_1 = 1$, $a_2 = 0$, $a_3 = -2/21$, and $a_4 = 0$, giving the Padé approximant as,

$$R_{4,4} = \frac{x - \dfrac{2}{21} x^3}{1 - \dfrac{3}{7} x^2 + \dfrac{1}{105} x^4}$$

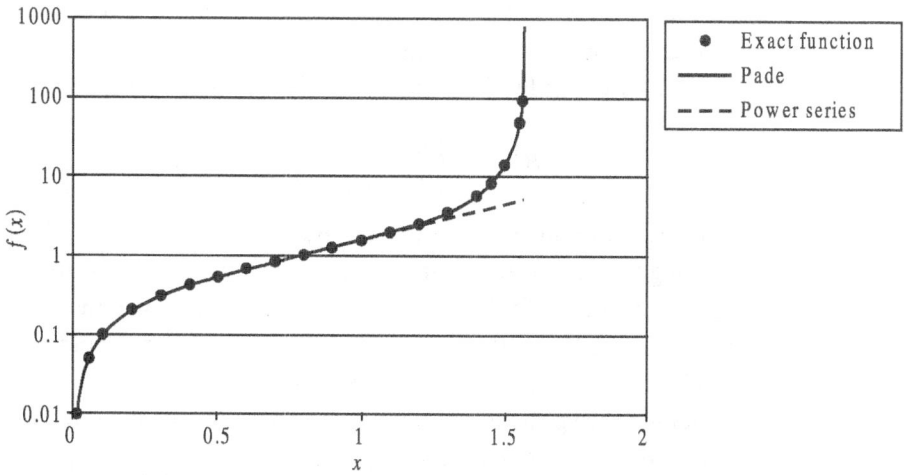

Fig. 8.3   Comparison of power series and Padé approximations.

Figure 8.3 shows the function, its power series, and the Padé approximant over the interval $(0, \pi/2)$ and demonstrates the superiority of the Padé approximation over the power series, although it is derived from the power series. The estimate at $x = 1.51$ is obtained as 16.3438, an error of about 0.5% (the power series results in a value of 4.67!).

8.3 B-SPLINES

In Section 4.5, we looked at spline interpolation and discussed the linear, quadratic, and cubic splines. Here, we describe a class of splines which may be used as a basis for forming splines of any order. These are called the *B-splines* and follow a simple recursive formula which relates a higher-order spline to a lower-order spline. Moreover, the derivatives and integral of these splines are easily obtainable. The basic idea is to generate a higher-order spline by a combination of splines which are one order lower multiplied by a linear function.

We concentrate on the case where the knots used for defining the B-spline are same as the nodes for interpolation (i.e., the points where the function values are given). In general, the knots and nodes may be different and B-splines may still be used as long as the nodes satisfy certain conditions with respect to the knots.

Let $\{(x_k, f(x_k)), k = 0, 1, \ldots, n\}$ denote the set of nodes where the function values are given. The interpolating spline of order m is written as,

$$S^m(x) = \sum_{j=-m}^{n-1} c_j B_j^m(x) \tag{8.15}$$

in which B_j^m represents the B-spline of order m and c_j are constants (the range of j is taken from $-m$ to $(n-1)$, and the reasoning would be explained a little later. However, for $m = 0$, the range of j is 0 to n).

The lowest-order B-splines, i.e., of order 0, are defined as

$$B_i^0(x) = \begin{cases} 1, & \text{if } x_i \leq x < x_{i+1} \\ 0, & \text{otherwise} \end{cases} \quad \text{for } i = 0, 1, 2, \ldots, n \tag{8.16}$$

See Fig. 8.4 and note the discontinuity at the right end of an interval. Further, although we are using a finite set of nodes here, it is customary to define the B-splines over an infinite set using additional nodes on either side as $x_{-1}, x_{-2}, \ldots$, and $x_{n+1}, x_{n+2}, \ldots$. We will assume that all these nodes are equidistant.

Clearly, the value of the splines at the nodes are given by $B_i^0(x_j) = \delta_{i,j}$ and matching the spline interpolant at the nodes results in $c_j = f(x_j)$ for $j = 0, 1, \ldots, n$. These splines form the building blocks for higher-order splines. For example, the B-spline of order 1 are obtained from,

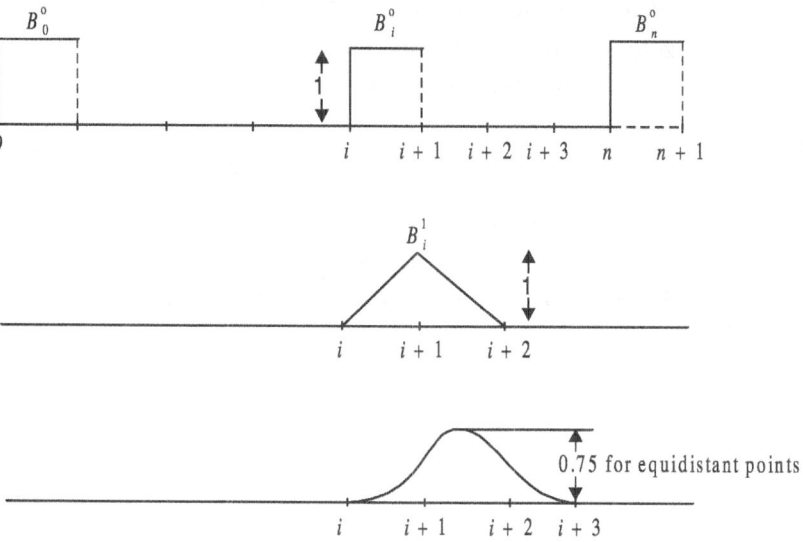

Fig. 8.4   B-splines of order 0, 1, and 2.

$$B_i^1(x) = \frac{x - x_i}{x_{i+1} - x_i} B_i^0(x) + \frac{x_{i+2} - x}{x_{i+2} - x_{i+1}} B_{i+1}^0(x), \text{ for } i = 0,1,2, \ldots, n-1 \qquad (8.17)$$

which is a linear function and its support is the interval (x_i, x_{i+2}), see Fig. 8.4. It is expressed as,

$$B_i^1(x) = \begin{cases} \dfrac{x - x_i}{x_{i+1} - x_i}, & \text{for } x_i \leq x < x_{i+1} \\[3mm] \dfrac{x_{i+2} - x}{x_{i+2} - x_{i+1}}, & \text{for } x_{i+1} \leq x < x_{i+2} \\[3mm] 0, & \text{otherwise} \end{cases} \qquad (8.18)$$

It is convenient at this stage to define a local variable (or shape function) as

$$l_i^h(x) = \frac{x - x_i}{x_{i+h} - x_i} \qquad (8.19)$$

Then, the B-spline of order 1 would be equal to $l_i^1(x)$ over the first interval and $1 - l_{i+1}^1(x)$ over the second. Note that the value of the B-spline of order 1 at any node is given by $B_i^1(x_j) = \delta_{i+1,j}$ and the coefficients in the interpolating spline would be given by $c_j = f(x_{j+1})$ for $j = -1, 0, 1, \ldots, n-1$. Also note that the spline $B_{-1}^1(x)$ would exist over the first interval (x_0, x_1) but the spline $B_n^1(x)$ would be zero over the entire domain (x_0, x_n).

We now write a recursive scheme for higher order B-splines as,

$$
\begin{aligned}
B_i^h(x) &= \frac{x - x_i}{x_{i+h} - x_i} B_i^{h-1}(x) + \frac{x_{i+h+1} - x}{x_{i+h+1} - x_{i+1}} B_{i+1}^{h-1}(x) \\
&= l_i^h(x) B_i^{h-1}(x) + \left[1 - l_{i+1}^h(x)\right] B_{i+1}^{h-1}(x)
\end{aligned}
\tag{8.20}
$$

For example, the second order B-spline is given by

$$
B_i^2(x) = \begin{cases}
l_i^2 l_i^1, & \text{for } (x_i \leq x < x_{i+1}) \\
l_i^2 (1 - l_{i+1}^1) + (1 - l_{i+1}^2) l_{i+1}^1, & \text{for } (x_{i+1} \leq x < x_{i+2}) \\
(1 - l_{i+1}^2)(1 - l_{i+2}^1), & \text{for } (x_{i+2} \leq x < x_{i+3}) \\
0, & \text{elsewhere}
\end{cases}
\tag{8.21}
$$

However, one should keep in mind that the actual computation of the interpolated value of the function would be done by the recursive relation [Eq. (8.20)] and explicit expressions for higher-order B-splines are not needed. The derivative and integral of the B-splines are given by

for $h \geq 2$,
$$
\frac{dB_i^h(x)}{dx} = \frac{h B_i^{h-1}(x)}{x_{i+h} - x_i} - \frac{h B_{i+1}^{h-1}(x)}{x_{i+h+1} - x_{i+1}}
\tag{8.22a}
$$

$$
\int_{-\infty}^{x} B_i^h(\chi) \, d\chi = \frac{x_{i+h+1} - x_i}{h+1} \sum_{j=i}^{\infty} B_j^{h+1}(x)
\tag{8.22b}
$$

The expression for the derivative may be used for $h = 1$ also except when x coincides with one of the nodes x_i, x_{i+1}, or x_{i+2}.

To obtain the coefficients, c_j, for interpolation with second-order B-splines, we note that the nodal values of this spline are given by,

$$
B_i^2(x_j) = \begin{cases}
\dfrac{x_{i+1} - x_i}{x_{i+2} - x_i}, & \text{for } j = i+1 \\[2ex]
\dfrac{x_{i+3} - x_{i+2}}{x_{i+3} - x_{i+1}}, & \text{for } j = i+2 \\[2ex]
0, & \text{otherwise}
\end{cases}
\tag{8.23}
$$

The equality of the interpolant and the function at the nodes then provides us the following relations:

$$\frac{x_{j+1} - x_j}{x_{j+1} - x_{j-1}} c_{j-2} + \frac{x_j - x_{j-1}}{x_{j+1} - x_{j-1}} c_{j-1} = f(x_j), \text{ for } j = 0, 1, 2, \ldots, n \qquad (8.24)$$

Thus, there are $(n + 1)$ equations in $(n + 2)$ unknowns $(c_j, j = -2, -1, \ldots, n-1)$. We, therefore, have a degree of freedom (compare with the discussion on quadratic spline in Section 4.5) and may choose one of the coefficients arbitrarily. For example, assuming that the spline is a straight line in the first segment, we should have

$$\left. \frac{dS^2(x)}{dx^2} \right|_{x=x_0^+} = 0$$

(note that the right-hand limit is used at $x = x_0$). The second derivative is easily obtained from Eq. (8.22a) as,

$$\frac{d^2 B_i^2(x)}{dx^2} = \frac{2}{(x_{i+2} - x_i)} \frac{dB_i^1(x)}{dx} - \frac{2}{(x_{i+3} - x_{i+1})} \frac{dB_{i+1}^1(x)}{dx}$$

and using the fact that

$$\left. \frac{dB_i^1(x)}{dx} \right|_{x_j^+} = \frac{\delta_{i,j}}{x_{i+1} - x_i} - \frac{\delta_{i+1,j}}{x_{i+2} - x_{i+1}}$$

Hence,

$$\left. \frac{d^2 S^2(x)}{dx^2} \right|_{x=x_0^+} = \sum_{i=-2}^{n-1} c_i \left\{ \begin{array}{l} \dfrac{2}{x_{i+2} - x_i} \left[\dfrac{\delta_{i,0}}{x_{i+1} - x_i} - \dfrac{\delta_{i+1,0}}{x_{i+2} - x_{i+1}} \right] \\[4mm] - \dfrac{2}{x_{i+3} - x_{i+1}} \left[\dfrac{\delta_{i+1,0}}{x_{i+2} - x_{i+1}} - \dfrac{\delta_{i+2,0}}{x_{i+3} - x_{i+2}} \right] \end{array} \right\} = 0$$

$$\Rightarrow \quad \frac{2c_{-2}}{(x_1 - x_{-1})(x_1 - x_0)} - \frac{2c_{-1}}{x_1 - x_0} \left[\frac{1}{x_1 - x_{-1}} + \frac{1}{x_2 - x_0} \right] + \frac{2c_0}{(x_2 - x_0)(x_1 - x_0)} = 0$$

$$\Rightarrow \quad (x_2 - x_0)c_{-2} - (x_2 + x_1 - x_0 - x_{-1})c_{-1} + (x_1 - x_{-1})c_0 = 0 \qquad (8.25)$$

(we typically assume $x_{-1} = 2x_0 - x_1$). The following example shows the application of quadratic spline interpolation using the B-splines.

Example 8.6   Obtain the quadratic spline which fits $f(x) = x^2$ sampled at points $x = 0, 1, 2$, and 3 assuming that the first segment is a straight line and estimate the function value at $x = 1.5$.

Solution

Here, the order m is 2 and the number of segments is 3. The interpolating spline is written as,

$$S^2(x) = \sum_{j=-2}^{2} c_j B_j^2(x)$$

The five coefficients are obtained by matching the four function values at the nodes [Eq. (8.24) for equidistant nodes] and the additional condition $c_{-2} - 2c_{-1} + c_0 = 0$ [from Eq. (8.25) for equidistant nodes]. These are written as,

$$\frac{1}{2}c_{-2} + \frac{1}{2}c_{-1} = 0$$

$$\frac{1}{2}c_{-1} + \frac{1}{2}c_0 = 1$$

$$\frac{1}{2}c_0 + \frac{1}{2}c_1 = 4$$

$$\frac{1}{2}c_1 + \frac{1}{2}c_2 = 9$$

$$c_{-2} - 2c_{-1} + c_0 = 0$$

and the solution is

$$c_{-2} = -1/2, c_{-1} = 1/2, c_0 = 3/2, c_1 = 13/2, c_2 = 23/2$$

The value of the B-splines at $x = 1.5$ is obtained by using the recursive relationship as shown in Table 8.2 (note the extension of domain on either side by using equispaced fictitious nodes).

Table 8.2

i	-2	-1	0	1	2	3	4
x_i	-2	-1	0	1	2	3	4
$B_i^0(1.5)$	0	0	0	1	0	0	0
$B_i^1(1.5)$	0	0	$0.5 \times 0 + 0.5 \times 1 = 0.5$	$0.5 \times 1 + 0.5 \times 0 = 0.5$	0	0	
$B_i^2(1.5)$	0	$0 + 0.25 \times 0.5$ $= 0.125$	$0.75 \times 0.5 + 0.75 \times 0.5$ $= 0.75$	$0.25 \times 0.5 + 0$ $= 0.125$	0		

The interpolated function value is therefore,

$$S^2(1.5) = \sum_{j=-2}^{2} c_j B_j^2(1.5) = \frac{1}{2} \times 0.125 + \frac{3}{2} \times 0.75 + \frac{13}{2} \times 0.125 = 2$$

Similarly, at $x = 2.5$, following an identical procedure, the interpolated value is

$$S^2(2.5) = \sum_{j=-2}^{2} c_j B_j^2(1.5) = \frac{3}{2} \times 0.125 + \frac{13}{2} \times 0.75 + \frac{23}{2} \times 0.125 = 6.5$$

These values match, as they should, with those obtained from the quadratic spline interpolation in Example 4.9 (see the dotted line in Fig. 4.9).

An alternative method of computing the interpolated value is to define a recursive relationship for the coefficients c as $c_i^{h-1} = l_i^h(x)c_i^h + \left[1 - l_i^h(x)\right]c_{i-1}^h$ and obtaining c_i^0.

8.3.1 Cubic Spline

As discussed earlier, the most commonly used spline is a cubic spline. To extend the B-splines to this case, we write,

$$S^3(x) = \sum_{j=-3}^{n-1} c_j B_j^3(x) \tag{8.26}$$

To obtain the coefficients, c_j, for interpolation with third-order B-splines, we note that the nodal values of this spline are given by,

$$B_i^3(x_j) = \begin{cases} \dfrac{(x_{i+1} - x_i)^2}{(x_{i+3} - x_i)(x_{i+2} - x_i)}, & \text{for } j = i+1 \\[2ex] \dfrac{(x_{i+4} - x_{i+2})(x_{i+2} - x_{i+1})}{(x_{i+4} - x_{i+1})(x_{i+3} - x_{i+1})} + \dfrac{(x_{i+2} - x_i)(x_{i+3} - x_{i+2})}{(x_{i+3} - x_i)(x_{i+3} - x_{i+1})}, & \text{for } j = i+2 \\[2ex] \dfrac{(x_{i+4} - x_{i+3})^2}{(x_{i+4} - x_{i+1})(x_{i+4} - x_{i+2})}, & \text{for } j = i+3 \\[2ex] 0, & \text{otherwise} \end{cases} \tag{8.27}$$

The equality of the interpolant and the function at the nodes then provides us the following relations:

$$\frac{(x_{j+1} - x_j)^2}{(x_{j+1} - x_{j-2})(x_{j+1} - x_{j-1})} c_{j-3} + \left[\frac{(x_{j+2} - x_j)(x_j - x_{j-1})}{(x_{j+2} - x_{j-1})(x_{j+1} - x_{j-1})} + \frac{(x_j - x_{j-1})(x_{j+1} - x_j)}{(x_{j+1} - x_{j-2})(x_{j+1} - x_{j-1})} \right] c_{j-2}$$

$$+ \frac{(x_j - x_{j-1})^2}{(x_{j+2} - x_{j-1})(x_{j+1} - x_{j-1})} c_{j-1} = f(x_j) \tag{8.28}$$

$$\text{for } j = 0, 1, 2, \ldots, n.$$

Thus there are $(n + 1)$ equations in $(n + 3)$ unknowns $(c_j, j = -3, -2, \ldots, n-1)$. We, therefore, have two degrees of freedom (compare with the discussion on cubic spline in Section 4.5) and may choose two of the coefficients arbitrarily. For the natural cubic spline, we should have

$$\left.\frac{dS^3(x)}{dx^2}\right|_{x=x_0^+} = \left.\frac{dS^3(x)}{dx^2}\right|_{x=x_n^-} = 0$$

Following a procedure similar to that used for quadratic B-splines, these boundary conditions reduce to,

$$(x_2 - x_{-1})c_{-3} - (x_2 + x_1 - x_{-1} - x_{-2})c_{-2} + (x_1 - x_{-2})c_{-1} = 0$$
$$(x_{n+2} - x_{n-1})c_{n-3} - (x_{n+2} + x_{n+1} - x_{n-1} - x_{n-2})c_{n-2} + (x_{n+1} - x_{n-2})c_{n-1} = 0 \quad (8.29)$$

The following example shows the application of cubic spline interpolation using the B-splines.

Example 8.7

Obtain the natural cubic spline which interpolates $f(x) = 1/(1 + x^2)$ sampled at points $x = -2, -1, 0, 1$, and 2. Estimate the function value at $x = 1.6$.

Solution

Here, the order m is 3 and the number of segments is 4.

The interpolating spline is written as,

$$S^3(x) = \sum_{j=-3}^{3} c_j B_j^3(x)$$

The seven coefficients are obtained by matching the five function values at the nodes [Eq. (8.28)] and the additional boundary conditions $c_{-3} - 2c_{-2} + c_{-1} = 0$ and $c_{n-3} - 2c_{n-2} + c_{n-1} = 0$ [Eq. (8.29)]. These are written as,

$$\frac{1}{6}c_{-3} + \frac{2}{3}c_{-2} + \frac{1}{6}c_{-1} = 0.2$$

$$\frac{1}{6}c_{-2} + \frac{2}{3}c_{-1} + \frac{1}{6}c_0 = 0.5$$

$$\frac{1}{6}c_{-1} + \frac{2}{3}c_0 + \frac{1}{6}c_1 = 1.0$$

$$\frac{1}{6}c_0 + \frac{2}{3}c_1 + \frac{1}{6}c_2 = 0.5$$

$$\frac{1}{6}c_1 + \frac{2}{3}c_2 + \frac{1}{6}c_3 = 0.2$$

$$c_{-3} - 2c_{-2} + c_{-1} = 0$$

$$c_1 - 2c_2 + c_3 = 0$$

and the solution is $c_{-3} = c_3 = 1/35$, $c_{-2} = c_2 = 7/35$, $c_{-1} = 13/35$, $c_0 = 46/35$. The value of the B-splines at $x = 1.6$ is obtained by using the recursive relationship as shown in Table 8.3.

Table 8.3

i	-3	-2	-1	0	1	2	3	4	5	6
x_i	-5	-4	-3	-2	-1	0	1	2	3	4
$B_i^0(1.6)$	0	0	0	0	0	0	1	0	0	0
$B_i^1(1.6)$	0	0	0	0	0	$0 + 0.4 \times 1 = 0.4$	$0.6 \times 1 + 0 = 0.6$	0	0	
$B_i^2(1.6)$	0	0	0	0	$0 + 0.2 \times 0.4$ $= 0.08$	$0.8 \times 0.4 + 0.7 \times 0.6$ $= 0.74$	$0.3 \times 0.6 + 0$ $= 0.18$	0		
$B_i^3(1.6)$	0	0	0	$0 + 0.4/3$ $\times 0.08 = 0.01067$	$2.6/3 \times 0.08$ $+ 1.4/3 \times 0.74$ $= 0.4147$	$1.6/3 \times 0.74$ $+ 0.8 \times 0.18$ $= 0.5387$	$0.2 \times 0.18 + 0$ $= 0.036$			

The interpolated function value is, therefore,

$$S^3(1.6) = \sum_{j=-3}^{3} c_j B_j^3(1.6) = \frac{46}{35} \times 0.01067 + \frac{13}{35} \times 0.4147$$

$$+ \frac{7}{35} \times 0.5387 + \frac{1}{35} \times 0.036 = 0.2768$$

which is the same as that obtained in Example 4.10.

The B-spline could be used to generate spline interpolation of any order. However, an alternative formulation, known as a *tension spline*, is often used to unify the concepts of linear and cubic splines through a *tension* parameter. This parameter indicates the force applied to the spline between any two knots. For high tension, we would get close to a linear spline and for small tension, we approach the cubic spline. The tension parameter, τ, is defined in such a way that over each interval, the interpolating function, which has C^2 continuity, satisfies the equation,

$$f^{iv}(x) = \tau^2 f''(x) \tag{8.30}$$

Obviously, when $\tau = 0$, the fourth derivative is equal to zero, and we get the cubic spline. For very large τ, the second derivative approaches zero, and we get piecewise linear spline. The general solution involves the tanh function and results in a *tridiagonal* system of equations. Similarly, a *taut spline* is sometimes used for the same purpose. These are not described here. Interested readers are referred to de Boor (2001).

8.4 MULTIVARIATE FUNCTIONS

Except for a few places (e.g., multiple linear regression, nonlinear simultaneous equations, partial differential equations), we have assumed the function to be dependent on a single variable only. A number of times, we would require to approximate, differentiate, or integrate, a function of several variables. Although the basic philosophy is similar to that for the function of a single variable, we did not discuss these cases earlier. In this section, we provide a brief discussion on how to extend the methods to multivariate functions.

8.4.1 Interpolation

Let the function $f(x_1, x_2, ..., x_n)$ be sampled at $(k + 1)$ locations $\{x\}_0, \{x\}_1, ..., \{x\}_k$ and the values be denoted by $f_0, f_1, ..., f_k$. We now need to interpolate the function value at a point (target) which does not coincide with the sample locations (nodes). Various options could be tried for this purpose. For example, probably the simplest way would be to use the *nearest* nodal value as the target value. However, this would not be very accurate. We may think of using a weighted average of function values at some neighbouring points (or may be all points) with the weight dependent on the distance between the target and the corresponding node (a weight proportional to the inverse of the square of the distance is very commonly used). However, if the nodes are arranged in a Cartesian grid, we may extend the methods discussed for a single variable (Chapter 4). Since most of the practical cases would involve a regular grid, we assume in this section that the nodes are not randomly spaced. Although extension to 3 or more dimensions is straightforward, we describe only two-dimensional cases here.

Let the nodes be located in a grid of $(k_1 + 1)$ columns and $(k_2 + 1)$ rows such that $x_1 = x_{1,0}, x_{1,1}, x_{1,2}, ..., x_{1,k1}$ and $x_2 = x_{2,0}, x_{2,1}, x_{2,2}, ..., x_{2,k2}$, and the function value at the node $(x_{1,i}, x_{2,j})$ be denoted by $f_{i,j}$. The two extremes of interpolation methods are (Fig. 8.5): (i) the minimum possible continuity approach (i.e., the nearest neighbour approach) which assumes the function to be constant within a neighbourhood of the nodes, and (ii) the maximum possible continuity approach which assumes the function to be of highest possible degree, i.e., of the form

$$\sum_{i=0}^{k_1} \sum_{j=0}^{k_2} c_{i,j} x_1^i x_2^j$$

As before, we assume polynomial interpolating functions only.

While the first approach is not very accurate, the second would be computationally intensive (the coefficients $c_{i,j}$ could be obtained by first performing a single variable interpolation along, say, x_1, using a k_1 degree polynomial. The resulting (k_1+1) coefficients would naturally be different along each of the k_2 rows.

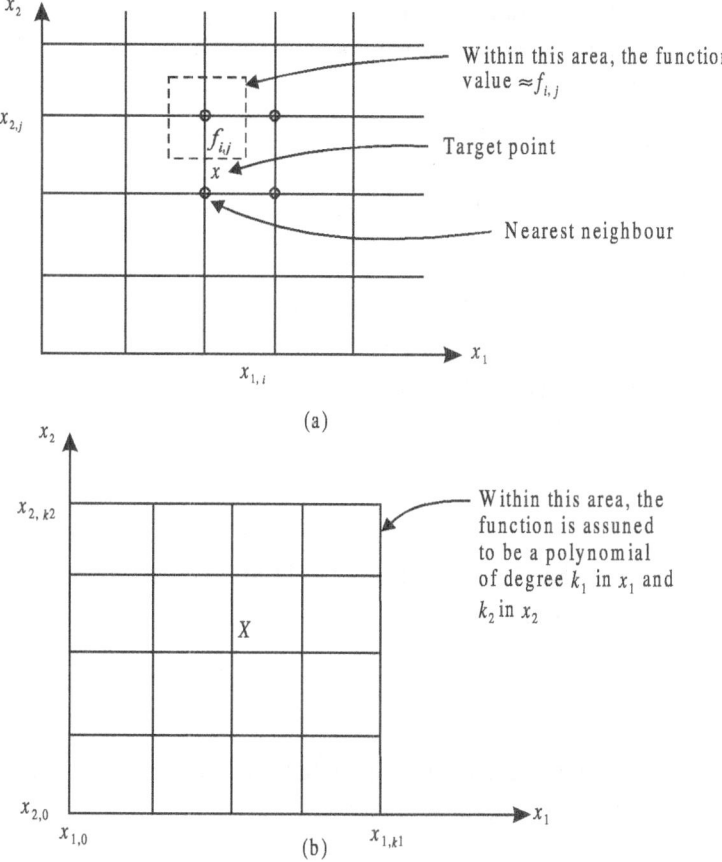

Fig. 8.5 Interpolation on two-dimensional regular grid.

An interpolation along x_2 would then result in expressions for these coefficients as k_2 degree polynomials in x_2. In case, we want only the function value, and not the interpolating polynomial expression, similar philosophy could be used to first interpolate the function values corresponding to the target x_1 along all rows and then performing a single interpolation along x_2. Therefore, typically we use a predefined order of polynomial depending on the requirements of accuracy and use the appropriate number of nodes *near* the target for interpolation.

Since the interpolation error is least near the middle of the range of node-span, it is preferred that the target point be close to the centre of the selected span (sometimes, the target point may be close to the boundary of the sampled area and it may not be possible to do so).

Although a constant function is the simplest interpolant, it implies discontinuity in the function values which is not desirable. Hence, most commonly, a linear interpolation in both directions (called a bilinear interpolation) is used. Clearly, it would need to use two nodes in each direction around the target point. Figure 8.5 shows the arrangement of the nodes and the target point. Performing linear interpolation along the rows, we get

$$\text{Along } x_2 = x_2^l : \quad f_1(x_1, x_2) = \frac{x_1^u f_{l,l} - x_1^l f_{u,l}}{x_1^u - x_1^l} + \frac{f_{u,l} - f_{l,l}}{x_1^u - x_1^l} x_1$$

$$\text{Along } x_2 = x_2^u : \quad f_1(x_1, x_2) = \frac{x_1^u f_{l,u} - x_1^l f_{u,u}}{x_1^u - x_1^l} + \frac{f_{u,u} - f_{l,u}}{x_1^u - x_1^l} x_1 \qquad (8.31)$$

in which the subscripts (and superscripts) l and u denote the lower and upper values, respectively. At this stage, it is convenient to define a local coordinate system as,

$$\xi = \frac{x_1 - x_1^l}{x_1^u - x_1^l} \quad \text{and} \quad \eta = \frac{x_2 - x_2^l}{x_2^u - x_2^l} \qquad (8.32)$$

such that these vary from 0 to 1 within the relevant domain. Equation (8.31) is then written as,

$$\text{Along } \eta = 0: \quad f_1(\xi, \eta) = f_{0,0} + \xi(f_{1,0} - f_{0,0})$$

$$\text{Along } \eta = 1: \quad f_1(\xi, \eta) = f_{0,1} + \xi(f_{1,1} - f_{0,1}) \qquad (8.33)$$

Linear interpolation of the coefficients then provides the final form of the interpolating linear polynomial as,

$$f_1(\xi, \eta) = \left[f_{0,0} + \eta(f_{0,1} - f_{0,0}) \right] + \xi \left[\begin{array}{c} (f_{1,0} - f_{0,0}) \\ + \eta(f_{1,1} - f_{0,1} - f_{1,0} + f_{0,0}) \end{array} \right]$$

$$= (1-\xi)(1-\eta) f_{0,0} + \xi(1-\eta) f_{1,0} + \xi\eta f_{1,1} + (1-\xi)\eta f_{0,1} \qquad (8.34)$$

Note that the expression is a two-dimensional counterpart of the Lagrange interpolating polynomial, and it is readily verified that these polynomials (i.e., the multiplying coefficients to each function value) are unity at the corresponding node and zero at all other nodes. Also, these vary linearly along the axis but quadratically along any other straight line. For example, $L_{0,0}$ would be equal to $1 - \xi$ along $\eta = 0$ and $1 - \eta$ along $\xi = 0$; and it would vary as $(1 - \xi)^2$ along the line $\xi = \eta$.

Example 8.8   Table 8.4 shows the values of a function at the nodes of a regular grid. Estimate the value of the function at the point (0.2, 0.3).

Table 8.4

x_2	x_1				
	-1	-0.5	0	0.5	1
1	0.0196078	0.0310078	0.0384615	0.0310078	0.0196078
0.5	0.0310078	0.0740741	0.137931	0.0740741	0.0310078
0	0.0384615	0.137931	1.000000	0.137931	0.0384615
-0.5	0.0310078	0.0740741	0.137931	0.0740741	0.0310078
-1	0.0196078	0.0310078	0.0384615	0.0310078	0.0196078

Solution

The simplest approach, using the nearest neighbour, results in the estimated value as that at the node (0,0.5), i.e., 0.137931. Note that, had the target point been at (0.25,0.3), we could choose the nearest node as either (0,0.5) or (0.5,0.5). Similarly, if the target point is (0.25,0.25), any of the four surrounding nodes could be chosen.

The polynomial interpolant of the highest possible degree could be written as

$$\sum_{i=0}^{4}\sum_{j=0}^{4} c_{i,j} x_1^i x_2^j$$

and the 25 coefficients could be obtained by the equality condition at the nodes. However, it would be simpler to perform unidirectional interpolations as described below.

Along the top and bottom rows ($x_2 = \pm 1$) (note the symmetry in the function), the interpolating polynomial is obtained as

$$f_{4,4}(x_1, x_2 = \pm 1) = 0.0384615 - 0.0334690 x_1^2 + 0.0146153 x_1^4$$

Similarly, the other rows provide the following:

$$f_{4,4}(x_1, x_2 = \pm 0.5) = 0.137931 - 0.304929 x_1^2 + 0.198006 x_1^4$$

$$f_{4,4}(x_1, x_2 = 0) = 1.00000 - 4.27719 x_1^2 + 3.31565 x_1^4$$

An interpolation on these coefficients then results in the final form of the interpolant as

$$f_{4,4}(x_1, x_2) = (1.00000 - 4.27719 x_2^2 + 3.31565 x_2^4)$$
$$+ (-4.27719 + 19.7708 x_2^2 - 15.5271 x_2^4) x_1^2$$
$$+ (3.31565 - 15.5271 x_2^2 + 12.2261 x_2^4) x_1^4$$

The interpolated value at the point (0.2,0.3) is computed as 0.540194.

If the estimate of the function value is required only at a single point, it would be more efficient to perform five interpolations along the rows to obtain the estimates at the points (0.2, ±1), (0.2, ±0.5), and (0.2, 0). An interpolation along this column would then provide the answer. The interpolated values along the rows are obtained as 0.0371462 at (0.2, ± 1), 0.126051 at (0.2, ± 0.5), and 0.834218 at (0.2,0). A fourth degree polynomial interpolation along this column provides the estimate at (0.2,0.3) as 0.540194 (same as above).

If we use the bilinear interpolation using the nodes (0,0), (0.5,0), (0.5,0.5), (0,0.5) to enclose the target point, the local coordinates of the target points would be (0.4, 0.6). The interpolated function value, using Eq. (8.34), is therefore,

$$0.6 \times 0.4 \times 1.00000 + 0.4 \times 0.4 \times 0.137931 + 0.4 \times 0.6 \times 0.074074$$
$$+ 0.6 \times 0.6 \times 0.137931 = 0.329502$$

The same answer could, of course, be obtained by performing linear interpolation along the two axis separately. For example, a linear interpolation along the x_1 axis results in the values 0.655172 at (0.2,0) and 0.112388 at (0.2,0.5). Linear interpolation along x_2 provides the value of 0.329502 at (0.2,0.3).

This example shows the two extremes of possible accuracy for interpolation with a given set of data points and also described an *intermediate* accuracy by using bilinear interpolation. Due to their computational simplicity and reasonable accuracy, the bilinear (or, more generally, multilinear) interpolation is the method of choice for multi-dimensional interpolation. However, as mentioned earlier, the function is assumed to vary linearly along each axis. This means that its derivative along the axis in any direction (derivative here implies partial derivative with respect to that direction) would be a constant within an element and, in general, would be discontinuous at the element boundaries. To achieve a higher continuity than the C^0 continuity achieved in bilinear interpolation, we may either use more number of grid points around the target point or use higher derivatives also at the four surrounding nodes (comparable to the Hermite interpolation, Section 4.5). Both these techniques are described hereunder.

8.4.1.1 *Use of More Points*

If we use a subgrid of $n1 + 1$ nodes along the x_1 direction and $n2 + 1$ along the x_2 direction, we could write an interpolating function

$$\sum_{i=0}^{n1} \sum_{j=0}^{n2} c_{i,j} x_1^i x_2^j$$

and proceed as described in Example 8.8. Another option could be to use a, say, cubic spline along each direction using a subset of the grid or even the entire grid. We prefer the first method which is computationally less intensive and illustrate its use by an example.

Example 8.9   For the function given in Example 8.8, estimate the value of the function at the point (0.2,0.3) by using a bi-quadratic interpolation.

Solution   We choose the sub-grid bounded by the lines $x_1 = -0.5$ and 0.5 and $x_2 = 0$ and 1.0 ($n1 = n2 = 2$). Note that these are chosen so as to be nearest to the target point.

Quadratic interpolations along $x_2 = 0$, 0.5, and 1.0, provide the function values as

0.862069 at (0.2,0), 0.127714 at (0.2,0.5), and 0.0372689 at (0.2,1)

Another quadratic interpolation results in an estimate of 0.344187 at (0.2,0.3).

8.4.1.2 *Use of Higher Derivatives*

If we use the same four nodes as in the bilinear interpolation, but have additional information about the function derivatives at these nodes, we could write an interpolating function

$$\sum_{i=0}^{n1} \sum_{j=0}^{n2} c_{i,j} x_1^i x_2^j$$

and obtain the coefficients by matching the function values and the derivatives at the nodes [compare with Hermite interpolation, Section 4.5, especially Eqs (4.48) and (4.49)]. The derivatives may be available analytically or we may compute them numerically from the known function values at the grid points.

The degrees of the interpolating polynomial, $n1$ and $n2$, naturally depend on the amount of information available to us. For example, if we want to perform a bi-quadratic interpolation ($n1 = n2 = 2$), we would have nine coefficients implying that five additional (other than the function values at the four nodes) *values* are needed. These values may be in the form of the partial derivatives (with respect to, say, x_1) at each node and another partial derivative (say, with respect to x_2) at one of the nodes. However, one should prefer a *symmetrical* scheme and, a bi-cubic ($n1 = n2 = 3$) interpolation is the method of choice.

Since there are 16 coefficients, 12 additional conditions are needed. Typically, these are in terms of $\partial f/\partial x_1$, $\partial f/\partial x_2$, and $\partial^2 f/\partial x_1 \partial x_2$ at each of the four nodes. We then write the interpolating bi-cubic polynomial in terms of the Hermite polynomials as (note that we have switched to the local coordinates, ξ and η),

$$f_{3,3} = \sum_{i=1}^{4} H_{00,i} f_i + H_{10,i} \left.\frac{\partial f}{\partial \xi}\right|_i + H_{01,i} \left.\frac{\partial f}{\partial \eta}\right|_i + H_{11,i} \left.\frac{\partial^2 f}{\partial \xi \partial \eta}\right|_i \qquad (8.35)$$

where i represents the nodal point (we follow the convention of taking the first point, $i = 1$, at the bottom left corner, i.e., $\xi = \eta = 0$, and then moving counter clockwise such that $i = 4$ corresponds to $\xi = 0$, $\eta = 1$). The H's are the Hermite polynomials, all of which are bi-cubic and follow constraints of the value or its derivative being 1 at the corresponding node for the appropriate function / derivative and 0 at other nodes and for other function / derivatives. For example, $H_{00,1}$ would be 1 at the first node and 0 at other three nodes and its partial derivatives, $\partial H_{00,1}/\partial \xi$, $\partial H_{00,1}/\partial \eta$, and $\partial^2 H_{00,1}/\partial \xi \partial \eta$, would be 0 at all four nodes. This provides us with 16 linear equations to obtain the 16 coefficients.

Similarly, $H_{11,4}$, $\partial H_{11,4}/\partial \xi$, and $\partial H_{11,4}/\partial \eta$, would be zero at all four nodes, and $\partial^2 H_{11,4}/\partial \xi \partial \eta$ would be 1 at the fourth node and 0 at other three nodes. To illustrate the method, we describe the determination of $H_{00,3}$ (note that Node 3 corresponds to $\xi = \eta = 1$). Others are obtained similarly, though the process is rather tedious. Assuming

$$H_{00,3} = \sum_{i=0}^{3} \sum_{j=0}^{3} c_{ij} \xi^i \eta^j$$

the 16 equations are written as (in each row, the four equations result by application at Nodes 1, 2, 3, and 4, respectively),

For f: $c_{00} = 0; \quad \sum_{i=0}^{3} c_{i0} = 0; \quad \sum_{i=0}^{3} \sum_{j=0}^{3} c_{ij} = 1; \quad \sum_{j=0}^{3} c_{0j} = 0$

For $\partial f / \partial \xi$: $c_{10} = 0; \quad \sum_{i=1}^{3} i c_{i0} = 0; \quad \sum_{i=1}^{3} \sum_{j=0}^{3} i c_{ij} = 0; \quad \sum_{j=0}^{3} c_{1j} = 0$

For $\partial f / \partial \eta$: $c_{01} = 0; \quad \sum_{i=0}^{3} c_{i1} = 0; \quad \sum_{i=0}^{3} \sum_{j=1}^{3} j c_{ij} = 0; \quad \sum_{j=1}^{3} j c_{0j} = 0$ $\qquad (8.36)$

For $\partial^2 f / \partial \xi \partial \eta$: $c_{11} = 0; \quad \sum_{i=1}^{3} i c_{i1} = 0; \quad \sum_{i=1}^{3} \sum_{j=1}^{3} i j c_{ij} = 0; \quad \sum_{j=1}^{3} j c_{1j} = 0$

which provide the solution $c_{22} = 9, c_{23} = c_{32} = -6, c_{33} = 4$ and all other coefficients as zero. We, therefore, have

$$H_{00,3} = 9\xi^2\eta^2 - 6\xi^2\eta^3 - 6\xi^3\eta^2 + 4\xi^3\eta^3 \qquad (8.37)$$

It is straightforward to verify that this polynomial is 1 at Node 3 and 0 at the other three nodes and that its relevant partial derivatives are 0 at all four nodes. All the Hermite polynomials are listed below:

$$H_{00,1} = (1 - 3\xi^2 + 2\xi^3)(1 - 3\eta^2 + 2\eta^3) \ ; \ H_{00,2} = (3\xi^2 - 2\xi^3)(1 - 3\eta^2 + 2\eta^3)$$

$$H_{00,3} = (3\xi^2 - 2\xi^3)(3\eta^2 - 2\eta^3) \qquad\quad ; \ H_{00,4} = (1 - 3\xi^2 + 2\xi^3)(3\eta^2 - 2\eta^3)$$

$$H_{10,1} = (\xi - 2\xi^2 + \xi^3)(1 - 3\eta^2 + 2\eta^3) \ ; \ H_{10,2} = (-\xi^2 + \xi^3)(1 - 3\eta^2 + 2\eta^3)$$

$$H_{10,3} = (-\xi^2 + \xi^3)(3\eta^2 - 2\eta^3) \qquad\quad ; \ H_{10,4} = (\xi - 2\xi^2 + \xi^3)(3\eta^2 - 2\eta^3) \qquad (8.38)$$

$$H_{01,1} = (1 - 3\xi^2 + 2\xi^3)(\eta - 2\eta^2 + \eta^3) \ ; \ H_{01,2} = (3\xi^2 - 2\xi^3)(\eta - 2\eta^2 + \eta^3)$$

$$H_{01,3} = (3\xi^2 - 2\xi^3)(-\eta^2 + \eta^3) \qquad\quad ; \ H_{01,4} = (1 - 3\xi^2 + 2\xi^3)(-\eta^2 + \eta^3)$$

$$H_{11,1} = (\xi - 2\xi^2 + \xi^3)(\eta - 2\eta^2 + \eta^3) \ ; \ H_{11,2} = (-\xi^2 + \xi^3)(\eta - 2\eta^2 + \eta^3)$$

$$H_{11,3} = (-\xi^2 + \xi^3)(-\eta^2 + \eta^3) \qquad\quad ; \ H_{11,4} = (\xi - 2\xi^2 + \xi^3)(-\eta^2 + \eta^3)$$

The following example describes the methodology.

Example 8.10 For the function given in Example 8.8, estimate the value of the function at the point (0.2,0.3) using a bi-cubic interpolation by estimating the partial derivatives of the function at the four nodes with central difference scheme.

Solution Table 8.5 shows the central difference approximations of the partial derivatives at the four nodes:

Table 8.5

Node	$\partial f / \partial x_1$	$\partial f / \partial x_2$	$\partial^2 f / \partial x_1 \partial x_2$
(0, 0)	0	0	0
(0.5, 0)	– 0.961538	0	0
(0.5, 0.5)	– 0.106923	– 0.106923	0.942685
(0, 0.5)	0	– 0.961538	0

Since we need the partial derivatives with respect to the local coordinates, we use the fact that $\partial / \partial \xi = \Delta x_1 \partial / \partial x_1$ and $\partial / \partial \eta = \Delta x_2 \partial / \partial x_2$ in which Δx_1 and Δx_2 represent the grid size along the x_1 and x_2 axis (here, 0.5). The values of the Hermite interpolating polynomials at the point (0.2,0.3), i.e., $\xi = 0.4$, $\eta = 0.6$, is obtained from Eq. (8.38) as,

$$H_{00,1} = 0.228096; \ H_{00,2} = 0.123904; \ H_{00,3} = 0.228096; \ H_{00,4} = 0.419904$$

$$H_{10,1} = 0.050688; \ H_{10,2} = -0.033792; \ H_{10,3} = -0.062208; \ H_{10,4} = 0.093312$$

$$H_{01,1} = 0.062208; \ H_{01,2} = 0.033792; \ H_{01,3} = -0.050688; \ H_{01,4} = -0.093312$$

$$H_{11,1} = 0.013824; \ H_{11,2} = -0.009216; \ H_{11,3} = 0.013824; \ H_{11,4} = -0.020736$$

The interpolated function value at (0.2,0.3) is then obtained as,

$$(0.228096 \times 1) + (0.123904 \times 0.137931)$$

$$+ (0.228096 \times 0.0740741) + (0.419904 \times 0.137931)$$
$$+ (0.033792 \times 0.5 \times 0.961538) + (0.062208 \times 0.5 \times 0.106923)$$
$$+ (0.050688 \times 0.5 \times 0.106923) + (0.093312 \times 0.5 \times 0.961538)$$
$$+ (0.013824 \times 0.25 \times 0.942685) = 0.390401$$

8.4.2 Integration

Multivariable integration of a function is obviously more computationally intensive, since the number of function evaluations increases significantly. Moreover, whereas for a single variable the limits of integration are just two points, for multi dimensional integrals, these would, in general, be lines or surfaces. Sometimes, it is possible to reduce a multi-dimensional integral to a one-dimensional integral (e.g., integration of a radially symmetric function over a circular area). We assume that for the given function, it is not possible to do so.

As discussed in Chapter 5, the function to be integrated may not be known and could be given in a discrete form by sampling at a few locations. Or, the function may be known and we could be free to decide where to sample it. For convenience, we discuss only two-variable cases though extension to higher dimensions is straightforward. We denote the required integral by,

$$I = \int_{x_{1l}}^{x_{1u}} \int_{x_{2l}(x_1)}^{x_{2u}(x_1)} f(x_1, x_2) \, dx_2 dx_1$$

We assume that the region of integration is convex and simply connected (see Fig. 8.6). If it is not, methods based on random sampling would probably be the best choice.

8.4.2.1 *Discrete Case*

We assume that the domain of the integral is rectangular (Cartesian for higher dimensions) and the function is sampled at the nodes of a regular grid *spanning the entire domain* (Fig. 8.6). The integral which in one-dimensional case represented area under the function within the integration domain, now represents the volume enclosed by the function. The mid-point rule may be used to approximate this volume as the sum of the volumes of rectangular prisms over each element. In other words, we write,

$$\tilde{I} = \sum_{i=1}^{n1} \sum_{j=1}^{n2} \overline{f}_{i,j}(x_{1,i} - x_{1,i-1})(x_{2,j} - x_{2,j-1}) \qquad (8.39)$$

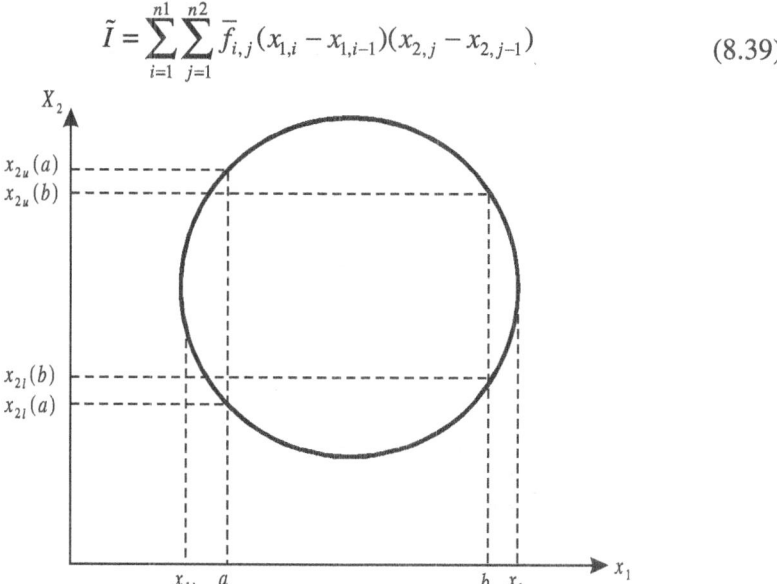

Fig. 8.6 Integration over a two-dimensional region.

in which $\overline{f}_{i,j}$ represents the average value of the function over the grid element bounded by $x_{1,i-1}, x_{1,i}, x_{2,j-1}$, and $x_{2,j}$, and could be taken as the arithmetic mean of the nodal values, $(f_{i-1,j-1} + f_{i,j-1} + f_{i,j} + f_{i-1,j})/4$. A higher-degree method could be obtained by using the bilinear interpolation [Eq. (8.34)] and integrating over an element to obtain,

$$\tilde{I} = \sum_{i=1}^{n1} \sum_{j=1}^{n2} \left(\begin{array}{l} f_{i-1,j-1}\int_0^1\int_0^1 (1-\xi)(1-\eta)d\xi d\eta + f_{i,j-1}\int_0^1\int_0^1 \xi(1-\eta)d\xi d\eta \\[2mm] + f_{i,j}\int_0^1\int_0^1 \xi\eta d\xi d\eta + f_{i-1,j}\int_0^1\int_0^1 (1-\xi)\eta d\xi d\eta \end{array} \right)$$
$$(x_{1,i} - x_{1,i-1})(x_{2,j} - x_{2,j-1}) \qquad (8.40)$$

which results in the same expression as the mid-point method (analogous to the one-dimensional equivalence of the mid-point rule and the trapezoidal method). Higher-order interpolation, as discussed in the previous subsection, may be used to improve the accuracy but becomes computationally intensive. If the nodal values of the derivatives are available, integration of the Hermite polynomials [Eq. (8.38)] could be performed to obtain,

$$
\tilde{I} = \sum_{i=1}^{n1}\sum_{j=1}^{n2}(x_{1,i}-x_{1,i-1})(x_{2,j}-x_{2,j-1})
\begin{bmatrix}
\dfrac{1}{4}(f_{i-1,j-1}+f_{i,j-1}+f_{i,j}+f_{i-1,j})+ \\[2ex]
\dfrac{x_{1,i}-x_{1,i-1}}{24}
\left(
\begin{array}{l}
\left.\dfrac{\partial f}{\partial x_1}\right|_{i-1,j-1} - \left.\dfrac{\partial f}{\partial x_1}\right|_{i,j-1} \\[2ex]
-\left.\dfrac{\partial f}{\partial x_1}\right|_{i,j} + \left.\dfrac{\partial f}{\partial x_1}\right|_{i-1,j}
\end{array}
\right)+ \\[3ex]
\dfrac{x_{2,j}-x_{2,j-1}}{24}
\left(
\begin{array}{l}
\left.\dfrac{\partial f}{\partial x_2}\right|_{i-1,j-1} + \left.\dfrac{\partial f}{\partial x_2}\right|_{i,j-1} \\[2ex]
-\left.\dfrac{\partial f}{\partial x_2}\right|_{i,j} - \left.\dfrac{\partial f}{\partial x_2}\right|_{i-1,j}
\end{array}
\right)+ \\[3ex]
\dfrac{(x_{1,i}-x_{1,i-1})(x_{2,j}-x_{2,j-1})}{144}
\left(
\begin{array}{l}
\left.\dfrac{\partial^2 f}{\partial x_1 \partial x_2}\right|_{i-1,j-1} \\[2ex]
-\left.\dfrac{\partial^2 f}{\partial x_1 \partial x_2}\right|_{i,j-1} \\[2ex]
+\left.\dfrac{\partial^2 f}{\partial x_1 \partial x_2}\right|_{i,j} \\[2ex]
-\left.\dfrac{\partial^2 f}{\partial x_1 \partial x_2}\right|_{i-1,j}
\end{array}
\right)
\end{bmatrix}
$$

$$(8.41)$$

For Cartesian grid, the desired integral may be easily evaluated by successive applications of one-dimensional methods. For example, we may write,

$$
I = \int_{x_{2l}}^{x_{2u}} \int_{x_{1l}}^{x_{1u}} f(x_1, x_2)\, dx_1 dx_2
$$

as
$$I = \int_{x_{2l}}^{x_{2u}} g(x_2)\,dx_2 \ \text{ with } g(x_2) = \int_{x_{1l}}^{x_{1u}} f(x_1, x_2)\,dx_1$$

and first evaluate $g(x_2)$ at each x_2 by performing a one-dimensional integration (using, say, Simpson's rule) along x_1. (If we are free to choose the location of the sampling points, we may be able to use the Gauss quadrature to obtain higher accuracy for the same number of sampling points by locating the sampling points at the zeros of the appropriate Legendre polynomial.) Another one-dimensional integration of these $g(x_2)$ values along x_2 provides the required estimate. If the boundary is not regular, the integration limits on, say, x_2, would be a function of x_1. Even for this case, successive one-dimensional integration may be used if the sample grid is regular along x_2 (i.e., the sampling locations correspond to a few selected values of x_2, although the x_1 ordinates may be arbitrarily located, see Fig. 8.7). If the function is sampled at the nodes of a non-regular grid and / or the boundary of the domain of integration is irregular, a triangular grid could be tried with the 'mid-point' rule.

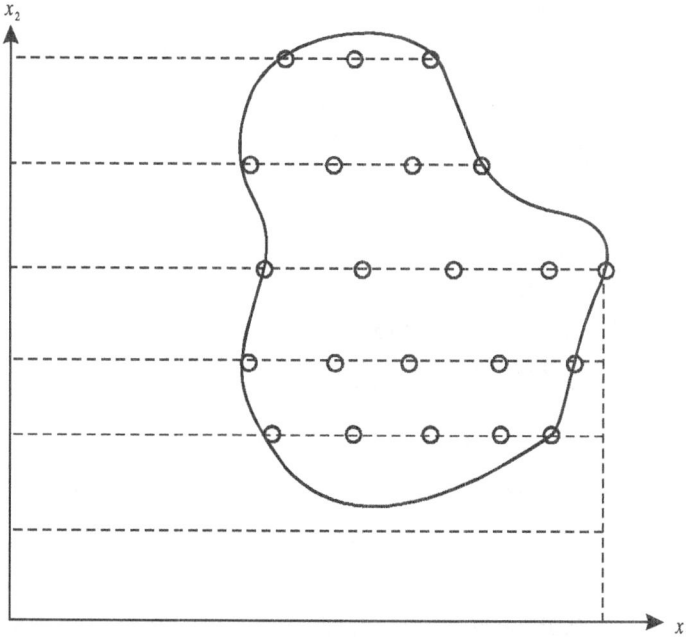

Fig. 8.7 Integration over a two-dimensional region with irregular boundary.

The following example illustrates the application of these methods.

Example 8.11 The table in Example 8.8 shows the values of a function, $f(x_1, x_2)$ at the nodes of a regular grid. Estimate the value of the integral

$$\int_{-1}^{1}\int_{-1}^{1} f(x_1, x_2)\, dx_1 dx_2$$

If additional information about the function derivatives is available as shown in the tables below, what would be the estimated value of the integral?

Values of $\partial f(x_1, x_2) \big/ \partial x_1$

x_2	x_1				
	-1	-0.5	0	0.5	1
1	0.0192234	0.0240370	0.0000000	-0.0240370	-0.0192234
0.5	0.0480740	0.137174	0.0000000	-0.137174	-0.0480740
0	0.0739645	0.475624	0.0000000	-0.475624	-0.0739645
-0.5	0.0480740	0.137174	0.0000000	-0.137174	-0.0480740
-1	0.0192234	0.0240370	0.0000000	-0.0240370	0.0192234

Values of $\partial f(x_1, x_2) \big/ \partial x_2$

x_2	x_1				
	-1	-0.5	0	0.5	1
1	-0.0192234	-0.0480740	-0.0739645	-0.0480740	-0.0192234
0.5	-0.0240370	-0.137174	-0.475624	-0.137174	-0.0240370
0	0.0000000	0.0000000	0.0000000	0.0000000	0.0000000
-0.5	0.0240370	0.137174	0.475624	0.137174	0.0240370
-1	0.0192234	0.0480740	0.0739645	0.0480740	0.0192234

Values of $\partial^2 f(x_1, x_2) \big/ \partial x_1 \partial x_2$

x_2	x_1				
	-1	-0.5	0	0.5	1
1	-0.0376929	-0.0745334	0.0000000	0.0745334	0.0376929
0.5	-0.0745334	-0.508053	0.0000000	0.508053	0.0745334
0	0.0000000	0.0000000	0.0000000	0.0000000	0.0000000
-0.5	0.0745334	0.508053	0.0000000	-0.508053	-0.0745334
-1	0.0376929	0.0745334	0.0000000	-0.0745334	-0.0376929

Solution   When only the function value is known, we may use the mid-point method (or the bilinear interpolation which provides the same estimate) to estimate the integral as,

$$\tilde{I} = \int\limits_{-1}^{1}\int\limits_{-1}^{1} f(x_1, x_2)\,dx_1 dx_2 = \sum_{i=1}^{4}\sum_{j=1}^{4} \frac{f_{i-1,j-1} + f_{i,j-1} + f_{i,j} + f_{i-1,j}}{4}(x_{1,i} - x_{1,i-1})(x_{2,j} - x_{2,j-1})$$

For constant grid size, as used in this case, this amounts to,

$$\tilde{I} = \Delta x_1 \Delta x_2 \left[\frac{1}{4}\sum f(\text{corner nodes}) + \frac{1}{2}\sum f(\text{other boundary nodes})\right] + \sum f(\text{internal nodes})$$

and provides the estimate as 0.517146.

Application of the Simpson's 1/3 rule along the x_1 direction provides the following estimates for

$$g(x_2) = \int\limits_{-1}^{1} f(x_1, x_2)\,dx_1$$

$g(1) = g(-1) = 0.0607000$; $g(-0.5) = g(0.5) = 0.155078$; and $g(0) = 0.530062$.

Another application in the x_2 direction provides the estimate of the desired integral as (taking advantage of the symmetry of the function),

$$\tilde{I} = \int\limits_{-1}^{1}\int\limits_{-1}^{1} f(x_1, x_2)\,dx_1 dx_2$$

$$= \int\limits_{-1}^{1} g(x_2)\,dx_2 = 2 \times \frac{0.5}{3}(0.060700 + 4 \times 0.155078 + 0.530062)$$

$= 0.403692$ Note the difference between the two estimates (the true value of the integral is 0.436193).

Using the Hermite interpolation and its integration [Eq. (8.41)], it is readily seen that for equal node spacing, we get

$\tilde{I} = $ Mid-point estimate

$$+ \frac{\Delta x_1^2 \Delta x_2}{24}\left[\frac{1}{4}\sum \begin{array}{l} \dfrac{\partial f}{\partial x_1}(\text{left boundary corner nodes}) \\ + \dfrac{1}{2}\sum \dfrac{\partial f}{\partial x_1}(\text{other left boundary nodes}) \end{array} - \frac{1}{4}\sum \begin{array}{l} \dfrac{\partial f}{\partial x_1}(\text{right boundary corner nodes}) \\ - \dfrac{1}{2}\sum \dfrac{\partial f}{\partial x_1}(\text{other right boundary nodes}) \end{array}\right]$$

$$+ \frac{\Delta x_1 \Delta x_2^2}{24} \left[\frac{1}{4} \sum \begin{array}{l} \dfrac{\partial f}{\partial x_2} \text{(lower boundary corner nodes)} \\[1.2em] +\dfrac{1}{2} \sum \dfrac{\partial f}{\partial x_2} \text{(other lower boundary nodes)} \\[1.2em] \dfrac{\partial f}{\partial x_2} \text{(upper boundary corner nodes)} \\[1.2em] -\dfrac{1}{2} \sum \dfrac{\partial f}{\partial x_2} \text{(other upper boundary nodes)} \end{array} \right.$$

$$+ \frac{\Delta x_1^2 \Delta x_2^2}{144} \left[\frac{1}{4} \sum \begin{array}{l} \dfrac{\partial^2 f}{\partial x_1 \partial x_2} \text{(lower left corner node)} \\[1.2em] +\dfrac{1}{4} \sum \dfrac{\partial^2 f}{\partial x_1 \partial x_2} \text{(upper right corner node)} \\[1.2em] \dfrac{\partial^2 f}{\partial x_1 \partial x_2} \text{(lower right corner node)} \\[1.2em] -\dfrac{1}{4} \sum \dfrac{\partial^2 f}{\partial x_1 \partial x_2} \text{(upper left corner node)} \end{array} \right.$$

and its value as 0.525100.

8.4.2.2 *Continuous Case*

If the function to be integrated is known in an analytic form, we could use any of the methods described in the previous sub-section by generating a grid of nodal values. The successive applications of one-dimensional Gauss quadrature, if possible to apply, would probably be the best. Sometimes, however, the integration domain may be so complex as to preclude the application of any of the previously discussed methods. The Monte Carlo method, based on relating the integral to the probability, could be applied in such cases. In this method, a volume (for two-dimensional integrals) is selected which encloses the volume represented by the desired integral.

For example, taking the minimum and maximum values of x_1 and x_2, and the minimum and maximum values of the function over this rectangle, we obtain a rectangular prism which will completely contain the desired volume. All that is left to do now is to generate a large number of points randomly and see whether they fall *inside the integral* or not. The fraction of the points lying within the integral multiplied by the volume of the encompassing rectangular prism provides an estimate of the required integral.

8.5 FAST FOURIER TRANSFORM

In Section 4.7, we discussed the interpolation of a periodic function, sampled at a uniform interval over a time period, using the partial sum of a Fourier series [see Eq. (4.69)]. Various issues related to this method were deliberately avoided at that time since these are slightly advanced. For example, *how closely does the interpolation represent the sampled function, is the sampling interval adequate to capture the essence of the function, is there a better way to approximate the data, what happens when the function is complex rather than real?* In this section, we look at some of these questions, introduce the *discrete fourier transform* (DFT), and describe an efficient way of evaluation of the DFT using the *fast fourier transform* (FFT).

We use the same notation as in Section 4.7 and assume that we have sampled a *real* function at $(n + 1)$ points,

$$x_l = -\pi + \frac{2\pi l}{n+1}, \quad l = 0, 1, 2, ..., n$$

As seen in Section 4.7, we may interpolate this data by using a partial sum of a Fourier series including terms up to mth harmonic (m is equal to $n/2$ or, if n is odd, $(n-1)/2$). Through an example, we also saw that the terms of these series were different from the Fourier series of the actual function. However, if the Fourier series of the function contained frequencies only up to the mth harmonic, the interpolation would reproduce the function *exactly*.

The next higher frequency, i.e., $(m + 1)/2\pi$, is called the *Nyquist critical frequency*, f_c, such that if the function is *band-limited* to this frequency, i.e., all components with frequencies f_c and higher vanish, it would be *completely determined* by the discretely sampled values. (For a function which has complex values, we need to use *negative frequencies* also in the Fourier transform or Fourier series. In that case, the Nyquist critical frequency is described in terms of its magnitude and the function would be captured completely if its bandwidth is limited to frequencies between $-f_c$ and f_c, i.e., all components with frequencies greater than or equal to f_c and less than or equal to $-f_c$ vanish.) Clearly, f_c depends on the sampling interval, $2\pi/(n + 1)$, and is obtained from,

$$f_c = \frac{m+1}{2\pi} = \frac{n+2}{2 \times 2\pi}, \text{ for even } n$$

$$= \frac{n+1}{2 \times 2\pi}, \text{ for odd } n$$

Thus, for odd n, f_c is half of the inverse of the sampling interval and; for even n, it is slightly larger. However, considering the definition of f_c and the fact that for even n, the mth harmonic would have a lower frequency than $(n + 1)/2\pi$, it is customary to use f_c as half of the inverse of the sampling interval (i.e., half the sampling frequency) for all cases. If the function does contain frequencies higher than the critical frequency, these higher frequencies are manifested as lower frequency terms, a phenomenon known as *aliasing* as depicted in the example below.

Example 8.12   A function is periodic with a time period of 5 sec and is given as,

$$f(t) = -0.064t^3 + 0.64t^2 - 1.6t + 1, \quad t = 0 \text{ to } 5 \text{ sec}$$

It is sampled at a 1 sec interval (t = 0, 1, 2, 3, 4) under the following conditions:

(a) after passing it through a filter which eliminates all components with frequency equal to or higher than the critical frequency,

(b) after passing it through a filter which retains an additional frequency over that of signal in (a), and

(c) with no filtering.

Find the interpolating trigonometric series in each case and comment.

Solution   We first transform the function from period $(0,5)$ to $(-\pi,\pi)$ by using $t = 5/2(x/\pi + 1)$ as,

$$f(x) = -\left(\frac{x}{\pi}\right)^3 + \left(\frac{x}{\pi}\right)^2 + \frac{x}{\pi}; \qquad x = -\pi \text{ to } \pi \text{ radians}$$

The Fourier series is obtained as (see Chapter 4)

$$f(x) = \frac{1}{3} + \sum_{j=1}^{\infty} \left[\frac{4(-1)^j}{j^2\pi^2} \cos jx + \frac{12(-1)^{j+1}}{j^3\pi^3} \sin jx \right]$$

Since the sampling interval is 1 sec, the Nyquist critical frequency would be 1/2 Hz. Or, since the sampling frequency is 5 per time period, the critical frequency would be 5/2 cycles per time period.

Thus in (a), we truncate the Fourier series at $j = 2$ and in (b), we use terms up to $j = 3$. For (c), of course, we use the entire series which represents the given function itself.

(a) We use,

$$f_a(x) = \frac{1}{3} - \frac{4}{\pi^2}\cos x + \frac{12}{\pi^3}\sin x + \frac{1}{\pi^2}\cos 2x - \frac{3}{2\pi^3}\sin 2x$$

and obtain the following values:

x	$-\pi$	$-3\pi/5$	$-\pi/5$	$\pi/5$	$3\pi/5$
$f_a(x)$	0.839939	-0.0199092	-0.144713	0.218235	0.773114

Writing the interpolating polynomial as,

$$f_2(x) = \frac{c_0}{2} + \sum_{j=1}^{2}(c_j \cos jx + s_j \sin jx)$$

we obtain the coefficients as, $c_0 = 0.666667$, $c_1 = -0.405285$, $s_1 = 0.387018$, $c_2 = 0.101321$, and $s_2 = -0.0483773$, which exactly match the function.

(b) We use,

$$f_b(x) = \frac{1}{3} - \frac{4}{\pi^2}\cos x + \frac{12}{\pi^3}\sin x + \frac{1}{\pi^2}\cos 2x$$

$$- \frac{3}{2\pi^3}\sin 2x - \frac{4}{9\pi^2}\cos 3x + \frac{4}{9\pi^3}\sin 3x$$

and get the following values:

x	$-\pi$	$-3\pi/5$	$-\pi/5$	$\pi/5$	$3\pi/5$
$f_b(x)$	0.884971	-0.0479152	-0.144430	0.245783	0.728258

The interpolating polynomial is again written as in (a) and we get $c_0 = 0.666667$, $c_1 = -0.405285$, $s_1 = 0.387018$, $c_2 = 0.146353$, and $s_2 = -0.0340433$

Note that the fundamental frequency is not affected but the amplitude of the second harmonic has changed. It is readily seen that c_2 is the sum of the coefficient of $\cos(2x)$ term and (negative of) the coefficient of $\cos(3x)$ term. Similarly, s_2 is the sum of the coefficient of $\sin(2x)$ term and the coefficient of $\sin(3x)$ term. In other words, the third harmonic manifests itself as a second harmonic if five sampling point per time period are chosen. This can be explained by the graphs of sine and cosine functions as shown in Fig. 8.8.

Thus, at the sampling points ($x/\pi = -1$, -0.6, -0.2, 0.2, and 0.6), $\sin(2x)$ and $\sin(3x)$ are equal while $\cos(2x)$ and $\cos(3x)$ have the same magnitude but are opposite in sign (it is also clear that at points other than the sampling points, the two frequencies components are not equal). Though not shown here, the same

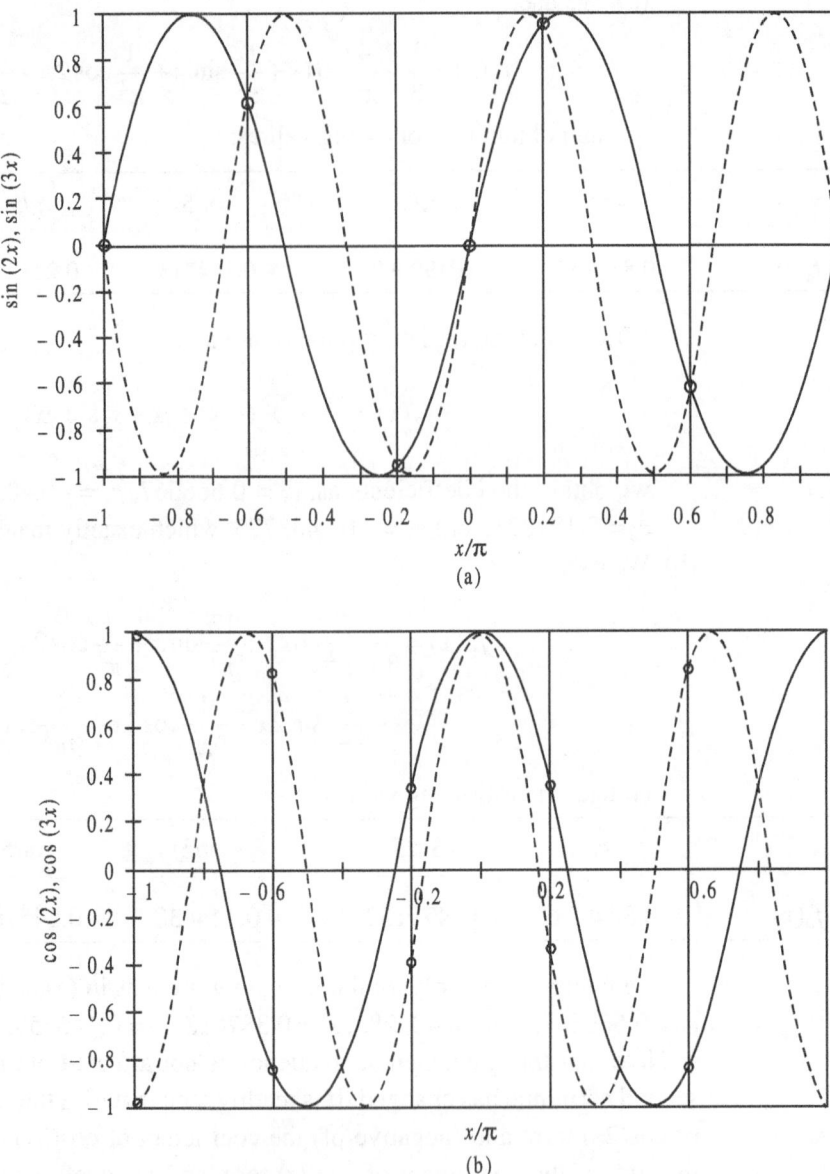

Fig. 8.8 Plots of sine and cosine of $2x$ and $3x$ showing the aliasing effect.

thing holds for the arguments x and $4x$; and 0 and $5x$. For higher frequencies, the *aliased* frequency is one of these three such that the magnitudes of the sine and cosine terms are equal at the sampling locations but the signs may be same or opposite (e.g., $6x$ is aliased to x and $7x$ is aliased to $2x$ such that both sine and cosine are opposite in sign, while $8x$ is aliased to $2x$ such that cosine has same sign but sine is opposite in sign. If we take the interval as $(0, 2\pi)$ instead of $(-\pi, \pi)$, we observe the same behaviour with respect to the magnitude but the signs of various terms may be different. This is sometimes known as the *folding* of frequencies higher than the critical frequency onto those smaller than critical (and the Nyquist frequency is, therefore, also called *folding frequency*).

(c) If we use the entire Fourier series, i.e., the exact function, we get the following:

x	$-\pi$	$-3\pi/5$	$-\pi/5$	$\pi/5$	$3\pi/5$
$f(x)$	1.000	−0.024	−0.152	0.232	0.744

The coefficients of the interpolating trigonometric series are obtained as: $c_0 = 0.72$, $c_1 = -0.463108$, $s_1 = 0.382448$, $c_2 = 0.176892$, and $s_2 = -0.0344853$.

Figure 8.9 shows the function and the interpolant for each of these cases.

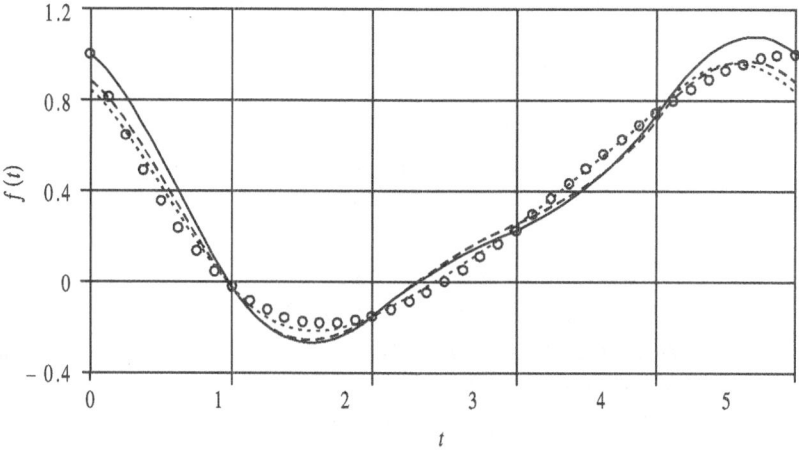

Fig. 8.9 Interpolation using the data generated by Fourier series up to second harmonic (dotted line), up to third harmonic (dashed line) and the entire function (solid line). The circles represent the exact function value.

We see that the interpolation using the function values at five points has significant errors implying that we need to reduce the sampling interval to get reasonably close to the function. However, this would increase the computational burden, since the computational effort is of the order of n^2 [see Eqs (4.69) and (4.70) in which evaluation of each coefficient requires $O(n)$ multiplications]. Efforts have been made (the earliest by Gauss) to take advantage of symmetries in order to increase the computational efficiency, and here we describe one such method of *fast Fourier transform* (FFT). To do that, we would have a re-look at the interpolation problem and cast it in terms of a *discrete Fourier transform* (DFT).

8.5.1 Discrete Fourier Transform

Equation (B4.5.3) expresses the function in the complex form of the Fourier series and the coefficients are expressed as complex integrals involving the function. Now, if the function is not known but its values are known at sampling interval of, say Δt, we may approximate these integrals by using a quadrature scheme and estimate the Fourier coefficients.

For a function sampled at $t_l = l\Delta t$; $l = 0, 1, 2,..., N-1$, using the left-end rectangular rule to estimate the integral, we have

$$
\begin{aligned}
C_j &= \frac{1}{T}\int_0^T f(t)e^{-i2\pi jt/T}\,dt, \qquad \text{for } j = -\infty \text{ to } \infty \\[2mm]
&\approx \frac{1}{N\Delta t}\sum_{l=0}^{N-1} f(t_l)e^{-i2\pi jt_l/T}\Delta t \\[2mm]
&\approx \frac{1}{N}\sum_{l=0}^{N-1} f(t_l)e^{-i2\pi jl/N}
\end{aligned}
\tag{8.42}
$$

Note that to simplify the analysis, we now use the variable t, the sampling domain as $(0, N\Delta t)$ rather than $(-\pi, \pi)$ such that the period T is equal to $N\Delta t$ and there are N samples starting at $t = 0$ but not including $t = T$ as the function value is assumed to be same at $t = 0$ and $t = T$.

For now, we assume that the function may take on complex values. However, since we are mostly concerned with real functions and want to avoid complex computations, we will later switch to the discrete sine and cosine transforms. The other assumption we make is that the sampling domain of $(0, T)$ represents the function completely. This means that if the function is periodic and infinite in length, $(0, T)$ represents one period; and if the function is zero everywhere except over a finite length, $(0, T)$ includes that length. It should also be noted that in the latter case of the function being zero everywhere outside the interval $(0, T)$, we may arbitrarily define the function over the interval $(-T, 0)$ as a mirror image of the given function to make it an even function over $(-T, T)$ or reverse the mirror image to make it an odd function over $(-T, T)$. This will result in the function being described by a Fourier cosine series or a Fourier sine series, respectively. In this book, we do not discuss these.

The sum on the right-hand side of Eq. (8.42) is called the N-point DFT of the N sample values $f(t_l)$ and will be denoted by F_j (note that $C_j \cong F_j/N$). One should note that

$$F_{j+N} = \sum_{l=0}^{N-1} f(t_l) e^{-i2\pi(j+N)l/N} = \sum_{l=0}^{N-1} f(t_l) e^{-i2\pi jl/N} e^{-i2\pi} = F_j \qquad (8.43)$$

showing that the N-point DFT is periodic with a period of N (one must keep in mind that the coefficients of the Fourier series are, in general, not periodic. This issue of closeness of the DFT and the Fourier coefficients would be discussed later in this section). Also,

$$\sum_{k=0}^{N-1} F_k e^{i2\pi jk/N} = \sum_{k=0}^{N-1} \left[\sum_{l=0}^{N-1} f(t_l) e^{-i2\pi kl/N} \right] e^{i2\pi jk/N} = \sum_{k=0}^{N-1} \sum_{l=0}^{N-1} f(t_l) e^{-i2\pi k(j-l)/N} \qquad (8.44)$$

$$\text{for } j = 0, 1, 2, \ldots, N-1$$

Changing the order of summation and using the properties of a geometric series (see Box 4.5) to show that

$$\sum_{k=0}^{N-1} e^{-i2\pi k(j-l)/N} = \delta_{jl}$$

we get

$$\sum_{k=0}^{N-1} F_k e^{i2\pi jk/N} = Nf(t_j) \qquad (8.45)$$

The sum on the left-hand side is called the N-point inverse DFT of the N (generally complex) numbers, F_k, $k = 0$ to $N-1$, and may be denoted by f_j. Clearly, if we obtain the DFT of a sampled function, and then the inverse DFT of the *truncated* DFT (note that the DFT, F_k, has a range of k from $-\infty$ to ∞, but the inverse uses only $k = 0$ to $N-1$), we will get back the function value at all the sampling points on dividing the inverse DFT by N, the number of sampling locations.

Comparing with the technique used for interpolation (Section 4.7), we see that we have not gained in computational efficiency by expressing the procedure in terms of DFT. However, there is a simple technique which reduces the computational time significantly as described next. We describe the DFT and FFT using the complex form of the Fourier series. For real functions, an alternative would be to define a discrete sine transform (DST) and a discrete cosine transform (DCT) on similar lines. Interested readers should see Walker (1996).

If we assume that N is even, the N-point DFT may be split into two parts as,

$$F_j = \sum_{l=0}^{N-1} f(t_l) e^{-i2\pi jl/N} = \sum_{l=0}^{N/2-1} \left[f(t_{2l}) e^{-i2\pi j 2l/N} + f(t_{2l+1}) e^{-i2\pi j(2l+1)/N} \right]$$

$$= \sum_{l=0}^{N/2-1} f(t_{2l}) e^{-i2\pi jl/(N/2)} + e^{-i2\pi j/N} \sum_{l=0}^{N/2-1} f(t_{2l+1}) e^{-i2\pi jl/(N/2)}$$

(8.46)

The summations on the right-hand side may be thought of as N/2-point DFT of the even-numbered and odd-numbered samples, respectively. If we further assume that N/2 is even, we may split each of these N/2-point DFTs into two N/4-point DFTs. The *even set* would be split into an even-even and an even-odd set and the odd set would be split into an odd-even and an odd-odd set.

For example, if we start with eight points, numbered from 0 to 7, the first subdivision would result in 4-point DFT of points numbered 0, 2, 4, 6 and 1, 3, 5, 7 and the second subdivision would result in 2-point DFT of points 0, 4; 2, 6; 1, 5; and 3, 7. Carrying on this process (which necessitates the assumption of *N being an integer power of* 2), we ultimately reach a 1-point DFT which is the function value itself! This provides an efficient method of evaluating the N-point DFT.

Denoting even by the superscript zero and odd by one, and using the symbol ζ to represent the Nth root of unity ($\zeta = e^{-i2\pi/N}$) we have,

(for $j = 0$ to $N - 1$):

$$F_j = F_j^0 + e^{-i2\pi j/N} F_j^1 = (F_j^{00} + e^{-i2\pi j/(N/2)} F_j^{01}) + e^{-i2\pi j/N} (F_j^{10} + e^{-i2\pi j/(N/2)} F_j^{11})$$

$$= (F_j^{000} + \zeta^{4j} F_j^{001}) + \zeta^{2j} (F_j^{010} + \zeta^{4j} F_j^{011}) + \zeta^{j} \left[\begin{matrix} (F_j^{100} + \zeta^{4j} F_j^{101}) \\ + \zeta^{2j} (F_j^{110} + \zeta^{4j} F_j^{111}) \end{matrix} \right]$$

$$= \ldots$$

(8.47)

Using the fact that both F_j^0 and F_j^1, which are N/2-point DFTs, are periodic with a period N/2 and $\zeta^{N/2} = -1$, we write the first line of the equation above as. (The terms ξ^j which multiply the 'odd' components are sometimes called the twiddle factors.)

$$j = 0 \text{ to } N/2 - 1: \quad F_j = F_j^0 + \xi^j F_j^1; \quad F_{j+N/2} = F_j^0 - \xi^j F_j^1 \quad (8.48a)$$

Similarly, subsequent steps could be written as,

$$j = 0 \text{ to } N/4 - 1: \quad F_j^0 = F_j^{00} + \xi^{2j} F_j^{01}; \quad F_{j+N/4}^0 = F_j^{00} - \xi^{2j} F_j^{01} \quad (8.48b)$$

$$j = 0 \text{ to } N/4 - 1: \quad F_j^1 = F_j^{10} + \xi^{2j} F_j^{11}; \quad F_{j+N/4}^1 = F_j^{10} - \xi^{2j} F_j^{11} \quad (8.48c)$$

and so on, till we get $F_j^{\text{binary string}} = f(t_m)$, in which the superscript binary string is of length $\log_2 N$. The only thing left to do is to find the value of m corresponding to a particular string. To do this, we write the subscript m also as a binary number of the same length. As an example, for $N = 8$, the sampling locations are written as 000,001,010,011,100,101,110, and 111 corresponding to $t_0, t_1, \ldots, t_7$.

Clearly, F^0 uses all even points, i.e., those ending in 0, and F^1 uses all odd points, i.e., those ending in 1. At the next step, F^{00} uses all even-even points, i.e., those ending in 00; F^{01} uses all even-odd points, i.e., those ending in 10; F^{10} uses all even-odd points, i.e., those ending in 01; and F^{11} uses all odd-odd points, i.e., those ending in 11 (note the reversal of order in the superscript and the *end-bits* of the corresponding sampling locations).

Proceeding in this way, it is easily shown that m would be the bit-reversed value of the superscript binary string. Thus, if we re-arrange the observed values in bit-reversed order, we get the 1-point DFTs. The recursive relations are then used to march forward to 2-point DFTs, 4-point DFTs, ..., N-point DFT. The computational effort is only of the order of $N\log_2 N$ as compared to N^2 for regular DFT computation.

Example 8.13   A function is periodic with a time period of 2 sec and is given as,

$$f(t) = -t^3 + 4t^2 - 4t + 1; \qquad t = 0 \text{ to } 2 \text{ sec}$$

It is sampled at a 1/4 sec interval ($l = 0, 1, 2, \ldots, 7$). Find the DFT.

Solution   Table 8.6 shows the sampled values and the bit-reversed 1-point DFTs:

Table 8.6

l	0	1	2	3	4	5	6	7
Binary subscript	000	001	010	011	100	101	110	111
t_l	0	0.25	0.5	0.75	1.0	1.25	1.5	1.75
$f(t_l)$	1.000000	0.234375	−0.125000	−0.171875	0.000000	0.296875	0.625000	0.890625
Reverse superscript	000	100	010	110	001	101	011	111

The recursive computations are shown in Table 8.7 below starting from the 1-point DFTs (which are same for all j's as they are periodic with period 1) to 2-point DFTs:

Table 8.7

	1-point DFT			2-point DFT	
	Real	Imaginary		Real	Imaginary
F_j^{000}	1.00000	0.000000	F_0^{00}	1.00000	0.000000
F_j^{001}	0.000000	0.000000	F_0^{01}	0.500000	0.000000
F_j^{010}	-0.125000	0.000000	F_0^{10}	0.531250	0.000000
F_j^{011}	0.625000	0.000000	F_0^{11}	0.718750	0.000000
F_j^{100}	0.234375	0.000000	F_1^{00}	1.00000	0.000000
F_j^{101}	0.296875	0.000000	F_1^{01}	-0.750000	0.000000
F_j^{110}	-0.171875	0.000000	F_1^{10}	-0.062500	0.000000
F_j^{111}	0.890625	0.000000	F_1^{11}	-1.06250	0.000000

Table 8.8

	2-point DFT			4-point DFT			8-point DFT	
	Real	Imaginary		Real	Imaginary		Real	Imaginary
F_0^{00}	1.00000	0.000000	F_0^0	1.50000	0.000000	F_0	2.750000	0.000000
F_0^{01}	0.500000	0.000000	F_0^1	1.25000	0.000000	F_1	1.707107	1.545495
F_0^{10}	0.531250	0.000000	F_1^0	1.00000	0.750000	F_2	0.500000	0.187500
F_0^{11}	0.718750	0.000000	F_1^1	-0.062500	1.06250	F_3	0.292893	0.045495
F_1^{00}	1.00000	0.000000	F_2^0	0.500000	0.000000	F_4	0.250000	0.000000
F_1^{01}	-0.750000	0.000000	F_2^1	-0.187500	0.000000	F_5	0.292893	-0.045495
F_1^{10}	-0.062500	0.000000	F_3^0	1.00000	-0.750000	F_6	0.500000	-0.187500
F_1^{11}	-1.06250	0.000000	F_3^1	-0.062500	-1.06250	F_7	1.707107	-1.545495

Note that the 2-point DFTs are periodic, and we do not need to store these for $j = 2, 3,\ldots, 7$. The computation of 2-point DFT uses $F_j^{00} = F_j^{000} + \zeta^{4j} F_j^{001}$ (and similar expressions) and the fact that $\zeta^4 = -1$. Going from 2-point DFTs to 4-point DFTs, we use $F_j^0 = F_j^{00} + \xi^{2j} F_j^{01}$ and then obtain the required 8-point DFTs using $F_j = F_j^0 + \xi^j F_j^1$ as shown in Table 8.8. From the periodicity of F_j, we have,

$$F_{-3} = F_5 = 0.292893 - i\,0.045495$$

$$F_{-2} = F_6 = 0.500000 - i\,0.187500$$

$$F_{-1} = F_7 = 1.707107 - i\,1.545495$$

and using the relationship $C_j = F_j/N$, we could write the Fourier series in terms of the sines and cosines as,

$$
\begin{aligned}
f(t) &\approx 0.3437500 + 0.4267768 \, \cos(\pi t) - 0.3863738 \sin(\pi t) \\
&+ 0.1250000 \cos(2\pi t) - 0.0468750 \sin(2\pi t) \\
&+ 0.0732233 \cos(3\pi t) - 0.0113738 \sin(3\pi t) + 0.0312500 \cos(4\pi t)
\end{aligned}
$$

Comparing with the exact coefficients (see Example 8.12 and note that the function is same except that the domain has been changed from $x = (-\pi, \pi)$ to $t = (0, 2)$ and $x = -\pi + t\pi$),

$$a_0 = 1/3; \qquad a_1 = 0.405285; \quad b_1 = -0.387018; \quad a_2 = 0.101321;$$
$$b_2 = -0.0483773; \quad a_3 = 0.045032; \quad b_3 = -0.014334; \quad a_4 = 0.0253302$$

we see that the coefficients obtained from the DFT are slightly different from those of the full Fourier series. It has been suggested that for a N-point DFT, reasonably accurate coefficients may be obtained till order j which is roughly 1/8th of N. It should also be noted that this partial Fourier series interpolates the function value at all nodes, and we would get the same series by the interpolation method discussed in Chapter 4. However, the computational effort in DFT is much smaller than that in the interpolation.

Another point worth noting for this example is that even though the data is real, the DFTs are complex. However, we have complex conjugate pairs of DFTs resulting in real coefficients of the Fourier series written in terms of sines and cosines. The obvious question arises that *is there a method which could perform the discrete transform without using complex data storage and operations*. One answer, as shown in Chapter 4, is to perform an interpolation using the sine and cosine functions but it would be computationally intensive.

There are several alternatives for performing a FFT of real-valued data (Sorensen et al. 1987) and the most commonly used technique uses the fact that the DFT of a real sequence has an *even* real part and an *odd* imaginary part (as is clear from the Table 8.8). Then, if transforms of two different sets of real sequences are needed, the complex FFT is used with one sequence as the real part and the other as imaginary part. However, if only a single real sequence needs to be transformed, the 2-point DFTs (which are also real) are used as the real and imaginary parts of a $N/2$-point complex FFT.

A detailed description of various techniques and their relative computational efficiency may be seen in Sorensen et al. (1987) while Walker (1996) should be consulted for a thorough discussion of the discrete sine and cosine transforms and fast computation of these.

8.6 UNCERTAINTY IN PARAMETERS ESTIMATED FROM REGRESSION

In the discussion on regression, we mentioned that the data may contain certain errors and we may not want to perform an interpolation (Section 4.6). In such cases, we would prefer to obtain an interpolating polynomial (or some other function) which usually leads to an estimate of some physical parameter (e.g., the half-life of a substance, the velocity of an object). However, we did not touch upon the subject of the reliability of these estimates. In other words, since the data is not error-free, we would expect the estimated parameters also to have some uncertainty associated with them. In this section, we discuss this aspect in terms of a *confidence interval* for the estimated parameters. The reader is expected to be familiar with elementary statistics. However, some relevant information is presented briefly in Box 8.1.

BOX 8.1 Some Basic Statistical Concepts

In most of this section, we assume that x is a normally distributed continuous random variable which may take any value between $-\infty$ to ∞.

The probability density function (pdf), $f(x)$, is defined such that the probability of the variable x being in the range X and $X + dx$ is $f(X)\ dx$. The cumulative pdf, $F(x)$, is defined such that the probability of x not exceeding X is $F(X)$, implying

$$F(X) = \int_{-\infty}^{X} f(x)dx$$

The mean (or expected value) of x is written as,

$$E[x] \equiv \mu = \int_{-\infty}^{\infty} x f(x)\, dx$$

and the variance of x is given by,

$$E\left[(x-\mu)^2\right] \equiv \sigma^2 = \int_{-\infty}^{\infty} (x-\mu)^2 f(x)\, dx$$

(the square root of the variance is the standard deviation, σ). Another useful relationship obtained from this expression is

$$\sigma^2 = E\left[x^2\right] - 2\mu E[x] + \mu^2 \Rightarrow E\left[x^2\right] = \sigma^2 + \mu^2 \tag{B8.1.1}$$

For a random variable, z, which is a linear combination of k mutually independent random variables, $x_1, x_2, \ldots, x_k$, i.e.,

$$z = \sum_{i=1}^{k} a_i x_i$$

we have

$$E[z] \equiv \mu_z = \sum_{i=1}^{k} a_i \mu_i$$

$$E\left[(z-\mu_z)^2\right] \equiv \sigma_z^2 = \sum_{i=1}^{k} a_i^2 \sigma_i^2 \tag{B8.1.2}$$

in which μ_i is the mean of x_i and σ_i^2 is its variance.

For a sample consisting of m independent observations $(x_1, x_2, \ldots, x_m)$ of a random variable, x, which has a mean of μ and variance σ^2, the sample mean is given by,

$$\bar{x} = \frac{1}{m} \sum_{i=1}^{m} x_i$$

and its expected value is [The expected value of the sample mean denotes the average of the means of *several* samples of size k]

$$E[\bar{x}] = \frac{1}{m} \sum_{i=1}^{m} E[x_i] = \mu$$

indicating that the sample mean is an unbiased estimate of the population mean, μ. (The bias of an estimator is the expected value of the difference between the estimator and the true value of the parameter being estimated. In this case, the sample mean is an estimate of the population mean and, since the expected value of the sample mean is equal to the population mean, the bias is zero.) The variance of the sample mean is obtained from

$$E\left[(\bar{x}-\mu)^2\right] \equiv \sigma_{\bar{x}}^2 = E\left[\bar{x}^2\right] - 2\mu E\left[\bar{x}\right] + E\left[\sum_{i=1}^{m}\mu^2\right]$$

$$= \frac{1}{m^2}E\left[\left(\sum_{i=1}^{m}x_i\right)^2\right] - \mu^2$$

$$= \frac{1}{m^2}E\left[\sum_{i=1}^{m}x_i^2 + \sum_{i=1}^{m}\sum_{j=1, j\neq i}^{m}x_i x_j\right] - \mu^2$$

Now, since x_i and x_j are independent

$$E\left[x_i x_j\right] = E\left[x_i\right]E\left[x_j\right] = \mu^2$$

Also, from Eq. (B8.1.1)

$$E\left[x_i^2\right] = \sigma^2 + \mu^2$$

Therefore,

$$\sigma_{\bar{x}}^2 = \frac{1}{m^2}\left[m(\sigma^2 + \mu^2) + m(m-1)\mu^2\right] - \mu^2 = \frac{\sigma^2}{m} \tag{B8.1.3}$$

The expected value of the sum of squares of the deviations about the sample mean is

$$E\left[\sum_{i=1}^{m}(x_i-\bar{x})^2\right] = E\left[\sum_{i=1}^{m}x_i^2\right] - 2E\left[\bar{x}\sum_{i=1}^{m}x_i\right] + E\left[\sum_{i=1}^{m}\bar{x}^2\right]$$

$$= E\left[\sum_{i=1}^{m}x_i^2\right] - mE\left[\sum_{i=1}^{m}\bar{x}^2\right] \tag{B8.14}$$

$$= m(\sigma^2 + \mu^2) - m\left(\frac{\sigma^2}{m} + \mu^2\right) = (m-1)\sigma^2$$

Which indicates that the expected value of the average of the sum of squares of deviation, i.e., $\sum_{i=1}^{m}(x_i-\bar{x})^2/m$ would be smaller than the population variance and hence is not an unbiased estimate. We define a sample variance as

$$s^2 = \frac{\sum_{i=1}^{m}(x_i-\bar{x})^2}{m-1} \tag{B8.1.5}$$

which, clearly, is an unbiased estimate of the population variance. The reason for having only $m - 1$ degrees of freedom is that the deviations are taken about the sample mean and not the population mean. Since the sample mean is obtained from the m observations, it effectively implies that only $m - 1$ observations are independent (as $\bar{x}$ acts as a constraint on these values).

BOX 8.2 Some Probability Distributions

Normal Distribution

The pdf for a normal (or Gaussian) distribution is given by,

$$f(x) = \frac{1}{\sigma\sqrt{2\pi}} e^{-\frac{(x-\mu)^2}{2\sigma^2}}$$

(B8.1.6)

The variable x is normally distributed with a mean μ and variance σ^2 [commonly written as $N(\mu, \sigma^2)$] and is shown in Fig. 8.10 [for $\mu = 0, \sigma^2 = 1$, which is called the *standard normal distribution*, $N(0, 1)$]:

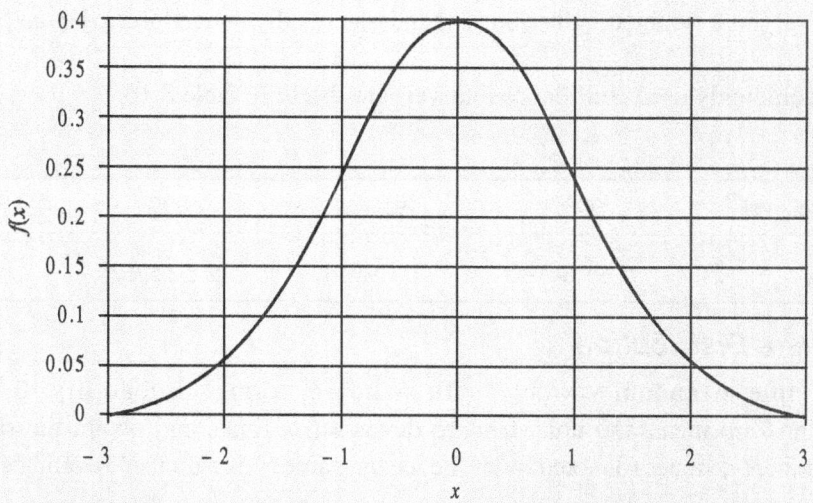

Fig. 8.10 Probability density function for the normal distribution.

Note that the distribution is symmetric about the mean, the cdf is maximum at the mean, and the probability of the random variable attaining values less than $\mu - 3\sigma$ or more than $\mu + 3\sigma$ is very small. Tables of the cdf and the cumulative cdf for the normal distribution are available in a number of books (e.g., Abramowitz and Stegun 1964), spreadsheet software (e.g., Microsoft Excel), and on the internet (e.g., http://www.itl.nist.gov/div898/handbook/eda/section3/eda3671.htm). A few typical values are shown in Table 8.9.

Table 8.9

X	μ	$\mu \pm \sigma$	$\mu \pm 2\sigma$	$\mu \pm 3\sigma$
$f(X)$	0.3989	0.2420	0.0540	0.0044
$F(X)$	0.5000	0.8413, 0.1587	0.9772, 0.0228	0.9986, 0.0014

Confidence Interval

From the table, we see that the probability of a normally distributed random variable attaining a value between $\mu - \sigma$ and $\mu + \sigma$ is $0.8413 - 0.1587$, i.e., 0.683. In other words, if we sample a random variable, we would expect its value to be within one standard deviation of the mean about 68% of the time. This leads us to the definition of a *confidence interval*, and we say that the interval $(\mu - \sigma, \mu + \sigma)$ is a 68% confidence interval. It should be noted that this is not the *only* 68% confidence interval. For example, the interval $(\mu - 0.7\sigma, \mu + 1.5\sigma)$ is also a nearly 68% confidence interval. However, the interval centered on mean is the smallest such interval (for a symmetric distribution) and is generally understood to be *the* confidence interval.

Some commonly used confidence intervals are listed in Table 8.10.

Table 8.10

Confidence level	90%	95%	99%	99.9%
Interval	$\mu \pm 1.645\sigma$	$\mu \pm 1.960\sigma$	$\mu \pm 2.576\sigma$	$\mu \pm 3.291\sigma$

Chi-square Distribution

If we sample a random variable with standard normal probability distribution (i.e., having zero mean and unit standard deviation) k times and obtain the sum of the squares of these k values (denoted by χ^2, hence the name) this sum itself would be a random variable:

$$\chi^2 = \sum_{i=1}^{k} x_i^2 \tag{B8.1.7}$$

It can be shown that the pdf for χ^2 is given by

$$f(x) = \frac{1}{2^{k/2}\Gamma(k/2)} x^{\frac{k}{2}-1} e^{-x/2} \tag{B8.1.8}$$

with k known as the degree of freedom of the χ^2 distribution. The Gamma function for a positive integer i is given by (since k is an integer, $k/2$ would be either an integer or an integer $+ 1/2$)

$$\Gamma(i) = (i-1)!$$

$$\Gamma\left(i+\frac{1}{2}\right) = \frac{1.3.5...(2i-1)}{2^i}\sqrt{\pi}$$

Figure 8.11 shows the distribution for a few selected values of k [note that $k = 1$ implies the pdf of the *square* of a $N(0,1)$ random variable]:

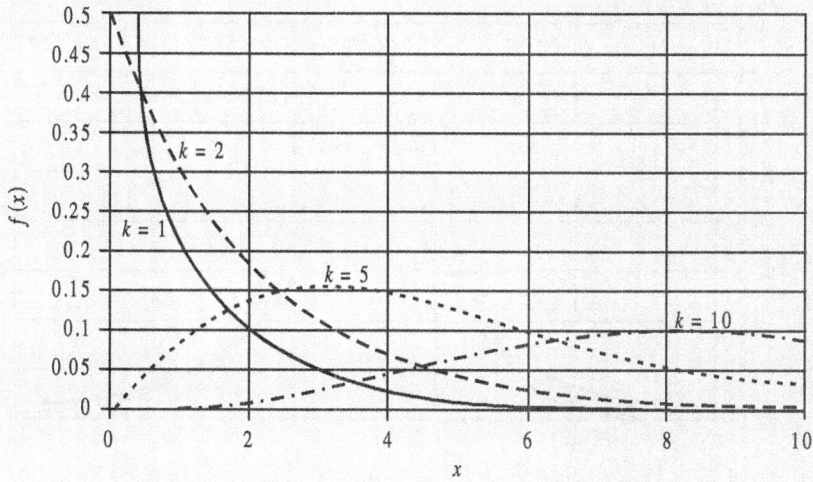

Fig. 8.11 Chi-square distribution.

The mean of the χ^2 distribution is k and its variance is $2k$, and it approaches the normal distribution $N(k, 2k)$ as k becomes very large.

Student's t-Distribution

If x_a is a random variable with standard normal probability distribution $N(0, 1)$, and x_b is another (independent) random variable which has a Chi-square distribution with k degrees of freedom, then the random variable t, defined by

$$t = \frac{x_a}{\sqrt{x_b/k}}$$

(B8.1.9)

has the following Student's *t*-distribution (Published by W.S. Gosset in 1908 under the pseudonym of *Student*)

$$f(t) = \frac{\Gamma\left(\dfrac{k+1}{2}\right)}{\sqrt{k\pi}\,\Gamma\left(\dfrac{k}{2}\right)}\left(1+\frac{t^2}{k}\right)^{-\frac{k+1}{2}}$$

(B8.1.10)

The mean of the *t*-distribution is 0 for $k > 1$ and its variance is $k/(k-2)$ for $k > 2$ (the mean is undefined for $k = 1$ and the variance is undefined for $k = 1$ and 2). The *t*-distribution approaches $N(0, 1)$ as k becomes large. Its plot for a few selected k values is shown in Fig. 8.12.

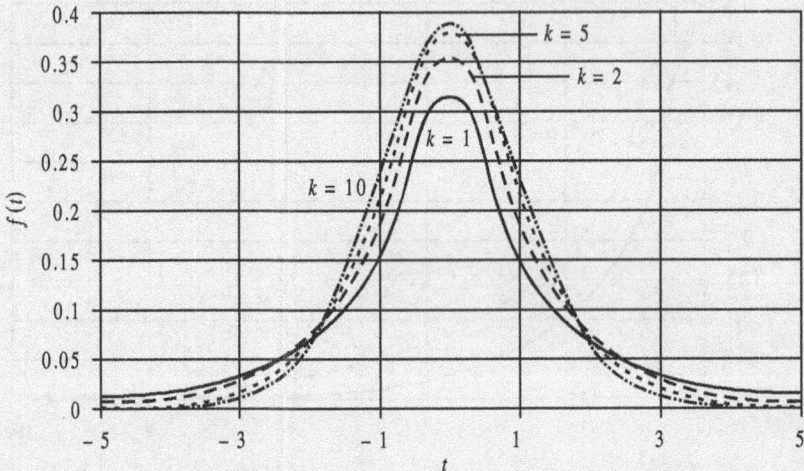

Fig. 8.12 Student's *t*-distribution.

Similar to the normal distribution, the confidence interval may be defined for the *t*-distribution and a few typical values are listed in Table 8.11.

Table 8.11

		Confidence level			
		90%	95%	99%	99.9%
Interval of *t*	$k = 1$	± 6.314	± 12.71	± 63.66	± 636.6
values for	$k = 2$	± 2.920	± 4.303	± 9.925	± 31.60
degree of	$k = 5$	± 2.015	± 2.571	± 4.032	± 6.869
freedom, *k*	$k = 10$	± 1.812	± 2.228	± 3.169	± 4.587
	$k = 100$	± 1.660	± 1.984	± 2.626	± 3.390

Let us assume that we measure the value of the dependent variable y at a few values of the independent variable x. We assume that x contains no errors and is precisely known (or, if it is not precisely known, the error in measurement of x is negligible compared to that in y). On the other hand, y is assumed to contain random measurement errors due to which it is normally distributed with a mean, $\mu(x)$ and a variance σ^2 (note that we assume the variance to be independent of x). Further, we assume that there is a linear relationship between y and x if both are precisely measured and it is given by

$$y = \alpha_0 + \alpha_1 x \tag{8.49}$$

Since the measurement of y is not precise, the plot of y vs. x will not, in general, be a straight line (see Fig. 8.13). The question of estimating the linear relationship between y and x has been addressed in Chapter 4. Here, we address the issue of how confident we are in the values of the coefficients (c_0 and c_1) estimated by a linear least square regression of y on x.

Let (x_i, y_i) denote a pair of the independent variable (x_i), and the corresponding measured value of y (y_i) and let there be a total of $n + 1$ observations ($i = 0, 1, 2, \ldots, n$). We follow the same convention as in Chapter 4 and start the index with 0. A number of times, the index starts at 1 so that the total number of points is n and not $n + 1$. The linear regression is given by the equation,

$$y = c_0 + c_1 x \tag{8.50}$$

and the coefficients are obtained by minimizing the sum of squared residuals [It can be shown that the least square regression also maximizes the probability of all the y values occurring simultaneously. However, we will follow the concept of minimizing the sum of square of residuals] as [see Eq. (4.55)],

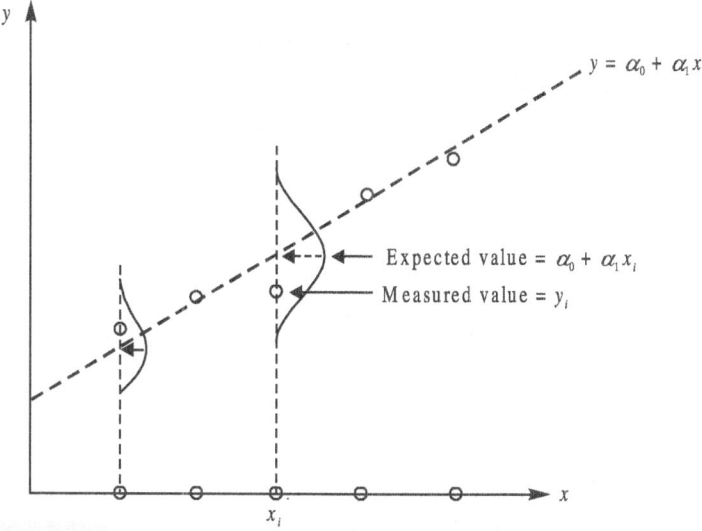

Fig. 8.13 A linear function of x with random measurement errors.

$$\begin{bmatrix} n+1 & \displaystyle\sum_{i=0}^{n} x_i \\[2ex] \displaystyle\sum_{i=0}^{n} x_i & \displaystyle\sum_{i=0}^{n} x_i^2 \end{bmatrix} \begin{Bmatrix} c_0 \\ c_1 \end{Bmatrix} = \begin{Bmatrix} \displaystyle\sum_{i=0}^{n} y_i \\[2ex] \displaystyle\sum_{i=0}^{n} x_i y_i \end{Bmatrix} \tag{8.51}$$

which is more conveniently written in terms of the sample means as,

$$\begin{bmatrix} 1 & \bar{x} \\[2ex] (n+1)\bar{x} & (n+1)\bar{x}^2 + \displaystyle\sum_{i=0}^{n}(x_i - \bar{x})^2 \end{bmatrix} \begin{Bmatrix} c_0 \\ c_1 \end{Bmatrix} = \begin{Bmatrix} \bar{y} \\[2ex] (n+1)\bar{x}\,\bar{y} + \displaystyle\sum_{i=0}^{n}(x_i - \bar{x})(y_i - \bar{y}) \end{Bmatrix} \tag{8.52}$$

resulting in,

$$c_1 = \frac{\displaystyle\sum_{i=0}^{n}(x_i - \bar{x})(y_i - \bar{y})}{\displaystyle\sum_{i=0}^{n}(x_i - \bar{x})^2}; \quad c_0 = \bar{y} - c_1\bar{x} \tag{8.53}$$

Estimate of the population variance about the regression line is obtained by,

$$s^2 = \frac{\displaystyle\sum_{i=0}^{n}(y_i - c_0 - c_1 x_i)^2}{n-1} \tag{8.54}$$

[see Eq. (B8.1.5) and note that the degrees of freedom has been reduced by two since c_0 and c_1 are estimated from the $n + 1$ observed values].

The expected values of c_0 and c_1 are shown to be equal to α_0 and α_1, respectively, implying that these are unbiased estimates of the true values of the intercept and the slope:

$$E[c_1] = E\left[\frac{\displaystyle\sum_{i=0}^{n}(x_i - \bar{x})(y_i - \bar{y})}{\displaystyle\sum_{i=0}^{n}(x_i - \bar{x})^2}\right] = E\left[\sum_{i=0}^{n}\left\{\frac{(x_i - \bar{x})}{\displaystyle\sum_{j=0}^{n}(x_j - \bar{x})^2} y_i\right\}\right] \tag{8.55}$$

Now, since y_i are independent and their expected value is equal to $\alpha_0 + \alpha_1 x_i$, using Eq. (B8.1.2), we get

$$E[c_1] = \frac{\sum_{i=0}^{n}(x_i - \bar{x})(\alpha_0 + \alpha_1 x_i)}{\sum_{i=0}^{n}(x_i - \bar{x})^2} = \alpha_1 \frac{\sum_{i=0}^{n}(x_i - \bar{x})x_i}{\sum_{i=0}^{n}(x_i - \bar{x})^2} = \alpha_1 \qquad (8.56a)$$

$$E[c_0] = E[\bar{y} - c_1\bar{x}] = \sum_{i=0}^{n}\frac{\alpha_0 + \alpha_1 x_i}{n+1} - \bar{x}\alpha_1 = \alpha_0 \qquad (8.56b)$$

The variances are obtained as,

$$\sigma_{c_1}^2 = \frac{\sum_{i=0}^{n}(x_i - \bar{x})^2 \sigma^2}{\left[\sum_{i=0}^{n}(x_i - \bar{x})^2\right]^2} = \frac{\sigma^2}{\sum_{i=0}^{n}(x_i - \bar{x})^2} \qquad (8.57a)$$

$$\sigma_{c_0}^2 = \sum_{i=0}^{n}\frac{\sigma^2}{(n+1)^2} + \bar{x}^2 \sigma_{c_1}^2 = \frac{\sigma^2 \sum_{i=0}^{n}x_i^2}{(n+1)\sum_{i=0}^{n}(x_i - \bar{x})^2} \qquad (8.57b)$$

In order to get a confidence interval, if σ^2 is known, the variance of both c_0 and c_1 could be obtained and the normal distribution could be used. Or, we may estimate σ^2 by the sample variance, s^2, and obtain the confidence interval. However, noting that $(y_i - c_0 - c_1 x_i)/\sigma$ follows a standard normal distribution, $N(0,1)$, we conclude that

$$\sum_{i=0}^{n}\left(\frac{y_i - c_0 - c_1 x_i}{\sigma}\right)^2$$

follows the Chi-square distribution with $(n-1)$ degrees of freedom. Moreover $\left((c_1 - \alpha_1)/\sigma_{c_1}\right)$ follows a standard normal distribution. Hence, [see Eq. (B8.1.9)]

$$\frac{(c_1 - \alpha_1)/\sigma_{c_1}}{\sqrt{\sum_{i=0}^{n}\left(\frac{y_i - c_0 - c_1 x_i}{\sigma}\right)^2 / (n-1)}}, \text{ i.e., } t_1 = \frac{(c_1 - \alpha_1)\sqrt{\sum_{i=0}^{n}(x_i - \bar{x}^2)}}{\sqrt{\sum_{i=0}^{n}(y_i - c_0 - c_1 x_i)^2 / (n-1)}}$$

follows a Student's t-distribution with $n - 1$ degrees of freedom (and does not involve σ^2). Similarly,

$$t_0 = \frac{(c_0 - \alpha_0)(n+1)\sum_{i=0}^{n}(x_i - \bar{x}^2)}{\sum_{i=0}^{n} x_i^2 \sqrt{\sum_{i=0}^{n}(y_i - c_0 - c_1 x_i)^2 /(n-1)}}$$

also follows a t-distribution. From these, the confidence intervals on the estimates of the slope and intercept are readily obtained. The following example illustrates the application.

Example 8.14 There are 12 boxes numbered 0 to 11 and the mass of sand stored in each is linearly related to the number on the box. The first two columns of the table below show the observed values of the mass in these boxes. Estimate the 95% confidence interval for the slope (α_1) and intercept (α_0) of the linear relationship.

Solution The sample means of x and y are obtained from the observed data in the first two columns of Table 8.12. Columns 3 to 6 are computed using these means. Column 8 is computed after obtaining the regression coefficients c_0 and c_1.

Table 8.12

Number on box (x)	Mass of sand (y)	$x - \bar{x}$	$y - \bar{y}$	$(x-\bar{x})(y-\bar{y})$	$(x-\bar{x})^2$	x^2	$(y - c_0 - c_1 x)^2$
0	1.01174	−5.5	−11.0142	60.5783	30.25	0	4.01094E−04
1	2.95054	−4.5	−9.07544	40.8395	20.25	1	2.24773E−03
2	5.03842	−3.5	−6.98755	24.4564	12.25	4	1.17301E−03
3	7.07363	−2.5	−4.95235	12.3809	6.25	9	3.99739E/0−03
4	8.91799	−1.5	−3.10799	4.66198	2.25	16	9.73036E−03
5	11.0289	−0.5	−0.997086	0.498543	0.25	25	3.63413E− 05
6	13.0574	0.5	1.03144	0.515720	0.25	36	8.02326E−04
7	15.0223	1.5	2.99634	4.49450	2.25	49	1.69196E−04
8	17.0326	2.5	5.00664	12.5166	6.25	64	7.97735E−05
9	19.0519	3.5	7.025940	24.5908	12.25	81	1.71480E−05
10	21.0770	4.5	9.051031	40.7296	20.25	100	5.29152E−04
11	23.0492	5.5	11.023249	60.6279	30.25	121	1.21166E−04

$\bar{x} = 5.5$ $\bar{y} = 12.0260$ $\Sigma = 286.891$ $\Sigma = 143$ $\Sigma = 506$ $\Sigma = 1.93047E−02$

The estimates are, therefore, $c_1 = 286.891/143 = 2.00623$ and $c_0 = 12.0260 - 2.00623 \times 5.5 = 0.991717$. The last two columns in Table 8.12 show the other parameters required for the t-distribution. From the confidence intervals for the t-distribution for 10 degrees of freedom, we get $t = \pm 2.228$. Thus, for the intercept, the confidence limits are given by

$$\pm 2.228 = \frac{(c_0 - \alpha_0)(n+1)\sum_{i=0}^{n}\left(x_i - \bar{x}^2\right)}{\sum_{i=0}^{n} x_i^2 \sqrt{\sum_{i=0}^{n}(y_i - c_0 - c_1 x_i)^2/(n-1)}} = \frac{(0.991717 - \alpha_0)12 \times 143}{506\sqrt{\dfrac{1.93047E - 02}{10}}}$$

which results in the 95% confidence interval on α_0 as $(0.962852, 1.02058)$. Similarly, the 95% confidence interval for the slope is given by

$$\pm 2.228 = \frac{(c_1 - \alpha_1)\sqrt{\sum_{i=0}^{n}\left(x_i - \bar{x}^2\right)}}{\sqrt{\sum_{i=0}^{n}(y_i - c_0 - c_1 x_i)^2/(n-1)}} = \frac{(2.00623 - \alpha_1)\sqrt{143}}{\sqrt{\dfrac{1.93047E - 02}{10}}}$$

which results in the 95% confidence interval on α_1 as $(1.99804, 2.01441)$.

It should be mentioned that the values of y were generated using $\alpha_0 = 1$ and $\alpha_1 = 2$ and adding a normally distributed random component with mean 0 and variance 0.0025. The variance of c_0 and c_1 [from Eq. (8.57)] are 7.37E–04 and 1.75E–05, respectively. Generally the variance of y about the regression line is not known. If we know its value, we do not need to use the chi-square analysis to estimate the confidence interval since the mean and variance of c_0 and c_1 are easily obtained and the confidence interval for a normal distribution could be used.

8.7 METHOD OF WEIGHTED RESIDUALS

In Chapters 6 and 7, we discussed various ways of solving ordinary and partial differential equations. The general technique was to substitute the derivatives by *finite differences* and then solve the resulting *algebraic* equations to obtain the *nodal* values. However, as we did for approximating a function, we may assume an approximate solution containing some *undetermined coefficients* (this assumed function will generally not satisfy the given differential equation), define a residual (which will represent how *far* the approximate solution is from satisfying the differential equation), and then minimise some norm of the residual by adjusting the coefficients.

Here, we describe this *method of weighted residuals*, and its extension to the *finite element method*. Another technique known as the *variational method* may be used in some cases (see Box 8.3) but is not discussed here.

Let the given differential equation be written as

$$\mathfrak{L}y - f = 0, \quad a \le x \le b \tag{8.58}$$

in which $\mathfrak{L}$ is a linear differential operator operating on the function y and f is a specified function of x. For easier presentation, we assume that $\mathfrak{L}$ is second order

$$\text{(specifically, } \mathfrak{L}y = \frac{d}{dx}\left[A_2(x)\frac{dy}{dx}\right] + \frac{d}{dx}[A_1(x)y] + A_0(x)y)$$

so that two boundary conditions are needed and that these boundary conditions are specified as

$$y\big|_{x=a} = y_a \tag{8.59a}$$

$$\frac{dy}{dx}\bigg|_{x=b} = d_b \tag{8.59b}$$

The boundary where the function value is specified is called the Dirichlet boundary and the boundary at which the gradient (or flux) is specified is known as the Neumann boundary. Sometimes, a Robin boundary condition is specified at which a relationship involving the function and its derivative is specified. Also note that we have used one condition at each of the two boundaries. Sometimes both the conditions may be specified at the same boundary. Similarly, for multidimensional problems, there are several possible boundary conditions. These are not discussed here.

In the weighted residual method, we assume a trial solution of the form

$$\tilde{y} = \tilde{y}(x, c_0, c_1, \ldots, c_n) = \sum_{j=0}^{n} c_j \phi_j(x) \tag{8.60}$$

in which the c's are undetermined coefficients and the ϕ's are basis functions. We will generally assume that the trial solution is de-linked with the boundary conditions. Sometimes, however, we would assume, and explicitly mention, that the trial function satisfies one or more of the boundary conditions a priori. The residual is defined as

$$R(x) = \mathfrak{L}\tilde{y} - f \tag{8.61}$$

and will naturally not be zero everywhere in (a, b) unless the approximate solution matches with the exact solution. To determine the coefficients $c_0, c_1, \ldots, c_n$, we stipulate that some (here since there are $n + 1$ undetermined coefficients and two boundary conditions, we would need $n - 1$ additional conditions) weight functions, $w_0(x), w_1(x), \ldots, w_{n-2}(x)$, be chosen in such a way that

BOX 8.3 Variational Method for Solving Differential Equations

Let the given differential equation be $\mathcal{L}y = f$ over the domain $x = (a, b)$, in which $\mathcal{L}$ is a linear differential operator operating on the function y and f is a specified function of x. Obviously, the required number of boundary conditions have to be specified (depending on the order of operator $\mathcal{L}$). The direct (analytical) solution of this equation may or may not be possible.

However, if $\mathcal{L}$ is self-adjoint (meaning that $\int_a^b (\mathcal{L}y_1) y_2 dx = \int_a^b (\mathcal{L}y_2) y_1 dx$, or using the inner product notation $\langle \mathcal{L}y_1, y_2 \rangle = \langle \mathcal{L}y_2, y_1 \rangle$) and positive definite (meaning that $\langle \mathcal{L}y, y \rangle > 0, \ \forall \ y \neq 0$), then the solution y of $\mathcal{L}y = f$ will minimize the functional

$$F(y) = \frac{1}{2}\langle \mathcal{L}y, y \rangle - \langle f, y \rangle$$

Proof:
From the definition of the function, we have

$$F(y + \alpha) = \frac{1}{2}\langle \mathcal{L}(y + \alpha), y + \alpha \rangle - \langle f, y + \alpha \rangle$$

$$= \frac{1}{2}\left(\langle \mathcal{L}y, y \rangle + \langle \mathcal{L}y, \alpha \rangle + \langle \mathcal{L}\alpha, y \rangle + \langle \mathcal{L}\alpha, \alpha \rangle\right) - \langle f, y \rangle - \langle f, \alpha \rangle$$

$$= \left(\frac{1}{2}\langle \mathcal{L}y, y \rangle - \langle f, y \rangle\right) + \frac{1}{2}\left(\langle \mathcal{L}y, \alpha \rangle + \langle \mathcal{L}\alpha, y \rangle\right) + \frac{1}{2}\langle \mathcal{L}\alpha, \alpha \rangle - \langle f, \alpha \rangle$$

The first term on the right-hand side is $F(y)$, the second term, using the self-adjoint property, is $\langle \mathcal{L}y, \alpha \rangle$, which combines with the last term and vanishes since $\mathcal{L}y - f$ is identically zero. Therefore,

$$F(y + \alpha) = F(y) + \frac{1}{2}\langle \mathcal{L}\alpha, \alpha \rangle$$

and, using the positive-definiteness of $\mathcal{L}$, it is established that $F(y)$ would be a minimum. Moreover, the uniqueness of the solution may be shown by taking two different solutions, y_a and y_b, which would imply $\mathcal{L}(y_a - y_b) = 0$ and consequently, $\langle \mathcal{L}(y_a - y_b), (y_a - y_b) \rangle = 0$. The positive definiteness then necessitates $y_a - y_b = 0$ proving the uniqueness of the solution.

  An advantage of using the variational formulation is that the continuity requirement on the trial function (i.e., the assumed solution) may be relaxed by an order. For example, if the operator $\mathcal{L}$ is a second derivative, the trial solution should have C^1 continuity so that the second derivatives are defined at every point within the domain. However, in the variational formulation, integration by parts will convert the integral of the term involving the *second derivative of the trial solution multiplied with the trial solution* to an integral involving *the square of the first derivative of the trial solution*, thereby enabling us to use only a C^0 continuity for the trial solution. Since this characteristic is similar to that of the method of weighted residuals, and the method of weighted residual does not have stringent requirement of self-adjoint property of the operator $\mathfrak{L}$, we do not discuss the variational methods further.

$$\int_a^b w_i(x)R(x)dx = 0, \qquad \text{for } i = 0,1,2,...,n-2 \tag{8.62}$$

which amounts to making the residual orthogonal to the weight functions. Sometimes the w's are called the *test* functions. These should be from a function space which is large enough to satisfy the requirements of the problem but is also small enough to be useful. The two boundary conditions provide the following equations:

$$\sum_{j=0}^{n} c_j \phi_j(a) = y_a \tag{8.63a}$$

$$\sum_{j=0}^{n} c_j \phi_j'(b) = d_b \tag{8.63b}$$

Thus we have $(n + 1)$ linear equations which may be solved to obtain the $(n + 1)$ coefficients. Different choices of the weight function lead to different methods, some of which are described here.

8.7.1 Point Collocation (or Collocation) Method

The simplest choice of the weights is the *Dirac delta function* at $(n-1)$ judiciously chosen points, i.e.,

$$w_i(x) = \delta(x - x_i), \quad i = 0,1,2,...,n-2$$

The Dirac delta function, $\delta(x)$ is defined such that it is zero everywhere except at $x = 0$, where it is infinite in such a way that

$$\int_{-\infty}^{\infty} \delta(x)\,dx = 1$$

A consequence of this definition is that

$$\int_{-\infty}^{\infty} f(x)\delta(x-x_i)\,dx = f(x_i)$$

Equation (8.62) then reduce to

$$R(x_i) = 0, \qquad i = 0,1,2,...,n-2 \tag{8.64}$$

8.7.2 Subdomain Collocation (or Subdomain) Method

In this method, the domain (a, b) is divided into $(n-1)$ subdomains $\Omega_0, \Omega_1,...,\Omega_{n-2}$ and the weights are taken as unity over the corresponding subdomain and zero everywhere else, i.e.,

$$w_i(x) = 1 \text{ over } \Omega_i \text{ and zero everywhere else}, \quad i = 0,1,2,...,n-2$$

Equation (8.62) then reduces to

$$\int_{\Omega_i} R(x)\,dx = 0, \quad i = 0, 1, 2, ..., n-2 \tag{8.65}$$

8.7.3 Least Squares Method

In this method, the L_2 norm of the residual

$$\left(\int_a^b R^2(x)dx\right)$$

is minimized. Using the boundary conditions, we could express two of the coefficients, say c_{n-1} and c_n in terms of others and the range of i in Eq. (8.62) would again be from 0 to $n-2$ (equivalently, we could solve it as a constrained optimization problem with the two constraints coming from the boundary conditions. However, we do not describe this option here). From the stationarity condition, it is readily seen that this corresponds to using weights which are partial derivatives of the residual with each of the undetermined coefficients, c_i, i.e.,

$$w_i(x) = \frac{\partial R(x)}{\partial c_i}, \quad i = 0,1,...,n-2$$

and Eq. (8.62) reduces to

$$\int_a^b \frac{\partial R(x)}{\partial c_i} R(x)dx = 0, \qquad i = 0,1,...,n-2 \tag{8.66}$$

8.7.4 Method of Moments

In this method, the weight functions are taken as a family of polynomials, i.e.,

$$w_i(x) = x^i, \qquad i = 0,1,2,\ldots,n-2$$

which amounts to taking the ith moment of the residual, and Eq. (8.62) reduce to

$$\int_a^b x^i R(x)\,dx = 0, \qquad i = 0,1,2,\ldots,n \tag{8.67}$$

The method works well if n is small but not for large n, since high powers of x are likely to become linearly dependent due to round-off errors.

8.7.5 Galerkin Method

This is the most widely used method in the family of the weighted residual methods. In this method, the trial solution is written in terms of basis functions and the test functions are either the basis functions (Bubnov-Galerkin or Galerkin method) or the sum of the basis function and an additional term (Petrov-Galerkin method) which may improve the solution accuracy in some cases. Here, we will discuss only the Bubnov-Galerkin method.

We start by assuming that the trial solution satisfies the boundary conditions *a priori*. More discussion on boundary condition follows later in this subsection. We would then need as many equations as there are undetermined coefficients. The weight functions are taken as the basis functions:

$$w_i(x) = \phi_i(x), \qquad i = 0,1,2,\ldots,n$$

and Eq. (8.62) reduces to

$$\int_a^b \phi_i(x)R(x)\,dx = 0, \qquad i = 0,1,2,\ldots,n \tag{8.68}$$

which is a set of $(n+1)$ linear equations in the coefficients. It should be noted that, written in this form and assuming the operator $\mathcal{L}$ to be of second order, the basis functions should satisfy the condition of *twice differentiability*. However, it is readily seen that integration by parts reduces the highest-order term as follows:

$$\int_a^b \phi_i \frac{d}{dx}\left(A_2 \frac{d}{dx}\sum_{j=0}^n c_j \phi_j\right) dx = \left[\phi_i A_2 \frac{d\tilde{y}}{dx}\right]_a^b - \sum_{j=0}^n c_j \int_a^b \frac{d\phi_i}{dx} A_2 \frac{d\phi_j}{dx} d \tag{8.69}$$

Thus, the more strong *twice differentiability* requirement is replaced by a *weaker square integrability of the first derivative* requirement, admitting basis functions from a larger function space. This implies that we may choose basis functions which are not twice-differentiable at a finite number of points. Also note the presence of the boundary term in the above expression, which brings us to the question—

what boundary conditions should be satisfied by the basis functions used in the trial solution in the weak formulation. We discuss this aspect for a simpler case in which the operator $\mathcal{L}$ is defined such that

$$\mathcal{L}y = \frac{d}{dx}\left[A_2(x)\frac{dy}{dx}\right]$$

Similar analysis could be performed for cases where the operator includes additional higher or lower order terms.

The weighted residual formulation is written as

$$\int_a^b \phi_i(x)\left[\mathcal{L}\tilde{y} - f(x)\right]dx = 0, \qquad i = 0,1,2,...,n$$

which, on integration by parts, becomes

$$\left[\phi_i A_2 \frac{d\tilde{y}}{dx}\right]_a^b - \int_a^b \frac{d\phi_i}{dx}A_2\frac{d\tilde{y}}{dx}dx - \int_a^b \phi_i f\,dx = 0 \qquad (8.70)$$

The first term on the left-hand side, using the boundary conditions [Eq. (8.63)], may be written as

$$\phi_i(b)A_2(b)d_b - \phi_i(a)A_2(a)\frac{d\tilde{y}}{dx}\bigg|_{x=a}$$

implying that the specified gradient at $x = b(d_b)$ is naturally incorporated in the formulation leading to the nomenclature of *natural boundary condition* (or, sometimes, *weak* boundary condition) for it. However, the specified function value at $x = a$ (y_a) is not accounted for and also we do not know the gradient at $x = a$. To get around this problem, the easiest approach (as suggested by the expression above) would be to stipulate that all the basis functions should vanish at $x = a$, thus eliminating the second term. However, since $y = y_a$, at $x = a$, we must have one of the basis functions, say ϕ_0, equal to 1 at $x = a$, and its coefficient c_0 equal to y_a.

We would need n additional equations to solve for the remaining coefficients $(c_1, c_2,...,c_n)$, which are obtained by using the n weights as the basis functions ϕ_1, $\phi_2,...,\phi_n$ (note that we would not be able to use ϕ_0 as a weight since dy/dx at $x = a$ is not known). Since the weak form requires the basis functions to essentially assume a fixed value at the boundary where the function value is specified, it is also called an *essential boundary condition* or, sometimes, a *strong boundary condition*.

The following example illustrates the use of these methods of weighted residuals.

Example 8.15   Solve the differential equation

$$\frac{d^2y}{dx^2} + \frac{8}{(1+x)^3} = 0 \quad \text{over } x = (0,1)$$

with the boundary conditions

$$y(0) = 0; \quad \left.\frac{dy}{dx}\right|_{x=1} = 1$$

using the method of weighted residuals and assuming the trial solution to be a cubic polynomial.

Solution   We do not require the trial solution to satisfy the boundary conditions a priori, and assume the trial solution to be

$$\tilde{y} = c_0 + c_1 x + c_2 x^2 + c_3 x^3$$

(i) *Point collocation method*

Application of the boundary conditions results in $c_0 = 0$ and $c_1 + 2c_2 + 3c_3 = 1$. The other two equations are obtained by choosing the collocation points as $x_0 = 1/3$ and $x_1 = 2/3$ (while any two points could be chosen, in absence of any other information, we prefer these to be symmetrically located) and give rise to

$$R\left(\frac{1}{3}\right) = 2c_2 + 6c_3\frac{1}{3} + \frac{8}{\left(1+\frac{1}{3}\right)^3} = 0 \Rightarrow c_2 + c_3 + \frac{27}{16} = 0$$

$$R\left(\frac{2}{3}\right) = 2c_2 + 6c_3\frac{2}{3} + \frac{8}{\left(1+\frac{2}{3}\right)^3} = 0 \Rightarrow c_2 + 2c_3 + \frac{108}{125} = 0$$

which results in $c_2 = -2.511$ and $c_3 = 0.823500$. c_1 is then obtained as 3.5515 and the approximate solution as

$$\tilde{y} = 3.55150x - 2.51100x^2 + 0.8235x^3$$

(ii) *Subdomain collocation method*

We again have $c_0 = 0$ and $c_1 + 2c_2 + 3c_3 = 1$. The other two equations are obtained by choosing the collocation domains as (0, 1/2) and (1/2, 1) and give rise to

$$\int_0^{1/2} 2c_2 + 6c_3 x + \frac{8}{(1+x)^3}dx = 0 \quad \Rightarrow \quad c_2 + \frac{3}{4}c_3 + \frac{20}{9} = 0$$

$$\int_{1/2}^1 2c_2 + 6c_3 x + \frac{8}{(1+x)^3}dx = 0 \quad \Rightarrow \quad c_2 + \frac{9}{4}c_3 + \frac{7}{9} = 0$$

which results in $c_2 = -2.94444$ and $c_3 = 0.962963$. c_1 is then obtained as 4.00000 and the approximate solution as

$$\tilde{y} = 4.00000x - 2.94444x^2 + 0.962963x^3$$

(iii) *Least squares method*

Applying the boundary conditions, the trial solution is written as

$$\tilde{y} = x + c_2(x^2 - 2x) + c_3(x^3 - 3x)$$

and the residual is obtained as

$$\frac{d^2\tilde{y}}{dx^2} + \frac{8}{(1+x)^3} = 2c_2 + 6c_3 x + \frac{8}{(1+x)^3}$$

Minimization of the L_2 norm of the residual results in

$$\int_0^1 2\left[2c_2 + 6c_3 x + \frac{8}{(1+x)^3}\right] dx = 0 \;\Rightarrow\; c_2 + \frac{3}{2}c_3 + \frac{3}{2} = 0$$

$$\int_0^1 6x\left[2c_2 + 6c_3 x + \frac{8}{(1+x)^3}\right] dx = 0 \;\Rightarrow\; c_2 + 2c_3 + 1 = 0$$

which provides $c_2 = -3$, $c_3 = 1$, and the approximate solution as

$$\tilde{y} = 4x - 3x^2 + x^3$$

(iv) *Method of moments*

We again have $c_0 = 0$ and $c_1 + 2c_2 + 3c_3 = 1$. The other two equations are obtained by equating the first two moments of the residual to zero and results in

$$\int_0^1 1\left[2c_2 + 6c_3 x + \frac{8}{(1+x)^3}\right] dx = 0 \;\Rightarrow\; c_2 + \frac{3}{2}c_3 + \frac{3}{2} = 0$$

$$\int_0^1 x\left[2c_2 + 6c_3 x + \frac{8}{(1+x)^3}\right] dx = 0 \;\Rightarrow\; c_2 + 2c_3 + 1 = 0$$

which provides $c_2 = -3$, $c_3 = 1$ (and $c_1 = 4$), and the approximate solution as

$$\tilde{y} = 4x - 3x^2 + x^3$$

Note that the use of polynomial test functions implies that the method of moments and the method of least squares are virtually identical.

(v) *Galerkin method*

We use the weak form of the equations and note that the essential boundary condition is satisfied by the basis functions when the trial solution is written as $\tilde{y} = c_0 + c_1 x + c_2 x^2 + c_3 x^3$ with $c_0 = 0$ [if the function value is specified at $x = 1$, we could use the basis functions as 1, $x - 1$, $(x - 1)^2$, $(x - 1)^3$]. The remaining three coefficients are obtained from

$$\int_0^1 \phi_i(x)\left[\frac{d^2\tilde{y}}{dx^2} + \frac{8}{(1+x)^3}\right] dx = 0, \quad i = 1, 2, 3$$

The weak form is written as

$$\phi_i(1)\times 1 - \int_0^1 \frac{d\phi_i}{dx}\frac{d\tilde{y}}{dx} dx + \int_0^1 \phi_i \frac{8}{(1+x)^3} dx = 0$$

which results in the equations (with the basis functions as x, x^2, and x^3)

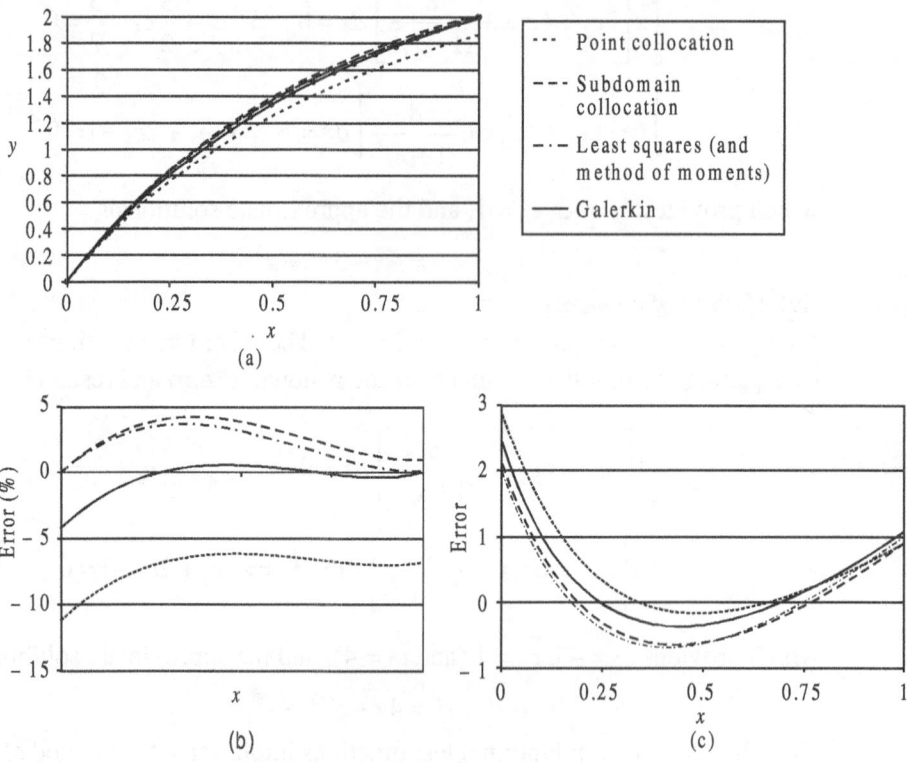

(a)

(b)   (c)

Fig. 8.14 (a) Comparison of approximate solution (lines) and the exact solution (circles).
(b) Percent error in approximate solution, and (c) error in the differential equation, i.e.,

$$\frac{d^2\tilde{y}}{dx^2} + \frac{8}{(1+x)^3}.$$

$$1 \times 1 - \int_0^1 1(c_1 + 2c_2 x + 3c_3 x^2)dx + \int_0^1 x \frac{8}{(1+x)^3}dx = 0$$

$$\Rightarrow \qquad c_1 + c_2 + c_3 = 2$$

$$1 \times 1 - \int_0^1 2x(c_1 + 2c_2 x + 3c_3 x^2)dx + \int_0^1 x^2 \frac{8}{(1+x)^3} dx = 0$$

$$\Rightarrow \qquad c_1 + \frac{4}{3}c_2 + \frac{3}{2}c_3 = 1.54518$$

$$1 \times 1 - \int_0^1 3x^2(c_1 + 2c_2 x + 3c_3 x^2)dx + \int_0^1 x^3 \frac{8}{(1+x)^3}dx - 0$$

$$\Rightarrow \qquad c_1 + \frac{3}{2}c_2 + \frac{9}{5}c_3 = 1.36447$$

from which we get $c_1 = 3.83148$, $c_2 = -2.76551$, and $c_3 = 0.934030$ and the approximate solution as

$$\tilde{y} = 3.83148x - 2.76551x^2 + 0.934030x^3$$

It should be noted that this approximate solution does not satisfy the gradient boundary condition at $x = 1$ (the gradient is equal to 1.10255 as against the specified value of 1). In other words, by relaxing the condition at this boundary, we will have some error in the boundary condition but hope that the overall solution accuracy would be higher. It is instructive to compare the approximate solutions obtained from these methods as shown in Fig. 8.14.

The Galerkin method appears to be the best. However, we must keep in mind that we have relaxed one boundary condition and therefore have an additional degree of freedom. Also note from Fig. 8.14(c) that the residual is zero at the collocation points ($x = 1/3$ and $2/3$) for the point collocation method.

For this simple problem, the computations were not very involved. However, in practice, we would require a much higher accuracy implying that the number of undetermined coefficients would be very large. The system of equations to be solved would be of order n and would typically not be sparse or banded since the basis functions (or their derivatives) do not vanish. Clearly, a more efficient scheme could be devised in which the basis functions would have a *local support* and the resulting system of equation would be banded. This leads us to the *finite element method* which is described in the next subsection.

8.7.6 Finite Element Method

In this method, we divide the computational domain in a number of *finite elements*. Each element has a number of *nodes*, the most common being a two-noded element having nodes at either end (a three-noded element with an additional node at the centre is sometimes used to provide a higher-order local continuity). We describe only one-dimensional cases. For two-dimensional problems, the elements would be triangles, rectangles, or other polygons, and not line segments. Let the domain be divided into n elements with $n + 1$ nodes (Fig. 8.15). The approximate solution is written as

$$\tilde{y} = \sum_{j=0}^{n} L_j(x)\tilde{y}_j \tag{8.71}$$

in which $\tilde{y}_j$ denotes the nodal value obtained from the approximate solution and L is similar to (but not same as) the Lagrange interpolating polynomial (see Section 4.4)

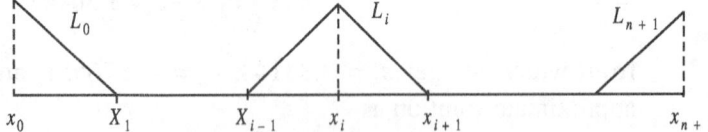

Fig. 8.15 Discretisation into finite elements and some shape functions.

and is commonly called the *shape function* (since it describes the shape of the approximate solution). Clearly L_j must be unity at $x = x_j$ and zero at all other nodes.

The most commonly used shape functions are piecewise linear, defined as (Fig. 8.15)

$$L_j = \begin{cases} \dfrac{x - x_{j-1}}{x_j - x_{j-1}}, & x_{j-1} \le x \le x_j \\[2ex] \dfrac{x_{j+1} - x}{x_{j+1} - x_j}, & x_j \le x \le x_{j+1} \\[2ex] 0, & \text{elsewhere} \end{cases} \tag{8.72}$$

As a result, the integrals involving the shape function L_j (and its derivatives) would vanish everywhere except over the jth and $(j + 1)$th element and we would get a *tridiagonal* system of equations which may be efficiently solved to obtain the nodal values, $\tilde{y}_j$. Note that the essential boundary condition is automatically satisfied by the shape functions. Obviously, the assumption of piecewise linear variation would not work with the second-order differential equation since it would not be twice-differentiable at the nodes. However, since the weak form requires only the square integrability of the first derivative, this should work. For higher-order equations, we would use, for example, three-noded elements with piecewise parabolic variation or a two-noded Hermitian element. These are not discussed here. The following example illustrates the application of the finite element method.

Example 8.16   Solve the differential equation

$$\frac{d^2y}{dx^2} + \frac{8}{(1+x)^3} = 0 \ \text{ over } x = (0,1)$$

with the boundary conditions

$$y(0) = 0, \ \left.\frac{dy}{dx}\right|_{x=1} = 1$$

using the finite element method with four elements.

Solution   We use four elements of equal length ($\Delta x = 0.25$) and write the approximate solution as

$$\tilde{y} = \sum_{j=0}^{4} L_j(x)\tilde{y}_j$$

Our aim is to solve for the nodal values which, with the assumed piecewise linear variation, would provide the solution throughout the domain.

The Galerkin method applied to the weak form of the equation is written as

$$L_i(1)\times 1 - L_i(0)\left.\frac{d\tilde{y}}{dx}\right|_{x=0} - \int_0^1 \frac{dL_i}{dx}\frac{d\sum_{j=0}^{4}L_j\tilde{y}_j}{dx}\,dx + \int_0^1 L_i\frac{8}{(1+x)^3}\,dx = 0, \quad i=0,1,2,3,4$$

(8.73)

Note that all the shape functions, except L_0, are zero at $x = 0$, satisfying the essential boundary condition. Moreover, $\tilde{y}_0 = 0$, to satisfy this boundary condition. The other four nodal values need to be obtained from the weighted residual formulation and using $i = 1, 2, 3, 4$, would be sufficient to get the four equations. However, we would write the full set of five equations and apply the essential boundary condition at the end. Equation (8.73) is written as

$$\int_0^1 \frac{dL_i}{dx}\frac{d\sum_{j=0}^{4}L_j\tilde{y}_j}{dx}\,dx = L_i(1)\times 1 - L_i(0)\left.\frac{d\tilde{y}}{dx}\right|_{x=0} + \int_0^1 L_i\frac{8}{(1+x)^3}\,dx, \quad i=0,1,2,3,4$$

or, since the nodal values are not functions of x,

$$\sum_{j=1}^{4}\left(\int_{0}^{1}\frac{dL_i}{dx}\frac{dL_j}{dx}dx\right)\tilde{y}_j = L_i(1)\times 1 - L_i(0)\frac{d\tilde{y}}{dx}\bigg|_{x=0} + \int_{0}^{1}L_i\frac{8}{(1+x)^3}dx, \quad i=0,1,2,3,4$$

These equations are written in a compact form as

$$[A]\{\tilde{y}\} = \{b\}$$

where the 5×5 coefficient matrix A is given by

$$a_{ij} = \int_{0}^{1}\frac{dL_i}{dx}\frac{dL_j}{dx}dx$$

and the 5×1 vector b is given by

$$b_i = L_i(1)\times 1 - L_i(0)\frac{d\tilde{y}}{dx}\bigg|_{x=0} + \int_{0}^{1}L_i\frac{8}{(1+x)^3}dx$$

One should note from the definition of A that, in this case, it is symmetric. Moreover, although the integral is over the entire domain, it would effectively involve only those elements over which the shape functions are non-zero, i.e., only the element before the corresponding node and the element after it. It is, therefore, expedient to write the integral as the sum of integral over the finite elements and add the contribution of the appropriate elements. A local element matrix (2×2) is defined for this purpose as (see Fig. 8.16)

$$a_{ij}^e = \int_{0}^{\Delta x}\frac{dl_i}{dx}\frac{dl_j}{dx}dx, \quad i,j=1,2 \tag{8.74}$$

which, on using $l_1 = 1 - x/\Delta x$ and $l_2 = x/\Delta x$, becomes

$$[A]^e = \frac{1}{\Delta x}\begin{bmatrix} 1 & -1 \\ -1 & 1 \end{bmatrix} \tag{8.75}$$

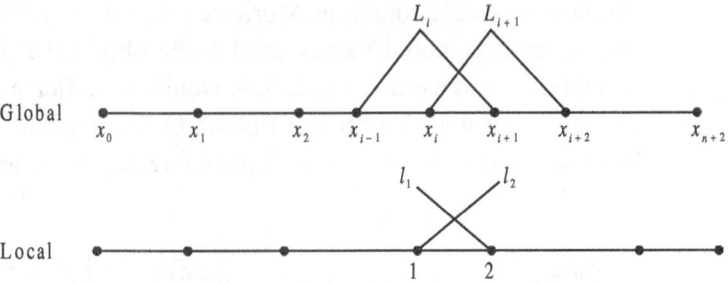

Fig. 8.16   Local and global shape functions.

Here, since all elements are of the same length, the element matrices are identical as $\begin{bmatrix} 4 & -4 \\ -4 & 4 \end{bmatrix}$.

We now detail the procedure of obtaining the equations by considering three rows of the matrix ($i = 0$, 1, and 4) and *assembling* the element matrices.

For $i = 0$, the weight is L_0 which is non-zero only for the first element. Also note that L_0 corresponds to the local shape function l_1 for this element. Similarly, the jth column of this row involves the shape function L_j, which would vanish over the first element for $j = 2, 3$, and 4. Moreover, L_1 corresponds to the local shape function l_2 for the first element. Therefore, the top row of the coefficient matrix will have only two entries, a_{00} and a_{01}, which are given by

$$a_{00} = \int_{\Sigma\text{element 1}} \frac{dL_0}{dx} \frac{dL_0}{dx}\, dx = a_{11}^e \Big|_{\text{element 1}} = 4$$

$$a_{01} = \int_{\Sigma\text{element 1}} \frac{dL_0}{dx} \frac{dL_1}{dx}\, dx = a_{12}^e \Big|_{\text{element 1}} = -4$$

The top element of the right-hand side vector is given by

$$b_0 = L_0(1) \times 1 - L_0(0) \frac{d\tilde{y}}{dx}\Big|_{x=0} + \int_0^{0.25} \frac{0.25 - x}{0.25} \frac{8}{(1+x)^3}\, dx$$

$$= 0 \qquad - \qquad \frac{d\tilde{y}}{dx}\Big|_{x=0} + \frac{4}{5}$$

For $i = 1$, the weight is L_1 which is non-zero only for the first and second elements. L_1 corresponds to the local shape function, l_2 for the first element, and the local shape function l_1 for the second element. Similarly, the jth column of this row involves the shape function L_j, which would vanish over both these elements for $j = 3$ and 4, over the second element for $j = 0$, and over the first element for $j = 2$. Moreover, L_0 corresponds to the local shape function, l_1 for the first element, and L_2 corresponds to the local shape function l_2 for the second element. Therefore, the next row of the coefficient matrix will have only three entries, a_{10}, a_{11}, and a_{12}, which are given by

$$a_{10} = \int_{\Sigma\text{element 1}} \frac{dL_1}{dx} \frac{dL_0}{dx}\, dx = a_{21}^e \Big|_{\text{element 1}} = -4$$

$$a_{11} = \int_{\Sigma\text{element 1, element 2}} \frac{dL_1}{dx} \frac{dL_1}{dx}\, dx = a_{22}^e \Big|_{\text{element 1}} + a_{11}^e \Big|_{\text{element 2}} = 8$$

$$a_{12} = \int_{\Sigma\text{element 2}} \frac{dL_1}{dx} \frac{dL_2}{dx}\, dx = a_{12}^e \Big|_{\text{element 2}} = -4$$

The corresponding element of the right-hand side vector is given by

$$b_1 = L_1(1) \times 1 - L_1(0) \left. \frac{d\tilde{y}}{dx} \right|_{x=0} + \int_0^1 L_1 \frac{8}{(1+x)^3} \, dx$$

$$= 0 - 0 + \int_0^{0.25} \frac{x}{0.25} \frac{8}{(1+x)^3} \, dx + \int_{0.25}^{0.5} \frac{0.5-x}{0.25} \frac{8}{(1+x)^3} \, dx$$

$$= 0 - 0 + \frac{16}{25} + \frac{32}{75}$$

For $i = 4$, the weight is L_4 which is non-zero only for the last element. L_4 corresponds to the local shape function l_2 for this element. The jth column of this row involves the shape function L_j, which would vanish over this element for $j = 0$, 1 and 2. L_3 corresponds to the local shape function l_1 for the first element. Therefore, the bottom row of the coefficient matrix will have only two entries, a_{43} and a_{44}, which are given by

$$a_{43} = \int_{\Sigma \text{element } 4} \frac{dL_4}{dx} \frac{dL_3}{dx} \, dx = a_{21}^e \Big|_{\text{element } 4} = -4$$

$$a_{44} = \int_{\Sigma \text{element } 4} \frac{dL_4}{dx} \frac{dL_4}{dx} \, dx = a_{22}^e \Big|_{\text{element } 4} = 4$$

The bottom element of the right-hand side vector is given by

$$b_4 = L_4(1) \times 1 - L_4(0) \left. \frac{d\tilde{y}}{dx} \right|_{x=0} + \int_{0.75}^{1.0} \frac{x-0.75}{0.25} \frac{8}{(1+x)^3} \, dx$$

$$= 1 - 0 + \frac{1}{7} = \frac{8}{7}$$

The assembled equations are written in the following detailed form to explain the process:

$$\begin{bmatrix} 4 & -4 & & & \\ -4 & 4+4 & -4 & & \\ & -4 & 4+4 & -4 & \\ & & -4 & 4+4 & -4 \\ & & & -4 & 4 \end{bmatrix} \begin{Bmatrix} \tilde{y}_0 \\ \tilde{y}_1 \\ \tilde{y}_2 \\ \tilde{y}_3 \\ \tilde{y}_4 \end{Bmatrix} = \begin{Bmatrix} -\tilde{y}'(0) + {}^4\!/\!_5 \\ \left({}^{16}\!/\!_{25}\right) + \left({}^{32}\!/\!_{75}\right) \\ \left({}^{16}\!/\!_{45}\right) + \left({}^{16}\!/\!_{63}\right) \\ \left({}^{32}\!/\!_{147}\right) + \left({}^{8}\!/\!_{49}\right) \\ 8/7 \end{Bmatrix}$$

$$\Rightarrow \begin{bmatrix} 4 & -4 & & & \\ -4 & 8 & -4 & & \\ & -4 & 8 & -4 & \\ & & -4 & 8 & -4 \\ & & & -4 & 4 \end{bmatrix} \begin{bmatrix} \tilde{y}_0 \\ \tilde{y}_1 \\ \tilde{y}_2 \\ \tilde{y}_3 \\ \tilde{y}_4 \end{bmatrix} = \begin{Bmatrix} -\tilde{y}'(0)+(4/5) \\ 16/15 \\ 64/105 \\ 8/21 \\ 8/7 \end{Bmatrix}$$

By applying the essential boundary condition ($\tilde{y}_0 = 0$), the tridiagonal system is solved to obtain $\tilde{y}_1 = 0.800000$, $\tilde{y}_2 = 1.33333$, $\tilde{y}_3 = 1.71429$, $\tilde{y}_4 = 2.00000$. It should be noted that these nodal values match the analytical solution exactly. However, the approximate solution is assumed to vary linearly within the elements and does not match the analytical solution (Fig. 8.17). Note that, the first derivative of the approximate solution at $x = 1$ is equal to $(\tilde{y}_4 - \tilde{y}_3)/0.25 = 1.14286$, which does not match the specified value of 1. Also, from the first equation, we get the approximate value of the first derivative at $x = 0$ as

$$\tilde{y}'(0) = \frac{4}{5} + 4\tilde{y}_1 = 4$$

which is the exact value.

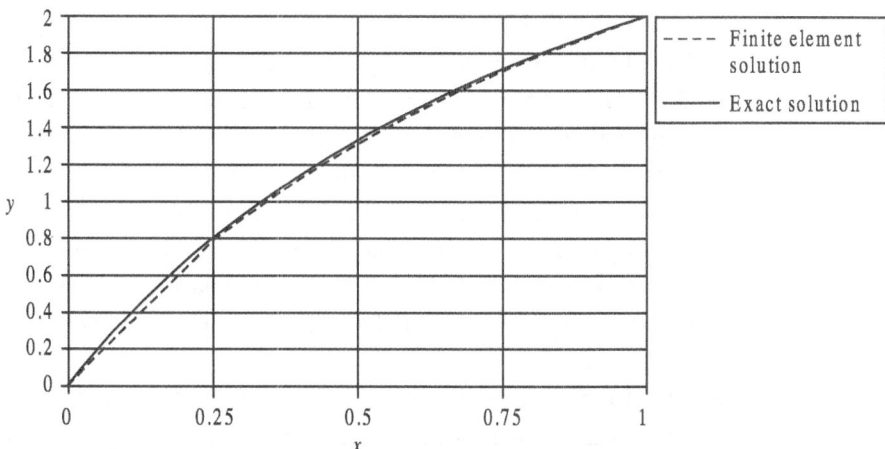

Fig. 8.17 Comparison of finite element solution and the exact solution.

The finite element method is particularly well suited for transient problems since the coefficient matrix is time-invariant (for linear equations) and needs to be computed only once. For subsequent time steps, only the right-hand side vector needs to be re-computed. The LU decomposition would be very efficient in such cases (see Chapter 2).

We have provided a very elementary introduction to the finite element method. Issues like higher-order elements in one-dimensional problems or multi-dimensional problems have not been discussed. The interested reader should refer to Strang and Fix (1973) or Zienkiewicz and Taylor (2000).

8.8 SPATIAL DISCRETIZATION ON IRREGULAR BOUNDARY

Majority of the natural science and engineering problems deal with two or three spatial dimensions. In order to reduce the quantum of computation and geometric complications, three-dimensional problems are often converted (approximated) to two-dimensional problems through some practical or plausible and occasionally ridiculous assumptions (read spherical cow or cylindrical horse). In Chapter 7, we have shown how to obtain spatial discretization in two dimensions for both Lapace equation and two-dimensional diffusion equation. The same could have easily been expanded to the two-dimensional wave equation. However, we always dealt with boundaries that aligned with Cartesian system (Fig. 7.4) or polar system (Exercise 7.3, Problem 7).

In this section, we shall explain how to write difference approximation for the spatial derivatives when the boundaries of the domains do not align with any particular co-ordinate systems. Typically, the problem that arise out of an irregular boundary can be broken down to three parts as follows:

- How to lay a grid on an irregular boundary?
- Difference approximation on irregular grid. We have seen in Chapter 7 that two-dimensional problem does not change the difference approximations derived for one-dimension in Chapter 5. Rather, the same one-dimensional approximations were applied independent in two different directions. However, in Chapter 5, we only derived finite different approximations where all the grid spacings were equal. The irregular boundary will require derivation of finite difference approximations for spatial derivatives when adjacent grid spacings are unequal.
- How to incorporate derivative boundary conditions on irregular boundary?

We will address each of these problems in the subsequent subsections.

8.8.1 Grid Layout

Basic principle of laying out a grid on any irregular geometry is to locate the outer rectangle, i.e., the rectangle whose sides are tangent to the farthest points of the irregular boundary on all four sides. If the equation to be solved is in the polar coordinate, draw the outer circle instead, i.e., segment of a disc that is tangent to the irregular boundary on at least three points. A regular grid will then be formed on the outer rectangle or circle. The boundary points will be located between two regular grid points from the geometry and actual measurements or map or contour, depending on the problem concerned. Let us start with the easiest example of Fig. 7.4 with one of the boundary at an offset (Fig. 8.18). The new domain is *ABED* in place of older *ABCD*. The first type boundary condition along *CD*, $\phi = c$, is now specified along *ED*. We have first drawn the same grid as in Fig. 7.4. The new boundary has excluded from the region, the interior nodes 15, 16, and 6 boundary nodes (including point *C*) along boundaries *CD* and *BC*. Next step is to identify the new boundary nodes at each point where the boundary intersects the grid in either directions. These are marked as 15–19 in Fig. 8.18.

The last job is to compute the distances of the boundary nodes from the nearest interior nodes such as, 13–15, 14–16, 14–17, 11–18, and 12–19. This can be easily computed from the geometry and measurements. For example, let us assume each side of the rectangle (square really!) *ABCD* is of unit length and the included angle $\angle EDC = \theta$. Since the grid is regular, $\Delta x = \Delta y = 0.2$, the following can be easily established

$$\tan(\theta) = \frac{0.4}{1} = 0.4 \tag{8.76}$$

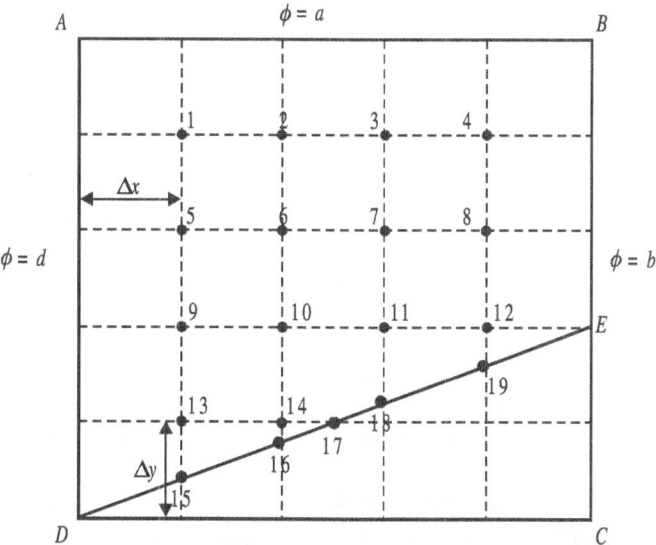

Fig. 8.18 A two-dimensional region with offset boundary.

Therefore, the distances could be calculated as follows:

$$[13-15] = 0.2 - 0.2\tan\theta = 0.12$$
$$[14-16] = 0.2 - 0.4\tan\theta = 0.04$$
$$[14-17] = 0.2\tan\left(\tfrac{\pi}{2}-\theta\right) - 0.4 = 0.1$$
$$[11-18] = 0.4 - 0.6\tan\theta = 0.16$$
$$[11-18] = 0.4 - 0.8\tan\theta = 0.08$$

(8.77)

Let us now consider approximation of the Laplace equation for an interior node adjacent to the boundary, for example Node 14. The approximation molecule is shown in Fig. 8.19 along with scaled distances between the nodes. Notice that the molecule is similar to Fig. 7.5 except for the fact that inter-nodal distances on two sides of the central node is not the same anymore. Actual distances are indicated. A central difference approximation similar to Eq. (7.69) cannot be written for either of the directions (x and y) because the approximation was derived based on the fact that the distances between the nodes (or grid sizes) were uniform. Therefore, we need to derive similar difference approximations for the spatial derivatives where grid sizes are not uniform. This will be done in Section 8.8.2.

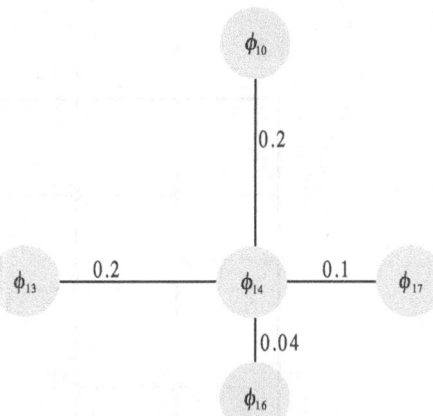

Fig. 8.19 Approximation molecule of Node 14 of Fig. 8.16.

While for Node 14 the distances are different on both the directions, this may not be the case for all the nodes adjacent to the irregular boundary. For example, the Nodes 11, 12, and 13, the irregular grid approximation applies only in y-direction while Eq. (7.69) can still be applied in x-direction.

Let us now look at another example (Fig. 8.20), where the distances adjacent to the boundary cannot be computed based on simple geometry.

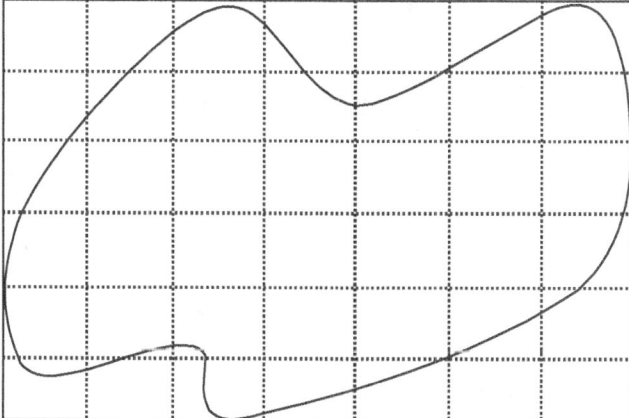

Fig. 8.20   Actual domain shown in dark boundary. The outer rectangle is drawn in solid lines. The dotted lines show the regular grid.

The arbitrary boundary shown in Fig. 8.20 can arise naturally in many engineering areas. For example, it may represent the boundary of a lake for water quality modeling or a plume of contaminant in the subsurface or shape of an object where temperature distribution has to be computed. In these cases, an exact scaled map is required for the domain on which the outer rectangle can be laid. The scaled regular grid is to be placed on top of the map. The exact grid spacing near the boundary or the location of the boundary points have to be transferred from the scaled map to the real program.

The grid may be uniform in each of the x and y directions but it is not necessary to have $\Delta x = \Delta y$, since the approximations are independent in each directions [Eq. (7.69)]. For smaller grid spacing, a large number of grid points will be placed in the interior. Therefore, uniform grid approximations similar to Eq. (7.69) are suitable for majority of the grid points. However, near the boundary, one requires the approximation on non-uniform grid, which we shall derive in the next section.

8.8.2 Finite Difference Approximation on a Non-uniform Grid

Our goal is to derive approximations for derivatives in terms of values at unequally spaced fixed points. Let us start with three adjacent points $(i-1, i, i+1)$ as shown in Fig. 8.21. Let us denote the interval between two adjacent nodes (say $i-1$ and i) as Δx. This may be the regular grid spacing of the irregular boundary problems described in the previous section. Then the other interval (between i and $i + 1$) can be represented as $\alpha \Delta x$, where $0 < \alpha < 1$.

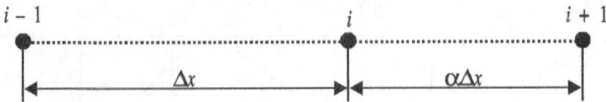

Fig. 8.21 A typical grid showing unequal grid spacing.

Let us now assume that the state variable is $f(x)$, and we intend to obtain a central difference approximation of

$$\frac{df}{dx} \quad \text{and} \quad \frac{d^2 f}{dx^2}$$

at node i as a function of values at the nodes $i - 1$, i, and $i + 1$. Such approximation can be written as

$$\frac{df_i}{dx} = a_1 f_{i-1} + a_2 f_i + a_3 f_{i+1} \tag{8.78}$$

$$\frac{d^2 f_i}{dx^2} = b_1 f_{i-1} + b_2 f_i + b_3 f_{i+1} \tag{8.79}$$

The goal is to estimate the coefficients a_j's and b_j's. We also have to make sure that the approximations are consistent. In order to establish that, we need to obtain the truncation error of the approximation and ensure that it approaches zero as $\Delta x \to 0$. We start by writing the following expansions using Taylor's series for the nodes shown in Fig. 8.21:

$$f_{i-1} = f(x_i - \Delta x) = f_i - \Delta x \frac{df_i}{dx} + \frac{\Delta x^2}{2} \frac{d^2 f_i}{dx^2} - \frac{\Delta x^3}{6} \frac{d^3 f_i}{dx^3} + \frac{\Delta x^4}{24} \frac{d^4 f_i}{dx^4} + \text{HOT}$$

$$\tag{8.80}$$

$$f_{i+1} = f(x_i + \alpha \Delta x) = f_i + \alpha \Delta x \frac{df_i}{dx} + \frac{(\alpha \Delta x)^2}{2} \frac{d^2 f_i}{dx^2} + \frac{(\alpha \Delta x)^3}{6} \frac{d^3 f_i}{dx^3}$$

$$+ \frac{(\alpha \Delta x)^4}{24} \frac{d^4 f_i}{dx^4} + \text{HOT} \tag{8.81}$$

Using these expansions in Eq. (8.78) and grouping similar terms yield

$$\frac{df_i}{dx} = (a_1 + a_2 + a_3) f_i + (-a_1 \Delta x + a_3 \alpha \Delta x) \frac{df_i}{dx} + \left\{ a_1 \frac{\Delta x^2}{2} + a_3 \frac{(\alpha \Delta x)^2}{2} \right\} \frac{d^2 f_i}{dx^2} +$$

$$\left\{ -a_1 \frac{\Delta x^3}{6} + a_3 \frac{(\alpha \Delta x)^3}{6} \right\} \frac{d^3 f_i}{dx^3} + \left\{ a_1 \frac{\Delta x^4}{24} + a_3 \frac{(\alpha \Delta x)^4}{24} \right\} \frac{d^4 f_i}{dx^4} + \text{HOT} \tag{8.82}$$

For a valid approximation of the left-hand side, we will now match the coefficients of each term of the right-hand side with the coefficient of the similar term on the left. Since, there are three adjustable parameters in a_j's, we can match a maximum of 3 terms. The rest will be residual to the approximation and lead to the truncation error. By matching three coefficients of the left-hand side with those of right leads to the following three equations:

$$a_1 + a_2 + a_3 = 0$$
$$-a_1 \Delta x + a_3 \alpha \Delta x = 1$$
$$a_1 \frac{\Delta x^2}{2} + a_3 \frac{(\alpha \Delta x)^2}{2} = 0$$

(8.83)

Solving the set of equations, we obtain

$$a_1 = -\frac{\alpha}{(\alpha + 1)\Delta x}, \quad a_2 = \frac{(\alpha - 1)}{\alpha \Delta x}, \quad \text{and } a_3 = \frac{1}{\alpha(\alpha + 1)\Delta x}$$

(8.84)

Using these values, the approximation of the first derivative we obtain from Eq. (8.78) as

$$\frac{df_i}{dx} = \frac{f_{i+1} + (\alpha^2 - 1)f_i - \alpha^2 f_{i-1}}{\alpha(\alpha + 1)\Delta x}$$

(8.85)

The truncation error can be obtained by using the values of Eq. (8.84) in Eq. (8.82) and deducting the approximation of Eq. (8.85) from it,

$$TE = \frac{\alpha \Delta x^2}{6} \frac{d^3 f_i}{dx^3} + \frac{\alpha(\alpha - 1)\Delta x^3}{24} \frac{d^4 f_i}{dx^4} + HOT$$

(8.86)

The reader can test that for $\alpha = 1$, the approximations degenerate to the uniform grid formula. For second derivative approximation shown in Eq. (8.79), we can proceed similarly to obtain the following set of equations for the constants:

$$b_1 + b_2 + b_3 = 0$$
$$-b_1 \Delta x + b_3 \alpha \Delta x = 0$$
$$b_1 \frac{\Delta x^2}{2} + b_3 \frac{(\alpha \Delta x)^2}{2} = 1$$

(8.87)

Solving these equations, we get the constants as

$$b_1 = \frac{2}{(\alpha + 1)\Delta x^2}, \quad b_2 = -\frac{2}{\alpha \Delta x^2}, \quad \text{and } b_3 = \frac{2}{\alpha(\alpha + 1)\Delta x^2}$$

(8.88)

The approximation and the truncation errors are obtained as follows:

$$\frac{d^2 f_i}{dx^2} = \frac{f_{i+1} - (\alpha + 1)f_i + \alpha f_{i-1}}{\frac{\alpha}{2}(\alpha + 1)\Delta x^2} \tag{8.89}$$

$$TE = \frac{(\alpha - 1)\Delta x}{3}\frac{d^3 f_i}{dx^3} + \frac{(\alpha^2 - \alpha + 1)\Delta x^2}{12}\frac{d^4 f_i}{dx^4} + HOT \tag{8.90}$$

It is interesting to note that the approximation is first order with unevenly spaced points but second order for evenly spaced points. Higher-order approximation would require involving more points, as there will be more number of adjustable constants and as a result, we will be able to equate more terms of Eq. (8.82). Similarly, approximations can also be obtained for higher-order derivatives involving more points. For example, an approximation of third order derivative would require equating at least four terms of right hand side of Eq. (8.82). This would mean at least four adjustable constants which would in turn translate to involving four nodal points in equations similar to Eqs (8.78) and (8.79).

So far, we have derived expressions for central difference approximations. Let us show one derivation for a backward difference approximation of first-order derivative using non-uniform grid, which may be required for simulating boundary conditions. The problem can be stated as

$$\frac{df_i}{dx} = a_1 f_i + a_2 f_{i-1} + a_3 f_{i-2} \tag{8.91}$$

The grid for a typical such problem is depicted in Fig. 8.22.

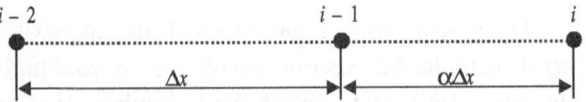

Fig. 8.22 The irregular grid for backward difference approximation.

We now write the following Taylor's series expansions:

$$f_{i-1} = f(x_i - \alpha \Delta x) = f_i - \alpha \Delta x \frac{df_i}{dx} + \frac{(\alpha \Delta x)^2}{2}\frac{d^2 f_i}{dx^2}$$

$$-\frac{(\alpha \Delta x)^3}{6}\frac{d^3 f_i}{dx^3} + \frac{(\alpha \Delta x)^4}{24}\frac{d^4 f_i}{dx^4} + HOT \tag{8.92}$$

$$f_{i-2} = f(x_i - \alpha \Delta x - \Delta x) = f_i - (\alpha + 1)\Delta x \frac{df_i}{dx}$$

$$+ \frac{\{(\alpha + 1)\Delta x\}^2}{2} \frac{d^2 f_i}{dx^2} - \frac{\{(\alpha + 1)\Delta x\}^3}{6} \frac{d^3 f_i}{dx^3} + \frac{\{(\alpha + 1)\Delta x\}^4}{24} \frac{d^4 f_i}{dx^4} + \mathrm{HOT}$$

$$(8.93)$$

Once again, putting the expansions of Eqs (8.92) and (8.93) in Eq. (8.91), grouping the similar terms and equating the coefficients of the first three terms, one obtains the following expression for the constants in Eq. (8.91):

$$a_1 = \frac{2\alpha + 1}{\alpha(\alpha + 1)\Delta x}, \quad a_2 = -\frac{(\alpha + 1)}{\alpha \Delta x}, \quad \text{and } a_3 = \frac{\alpha}{(\alpha + 1)\Delta x} \qquad (8.94)$$

Putting these values in Eq. (8.91), we obtain the required backward difference approximation as

$$\frac{df_i}{dx} = \frac{(2\alpha + 1)f_i - (\alpha + 1)^2 f_{i-1} + \alpha^2 f_{i-2}}{\alpha(\alpha + 1)\Delta x} \qquad (8.95)$$

Similarly, one can also derive a forward difference approximation of the form

$$\frac{df_i}{dx} = a_1 f_i + a_2 f_{i+1} + a_3 f_{i+2} \qquad (8.96)$$

This will be useful if the derivative boundary conditions is on the left side node or boundary. We leave it as exercise for the readers to derive. We now move on to describe the last item, i.e., implementation of derivative boundary conditions on an irregular boundary.

8.8.3 Second and Third Type Boundary Conditions on Irregular Boundary

Let us consider the problem depicted in Fig. 8.18 with a derivative boundary condition along *ED* similar to that on boundary *CD* in Fig. 7.7. The flux normal to line *ED* is equal to *c*. In the problem of Fig. 7.7, the normal to line *CD* was lying along *y*-axis and as a result, $\partial \phi / \partial y = c$ was the condition. In the present case, we will have to set $\partial \phi / \partial \eta = c$ where η is the direction normal to line *ED*. This is shown in Fig. 8.23 for any node lying on ED.

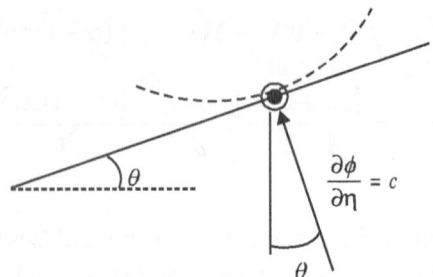

Fig. 8.23   The derivative boundary condition normal to the boundary line on an irregular boundary. The boundary can be straight line making an angle θ with the x-direction or tangent to the curved boundary at the nodal point could be at an angle θ with the x-direction.

Using the geometry, we can compute the component of flux (gradient) in two primary directions, x and y. From the figure, it is easy to see that the following holds:

$$\frac{\partial \phi}{\partial x} = c \sin \theta \quad \text{and} \quad \frac{\partial \phi}{\partial y} = c \cos \theta \qquad (8.97)$$

The sine and cosine of the angle can be easily computed from the geometry. Now, for example on Node 16 in Fig. 8.18, only y-grid line is intersecting. Therefore, a backward difference approximation of the y component of Eq. (8.97) can be written for Node 16 using Eq. (8.95). The derivative in y will be expressed in terms of nodal values at Nodes 16, 14, and 10. The equation can be used directly or used to replace the value at Node 16 in the equation for node 14, similar to Eqs (7.82–7.86). For Node 17 on the other hand, the x-direction component of the flux will be used for the backward difference approximation involving Nodes 17, 14, and 13.

If the node at the boundary is such that both x and y direction grid lines intersect, then both components of Eq. (8.97) have to be approximated with backward or forward difference approximations. In that case, the node will yield two equations.

For an arbitrary curved boundary, direction of the tangent at the node has to be established first. This will automatically give the direction of normal. Then, the same procedure can be followed as explained for the straight boundary *ED*. The curved boundary is also shown in Fig. 8.23. For arbitrary geometry, one will often have to work with a scaled map and work out the direction of tangent at each node. On other hand, if one can fit an equation (a polynomial) to the curved boundary, it is possible to evaluate the direction of tangent analytically or theoretically. However, taking help of geometry with actual measurements from the map is the most common way.

Use of the methods described in the three subsections enable one to handle boundaries of arbitrary shapes and sizes.

8.9 MULTI-GRID METHOD

In this section, we will present some basic concepts of multi-grid techniques used for accelerating convergence of iterative solution of linear systems of equation obtained by discretizing PDEs. It is especially useful for obtaining solution of PDEs of higher dimension where refinement of spatial grid results in multi-fold increase of total node numbers and as a result, dimension of the matrix equation to be solved. For example, solution of Laplace equation on the grid shown in Fig. 7.4 requires solution of a matrix equation of size 16. Halving the mesh size increases the size to 81, and a further halving will result in a matrix of size 361. The solution is typically computed using an iterative method such as Gauss-Siedel because the coefficient matrix is sparse.

If the solution is rapidly changing or has steep gradients, one is often interested in finding solution on a fine mesh. But the price paid is too heavy and most of the computation time is spent iterating for the solution using Gauss-Seidel or similar iterative procedures. The multi-grid acceleration technique helps in computing the solution much faster. The technique uses concepts from Chapters 2 and 4. Let us try to understand the multi-grid techniques intuitively. A full formal treatment is beyond the scope of this text. Let us recall that the initial error in an iterative method for the solution of linear system of algebraic equation could be expressed as linear combination of eigenvectors of the matrix Eq. (2.120):

$$e^{(0)} = \sum_{k=1}^{n} C_k v_k \qquad (8.98)$$

As the iteration progressed, the error in the mth iteration was given by

$$e^{(m)} = \sum_{k=1}^{n} C_k \lambda_k^m v_k \qquad (8.99)$$

In the context of the solution of PDE, the error vector contains the errors in the solution vector. Performing an analysis similar to Section 6.3.2, one can show that the error satisfies the original PDE similar to the case of ODE. Therefore, all the analysis and representation of the solution in Chapter 7 are also valid for the error vector.

Let us highlight the principles of multigrid methods using one-dimensional diffusion equation of Chapter 7. The principles are general and equally applicable to other PDEs as well. The component j in the error vector (represents error at the jth location in one-dimensional diffusion equation) can be represented as

$$e_j^{(m)} = \sum_{k=1}^{N} C_k \lambda_k^m v_{k,j} \qquad (8.100)$$

where $v_{k,j}$ is the jth element of vector v_k.

We have already seen that the λ_k of a tridiagonal matrix obtained from discretization of the PDE can be expressed by Eq. (7.127). The eigenvectors for the same matrix with $N + 1$ nodes are given by

$$v_{k,j} = \left(\frac{c}{b}\right)^{\frac{1}{2}j} \sin\left(\frac{k\pi j}{N}\right) \qquad (8.101)$$

Therefore, the error at any iteration is given by

$$e_j^{(m)} = \sum_{k=1}^{N} C_k \lambda_k^m \sin\left(\frac{k\pi j}{N}\right) \qquad (8.102)$$

You may want to compare this with Eqs (7.143) and (7.145) and see the similarity. Replace

$$x_j = j\Delta x = \frac{j}{N}$$

assuming a normalized spatial domain of length one. The λ_k behaves like an amplification factor at each iteration and C_k are the Fourier coefficients. The k is the wave number and determines how rapidly or smoothly the error is changing.

For convergence to the solution, we have to attenuate the errors at each wave number. Let us look at the sine component of the error. A wave that appears smooth when captured using a fine mesh will appear to be rapidly changing when represented on a coarse mesh. For this, visualize one complete sine wave captured by five points, a zero in the beginning, maxima, zero, minima, and zero again. This will essentially be two triangles and appear rapidly varying. This can also be demonstrated quantitatively.

Let us consider one sine wave with wave number k given by

$$\sin\left(\frac{k\pi j}{N}\right)$$

The complete range of k is 1 to N. Let us focus on a wave number in the lower half range, i.e., $1 \le k \le N/2$ (take $(N-1)/2$ if N is odd, it does not really change anything!). If we sample at the even numbered nodes, the wave gives

$$\sin\left(\frac{k\pi 2j}{N}\right) = \sin\left(\frac{k\pi j}{N/2}\right)$$

So, it is equivalent to looking at the same wave on a coarse grid of $(N/2 + 1)$ points. Therefore, a smooth wave of low wave number on a fine grid $(N + 1)$ appears as a high wave number on a coarse grid $(N/2 + 1)$. This shows that the coarse grid is better in representing the smoothly varying functions, and the fine grid is better for the rapidly varying functions.

Convergence on the other hand depends on λ_k which is inversely proportional to the square of the grid size [Eq. (7.128)]. Therefore, larger the grid size is (coarse grid), faster is the convergence. The relations between resolution capability, convergence, and grid size form the basis of multi-grid method. This essentially says that solution converges rapidly on the coarse grid but attenuates the smoothly varying part of the solution (low wave number).

The fine grid on the other hand attenuates the rapidly varying part of the solution but the convergence is slow. Therefore, an ideal solution will be to go back and forth between coarse and fine grids by alternatively attenuating the rapidly varying part and smoothly varying part of the solution such that no part grows unbounded. When all the parts of the solution (entire range of wave number) reach attenuation at different grid sizes, overall convergence is achieved.

In a regular grid, if one keeps on halving the grid sizes, the nodes of the largest grid size remain common in all the grid. In fact, between any two subsequent halving, the nodes of the larger grids are also present on the finer grid. Other points of the finer grid fall between the nodes of the coarse grid. If one estimates a solution at the coarse grid, it is then possible to transfer those values directly to the finer grid. For the intermediate points on the finer grid, one can then interpolate. These estimates serve as the initial guesses for the finer grid iterations, which then converges quickly. This way, one can keep refining the grids and obtain solutions with fewer Gauss-Seidel iterations. Let us now see how to operationalize an elementary multi-grid process from this conceptual description. The steps of a typical multi-grid process is as follows:

Step 1

Discretize the PDE and setup the matrix equations with different grid sizes. Let us take the example of one-dimensional diffusion equation. The matrix equation obtained for application of implicit methods can be represented as $Ax = b$ where, the matrix A contains the coefficients from the finite difference scheme used for discretization, the vector x contains the solution at time step $(n + 1)$ and vector b contains product of the vector x at time step n and a matrix similar to A, and a constant vector arising from boundary condition [right-hand side of Eq. (7.41)].

If we discretized the spatial dimension into five equally spaced intervals, we shall have four interior nodes and the size of the square matrix A is 4×4. Now, let us consider refining the grid size to half. This will give rise to 10 equal intervals and nine interior nodes. A similar set of equation can also be written for this grid size and the coefficient matrix A will be 9×9. Another refinement by further halving the grid size will give rise to a coefficient matrix of size 19×19. However, all of these matrix equations can be set up independently once a suitable discretization scheme is chosen. Notice that, for $(N + 1)$ equal intervals, the size of the coefficient square matrix is N. Each halving of the grid results in a coefficient matrix of size $(2N + 1)$ where N is size of the matrix for the coarse grid.

Similar matrix equation can also be setup for two-dimensional Laplace equation (Section 7.2.2). In this case, if a rectangular domain is divided into $(N + 1)$ equal intervals, the size of the coefficient matrix is $N^2 \times N^2$. For each halving of interval, the coefficient matrix becomes of size $(2N + 1)^2 \times (2N + 1)^2$. For a visual sense, the reader can look at Fig. 7.4 where five equal intervals in each direction resulted in a matrix of size 16×16. Now, visualize halving the intervals resulting in 9 interior grid points in each direction and a coefficient matrix of size 81×81.

Let us denote the smallest grid size as h and the corresponding matrix as A^h. Then the subsequent coarser grid sizes will be $2h$, $4h$, $8h$, etc. We shall denote the corresponding coefficient matrices as A^{2h}, A^{4h}, A^{8h}, etc. Therefore, for each problem, we will have the following set of matrix equations, set up using some discretization scheme chosen

$$A^h x^h = b^h \tag{8.103}$$

$$A^{2h} x^{2h} = b^{2h} \tag{8.104}$$

$$A^{4h} x^{4h} = b^{4h} \tag{8.105}$$

$$A^{8h} x^{8h} = b^{8h} \tag{8.106}$$

The solution vector x in each equation are of different lengths but some of the nodes are common because, by each halving of the intervals, essentially, one extra node is put in between two nodes of the coarser grid. For the one-dimensional diffusion problem with five equally spaced intervals, the size of the solution vector is 4, 9, 19, and 39 for grid sizes of $8h$, $4h$, $2h$, and h, respectively. Notice, that 4 nodes are common between vectors of $8h$ and $4h$, 9 nodes are common between vectors of $4h$ and $2h$, and 19 nodes are common between vectors of $2h$ and h.

For the two-dimensional Laplace equation of Fig. 7.4, the size of the vectors are 16, 81, 361, and 1521 for grid sizes of $8h$, $4h$, $2h$, and h, respectively. Therefore,

16 nodes are common between vectors of $8h$ and $4h$, 81 nodes are common between vectors of $4h$ and $2h$, and 361 nodes are common between vectors of $2h$ and h. Our final goal is to obtain a solution for the finest grid but that would require iterative solution of a matrix of very large size. The convergence may be slow, as the solution may not converge at the same rate in all parts of the domain. The convergence may be enhanced significantly if a better initial guess of the solution can be obtained. The principle of multi-grid method is based on this fact.

Through iteration of a smaller size matrix, values can be obtained at the coarse grid. These values can be transferred to the next finer grid and used as initial guess for the iteration. Instead of full convergence at any grid size, one can also go back and forth between multiple level grid sizes by performing only a few iterations at each grid size. This way, one modifies the solution at all grid sizes and overall, the convergence is faster. However, this would require relating the solution vector of the fine grid with the coarse grid at all levels. This is the topic of the next step.

Step 2

Since one has to go back and forth between coarse and fine grids, the transfer of information is two-way, i.e., coarse grid to fine grid and vice versa. Let us first discuss the transfer between coarse grid (say $2h$) to fine grid (h).

Intergrid Transfer: Coarse grid ($2h$) to Fine grid (h)
All nodes of the coarse grid are common to the fine grid. Therefore, the values at the coarse grid can be directly transferred to the fine grid for the common nodes. For the intermediate nodes on the fine grid, the value is obtained by linear interpolation between the neighbouring nodes on the coarse grid.

Since the interval is halved, the intermediate node is equidistant from two nodes on the coarse grid. Therefore, the value at the intermediate node on the fine grid is simply the average of values on the adjacent nodes of the coarse grid. Let us take the example of one-dimensional diffusion equation. If the nodes on the coarse grid are denoted as j; the nodes that are common on the fine grid are $2j$.

Therefore, we can write the following:

$$x_{2j}^h = x_j^{2h}$$

$$x_{2j+1}^h = \frac{1}{2}(x_j^{2h} + x_{j+1}^{2h}) \tag{8.107}$$

where

$$0 \le j \le \left(\frac{N}{2} - 1\right)$$

for even N and

$$0 \le j \le \left(\frac{N-1}{2}\right)$$

for odd N. Equation (8.107) can also be written in the matrix form and one can define a transfer matrix T_{2h}^h as follows:

$$x^h = T_{2h}^h x^{2h} \tag{8.108}$$

The transfer matrix T is not a square matrix. For example, let us take four equal intervals (three interior nodes) in the domain for $2h$. This will give eight equal intervals and seven interior nodes for h. The transfer matrix in this case transforms a vector of length 3 to a vector of length 7. Therefore, it must have a dimension of 7×3. In this case, Eq. (8.108) looks as follows:

$$
\begin{bmatrix} x_1^h \\ x_2^h \\ x_3^h \\ x_4^h \\ x_5^h \\ x_6^h \\ x_7^h \end{bmatrix} = \begin{bmatrix} 0.5 & 0 & 0 \\ 1 & 0 & 0 \\ 0.5 & 0.5 & 0 \\ 0 & 1 & 0 \\ 0 & 0.5 & 0.5 \\ 0 & 0 & 1 \\ 0 & 0 & 0.5 \end{bmatrix} \begin{bmatrix} x_1^{2h} \\ x_2^{2h} \\ x_3^{2h} \end{bmatrix} \tag{8.109}
$$

For the first and the last nodes on the fine grid, we assumed the value as half of the coarse grid nodal value. This is equivalent to assuming a zero boundary condition. If a first type boundary condition is known, one can take the average of the value at the boundary and the first node on the coarse grid. Needless to say that the same transfer protocol can also be applied between $4h$ and $2h$, $8h$ and $4h$, etc.

Intergrid Transfer: Fine grid (h) to Coarse grid (2h)

Since all nodes of the coarse grid is common to the fine grid, once a solution is available on the fine grid, it can be directly transferred to the coarse grid for the common points. However, since we never iterate to final solution at any of the grid level, we may improve the approximation on the coarse grids by taking a weighted average of the neighbouring points on the fine grid. These two concepts essentially lead to two different strategies for inter-grid transfer from fine to coarse grids. These are described below.

8.9.1 Direct Injection

In this, the values from the fine grid are directly transferred without modification. This can be expressed as:

$$x_j^{2h} = x_{2j}^h \tag{8.110}$$

where

$$0 \le j \le \left(\frac{N}{2} - 1\right)$$

for even N and

$$0 \le j \le \left(\frac{N-1}{2}\right)$$

for odd N. Once again, we can define a transfer matrix as follows:

$$x^{2h} = T_h^{2h} x^h \tag{8.111}$$

For example, problem of Eq. (8.109), the above matrix equation translates to

$$\begin{bmatrix} x_1^{2h} \\ x_2^{2h} \\ x_3^{2h} \end{bmatrix} = \begin{bmatrix} 0 & 1 & 0 & 0 & 0 & 0 & 0 \\ 0 & 0 & 0 & 1 & 0 & 0 & 0 \\ 0 & 0 & 0 & 0 & 0 & 1 & 0 \end{bmatrix} \begin{bmatrix} x_1^h \\ x_2^h \\ x_3^h \\ x_4^h \\ x_5^h \\ x_6^h \\ x_7^h \end{bmatrix} \tag{8.112}$$

8.9.2 Full Weighting

A weighted average of the neighbouring values are taken from the fine grid to approximate the values at the coarse grid. For the grid on one-dimension, this can be expressed as:

$$x_j^{2h} = \frac{1}{4}(x_{2j-1}^h + 2x_{2j}^h + x_{2j+1}^h) \tag{8.113}$$

where

$$0 \le j \le \left(\frac{N}{2} - 1\right)$$

for even N and

$$0 \le j \le \left(\frac{N-1}{2} \right)$$

for odd N. The matrix form is same as Eq. (8.111), but the transfer matrix is different.

For the one-dimensional diffusion problem, the transfer equation can now be written as:

$$\begin{bmatrix} x_1^{2h} \\ x_2^{2h} \\ x_3^{2h} \end{bmatrix} = \begin{bmatrix} 0.25 & 0.5 & 0.25 & 0 & 0 & 0 & 0 \\ 0 & 0 & 0.25 & 0.5 & 0.25 & 0 & 0 \\ 0 & 0 & 0 & 0 & 0.25 & 0.5 & 0.25 \end{bmatrix} \begin{bmatrix} x_1^h \\ x_2^h \\ x_3^h \\ x_4^h \\ x_5^h \\ x_6^h \\ x_7^h \end{bmatrix} \tag{8.114}$$

For a two-dimensional problem such as Laplace equation, the average in Eq. (8.113) can be taken in terms of neighbouring nodes on both directions. In this case, Eq. (8.113) becomes:

$$x_{i,j}^{2h} = \frac{1}{8}(x_{2i-1,2j}^h + x_{2i,2j-1}^h + 4x_{2i,2j}^h + x_{2i+1,2j}^h + x_{2i,2j+1}^h)$$

where, i and j are the indices for directions x and y, respectively.

Using the concepts developed in this section, we can now transfer information between various grids. Only thing left is the solution of Eqs (8.103) to (8.106) using an iterative method such as Gauss-Seidel.

Step 3

It is possible to start the iteration at the coarse grid, transfer the information at the finer grids, and then drop back to coarse grid, and finally again go up to the finer grid. Alternatively, one can perform a few iterations first at the finer grid, drop to the coarse grid, and finally go up to the fine grid with refined estimates. There is no unique optimized way to implement. However, one generally performs only 3-4 Gauss-Seidel iterations at each grid level at a time. Schematic of a few alternative schemes commonly found in the literature are shown in Fig. 8.24. These are termed as: *V-Cycle, W-Cycle, Full Multigrid V-Cycle (FMV Cycle)*, etc.

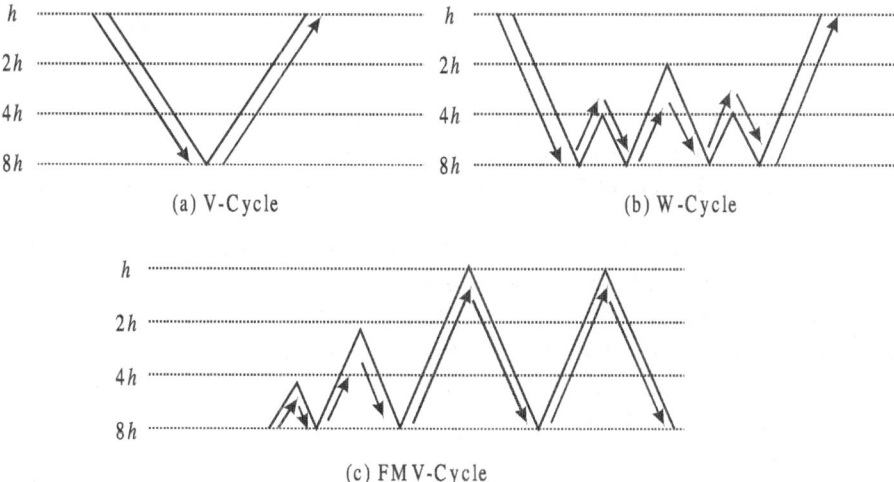

Fig. 8.24 Schematic of various multi-grid solution cycles.

The outline presented for multi-grid method in this chapter is elementary. There are significant amounts of literature that exist for the application of the multi-grid method. The background presented in this section should serve as the starting point for multi-grid application in the solution of PDEs.

References

Abramowitz, M. and I. A. Stegun (1964). *Handbook of Mathematical Functions*, Dover Publications, Inc., New York.

Butcher, J. C. (1965). On the attainable order of R-K methods, *Mathematics of Computation*, **19**, 408–417.

Celia, M. A. and W. G. Gray (1992). *Numerical Methods for Differential Equations*, Prentice-Hall International, Inc., Englewood Cliffs, NJ.

Chapra, S. C. and R. P. Canale (2006). *Numerical Methods for Engineers*, 5th Ed., McGraw Hill, Boston, MA.

Dahlquist, G. (1963). A special stability problem for linear multistep methods, *BIT*, **3**, 27–43.

Dahlquist, G. and Å. Björck (1969). *Numerical Methods*, Translated by N. Anderson, Prentice Hall, Englewood Cliffs, NJ.

Davis, P. J. and P. Rabinowitz (1967). *Numerical Integration*, Blaisdell Publishing Co., Waltham, MA.

de Boor, C. (2001). *A Practical Guide to Splines*, Springer, New York.

Enright, W. H., T. E. Hull, and B. Lindberg (1975). Comparing numerical methods for stiff systems of ODEs, *BIT*, **15**, 10–48.

Evans, L. C. (2010). *Partial Differential Equations*, 2nd Ed., American Mathematical Society, Providence.

Faddeev, D. and V. Faddeeva (1977). Corrections: Computational methods of linear algebra in *Zap. Naun. Sem. Leningrad. Otdel. Mat. Inst. Steklov.* (LOMI) **54** (1975), 3–228, *Zap. Naun. Sem. Leningrad. Otdel. Mat. Inst. Steklov.* (LOMI), **70**, 286–288.

Fadeeva, V. N. (1959). *Computational Methods of Linear Algebra*, Translated by C.D. Benster, Dover Publications, Inc., New York.

Gelfund, I. (1941). Normierte Ringe, *Mat. Sbornik (Recueil mathématique)*, N. S. **9** (51), 3–24.

Golub, G. H. and C. F. Van Loan (1996). *Matrix Computations*, 3rd Ed., Johns Hopkins University Press, Indian Ed. Published in 2007 by Hindustan Book Agency, New Delhi.

Gregory, R. T. and D. Karney (1969). *A Collection of Matrices for Testing Computational Algorithm*, Wiley-Interscience.

Hildebrand, F. B. (1974). *Introduction to Numerical Analysis*, 2nd Ed., Dover Publications, Inc., New York.

Hou, S. H. (1998). A simple proof of the Leverrier-Fadeev characteristic polynomial algorithm, *SIAM Rev.*, **40**(3), 706–709.

John, F. (1981). *Partial Differential Equations*, 4th Ed., Springer-Verlag, New York.

Kronrod, A. S. (1964). Integration with control of accuracy (in Russian), *Proc. USSR Academy of Sciences*, **154**, 283–286.

Psarakis, E. Z. and G. V. Moustakides (2003). A robust initialization scheme for the Remez exchange algorithm, *IEEE Signal Processing Letters*, **10**(1), 1–3.

Ralston, A. and H. S. Wilf (1960). *Mathematical Methods for Digital Computers*, John Wiley & Sons, New York.

Seinfeld, J. H. (1986). *Atmospheric Chemistry and Physics of Air Pollution*, John Wiley & Sons, Inc., USA.

Simpson, T. (1743). *Mathematical Dissertations on a Variety of Physical and Analytical Subjects: The Whole in a General and Perspicuous Manner*, T. Woodward (Half Moon), London.

Sorensen, H. V., D. L. Jones, M. T. Heideman, and C. S. Burrus (1987). Real-valued fast Fourier transform algorithms, *IEEE Trans. ASSP*, **35**(6), 849–863.

Steffensen, J. F. (1950). *Interpolation*, 2nd Ed., Chelsea, New York.

Stoer, J. and R. Bulirsch (2002). *Introduction to Numerical Analysis*, 3rd Ed., Springer, New York.

Strang, G. and G. J. Fix (1973). *An Analysis of the Finite Element Method*, Prentice Hall, Englewood Cliffs, NJ.

Van Barel, M. and A. Bultheel (1992). A parallel algorithm for discrete least squares rational approximation, *Numerische Mathematik*, **63**, 99–121.

Walker, J. S. (1996). *Fast Fourier Transforms*, 2nd Ed., CRC Press, New York.

Young, D. M. and R. T. Gregory (1988). *A Survey of Numerical Mathematics*, Vol. II, Dover Publications, Inc., New York.

Yueh, W. C. (2005). Eigenvalues of several tridiagonal matrices, *Applied Mathematics E-Notes*, **5**, 66–74.

Zienkiewicz, O. C. and R. L. Taylor (2000). *The Finite Element Method*, Vol. 1, The Basis, Butterworth-Heinemann, Oxford.

Index